PROBABILITY AND PROBABILITY DISTRIBUTIONS

PROBABILITY

Bayes' theorem

$$P(B_i \mid A) = \frac{P(B_i) \cdot P(A \mid B_i)}{P(B_1) \cdot P(A \mid B_1) + P(B_2) \cdot P(A \mid B_2) + \cdots + P(B_k) \cdot P(A \mid B_k)}$$

Conditional probability

$$P(A \mid B) = \frac{P(A \cap B)}{P(B)}$$

General addition rule

$$P(A \cup B) = P(A) + P(B) - P(A \cap B)$$

General multiplication rule

$$P(A \cap B) = P(B) \cdot P(A \mid B) \quad \text{or} \quad P(A \cap B) = P(A) \cdot P(B \mid A)$$

Mathematical expectation

$$E = a_1 p_1 + a_2 p_2 + \cdots + a_k p_k$$

PROBABILITY DISTRIBUTIONS

Binomial distribution

$$f(x) = \binom{n}{x} p^x (1 - p)^{n-x}$$

Mean of probability distribution

$$\mu = \sum x \cdot f(x)$$

Standard deviation of probability distribution

$$\sigma = \sqrt{\sum (x - \mu)^2 \cdot f(x)}$$

Continued inside back cover

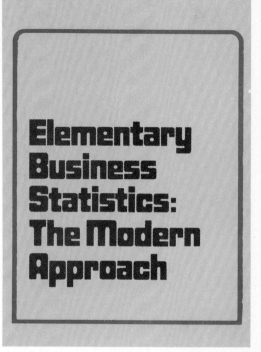

Elementary Business Statistics: The Modern Approach

PRENTICE-HALL, INC., Englewood Cliffs, New Jersey 07632

th EDITION

ELEMENTARY BUSINESS STATISTICS: THE MODERN APPROACH

JOHN E. FREUND

Arizona State University

FRANK J. WILLIAMS

San Francisco State University

Library of Congress Cataloging in Publication Data

FREUND, JOHN E.
  Elementary business statistics.

  Bibliography: p.
  Includes index.
  1. Commercial statistics.   2. Statistics.
I. Williams, Frank Jefferson.   II. Title.
HF1017.F73 1982   519.5′024658   81-12103
ISBN 0-13-253120-8   AACR2

10   9   8   7   6   5   4   3

Editorial/production supervision by Esther S. Koehn
Interior design and cover design by Walter A. Behnke
Manufacturing buyer: Ed O'Dougherty

ISBN 0-13-253120-8

PRENTICE-HALL INTERNATIONAL, INC., *London*
PRENTICE-HALL OF AUSTRALIA PTY. LIMITED, *Sydney*
PRENTICE-HALL OF CANADA, LTD., *Toronto*
PRENTICE-HALL OF INDIA PRIVATE LIMITED, *New Delhi*
PRENTICE-HALL OF JAPAN, INC., *Tokyo*
PRENTICE-HALL OF SOUTHEAST ASIA PTE. LTD., *Singapore*
WHITEHALL BOOKS LIMITED, *Wellington, New Zealand*

Contents

3

SUMMARIZING DATA: STATISTICAL DESCRIPTIONS 29

4

SUMMARIZING DATA: INDEX NUMBERS 71

5

POSSIBILITIES, PROBABILITIES, AND EXPECTATIONS 101

6

SOME RULES OF PROBABILITY 129

7

DECISION ANALYSIS 167

8

PROBABILITY DISTRIBUTIONS   209

9

BAYES' THEOREM
AND THE REVISION
OF PROBABILITIES   237

10

THE NORMAL DISTRIBUTION 255

11

SAMPLING AND SAMPLING DISTRIBUTIONS 277

12

DECISION MAKING: INFERENCES ABOUT MEANS 301

13

DECISION MAKING: INFERENCES ABOUT STANDARD DEVIATIONS 345

14

DECISION MAKING: INFERENCES ABOUT PROPORTIONS 357

15

DECISION MAKING:
ANALYSIS OF VARIANCE 389

16

DECISION MAKING:
NONPARAMETRIC TESTS 409

17

DECISION MAKING:
LINEAR REGRESSION 431

18

DECISION MAKING: CORRELATION 455

19

DECISION MAKING: TIME SERIES ANALYSIS 476

20

PLANNING BUSINESS RESEARCH 522

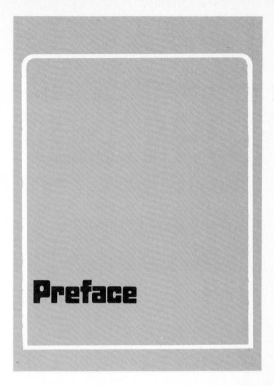

Preface

In the Preface to the first edition of this book we observed that, in the past few decades, the development and application of new mathematical, statistical, and computer techniques had brought about radical changes in virtually all areas of business. Statistics itself, we wrote, had clearly shifted from a backward-looking discipline, concerned with describing the past in numerical terms, to a forward- and outward-looking one—an action-oriented discipline—in which emphasis is placed on the decisions that must be made, and their consequences, for both current and future operations. Now, almost seventeen years after the first Preface was written, it is apparent that this new attitude toward statistics—what we call the "modern approach"—is pretty much the accepted way of life in both the practice and the teaching of business statistics.

Our aim in writing the first three editions of this book was to describe, as best we could, the modern approach to decision making in the face of uncertainty. This is again our aim, and in writing this fourth edition we now have years of valuable classroom experience to draw on. Over these years we, and hundreds of others, have discussed both formally and informally with thousands of students our earlier ideas on how modern business statistics could best be organized and presented at an elementary level of mathematical difficulty. Literally dozens of these teachers and hundreds of students have generously taken time to share their experiences and thoughts with us. The great extent to which we have relied on these friends will not be evident to every reader, but their counsel and advice are reflected everywhere in this book.

In this present version, there are substantial changes in content, organization, language, notation, and format. The chapter on nonparametric tests now includes small-sample tests with appropriate tables; in the discussion of sample spaces there is more emphasis on sets of outcomes than on geometrical configurations; there is an improved treatment of the relationship between probability and odds; cross-referencing among exercises is held to a minimum; statistical terminology and notation are updated to conform with current usage; sections are numbered; formulas are boxed with descriptions in the margins; some of the more important formulas are listed on the end papers; and each chapter has a list of key terms. We have also included in each chapter a set of review exercises. Altogether, the book contains more than 1,000 exercises—numbered consecutively throughout each chapter—many of which are new or updated. As in previous editions of this book, most of them are drawn from actual problems, but many of them have been modified and scaled down somewhat to simplify the computational burden.

In our judgment, this book contains enough material for a full year's work (six semester hours or nine quarter hours), and it permits a good deal of latitude in the selection of topics for shorter courses. Some recommendations for this are given in the Solutions Manual, published separately for the instructor. For these who are not familiar with the earlier editions of this book, let us point out that in view of the current emphasis on inference, the material on decision analysis is taken up as early as possible—in Chapters 7 and 9 immediately following the chapters on probability and probability distributions—and not as an afterthought at the end of the book. Also, the special topics on index numbers and time series are taken up at the appropriate places. Since we treat only the descriptive aspects of index number construction and not the sampling aspects of obtaining the necessary data, the index number chapter follows immediately the introductory chapters on descriptive statistics. Since time series analysis is essentially a problem of curve fitting, this subject is taken up following the chapters on regression and correlation. Finally, probability is introduced informally in Chapter 5 and formally in Chapter 6, thus making it possible to omit the formal treatment and continue into decision analysis and inference without going through the more rigorous study of probability.

We are greatly indebted to our many faculty colleagues and students who have contributed so much to this book; to them, to John Freund, Jr. and Rita Ewer who helped with the proofreading, and to Esther Koehn who was in charge of the production, we extend our special thanks. In addition, we would like to thank Professors John Gordon Foster, Jr., Montgomery College; Harold B. Hackett, Jr., State University of New York at Alfred; Daniel Sankowsky, Suffolk University; K. Goldstein, Miami Dade South; Joseph Perry, University of North Florida, Kathy Lewis Corriher, and Dr. Alan Oppenheim, Montclair State College, who reviewed the manuscript and offered many helpful suggestions.

We are also indebted to Professor E. S. Pearson and the *Biometrika* trustees for permission to reproduce parts of Tables 8 and 18 from their *Biometrika Tables for Statisticians*; to the Addison-Wesley Publishing Co. to base Table VII on Table 11.4 of D. B. Owen's *Handbook of Statistical Tables*; to the editor of the *Annals of Mathematical Statistics* to reproduce the material in Table VIII; and to the Literary Executor of the late Sir Ronald A. Fisher, F.R.S., and to the Macmillan Publishing Company to reprint part of Table IV from their *Statistical Methods for Research Workers* (14th edition).

JOHN E. FREUND

FRANK J. WILLIAMS

Everything dealing even remotely with the collection, processing, analysis, and interpretation of numerical data belongs to the domain of statistics. This includes, for example, calculating the average profit growth of a major chemical company over the past ten years; collecting and presenting annually, for all companies listed on the New York Stock Exchange, their short-term debt as a percentage of their long-term debt; evaluating the effectiveness of two different safety programs aimed at reducing the number of lost-time injury accidents on a hazardous job; and analyzing the variations which occur from time to time in series of economic data (retail sales, consumer and wholesale prices, money supply, common stock prices, farm labor productivity, and so on).

The word "statistics" itself is used in several ways. In one connection it means a collection of data such as may be found in the financial pages of newspapers, or in the *Statistical Abstract of the United States*. And in another connection it means the totality of the methods employed in the collection, processing, analysis, and interpretation of any kind of data. In this latter sense, statistics is a branch of applied mathematics, and it is this field of mathematics which is the subject matter of this book.

In Section 1.1 we discuss the recent growth of statistics and the distinction between descriptive statistics and statistical inference. Section 1.2 contains a brief introduction to the role of statistics in business management.

Introduction

1

1.1

Modern Statistics

One of the most remarkable phenomena of the past few decades has been the growth of statistical methods and statistical ideas. For many years, statistics was concerned mostly with the collection of data and their presentation in tables and charts; today, it has evolved to the extent that its impact is felt in almost every area of human endeavor. This is because modern statistics is directly concerned with the problem of decision making under uncertainty—needless to say, there are elements of uncertainty in nearly everything we do.

The most important feature of the recent growth of statistics has been the shift in emphasis from methods which merely describe to methods which serve to make generalizations, or in other words, a shift in emphasis from **descriptive statistics** to **statistical inference.** By descriptive statistics we mean any treatment of data which is designed to summarize or describe some of their important features without attempting to infer anything that goes beyond the data. For instance, if we compile the data from appropriate sources and report that at a recent early July auction sale at Keeneland Race Course 342 yearlings were auctioned off for a total of $18,344,000 and that 32 of them sold for $100,000 or more, our work belongs to the domain of descriptive statistics. This would also be the case if we determine that the average price per horse was $18,344,000/342 = $53,637, but not if we use these data to estimate, say, the average price of yearlings sold in that year at other auction centers or in private negotiations, or to predict the total or average sales price of yearlings at Keeneland in 1990.

Descriptive statistics is an important branch of statistics and it continues to be widely used in business and other areas of activity. In most cases, however, statistical information arises from samples (from observations made only on some of a large set of items), or from observations made on past happenings. Time, cost, or the impossibility of doing otherwise usually requires such a procedure, even though our real interest lies in the whole large set of items from which the sample came, and in future happenings, not in the past. Since generalizations of any kind lie outside the scope of descriptive statistics, we are led to the use of statistical inference in solving many problems of day-to-day operations and in making both short- and long-range plans. For instance, the methods of statistical inference are required to decide whether a large lot of 9-volt transistor batteries meets its manufacturer's guaranteed average useful life, to decide upon the minimum effective and maximum safe dose of a new anti-inflammatory agent in the treatment of painful local inflammations, to estimate the demand for new snow tires during the next snow season, or to predict the 1995 total demand for wood fibers for all uses in the United States.

At the very beginning of our study of statistics, however, we hasten to add that whenever we make a statistical inference—a generalization that goes beyond the limits of our observations—we must proceed with great caution. In fact, we must consider carefully whether it is possible to make any valid generalizations at all, and if it is, just how far we can go in generalizing. And yet, no matter how carefully we proceed, we may be entirely wrong in the generalizations and find ourselves in trouble. Indeed, one of the most basic problems of statistical inference is that of appraising the risks of making wrong generalizations, and perhaps doing the wrong thing, on the basis of sample data. Calling attention here to statistical errors may seem like a negative way to begin our study, but in a very real way, the constant awareness of the possibility of erroneous conclusions and actions, and the desire to control them, direct the course of statistical investigations.

Reality is harsh and unyielding, and must be dealt with on its own terms. We live in a world filled with uncertainties and there is no way to eliminate completely the risks of wrong decisions. This being the case, our real problem is not how to eliminate them, but how to live with them intelligently, and the sooner we realize this the better off we are, and the better we can understand why statistics is a discipline worth studying. One of the main reasons for studying statistics is that it addresses itself directly to this universal problem of how to make intelligent decisions in the face of uncertainty, or more briefly, to the problem of decision making under uncertainty.

1.2

Statistics in Business Management

Modern statistics is highly refined and it is now making a great contribution to the solution of many problems of decision making in the face of uncertainty. Moreover, continuing substantial progress is being made in developing new methods to meet urgent practical needs in many different areas of activity. For example, in the fields of medicine and public health the recently developed and rapidly growing science of biostatistics is applying powerful mathematical and statistical methods to the study of such fundamental problems connected with the growth, development, illnesses, and deaths of human populations as the harmful effects of air pollution, the relationship between diet and heart disease, and between smoking and lung cancer. In these areas, as in many others, statistical methods provide a framework for looking at problems in a systematic and logical way. As a matter of fact, these modern methods are in many cases absolutely essential to orderly and continued progress toward important goals.

There is hardly any area in which the impact of statistics has been felt more strongly than in business, where, as a daily way of life, decisions which affect profitability and continuity must be made at all levels of all kinds of businesses. Indeed, it would be hard to overestimate the contributions that statistical methods have made to the effective planning, operation, and control of business activities of all sorts. In the past 30 to 35 years the application of statistical methods has brought about drastic changes in all the major areas of business management: general management, research and development, finance, production, sales, advertising, and the rest. Of course, not all problems in these areas are of a statistical nature, but the list of those which can be treated either partly or entirely by statistical methods is very long. To illustrate, let us mention a few which might face a large multinational manufacturer.

In the general management area where long-range planning is of great concern, population trends in various countries must be forecast and their effects on consumer markets must be analyzed. In research and engineering, costs must be estimated for various projects, and manpower, skill, equipment, and time requirements must be anticipated. In the area of finance, the profit potentials of alternative capital investments must be determined, overall financial requirements must be projected, and capital markets must be studied so that sound long-range financing and investment plans can be developed.

In production, problems of a statistical nature arise in connection with such matters as plant layout and structure, plant size and location, inventory, production scheduling and control, maintenance, traffic and materials handling, and quality assurance. Enormous strides have been made in recent years in the application of statistics to the last area, that is, to sampling inspection and quality control. In the area of sales, many problems arise which require statistical solutions. For instance, sales must be forecast for both present and new products and for existing as well as new markets, channels of distribution must be determined, and requirements for sales forces must be estimated. In advertising, building successful campaigns can be a troublesome task. Budgets must be determined, allocations must be made to various media, and the effectiveness of the campaign must be measured (or predicted) by means of survey samples of public response and other statistical techniques.

So far we have been speaking of problems of a statistical nature which might typically be encountered by a large manufacturer. However, similar problems are faced, say, by a large railroad trying to make the best use of its thousands of freight cars; by a large rancher trying to decide how to feed his cattle so that their nutritional needs will be met at the lowest possible cost; by an open-end investment company trying to decide how much of its total assets should be kept in working cash balances and how much should be invested in common stocks and short-term notes; and by a large integrated gas company (which produces, processes, and transports natural gas, crude oil, and petroleum products) in planning its future conservation practices, transportation system, and energy sources development.

It is not at all necessary to refer to large organizations to find business applications of statistics. For the small businesses, problems usually differ more in degree than in kind from those of their large competitors. Neither the largest supermarket nor the smallest neighborhood grocery store, for example, has unlimited capital or shelf space, and neither can afford to tie up these two assets in the wrong goods. The problem of utilizing capital and shelf space most effectively is as critical for the small store as it is for the large, and it is extremely shortsighted to think that modern management tools (including modern statistical techniques) are of value only to big business. In fact, they could hardly be needed more anywhere else than in small business, where each year thousands of operating units fail and many of the thousands of new units entering the field are destined to fail because of inadequate capital, overextended credit, overloading with the wrong stock, and, generally speaking, no knowledge of the market or the competition.

Although in this text our attention will be directed largely toward business statistics and our specific goal is to introduce the basic concepts and methods of statistics to the beginning student, the formal notions of statistics as a way of making rational decisions should really be part of any thoughtful person's equipment. After all, the employees and managers of business are not the only persons who must make decisions involving uncertainties and risks. Everyone must make decisions of this sort professionally or simply as part of everyday life. It is true that some of the choices we must make entail only matters of personal preference (say, whether to watch television or read a book), but in many instances there is the possibility of being wrong in the sense that there is an actual loss or penalty—possibly only a minor annoyance, but possibly something as serious as the loss of one's fortune, even one's life, or something between these extremes. The methods of modern statistics deal with decision problems involving risks not only in business and industry, and in everyday life, but also in such fields as medicine, physics, chemistry, agriculture, foods and nutrition, economics, psychology, education, politics, government, ecology, and so forth. Although the examples and exercises used in this book are drawn mostly from the area of business, we shall also refer to some of these other areas from time to time. In this way, we hope to make the point that although specialized techniques exist for handling particular kinds of problems, the underlying statistical principles and ideas are identical regardless of the field of application.

1.3

A Word of Caution

The amount of statistical information that is disseminated to the public for one reason or another is almost beyond comprehension, and what part of it is "good" statistics and what part is "bad" statistics is anybody's guess. Certainly, all of it cannot be accepted uncritically. Sometimes entirely erroneous conclusions are based on sound data. For instance, a certain city once claimed to be the "nation's healthiest city," since its death rate was the lowest in the country. Even if we go along with their definition that healthy means "not dead," there is another factor that was not taken into account: since the city had no hospital, its citizens had to be hospitalized elsewhere, and their deaths were recorded in the cities in which death actually occurred. The following are some other *non sequiturs* based on otherwise sound statistical data: "Statistics show that there were fewer airplane accidents in 1920 than in 1980; hence, flying was safer in 1920 than in 1980." "Since there are more automobile accidents in the daytime than there are at night, it is safer to drive at night." "Recent statistics show that the average income per person in a certain area is $2,800; thus, the average income for a family of five is $14,000."

Sometimes, identical data are made the basis for directly opposite conclusions, as in collective-bargaining disputes when the same data are used by one side to show that employees are getting rich and by the other side to show that they are on the verge of starvation. In view of examples such as these, it is understandable that some persons are inclined to feel that figures can be made to show pretty much what one wants them to show. Sadly enough, as we have just demonstrated, this is uncomfortably close to the truth, unless we develop the ability to distinguish between "good" statistics and "bad" statistics, between statistical methods properly applied and statistical methods shamefully misapplied, and between statistical information correctly analyzed and interpreted and statistical information either intentionally or unintentionally perverted. We shall repeatedly remind the reader of this problem in special sections titled "A Word of Caution," which are given at the ends of many chapters.

To make one final point before beginning the formal study of statistics, let us make it clear that the sound statistical treatment of a problem consists of a good deal more than making a few observations, performing some calculations, and drawing some sort of conclusion. Questions as to how the data are collected and how the whole experiment or survey is planned are of basic importance. As elsewhere, we get "nothing for nothing" in statistics, and unless proper care is taken in all phases of an investigation—from the conception and statement of the problem (sometimes, the hardest job of all) to the planning and design, through the stages of data collection, analysis, and interpretation—no useful or valid conclusion whatever may be reached. Generally speaking, no amount of fancy mathematical or statistical manipulation of data on the most expensive computer hardware in the world can salvage poorly designed surveys or experiments. Indeed, professional statisticians insist that even the simplest of sampling studies be rigidly conducted according to well-defined rules. There is no more justification for calling a study which does not conform to these rules "statistical" than there is for calling a barnacle a ship.

In recent years, business decisions have come to depend more and more on the analysis of very large sets of data. Small businessmen need information about the income patterns and population characteristics in the areas they serve; market research analysts must deal with the views expressed by thousands of shoppers; government statisticians must handle, analyze, and interpret voluminous data collected in censuses of various kinds; and the heads of large corporations must consider information relevant to many problems which would simply overwhelm them if it were not presented in a compact and usable form. This trend in the use of mass data is due partly to the increasing availability of high-speed computers (many current applications of statistical methods would have been practically impossible before the advent of modern data-processing techniques) and partly to an increasing awareness of the need for scientific methods in business management. Of course, we do not always deal with very large sets of data; in many cases, they are prohibitively costly and hard to collect, but the problem of putting mass data into a usable form is so important that it requires special attention.

The most common method of summarizing mass data is to present them in condensed form in tables or charts, and the study of statistical presentation once took up most of the time in elementary statistics courses. Nowadays, the scope of statistics has grown to such an extent that much less time is devoted to this kind of work—in fact, we shall discuss it here only in this brief chapter.

2

Summarizing Data: Frequency Distributions

2.1

Frequency Distributions

We can often gain much valuable information about a large set of data, and get a good overall picture of it by grouping the data into a number of classes. For instance, the total billings, rounded to the nearest dollar, of 8,644 selected law firms may be summarized as follows:

Total billings	Number of firms
Less than $100,000	1,406
$100,000 to $249,999	4,352
$250,000 to $499,999	1,833
$500,000 to $749,999	489
$750,000 to $999,999	163
$1,000,000 or more	401
Total	8,644

Also, 2,439 complaints about comfort-related characteristics of an airline's planes may be presented as follows:

Nature of complaint	Number of complaints
Inadequate leg room	719
Uncomfortable seats	914
Narrow aisles	146
Insufficient carry-on facilities	218
Insufficient rest rooms	58
Miscellaneous other complaints	384
Total	2,439

Tables such as these are called **frequency distributions** (or simply **distributions**). The first one shows how the law firms' billings are distributed among the chosen classes, and the second one shows how the complaints are distributed among the different categories. If, as in the first table, data are grouped according to numerical size, the resulting table is called a **numerical** or **quantitative distribution**. In contrast, if data are grouped into categories which differ in kind rather than in degree, as in the second table, the resulting table is called a **categorical** or **qualitative distribution**.

Frequency distributions present data in a relatively compact form, give a good overall picture, and contain information that is adequate for many purposes, but some things which can be determined from the original data cannot be determined from a distribution. For instance, from the first distribution we can find neither the exact size of the lowest and highest of the law firms' total billings, nor the exact total or average of the billings of the 8,644 firms. Similarly, from the second distribution we cannot tell how many of the complaints about uncomfortable seats were over the width of the seats, how many of the complaints about insufficient carry-on facilities were over particular size luggage, and so forth. Nevertheless, frequency distributions present **raw** (unprocessed) **data** in a more readily usable form, and the price we pay for this—the loss of certain information—is usually a fair exchange.

The construction of a numerical distribution consists essentially of three steps: (1) choosing the classes, (2) sorting (or tallying) the data into these classes, and (3) counting the number of items in each class. The last two of these steps are purely mechanical, so we shall discuss here only the first, the problem of choosing a suitable classification.

The two things that we must consider in choosing a classification scheme are the number of classes we should use to accommodate the data, and the range of values each class should cover, that is, from where to where each class should go. Both choices are essentially arbitrary, but the following rules are usually observed:

> **We seldom use fewer than 6 or more than 15 classes; the exact number we use in a given situation will depend on the nature, magnitude, and range of the data.**

Clearly, we would lose more than we gain if we group 5 observations into 12 classes with most of them empty, and we would lose a great deal of information if we group 10,000 measurements into 3 classes.

> **We always make sure that each item (measurement or observation) goes into one and only one class.**

To this end we must make sure that the smallest and largest values fall within the classification, that none of the values can fall into gaps between successive classes, and that successive classes do not overlap and contain some values in common.

> **Whenever possible, we make the classes the same length; that is, we make them cover equal ranges of values.**

If we can, we also make these ranges multiples of easy to work with numbers, such as 5, 10, or 100, for this facilitates constructing, reading, and using the distribution.

Since the law firm billings of this section were all rounded to the nearest dollar, only the last of these rules was violated in constructing their distribution. (Had the billings been given to the nearest cent, however, a billing of, say, $249,999.53 would have fallen between the second class and the third class, and we would also have violated the second rule.) Actually, the last rule was violated in two ways: The interval from $100,000 to $249,999 and the interval from $250,000 to $499,999 cover unequal ranges of values, and the first class has no specific lower limit and the last has no specific upper limit.

In general, we refer to any class of the "less than," "or less," "more than," or "or more" type as an **open class.** If a set of data contains a few values which are much greater than or much smaller than the rest, open classes are quite useful in reducing the number of classes required to accommodate the data. However, we try to avoid them because they make it hard, if not impossible, to calculate values of interest, such as an average or a total.

As we have suggested, the appropriateness of a classification may depend on whether the data are rounded to the nearest dollar or to the nearest cent. It may also depend on whether the data are given to the nearest inch or to the nearest hundredth of an inch, to the nearest second or to the nearest millisecond, or whether they are rounded to the nearest percent or to the nearest tenth of a percent, and so on.

For instance, if we want to group the sizes of all sales, in dollars and cents, made by an office supply store on one day, we might use the classification

Size of sale *(dollars)*
0.00– 4.99
5.00– 9.99
10.00–14.99
15.00–19.99
20.00–24.99
etc.

Also, for the price–earnings ratios of common stocks given to the nearest tenth, we might use the classification

Price–earnings ratio
4.0– 5.9
6.0– 7.9
8.0– 9.9
10.0–11.9
12.0–13.9
etc.

and for the number of calls received per hour at a department store's switchboard, we might use the classification

Number of calls
0– 24
25– 49
50– 74
75– 99
100–124
etc.

To illustrate the construction of a frequency distribution, let us go through the actual steps of grouping a set of data.

Construct a distribution of the following scores which 150 applicants for secretarial positions in a large company made on a clerical aptitude test:

27	79	69	40	51	88	55	48	36	61
53	44	94	51	65	42	58	55	69	63
70	48	61	55	60	25	47	78	61	54
57	76	73	62	36	67	40	51	59	68
27	46	62	43	54	83	59	13	72	57
82	45	54	52	71	53	82	69	60	35
41	65	62	75	60	42	55	34	49	45
49	64	40	61	73	44	59	46	71	86
43	69	54	31	56	51	75	44	66	53
80	71	53	56	91	60	41	29	56	57
35	54	43	39	56	27	62	44	85	61
59	89	60	51	71	53	58	26	77	68
62	57	48	69	76	52	49	45	54	41
33	61	80	57	42	45	59	44	68	73
55	70	39	58	69	51	85	46	55	67

Since the smallest of the scores is 13 and the largest is 94, it seems reasonable to choose the nine classes 10–19, 20–29, . . . , and 90–99, which satisfy the three rules for choosing a suitable classification. Performing the actual tally and counting the number of items in each class, we get the following frequency distribution:

Scores	Tally	Frequency
10–19	/	1
20–29	ᵀᕼᐱ /	6
30–39	ᵀᕼᐱ ᵀᕼᐱ	9
40–49	ᵀᕼᐱ ᵀᕼᐱ ᵀᕼᐱ ᵀᕼᐱ ᵀᕼᐱ ᵀᕼᐱ /	31
50–59	ᵀᕼᐱ ᵀᕼᐱ ᵀᕼᐱ ᵀᕼᐱ ᵀᕼᐱ ᵀᕼᐱ ᵀᕼᐱ ᵀᕼᐱ //	42
60–69	ᵀᕼᐱ ᵀᕼᐱ ᵀᕼᐱ ᵀᕼᐱ ᵀᕼᐱ ᵀᕼᐱ //	32
70–79	ᵀᕼᐱ ᵀᕼᐱ ᵀᕼᐱ //	17
80–89	ᵀᕼᐱ ᵀᕼᐱ	10
90–99	//	2
	Total	150

The numbers given in the right-hand column of this table, which show how many items fall into each class, are called the **class frequencies**. The smallest and largest values that can go into any given class are called its **class limits**, and for this distribution they are 10 and 19, 20 and 29, 30 and 39, . . . , and 90 and 99. More specifically, 10, 20, 30, . . . , and 90 are called the **lower class limits**, and 19, 29, 39, . . . , and 99 are called the **upper class limits**.

Using this terminology, we can now say that the choice of the class limits depends on the extent to which the numbers we want to group are rounded. If our data are prices rounded to the nearest dollar, the class $5–9 actually contains all prices between $4.50 and $9.50; and if our data are lengths rounded to the nearest tenth of an inch, the class 1.5–1.9 actually contains all lengths between 1.45 and 1.95 inches. These pairs of values are usually called **class boundaries** or **real class limits**.

For the distribution of aptitude scores, the class boundaries are 9.5 and 19.5, 19.5 and 29.5, 29.5 and 39.5, . . . , and 89.5 and 99.5, namely, the midpoints between the respective class limits. Of course, these values must be, by their nature, "impossible" values which cannot occur among the data being grouped. To make sure of this, we have only to observe the extent to which the data are rounded; for instance, for the size-of-sales classification on page 10 the class boundaries are the impossible values −0.005, 4.995, 9.995, 14.995, and so on.

Numerical distributions also have what are called **class marks** and **class intervals.** Class marks are just the midpoints of the classes, and they are found by adding the lower and upper limits of a class (or its upper and lower boundaries) and dividing by 2. A class interval is simply the length of a class, or the range of values it can contain, and it is given by the difference between its class boundaries. If the classes of a distribution are all equal in length, their common class interval, which we call the class interval of the distribution, is also given by the difference between any two successive class marks.

EXAMPLE   Find the class marks and the class interval of the distribution of aptitude scores.

SOLUTION   The class marks are $\frac{10 + 19}{2} = 14.5$, $\frac{20 + 29}{2} = 24.5$, ..., and $\frac{90 + 99}{2} = 94.5$, and the class interval of the distribution is $24.5 - 14.5 = 10$. Note that the class interval is not given by the difference between the upper and lower limits of a class, which here is 9, not 10.

Sometimes it is preferable to present data in what is called a **cumulative frequency distribution**, or simply a **cumulative distribution**, which shows directly how many of the items are less than, or greater than, various values.

EXAMPLE   Convert the distribution of aptitude scores into a cumulative "or less" distribution.

SOLUTION   Since none of the values is 9 or less, one of the values is 19 or less, $1 + 6 = 7$ of the values are 29 or less, $1 + 6 + 9 = 16$ of the values are 39 or less, ..., the "or less" distribution is:

Scores	Cumulative frequencies
9 or less	0
19 or less	1
29 or less	7
39 or less	16
49 or less	47
59 or less	89
69 or less	121
79 or less	138
89 or less	148
99 or less	150

Had we written "less than 10" instead of "9 or less," "less than 20" instead of "19 or less," "less than 30" instead of "29 or less," ..., and "less than 100" instead of "99 or less," we would have a cumulative "less than" distribution. If we cumulate the frequencies, beginning with the largest class, we get an "or more" distribution if we designate the classes "10 or more," "20 or more," "30 or more," . . . , and "100 or more"; or a "more than" distribution if we designate the classes "more than 9," "more than 19," "more than 29," . . . , and "more than 99."

Often, it is better to show what percentage of the items falls into each class of a distribution instead of showing the actual class frequencies. To convert a frequency distribution into a corresponding **percentage distribution**, we divide each class frequency by the total number of items grouped and multiply the quotient by 100.

EXAMPLE   Convert the distribution of the aptitude scores into a percentage distribution.

SOLUTION   The first class contains $\frac{1}{150} \cdot 100 = 0.7$ percent of the data, the second class contains $\frac{6}{150} \cdot 100 = 4.0$ percent of the data, . . . , and the ninth class contains $\frac{2}{150} \cdot 100 = 1.3$ percent of the data. Hence, the percentage distribution is

Scores	Percentages
10–19	0.7
20–29	4.0
30–39	6.0
40–49	20.7
50–59	28.0
60–69	21.3
70–79	11.3
80–89	6.7
90–99	1.3

So far we have discussed only the construction of numerical distributions, but the general problem of constructing categorical (or qualitative) distributions is somewhat the same. Here again we must decide how many categories (classes) to use and what kind of items each category is to contain, making sure that all the items are accommodated and that there are no ambiguities. Since the categories must often be chosen before any data are actually collected, it is prudent to include a category labeled "others" or "miscellaneous."

For categorical distributions, we do not have to worry about such mathematical details as class limits, class boundaries, and class marks. On the other hand, there is often a serious problem with ambiguities and we must be very careful and explicit in defining what each category is to contain. For this reason, it is advisable, where possible, to use standard categories developed by the Bureau of the Census and other government agencies.†

†For references to these see the book by P. M. Hauser and W. R. Leonard in the Bibliography at the end of this book.

2.1 The wages paid to the piecework employees of a cotton textile mill in a given week varied from $165.26 to $263.55. Indicate the limits of six classes into which these wages might be grouped.

2.2 A customs official groups the declared values of a number of packages mailed from a foreign country into a frequency distribution with the classes $0.00–9.99, $10.00–19.99, $20.00–29.99, $30.00–39.99, $40.00–49.99, $50.00–59.99, and $60.00 and over. Is it possible to determine from this distribution the number of packages valued at

(a) less than $40.00; (c) more than $30.00;

(b) $40.00 or less; (d) $30.00 or more?

2.3 A sample of 200 outstanding balances is grouped into a table having the classes $0.00–24.99, $25.00–49.99, $50.00–99.99, and $100.00–199.99. Find

(a) the class boundaries; (c) the class intervals.

(b) the class marks;

2.4 The class marks of a distribution of the daily number of cameras brought in to a manufacturer's regional service center for repairs are 4, 11, 18, and 25. Find

(a) the class boundaries; (b) the class limits.

2.5 A set of measurements of the length of a large number of pieces of binding tape given to the nearest tenth of an inch is grouped into a table with the class boundaries 10.95, 12.95, 14.95, 16.95, 18.95, 20.95, and 22.95. What are the lower and upper limits of the six classes and the class marks?

2.6 To group sales invoices ranging from $15.00 to $40.00, a clerk uses the following classifications: $15.00–19.99, $20.00–25.99, $25.00–29.99, $30.00–34.90, and $35.00–39.99. Explain where difficulties might arise.

2.7 The following are the numbers of customers a restaurant served for lunch on 120 weekdays:

50	64	55	51	60	41	71	53	63	64	46	59
66	45	61	57	65	62	58	65	55	61	50	55
53	57	58	66	53	56	64	46	59	49	64	60
58	64	42	47	59	62	56	63	61	68	57	51
61	51	60	59	67	52	52	58	64	43	60	62
48	62	56	63	55	73	60	69	53	66	54	52
56	59	65	60	61	59	63	56	62	56	62	57
57	52	63	48	58	64	59	43	67	52	58	47
63	53	54	67	57	61	65	78	60	66	63	58
60	55	61	59	74	62	49	63	65	55	61	54

(a) Group these figures into a table having the classes 40–44, 45–49, 50–54, 55–59, 60–64, 65–69, 70–74, and 75–79.

(b) Convert the distribution of part (a) into a cumulative "less than" distribution.

2.8 A group of 50 employees from the accounting department of a large company is given an intensive course in computer programming. Of the various exercises assigned during the course, the following are the numbers of exercises satisfactorily completed by the members of the group:

```
13    9    5   11   14    6    5    8   11   13
10   16   15    3   19   18    9    9    5   12
13   12   15    9   18   12   16    7   12   13
11   18   15    9   21    9   11    6   12   12
10   16    2   14   10   17    8   15   11   12
```

(a) Group these figures into a table having the classes 2–4, 5–7, 8–10, . . . , and 20–22.

(b) Convert the distribution of part (a) into a corresponding percentage distribution and also a cumulative "or more" percentage distribution.

2.9 During the instruction period referred to in the preceding exercise, the study group was given a test on number systems and Boolean algebra, and the following are the grades made by the members of the group:

```
73   65   82   70   45   50   70   54   32   75
75   67   65   60   75   87   83   40   72   64
58   75   89   70   73   55   61   78   89   93
43   51   59   38   65   71   75   85   65   85
49   97   55   60   76   75   69   35   45   63
```

(a) Group these grades into a distribution having the classes 30–39, 40–49, 50–59, . . . , and 90–99.

(b) Convert the distribution of part (a) into a percentage distribution.

2.10 With reference to the preceding exercise, convert the percentage distribution of part (b) into a cumulative "less than" percentage distribution.

2.11 Asked to rate the maneuverability of a car as excellent, very good, good, fair, poor, or very poor, 45 drivers responded as follows: very good, good, good, fair, excellent, good, good, good, very good, poor, good, good, good, good, very good, good, fair, good, good, very poor, very good, fair, good, good, excellent, very good, good, good, good, fair, fair, very good, good, very good, excellent, very good, fair, good, good, very good, excellent, fair, good, good, and very good. Construct a table showing the frequencies corresponding to these ratings of the maneuverability of the car.

2.2

Graphical Presentations

When frequency distributions are constructed primarily to condense large sets of data and display them in an "easy to digest" form, it is usually advisable to present them graphically. The most common form of graphical presentation of statistical data is the histogram, an example of which is shown in Figure 2.1. A histogram is constructed by representing the measurements or observations that are grouped (in Figure 2.1, the clerical aptitude scores) on a horizontal scale, the class frequencies on a vertical scale, and drawing rectangles whose bases equal the class interval and whose heights are determined by the corre-

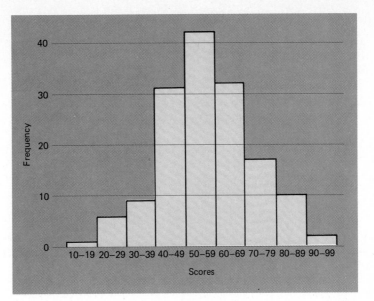

Histogram of the
distribution of the
aptitude scores.

sponding class frequencies. The markings on the horizontal scale can be the
class limits, as in Figure 2.1, the class boundaries, or arbitrary key values. For
easy readability, it is usually better to indicate the class limits, although the
rectangles actually go from one class boundary to the next. Histograms cannot
be used in connection with frequency distributions having open classes, and
they must be used with extreme care if the class intervals are not all equal (see
the discussion on page 22).

An alternative, although less widely used form of graphical presentation is
the **frequency polygon** (see Figure 2.2). Here the class frequencies are plotted at
the class marks and the successive points are connected by means of straight
lines. Note that we added classes with zero frequencies at both ends of the
distribution in order to "tie down" the graph to the horizontal scale. If we
apply the same technique to a cumulative distribution, we obtain what is called
an **ogive**. However, the cumulative frequencies are not plotted at the class
marks—it stands to reason that the frequency corresponding, say, to "29 or
less" should be plotted at 29.5, the class boundary, since "29 or less" actually
includes everything up to 29.5. Figure 2.3 shows an ogive corresponding to
the "or less" distribution of the aptitude scores.

Although the visual appeal of histograms, frequency polygons, and ogives
exceeds that of frequency tables, there are ways in which distributions can be
presented even more dramatically and often more effectively. Two kinds of such
pictorial presentations (often seen in newspapers, magazines, and reports of
various kinds) are illustrated by the **pictograms** of Figures 2.4 and 2.5.

Categorical (or qualitative) distributions are often presented graphically
as **pie charts** such as the one shown in Figure 2.6, where a circle is divided into

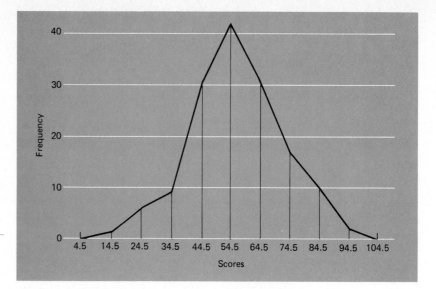

2.2

Frequency polygon of the distribution of the aptitude scores.

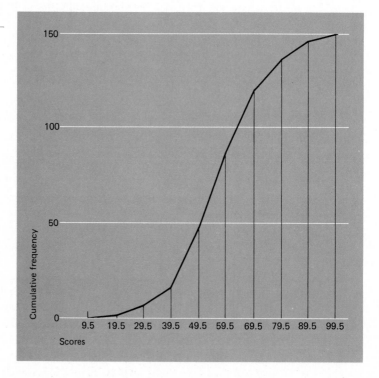

2.3

Ogive of the "or less" distribution of the aptitude scores.

sectors (pie-shaped pieces) which are proportional in size to the corresponding frequencies or percentages. To construct a pie chart, we first convert the distribution into a percentage distribution. Then, since a complete circle corresponds to 360 degrees, we get the central angles of the various sectors by multiplying the percentages by 3.6.

EXERCISES

2.12 The following is the distribution of the total finance charges which 200 customers paid on their budget accounts at a department store:

Amount (dollars)	Frequency
0–19	13
20–39	64
40–59	65
60–79	44
80–99	14

Draw a histogram of this distribution.

2.13 Convert the distribution of the preceding exercise into a cumulative "less than" distribution and draw its ogive.

2.14 The following distribution shows the scores made by 80 entering college freshmen on an English placement examination:

Score	Number of students
45–49	3
50–54	5
55–59	11
60–64	15
65–69	20
70–74	8
75–79	7
80–84	5
85–89	4
90–94	2
Total	80

(a) Draw a histogram of this distribution.
(b) Draw a frequency polygon of this distribution.

2.4

Pictogram of the population of the United States.

1920
1930
1940
1950
1960
1970
1980

Each symbol = 10 million persons

2.5

Electric energy production in the United States (billions of kilowatt-hours).

1955   629
1960   842
1965   1,158
1970   1,640

2.6

Pie chart of the source of the sales dollar of a large diversified paper company.

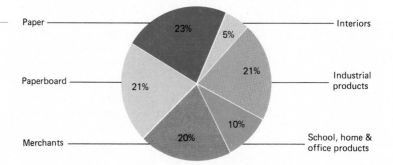

Paper — 23%
Interiors — 5%
Industrial products — 21%
School, home & office products — 10%
Merchants — 20%
Paperboard — 21%

2.15 Convert the distribution of the preceding exercise into an "or more" distribution and draw its ogive.

2.16 A bar chart is a form of graphical presentation which is very similar to a histogram. In this kind of chart (see Figure 2.7), the lengths of the bars are proportional to the class frequencies, but there is no pretense of having a continuous horizontal scale. Draw a bar chart of the distribution of
(a) Exercise 2.12; (b) Exercise 2.14.

2.7

Bar chart of the distribution of the aptitude scores.

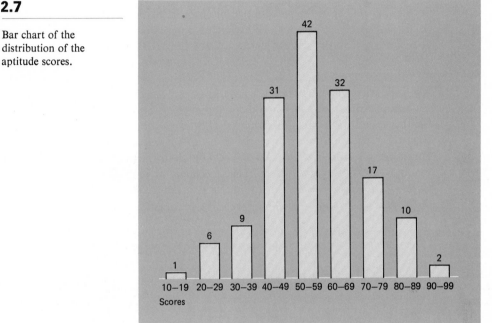

2.17 Draw a pie chart to display the information that a large hospital's expenses are as follows: 73 percent for salaries, professional medical fees, and employee benefits; 13 percent for medical and surgical supplies and equipment; 8 percent for maintenance, food, and power; and 6 percent for administrative services.

2.18 Of the $207 million total raised in a major university's fund drive, $117 million came from individuals and bequests, $24 million from industry and business, and $66 million from foundations and associations. Present this information in a pie chart.

2.19 With reference to Exercise 2.11, present the distribution obtained in that exercise in the form of a
(a) bar chart (see Exercise 2.16); (b) pie chart.

2.3

A Word of Caution

Intentionally or unintentionally, frequency tables, histograms, and other pictorial presentations are sometimes very misleading. Suppose, for instance, that when we grouped the clerical aptitude scores, we combined the two classes 60–69, and 70–79 into one class, the class 60–79. This new class has a frequency of 49, but in Figure 2.8, where we still use the heights of the rectangles to represent the class frequencies, we get the erroneous impression that this class contains about one-half of the scores (instead of one-third). This is due to the fact that when we compare the sizes of rectangles, triangles, and other plane figures, we instinctively compare their areas and not their sides. This does not matter when the class intervals are equal, but in Figure 2.8 the class 60–79 is twice as wide as the others, and we should compensate for this by dividing the height of the rectangle by 2. Figure 2.9 (where the vertical scale, which has lost its significance, has been omitted) shows the result of this adjustment. Now we get the correct impression that the class 60–79 contains nearly as many items as the two classes 50–59 and 80–89 combined. This is as it should be, since the frequency of the class 60–79 is 49 and the sum of the frequencies of the two classes 50–59 and 80–89 is $42 + 10 = 52$.

2.8

Incorrectly modified histogram of the distribution of the aptitude scores.

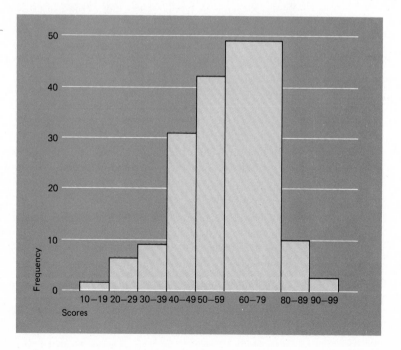

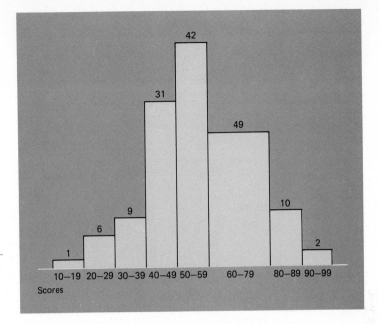

Correctly modified
histogram of the
distribution of the
aptitude scores.

The same difficulty arises in the construction of pictograms of the sort where the sizes of various objects are intended to illustrate and emphasize differences among the data. Suppose, for instance, we want to dramatize the fact that the total wine tax collected by the federal government increased from $91 million in 1959 to $182 million in 1972. Since the amount has doubled, we might be tempted to draw a picture such as the one in Figure 2.10, where the height

2.10

Misleading pictogram
of the 1959 and 1972
wine tax collected by
the federal government.

91 million tax dollars in 1959

182 million tax dollars in 1972

91 million tax dollars in 1959

2.11

Corrected pictogram of the 1959 and 1972 wine tax collected by the federal government.

182 million tax dollars in 1972

and width of the tax dollar representing 1972 collections are both twice the height and width of the tax dollar representing 1959 collections. However, this gives the false impression that 1972 collections were four times (instead of twice) those of 1959, and objectivity requires that we make the side of the 1972 dollar $\sqrt{2} = 1.41$ times the corresponding side of the 1959 dollar, as in Figure 2.11.

2.4

Check List of Key Terms

2.5

Review Exercises

2.20 The class marks of a distribution of the daily number of calls received by a landscaping service are 2, 7, 12, 17, and 22. What are the class limits of this distribution?

2.21 The number of empty seats on flights from Dallas to New Orleans are grouped into a table with the classes 0–9, 10–19, 20–29, 30–39, and 40 or more. Is it possible to determine from this table the number of flights on which there were
(a) at least 30 empty seats;   (c) at least 19 empty seats;
(b) more than 30 empty seats;   (d) at least 10 empty seats?

2.22 The year-end executive bonuses paid by a large manufacturer are all multiples of $100, the smallest being $8,200 and the largest being $24,700. Write both the class limits and the class boundaries of a distribution with six equal classes into which these payments might be grouped.

2.23 In a two-week study of the productivity of workers, the following data were obtained on the average number of acceptable pieces (rounded to the nearest tenth) which 100 workers produced per hour:

3.6	6.7	4.3	2.8	5.6	7.9	8.4	4.9	3.6	6.5
8.2	2.2	6.2	5.5	7.2	6.8	4.0	3.7	7.8	4.3
6.5	7.3	5.7	3.9	4.6	5.7	5.6	6.0	5.0	8.8
4.5	5.6	7.5	4.0	5.1	7.0	7.4	7.6	4.8	5.9
3.4	7.4	5.3	6.4	8.0	3.2	6.3	5.2	6.2	3.5
5.1	3.5	4.4	4.5	5.4	5.1	5.5	4.8	6.0	7.6
6.8	8.5	6.0	7.7	6.1	3.3	4.5	6.1	5.3	2.1
4.7	5.2	6.8	5.2	6.9	4.2	6.7	3.4	5.3	4.5
5.4	4.1	5.9	5.3	5.0	7.3	6.1	5.5	6.5	6.2
7.0	3.8	5.0	4.7	3.5	2.6	5.8	8.2	7.4	4.1

(a) Group these figures into a distribution having the classes 2.0–2.9, 3.0–3.9, 4.0–4.9, 5.0–5.9, 6.0–6.9, 7.0–7.9, and 8.0–8.9.
(b) Draw a histogram of the distribution of part (a).

2.24 Convert the distribution of part (a) of the preceding exercise into a cumulative "less than" distribution and draw its ogive.

2.25 The daily number of orders received by a wholesale hardware supplier are grouped into a table having the classes 100–139, 140–179, 180–219, 220–259, and 260–299. Find
(a) the class boundaries;   (c) the class interval.
(b) the class marks;

2.26 Use a daily newspaper listing prices on the American Stock Exchange to construct a table showing how many of the D, E, and F stocks traded on a certain day closed up, down, or unchanged from their previous close.

2.27 The following is the distribution of the actual shelf weight (in ounces) of a sample of 60 "one-pound" sacks of trisodium phosphate, which were filled from bulk stock by a part-time clerk in a hardware store:

Weight	Number of sacks
15.60–15.99	8
16.00–16.39	27
16.40–16.79	19
16.80–17.19	5
17.20–17.59	1

Draw a histogram of this distribution.

2.28 Convert the distribution of the preceding exercise into a cumulative "or more" distribution and draw its ogive.

2.29 To group data on the number of rainy days reported by a weather station for the month of June during the last 50 years, a meteorologist uses the classes 0–5, 6–10, 12–17, 18–23, and 23–30. Explain where difficulties might arise.

2.30 The distribution of revenues of a large diversified paper company in 1973 showed: raw materials, supplies, and services, 61 percent; salaries, wages, and employee benefits, 29 percent; retained earnings, 2 percent; dividends, 1.5 percent; depreciation, 2.5 percent; and taxes, 4 percent. Present this information in a pie chart.

2.31 A survey made at a resort city showed that 50 tourists arrived by the following means of transportation: car, train, plane, plane, plane, bus, train, car, car, car, plane, car, plane, train, car, car, bus, car, plane, plane, train, train, plane, plane, car, car, train, car, car, plane, car, car, plane, bus, plane, bus, car, plane, car, car, train, train, car, plane, bus, plane, car, car, train, and bus. Construct a categorical distribution showing the frequencies corresponding to the different means of transportation, and present it pictorially in the form of a pie chart.

2.32 The pictogram of Figure 2.12 is intended to illustrate the fact that the total value of corporation stock held by U.S. life insurance companies tripled from 1960 to 1970. How should it be modified so that it will convey a fair impression of the actual change?

2.33 The following is a distribution of the number of days which elapsed between the day on which 90 commercial property owners in a large city petitioned

2.12

Value of corporation stock held by U.S. life insurance companies.

5 billion dollars 1960

15 billion dollars 1970

for tax hearings and the day on which formal acknowledgment of the petitions was received from the appeals board:

Number of days	Number of petitioners
20–24	2
25–29	6
30–34	13
35–39	18
40–44	22
45–49	10
50–54	9
55–59	5
60–64	3
65–69	2

(a) Draw a histogram of this distribution.
(b) Draw a histogram of the modified distribution obtained by combining the cases from 30 through 44 into one class.

2.34 A set of measurements of the lengths of a large number of pieces of scrap metal given to the nearest tenth of an inch are grouped into a table whose classes have the boundaries 4.95, 6.95, 8.95, 10.95, and 12.95. What are the lower and upper limits of each class?

Summarizing Data: Statistical Descriptions

3

Descriptions of statistical data can be quite brief or quite elaborate, depending on what we intend to do with them. In any event, when we describe collections of data we try to say neither too little nor too much. On some occasions it may be satisfactory to present data just as they are, in raw form, and let them speak for themselves; on others, it may be necessary only to group a set of data and present its distribution in tabular or graphical form. Usually, though, data have to be summarized further by means of appropriate statistical descriptions, and in this chapter we shall concentrate on two special kinds of descriptive measures, called measures of location and measures of variation. The former are discussed in Sections 3.2 and 3.3, and the latter in Sections 3.5 and 3.6. The description of grouped data is treated in Section 3.8 and some further kinds of descriptions are presented in Section 3.9.

3.1

Populations and Samples

Before we study specific statistical descriptions, let us define the terms "population" and "sample." If a set of data consists of all conceivably possible (or hypothetically possible) observations of a certain phenomenon, we call it a **population**; if a set of data contains only a part of these observations, we call it a **sample**. We added the qualification "hypothetically possible" in the first definition to take care of such clearly hypothetical situations as where we look at the outcomes (heads or tails) of 10 tosses of a coin as a sample from the population which consists of all possible tosses of the coin, or where we look at five measurements of the length of a steel shaft as a sample from the population of all possible measurements of its length. In fact, we often look at the results of an experiment as a sample of what we would see if the experiment were repeated over and over again.

Although we are free to call any group of items a population, what we do in practice depends on the context in which the items are to be viewed. Suppose, for instance, that we are offered a lot of 500 ceramic tiles, which we may or may not buy depending on their strength. If we measure the breaking strength of 10 of these tiles in order to estimate the average breaking strength of all the tiles, these 10 measurements constitute a sample from the population which consists of the breaking strengths of all the tiles. Denoting the number of items in a finite population, the **population size,** by the letter N, and the number of items in a sample, the **sample size,** by the letter n, we could thus say that we have a sample of size $n = 10$ drawn from a population of size $N = 500$. In another context, however, if we were considering entering into a long-term contract calling for the delivery of tens of thousands of such tiles, we would look upon the breaking strengths of the original 500 tiles only as a sample. Similarly, the complete set of figures for a recent year, giving the elapsed time between the application and the issuance of residential building permits in Atlanta, can be looked on as either a population or a sample. If we were interested only in the city of Atlanta and that particular year, we would consider these data to constitute a population; on the other hand, if we wanted to generalize about the time required for the issuance of residential building permits in the entire United States, in some other large cities, or in some other years, these data would constitute only a sample.

As we have used it here, the word "sample" has very much the same meaning as it has in everyday language. A newspaper considers the attitudes of 100 readers toward a proposed city-county sports complex to be a sample of the attitudes of all its readers toward the complex; and a homemaker considers a trial-size box of a new heavy-duty laundry detergent a sample of all the boxes of this detergent. Later, we shall use the word "sample" to refer only to sets of data which can reasonably serve as the basis for valid generalizations about the populations from which they came, and in this more technical sense many sets of data which are popularly called samples are not samples at all.

In this chapter we shall describe things statistically, but not make any generalizations. For future reference, though, it is important even here to distinguish between populations and samples. Thus, we shall use different symbols depending on whether we are describing populations or samples, and sometimes even different formulas.

3.2

Measures of Location: The Mean, the Median, and the Mode

It is often necessary to represent a set of data in terms of a single number which, in its way, is descriptive of the entire set. Exactly what sort of number we choose depends on the particular characteristic we want to describe. In one study, for example, we might be interested in the extreme (smallest and largest) values among the data; in another, in the value which is exceeded by only 10 percent of the values; and in still another, in the total of all the values. In this section, we shall consider those measures which somehow describe the center or middle of a set of data—the **measures of central location**.

Of the different measures of central location, by far the best known and the most widely used is the **arithmetic mean**, or simply the **mean**, which we define as follows:

The mean of n values is the sum of the values divided by n.

In everyday speech this value is often called the "average," and on occasion we shall call it that ourselves. But, as we shall see, there are other "averages" in statistics, and we cannot afford to speak loosely when there is any danger of ambiguity.

EXAMPLE Given that during the 12 months of 1980 a state auditor charged 5, 2, 1, 3, 3, 8, 6, 7, 4, 1, 2, and 6 calls to her telephone credit card, find the mean, that is, the average number of charges per month.

SOLUTION The total for the twelve months is $5 + 2 + 1 + 3 + 3 + 8 + 6 + 7 + 4 + 1 + 2 + 6 = 48$, and therefore

$$\text{mean} = \frac{48}{12} = 4$$

Since we shall have occasion to calculate the means of many different sets of sample data, it will be convenient to have a simple formula that is always applicable. This requires that we represent the figures to be averaged by some general symbol such as x, y, or z. Choosing the letter x, we can refer to the n values in a sample as x_1 ("x sub-one"), $x_2, x_3, \ldots$, and x_n and write

$$\text{sample mean} = \frac{x_1 + x_2 + x_3 + \cdots + x_n}{n}$$

This formula is perfectly general and it will take care of any set of sample data, but it can be made more compact by assigning the sample mean the special symbol $\bar{x}$ ("x bar") and using the $\sum$ (Greek capital sigma) notation. In this notation we let $\sum x$ stand for "the sum of the x's," that is, for $x_1 + x_2 + x_3 + \ldots + x_n$. Thus, the mean, $\bar{x}$, of a set of sample values $x_1, x_2, x_3, \ldots$, and x_n is given by the formula

Sample mean

$$\bar{x} = \frac{\sum x}{n}$$

If we refer to the measurements as y's or z's, we write their mean as $\bar{y}$ or $\bar{z}$. In the formula above, the term $\sum x$ does not state explicitly which values of x are to be added; let it be understood, however, that $\sum x$ always refers to the sum of all the x's under consideration. In the technical note of Section 3.10 the sigma notation is discussed in more detail.

The mean of a population of N items is defined in the same way. It is the sum of the N items, $x_1 + x_2 + x_3 + \ldots + x_N$, or $\sum x$, divided by N. Giving the population mean the special symbol μ (mu, the Greek letter for lowercase m), we write

Population mean

$$\mu = \frac{\sum x}{N}$$

with the reminder that in this formula $\sum x$ is now the sum of all N values of x which constitute the population. Also, to distinguish between descriptions of populations and descriptions of samples, statisticians not only use different symbols such as μ and $\bar{x}$, but they refer to descriptions of populations as **parameters** and descriptions of samples as **statistics**. Parameters are usually denoted by Greek letters.

The terminology and notation introduced here is illustrated by the following example:

A sample of $n = 5$ high-precision spring-driven motors is taken from a production lot of $N = 100,000$ such motors. The motors are wound and started, and their running times clocked at 3.50, 3.65, 3.55, 3.58, and 3.52 minutes. Find their mean running time.

SOLUTION Substituting into the formula for $\bar{x}$, we have

$$\bar{x} = \frac{\sum x}{n} = \frac{3.50 + 3.65 + 3.55 + 3.58 + 3.52}{5} = 3.56 \text{ minutes}$$

and if these times constitute a sample in the technical sense (that is, a set of data from which valid generalizations can be made), we can estimate the mean running time μ of all the 100,000 motors as 3.56 minutes. Moreover, if the sample is a particular kind of sample called a "random sample," we can apply techniques we shall study later and say that we are reasonably sure that this estimate is in error by at most 0.073 minute (or about 4.4 seconds).

The widespread use of the mean to describe the "middle" of a set of data is not accidental. Aside from the fact that it is a simple, familiar measure, the mean has the following desirable properties:

1. It can be calculated for any set of numerical data, so it always exists;
2. A set of numerical data has one and only one mean, so it is always unique;
3. It lends itself to further statistical treatment (for instance, the means of several sets of data can be combined into the overall mean of all the data); and
4. It is relatively reliable in the sense that the means of many samples drawn from the same population usually do not fluctuate, or vary, as widely as other statistics used to estimate the population mean μ.

The fourth property is of fundamental importance in statistical inference.

There is another characteristic of the mean which, on the surface, seems desirable but may not be so:

5. It takes into account every item of the data.

However, samples sometimes contain very small or very large values which are so far removed from the main body of the data that the appropriateness of including them in the samples is questionable. When such values, called **outliers,** are averaged in with the other values, they can affect the mean to such an extent that its worth as a reasonable description of the "middle" of the data becomes debatable.

EXAMPLE Five light bulbs burned out after lasting for 867, 849, 840, 852, and 822 hours of continuous use. Find the mean and also determine what the mean would be if the second value is recorded incorrectly as 489 instead of 849.

SOLUTION For the original data we get

$$\bar{x} = \frac{867 + 849 + 840 + 852 + 822}{5} = \frac{4,230}{5} = 846$$

and with 489 instead of 849 we get

$$\bar{x} = \frac{867 + 489 + 840 + 852 + 822}{5} = \frac{3,870}{5} = 774$$

This illustrates how a careless error in recording a set of data can have a pronounced effect on the mean.

Outliers resulting from such things as gross errors in recording data, gross errors in calculations, malfunctioning of equipment, or contamination can sometimes be identified as to their source and simply eliminated from the data before they are averaged. However, instead of omitting outliers and calculating a sort of modified mean in an attempt to avoid this difficulty, we can use another measure of location, called the median, to describe the "middle" of the data. Unlike the mean, the median is not easily affected by extreme values. By definition,

> The median of a set of data is the value of the middle item (or the mean of the values of the two middle items) when the data are arrayed or ordered, that is, arranged in an increasing or decreasing order of magnitude.

Like the mean of a sample of n values of x, the sample median, which we denote $\tilde{x}$, can be used to estimate the population mean; in fact, this accounts for our main interest in this statistical measure.

If there is an odd number of items in a set of data, there is always a middle item whose value is the median.

EXAMPLE Eleven large corporations reported that in 1980 they made cash donations to 9, 16, 11, 19, 11, 10, 13, 12, 6, 9, and 12 colleges. Find the median number of donations.

SOLUTION Arranging these figures according to size, we get

$$6 \quad 9 \quad 9 \quad 10 \quad 11 \quad 11 \quad 12 \quad 12 \quad 13 \quad 16 \quad 19$$

and it can be seen that the median is 11.

Generally speaking, the median of a set of n items, where n is odd, is the value of the $\frac{n+1}{2}$th largest item. For instance, the median of 35 numbers is the value of the $\frac{35 + 1}{2} = $ 18th largest item, and the median of 199 numbers is the value of the $\frac{199 + 1}{2} = $ 100th largest item.

For a set of data containing an even number of items, there is no single middle item and the median is defined as the mean of the values of the two middle items, but the formula $\frac{n + 1}{2}$ still serves to locate the position of the median. For instance, for $n = 12$ we get $\frac{12 + 1}{2} = 6.5$, and the median is the mean of the values of the 6th and 7th largest items; also, for $n = 50$ we get $\frac{50 + 1}{2} = 25.5$, and the median is the mean of the values of the 25th and 26th largest items.

EXAMPLE   On ten days, a bank had 18, 13, 15, 12, 8, 3, 7, 14, 16, and 3 foreign currency transactions. Find the median.

SOLUTION   Arranging these figures according to size, we get

$$3 \quad 3 \quad 7 \quad 8 \quad 12 \quad 13 \quad 14 \quad 15 \quad 16 \quad 18$$

and it can be seen that the median is 12.5, the mean of 12 and 13.

The mean of the numbers in this example is 10.9, and it should not come as a surprise that it differs from the median, which is 12.5. Each of these averages describes the middle of the data in its own way. The median is typical (central or average) in the sense that it splits the data into two parts so that the values of half the items are less than or equal to the median, and the values of the other half are greater than or equal to the median. The mean, on the other hand, is typical in the sense that if all the values were the same size (but their total is the same), they would all be equal to the mean. (Since $\bar{x} = \frac{\sum x}{n}$, it follows that $n \cdot \bar{x} = \sum x$, and, hence, that n values, each of size $\bar{x}$, have the same total as the actual n values of x.)

Like the mean, the median always exists and is unique for any set of data. It can also be used to define the middle of a number of objects, properties, or qualities which are not really quantitative in nature. For instance, we can rank a number of tasks according to their difficulty and then describe the middle (or median) one as being of "typical" difficulty. On the less desirable side, ordering large sets of data manually can be a very tedious job, and what is more serious from the standpoint of statistical inference, a sample median is usually not so reliable an estimate of the population mean as the (arithmetic) mean of the same data. The medians of many samples drawn from the same population usually vary more widely than the corresponding sample means.

Another measure which is sometimes used to describe the center of a set of data is the **mode**, which is defined simply as the value that occurs with the highest frequency. Its two main advantages are that it requires no calculations, only counting, and that it can be determined for qualitative as well as quantitative data.

<table>
<tr><td>EXAMPLE</td><td>The 12 sessions of a seminar in personal financial planning were attended by 22, 16, 20, 20, 15, 16, 12, 14, 16, 14, 11, and 16 persons. Find the mode.</td></tr>
<tr><td>SOLUTION</td><td>Among these numbers 22, 15, 12, and 11 occur once, 20 and 14 occur twice, and 16 occurs four times. Thus, 16 is the modal attendance, or the mode.</td></tr>
</table>

Aside from the fact that the mode is an extremely poor measure of location in statistical inference, it also has the disadvantage that for some sets of data it may not exist and for others it may not be unique. For instance, there is no mode of the ages 19, 23, 29, 31, 25, and 22 (which are all different), and there are two modes, 9 and 14, of the dress sizes 7, 10, 14, 9, 9, 14, 9, 18, 16, 12, 11, 14, 14, 14, 9, 20, 9, and 11. The fact that a set of data has more than one mode (or is **bimodal**) is sometimes indicative of a lack of homogeneity in the data. For instance, the two modes of the dress sizes are accounted for by the fact that they are the sizes worn by nine mothers and their teenage daughters.

Besides the mean, the median, and the mode, there are various other measures which describe in some way the middle or the center of a set of data. In addition to the **weighted mean** discussed in Section 3.3, there are the **geometric** and **harmonic means** given in Exercises 3.23 and 3.24 on page 40, and the **midrange**, which is mentioned on page 65.

3.3

Measures of Location:
The Weighted Mean

In averaging quantities it is often necessary to account for the fact that not all of them are equally important in the phenomenon being described. For instance, if a person makes three investments which return 7, 8, and 9 percent, the average return is $\frac{7 + 8 + 9}{3} = 8$ percent only if the person puts the same amount in each of the investments. In order to give quantities being averaged their proper degree of importance, it is necessary to assign them (relative importance) **weights**, and then calculate a **weighted mean**. In general, the weighted mean $\bar{x}_w$ of a set of numbers $x_1, x_2, x_3, \ldots$, and x_n, whose relative importance is expressed numerically by a corresponding set of numbers $w_1, w_2, w_3, \ldots$, and w_n, is given by

Weighted mean

$$\bar{x}_w = \frac{w_1 x_1 + w_2 x_2 + \cdots + w_n x_n}{w_1 + w_2 + \cdots + w_n} = \frac{\sum w \cdot x}{\sum w}$$

If all the weights are equal, this formula reduces to that of the ordinary (arithmetic) mean.

EXAMPLE In a recent year, flounders, haddock, ocean perch, chinook salmon, and tuna brought commercial fishermen 26.5, 36.9, 8.1, 102.3, and 41.0 cents per pound. Given that the catches were 162 million pounds of flounders, 8 million pounds of haddock, 41 million pounds of ocean perch, 201 million pounds of chinook salmon, and 392 million pounds of tuna, what is the overall average price per pound received by these fishermen?

SOLUTION Substituting $x_1 = 26.5$, $x_2 = 36.9$, $x_3 = 8.1$, $x_4 = 102.3$, $x_5 = 41.0$, $w_1 = 162$, $w_2 = 8$, $w_3 = 41$, $w_4 = 201$, and $w_5 = 392$ into the formula for $\bar{x}_w$, we get

$$\bar{x}_w = \frac{162(26.5) + 8(36.9) + 41(8.1) + 201(102.3) + 392(41.0)}{162 + 8 + 41 + 201 + 392}$$

$$= \frac{41,554.6}{804}$$

$$= 51.7 \text{ cents}$$

The figure in the denominator, 804, is the total catch in millions of pounds, and the figure in the numerator, 41,554.6, is the total value of the catch in millions of cents, that is, in units of $10,000. Also, if we average 26.5, 36.9, 8.1, 102.3, and 41.0 without using weights, we get

$$\bar{x} = \frac{26.5 + 36.9 + 8.1 + 102.3 + 41.0}{5} = 43.0$$

which is much less than the actual average price of 51.7 cents. The reason for this should be apparent.

A special application of the formula for the weighted mean arises when we must find the overall mean, or **grand mean**, of k sets of data having the means $\bar{x}_1, \bar{x}_2, \bar{x}_3, \ldots$, and $\bar{x}_k$, and consisting of $n_1, n_2, n_3, \ldots$, and n_k measurements or observations. The result is given by

Grand mean of combined data

$$\bar{\bar{x}} = \frac{n_1 \bar{x}_1 + n_2 \bar{x}_2 + \cdots + n_k \bar{x}_k}{n_1 + n_2 + \cdots + n_k} = \frac{\sum n \cdot \bar{x}}{\sum n}$$

where the weights are the sizes of the samples, the numerator is the total of all the measurements or observations, and the denominator is the number of items in the combined samples.

EXAMPLE In three separate weeks a discount store chain sold 475, 310, and 420 microwave ovens at average prices of $490, $520, and $495. What is the average price of the ovens sold?

SOLUTION Substituting $n_1 = 475$, $n_2 = 310$, $n_3 = 420$, $\bar{x}_1 = 490$, $\bar{x}_2 = 520$, and $\bar{x}_3 = 495$ into the formula for the grand mean of combined data, we get

$$\bar{\bar{x}} = \frac{475(490) + 310(520) + 420(495)}{475 + 310 + 420}$$

$$= \frac{601,850}{1,205}$$

$$= \$499.46$$

or $499 rounded to the nearest dollar.

EXERCISES

3.1 Suppose that we are given complete information about the total sales tax collected by the 153 restaurants in a city during a given month. Give one illustration each of a problem in which we might consider these data to constitute
(a) a population; (b) a sample.

3.2 Suppose that we have for all the service companies in the states of Maine and Rhode Island the number of days' sales which their uncollected bills amounted to in the second quarter of a particular year. Give one example each of a situation in which these data might be looked upon as
(a) a population; (b) a sample.

3.3 The following are the ages of 20 persons empaneled for jury duty by a court: 48, 58, 33, 42, 57, 31, 52, 25, 46, 60, 61, 49, 38, 53, 30, 47, 52, 63, 41, and 34. Find the mean age.

3.4 During the years 1975 through 1979, the total rainfall in one city was 14.5, 14.2, 13.4, 13.2, and 10.7 inches, and in another city it was 13.9, 11.5, 14.1, 11.5, and 14.0 inches. Which city had the higher mean rainfall during these five years?

3.5 A passenger elevator has a rated maximum capacity of 4,000 pounds. Is the elevator overloaded if at one time it carries
(a) 25 passengers whose mean weight is 167 pounds;
(b) 10 children whose mean weight is 65 pounds and 20 adults whose mean weight is 150 pounds;
(c) 12 women whose mean weight is 115 pounds and 15 men whose mean weight is 180 pounds?

3.6 A sample of five universities in the Midwest in a certain year showed the following default rates on (federally funded) National Direct Student Loans: 40.5, 36.2, 42.1, 30.6, and 37.6 percent. Two years later, following intensive efforts at reducing these rates at the same universities, the rates were 20.2, 24.5, 28.3, 16.0, and 23.0 percent. Find the mean of the rates for each of the two years.

3.7 The "cut-out" syrup density in canned fruits is the percentage (or "degree") by weight of sugar in the syrup solution when a can is opened. Examination of samples of eight cans each of three grades of dried prunes yielded the following cut-out densities (measured in degrees on the Brix scale):

Fancy grade: 33, 35, 32, 32, 35, 30, 33, 34
Choice grade: 24, 27, 25, 30, 30, 28, 28, 30
Standard grade: 23, 22, 18, 18, 20, 24, 20, 20

Calculate the mean, the median, and the mode of each of the three samples.

3.8 In tests of the storage performance of a newly developed food starch in frozen fruit pie fillings, a sample of six specimens withstood 28, 25, 30, 25, 27, and 30 freeze–thaw cycles before evidence of curdling or syneresis (bleeding) appeared. Find the mean, median, and mode of these data.

3.9 A building materials company measures the performance of its 15 salespersons by expressing their total sales for the past year as percentages of quota with the following results: 80, 75, 98, 101, 90, 88, 94, 100, 110, 80, 95, 420, 98, 100, and 99 percent. Find the mean, median, and mode of these percentages and comment on the suitability of each as a way of expressing typical performance of the sales force.

3.10 An inspection of 18 newly completed home plumbing jobs in one city turned up 0, 1, 3, 0, 0, 0, 1, 0, 4, 0, 0, 1, 1, 2, 0, 1, 0, and 2 violations of the city building code. Find
(a) the mean; (c) the mode.
(b) the median;

3.11 Twenty-five power failures lasted 138, 142, 113, 126, 135, 142, 159, 157, 140, 157, 121, 128, 142, 164, 155, 139, 143, 158, 140, 118, 142, 146, 123, 130, and 137 minutes. Determine
(a) the median; (b) the mode.

3.12 Asked for their favorite color, fifty persons said: red, blue, blue, green, yellow, blue, brown, red, blue, red, red, green, white, blue, red, green, blue, red, green, green, purple, white, yellow, blue, blue, blue, red, red, brown, orange, red, green, blue, blue, red, black, blue, red, yellow, green, yellow, blue, blue, orange, red, green, white, purple, blue, and red. What is their modal choice?

3.13 It has been reported that "the typical embezzler is about 34, married and the father of two children, belongs to the middle-income group, and works on the average five years before he begins to rob the till." Comment on the statistical aspects of this statement.

3.14 A bill was once introduced in a state legislature to repeal the sales tax on prescription drugs. Comment on the argument of the state finance director that "the average per capita prescription bill for the past three years is a trifling $2.00, which is not really a burden to anyone."

3.15 If a person invests $2,000 at 7 percent, $5,000 at 8 percent, and $25,000 at 9 percent, what is the average return on these investments?

3.16 An instructor gives four tests during the semester, and in calculating the final grade, weights them in proportion to their length. If a student scores 66 on a 1-hour test, 72 on another 1-hour test, 85 on a 20-minute test, and 50 on the 2-hour final, what is her mean grade on the four tests?

3.17 With reference to Exercise 3.16, if a 65-point average on the tests is the minimum passing score, how many points would the student have needed on the final test in order to pass?

3.18 In a study concerned with the impact of rising home construction and replacement costs on the adequacy of home insurance policies, a sample of 12 type A homes, 24 type B homes, and 48 type C homes is taken in a certain region. Among them the type A homes were underinsured on the average by $4,000, the type B homes by $5,200, and the type C homes by $5,900. What is the average amount of underinsurance on all these homes?

3.19 On the first exam in a history course 5 freshmen averaged 62 points, 12 sophomores averaged 60 points, 15 juniors averaged 66 points, and 3 seniors averaged 70 points. What was the overall class average on the exam?

3.20 In one year a large national employer made plan A awards averaging $4,200 to 150 employees and plan B awards averaging $4,900 to 350 employees. What was the average award made to these employees?

3.21 In a study of insurance costs it was found that in one year 715,673 automobile property damage claims for $100 or less averaged $33.91, 157,879 claims for $101 to $1,000 averaged $216.89, and 1,707 claims for over $1,000 averaged $1,635.09. What is the overall average of these claims?

3.22 If some objects whose mean weight is 20 pounds weigh a total of 200 pounds, and some other objects whose mean weight is 35 pounds weigh a total of 700 pounds, what is the mean weight of all the objects combined?

3.23 The **geometric mean** of a set of n positive numbers is the nth root of their product. If the numbers are all the same, the geometric mean equals the arithmetic mean, but otherwise the geometric mean is always less than the arithmetic mean. For example, the geometric mean of the numbers 1, 1, 2, and 8 is $\sqrt[4]{1 \cdot 1 \cdot 2 \cdot 8} = 2$, but their (arithmetic) mean is 3. The geometric mean is used mainly to average ratios, rates of change, and index numbers (see Chapter 4), and in practice it is usually calculated by making use of the fact that the logarithm of the geometric mean of a set of numbers equals the arithmetic mean of their logarithms.
  (a) Find the geometric mean of 7 and 28.
  (b) Find the geometric mean of $\frac{1}{10}$, 2, and 625.
  (c) Find the geometric mean of 1, 2, 4, 8, and 16.
  (d) A company takes a sample of five items it buys and finds that they cost 112, 120, 104, 240, and 116 percent of what they cost a year earlier. Find the geometric mean of these percentages,

3.24 The **harmonic** mean of n numbers $x_1, x_2, x_3, \ldots,$ and x_n is defined as n divided by the sum of the reciprocals of the n numbers, or $\dfrac{n}{\sum 1/x}$. The harmonic mean has limited usefulness, but it is appropriate in some special situations. For instance, if a commuter drives 10 miles on the freeway at 60 miles per hour and the next 10 miles off the freeway at 30 miles per hour, he will not have averaged $\dfrac{60 + 30}{2} = 45$ miles per hour. He will have driven 20 miles in a total of 30 minutes, so his average speed is 40 miles per hour.
  (a) Verify that the harmonic mean of 60 and 30 is 40, so that it gives the appropriate "average" in this case.

(b) If a merchant spends $12 on novelty items costing 40 cents per dozen and another $12 on other novelty items costing 60 cents per dozen, what is her average cost per dozen?

(c) If a bakery buys $72 worth of honey at 60 cents per half pound, $72 worth at 72 cents per half pound, and $72 worth at 90 cents per half pound, what is the average cost per half pound of the honey?

3.4

The Concept of Variability

One of the most important characteristics of a set of data is that the values are usually not all alike; indeed, the precise extent to which they are unalike, or vary among themselves, is of basic importance in statistics. Measures of central location describe one important aspect of a set of data—their middle or their "average"—but they tell us nothing about this other basic characteristic. Hence, we require ways of measuring the extent to which data are dispersed, or spread out, and the statistical measures which provide this information are called **measures of variation.** We give below three examples which we hope will illustrate the importance of these measures.

Suppose that we are considering buying some common stock in one or the other of two food-service companies in a large metropolitan area. Among other things, we are interested in their operating results, as measured by the net profit on each $1 sale during the past six years, and we find that both firms averaged a net profit of 4.8 cents on each $1 sale. From this it might appear that the operating results of the two firms are equally good and that a choice between them must depend on other investment considerations. However, these averages have obscured an interesting and important fact. Over the six-year period the net profits of one firm were

$$5.2, \quad 4.5, \quad 3.9, \quad 4.8, \quad 5.0, \quad \text{and} \quad 5.4 \text{ cents}$$

varying from 3.9 to 5.4, while the net profits of the other were

$$7.9, \quad 7.0, \quad -5.3, \quad 14.2, \quad -11.0, \quad \text{and} \quad 16.0 \text{ cents}$$

varying from −11.0 to 16.0. These figures show clearly that the first firm has turned in a very consistent operating performance, but that the other is a highly inconsistent, erratic sort of "feast or famine" performer. This wide variability in the second firm's operating results introduces an element of risk into its investment potential which cannot be observed from a comparison of the mean profit figures alone.

The concept of variability or dispersion is of fundamental importance in statistical inference (estimation, tests of hypotheses, forecasting, and so on). Suppose that we have a somewhat bent and worn coin and we wonder whether it is still balanced or "fair" and will, in the long run, fall heads about 50 percent of the times it is tossed. What can we say if we actually toss the coin 100 times and get 29 heads and 71 tails? Is there anything unusual or out of the ordinary about this result? Specifically, does the shortage of heads—only 29 where we might have expected 50—suggest that, in fact, the coin is not fair?

To answer these questions, we must have some idea of how a fair coin behaves when it is tossed, that is, some idea about the magnitude of the fluctuations, or variations, produced by the action of chance in the number of times, in 100 tosses, that a fair coin falls heads. Suppose that, to get this information, we take a brand new coin in mint condition—presumably a balanced coin—and toss it 100 times, then again 100 times, and repeat this procedure until we have 10 sets of 100 tosses each. Suppose, furthermore, that in these 10 sets we get 51, 54, 58, 56, 41, 49, 58, 53, 47, and 56 heads. The number of heads varies from 41 to 58, so we might conclude from these results that a discrepancy of about 10 heads from the expected 50 heads is not unusual but that a discrepancy of 21 heads is so large that we would hesitate to attribute it to chance. On the basis of this small experiment, it seems more reasonable to conclude that the original coin is not balanced and is behaving accordingly, than that it is balanced and behaving in an unusual way.

For our third example, suppose that we want to estimate the true mean (net) weight of all cans of beef hash in a very large production lot put out by a food processor. To do this we take a random sample of three cans from the lot and find that their weights are 15.0, 14.8, and 15.2 ounces. The mean of this sample is $\bar{x} = 15.0$ ounces, and in the absence of any further information we may use this figure to estimate the actual mean weight of all the cans. But, obviously, the "goodness" of this estimate depends in a very real way on the population variability, that is, on the variability of the weights of all the cans in the entire lot. To illustrate this, let us consider the following two possibilities:

Case 1: The true mean weight is 15.1 ounces, the filling process is very consistent, and all the cans in the lot weigh somewhere between 14.8 and 15.4 ounces.

Case 2: The true mean weight is 15.1 ounces, but the filling process is very inconsistent, and the weights of the cans in the lot vary widely from 13.0 to 17.2 ounces.

If the population of weights whose mean we are estimating is the relatively homogeneous one described in case 1, we can be sure that the sample mean will not differ from the true mean by much, regardless of the size of the sample. In fact, a sample mean cannot possibly be off by more than 0.3 ounce, and off by that much only if the sample values are all 14.8 or all 15.4 ounces. The population described in case 2, however, is a much less homogeneous collection of weights, even though its mean is the same as that in case 1. If, by chance, the three sample weights chosen from the second population were all 13.0 or all 17.2 ounces, the sample mean would differ from the population mean (which we are estimating) by 2.1 ounces. Clearly, in order to judge the closeness of an estimate or the "goodness" of a generalization based on a sample, we must know something about the variability in the population from which the sample came.

The preceding three examples are intended to suggest how the concept of variability plays an important part in practically all aspects of statistics; specifically, we hope they illustrate the need for understanding and measuring chance variation. In later chapters we shall treat such problems more rigorously and in much more detail.

3.5

Measures of Variation: The Range

In the first of the examples above, we actually introduced one way of measuring variability when we gave the two extreme values of each set of data. More or less the same thing is accomplished by taking the difference between the two extremes; this statistical measure is called the range.

EXAMPLE   Four batteries have lifetimes of 6.2, 6.8, 6.0, and 6.4 hours. Find the range of these values.

SOLUTION   Since the largest value is 6.8 and the smallest value is 6.0, the range is $6.8 - 6.0 = 0.8$ hour.

The range is easy to calculate and easy to understand, but despite these advantages it is not a very useful measure of variation. Its main shortcoming is that it tells us nothing about the dispersion of the data which fall between the two extremes. For instance, each of the following sets of data

5	17	17	17	17	17	17	17	17	17
5	5	5	5	5	17	17	17	17	17
5	6	8	10	11	14	14	15	16	17

has a range of $17 - 5 = 12$, but the dispersion is quite different in each case.

In some cases, where the sample size is quite small, the range can be an adequate measure of variation. For instance, it is used widely in industrial quality control, where it is necessary to keep a close check on the quality of raw materials, semifinished, and finished products on the basis of many small samples taken at more or less regular intervals of time.

3.6

Measures of Variation: The Variance and the Standard Deviation

To define the standard deviation, by far the most useful measure of variation, let us observe that the dispersion of a set of data is small if the numbers are closely bunched about their mean, and that it is large if the numbers are scattered widely about their mean. Hence, it would seem reasonable to measure the variation of a set of data in terms of the amounts by which the various numbers deviate from their mean. If a set of numbers $x_1, x_2, x_3, \ldots$, and x_N, constituting a population, has the mean μ, the differences $x_1 - \mu$, $x_2 - \mu$, $x_3 - \mu, \ldots$, and $x_N - \mu$ are called the **deviations from the mean**, and it seems that we might use their arithmetic mean as a measure of the variation in the population. Unfortunately, this will not do. Unless the x's are all equal, some of the deviations will be positive, some will be negative, and it can be shown that their sum, $\sum (x - \mu)$, and consequently also their mean, is always zero.

Since we are really interested in the magnitude of the deviations, and not in their direction, we might simply ignore their signs and define a measure of variation in terms of the absolute values of the deviations from the mean. Indeed, adding the values of the deviations from the mean as if they were all positive and dividing by their number gives an intuitively appealing measure of variation called the **mean deviation**. However, using precisely the same deviations from the mean there is another way to proceed which does not require absolute values and, hence, is preferable on theoretical grounds. The squares of the deviations from the mean cannot be negative; in fact, they are all positive unless x happens to coincide with the mean, in which case both $x - \mu$ and $(x - \mu)^2$ are equal to zero. Therefore, it seems reasonable to measure the variability of a set of data in terms of the squared deviations from the mean, and this leads us to define the **population variance** in the following way:

Population variance

$$\sigma^2 = \frac{\sum (x - \mu)^2}{N}$$

This measure of variation, denoted by σ^2 (where σ, sigma, is the Greek letter for lowercase s), is simply the mean of the squared deviations from the population mean μ, and it is sometimes called the **mean-square deviation**.

The variance of a set of data is an extremely important measure of variation and it is used extensively in statistical work. By reason of squaring the deviations, however, the variance is not in the same unit of measurement as the data themselves and their mean—if the data are in inches, the variance is in inches squared; if the data are in pounds, the variance is in pounds squared; and so on. However, if we take the square root of the population variance, we get another measure of variability called the **population standard deviation**, or sometimes the **root-mean-square deviation**.

Population standard deviation

$$\sigma = \sqrt{\frac{\sum (x - \mu)^2}{N}}$$

The term "root-mean-square deviation" describes it precisely—it is the square root of the mean of the squared deviations from the population mean μ. Also, it is in the same unit of measurement as the original data.

It may seem logical to use the same formulas for a sample, with n and $\bar{x}$ substituted for N and μ, but this is not quite what we do. Instead of dividing the sum of the squared deviations from the sample mean $\bar{x}$ by n, we divide it by $n - 1$ and define the **sample standard deviation**, denoted by s, as

Sample standard deviation

$$s = \sqrt{\frac{\sum (x - \bar{x})^2}{n - 1}}$$

and its square, the **sample variance**, as

Sample variance

$$s^2 = \frac{\sum (x - \bar{x})^2}{n - 1}$$

In using $n - 1$ instead of n in the denominator of these two formulas, we are not just being arbitrary. There is a good reason for it, and it is explained in the technical note of Section 3.11.

To calculate the sample standard deviation by the definition formula, we must (1) find $\bar{x}$, (2) determine the n deviations from the mean $x - \bar{x}$, (3) square these deviations, (4) add the squared deviations, (5) divide by $n - 1$, and (6) take the square root of the quantity arrived at in step 5.†

EXAMPLE   The response times in a sample of six switches designed to activate an alarm system upon receiving a certain stimulus are 9, 8, 5, 11, 7, and 5 milliseconds. Calculate the standard deviation.

SOLUTION   We first calculate the mean

$$\bar{x} = \frac{9 + 8 + 5 + 11 + 7 + 5}{6}$$

$$= 7.5$$

then set up the work required to find $\sum (x - \bar{x})^2$ in the following table:

x	$x - \bar{x}$	$(x - \bar{x})^2$
9	1.5	2.25
8	0.5	0.25
5	−2.5	6.25
11	3.5	12.25
7	−0.5	0.25
5	−2.5	6.25
45	0.0	27.50

†Square roots may be looked up in Table XII at the end of the book.

Finally, we divide $\sum (x - \bar{x})^2$ by $6 - 1 = 5$ and take the square root of this quotient from Table XII. We get

$$s = \sqrt{\frac{27.50}{5}} = \sqrt{5.5} = 2.3$$

It was easy to calculate s in this example because the response times were whole numbers and their mean was exact to one decimal. Often, though, the calculations required by the formulas defining s and s^2 are quite cumbersome, and it may be better to use the following computing formula, which can be derived by applying the rules for summations given in the technical note of Section 3.10:

Computing formula
for the sample
standard deviation

$$s = \sqrt{\frac{n(\sum x^2) - (\sum x)^2}{n(n - 1)}}$$

This formula gives the exact value of s, not an approximation, and its advantage is that we do not actually have to find all the deviations from the mean. Instead we calculate $\sum x$, the sum of the x's, $\sum x^2$, the sum of their squares, and substitute into the formula. Aside from its advantage in manual calculations, this formula for s (or a slight modification of it) is the one usually preprogrammed into electronic statistical calculators, and it is the one most easily programmed for solution on a digital computer.†

EXAMPLE   Use the computing formula for s to rework the preceding example.

SOLUTION   First, we calculate the two sums

$$\sum x = 9 + 8 + 5 + 11 + 7 + 5 = 45$$

and

$$\sum x^2 = 81 + 64 + 25 + 121 + 49 + 25 = 365$$

Then, substituting these sums and $n = 6$ into the formula, we find that

$$s = \sqrt{\frac{6(365) - (45)^2}{6 \cdot 5}} = \sqrt{\frac{165}{30}} = \sqrt{5.5} = 2.3$$

and this agrees with the result we obtained before.

In Section 3.4 we showed that there are many ways in which knowledge of the variability of a set of data can be of importance. Another application arises in the comparison of numbers belonging to different sets of data. To illustrate, suppose that a large securities firm administers a battery of tests to all job ap-

†The computing formula for s can also be used to find σ, provided that we substitute n for the factor $n - 1$ in the denominator before we replace s and n with σ and N.

plicants, and a particular applicant, A, scores 135 on the General Information (GI) test and 265 points on the Accounting and Finance (AF) test. At first glance it may seem that A did much better (nearly twice as well) in accounting and finance than in general information. However, if the mean score which thousands of applicants made on the GI test was 100 points with a standard deviation of 15 points, and the mean score which all these applicants made on the AF test was 250 points with a standard deviation of 30 points, we can argue that A's score on the GI test is

$$\frac{135 - 100}{15} = 2\frac{1}{3} \text{ standard deviations}$$

above the mean of the distribution of all the scores on this test, while her score on the AF test was only

$$\frac{265 - 250}{30} = \frac{1}{2} \text{ standard deviation}$$

above the mean of the distribution of all the scores on this test. Whereas the original scores cannot be meaningfully compared, these new scores, expressed in terms of standard deviations, can. Clearly, A rates much higher on her command of general information than she does on her knowledge of accounting and finance.

What we did in the example is convert raw scores into standard units, or z-scores. If x is a measurement belonging to a set of data having the mean $\bar{x}$ (or μ) and the standard deviation s (or σ), then its value in standard units, denoted by the letter z, is given by

Formula for converting to standard units

$$z = \frac{x - \bar{x}}{s} \quad or \quad z = \frac{x - \mu}{\sigma}$$

depending on whether the data constitute a sample or a population. In these units, z tells us how many standard deviations a value lies above or below the mean of the set of data to which it belongs. Standard units will be used frequently in later chapters.

3.7

Chebyshev's Theorem

In the argument that led to the definition of the standard deviation, we observed that the dispersion of a set of data is small if the values are bunched closely about their mean and that it is large if the values are scattered widely about their mean. Correspondingly, we can now say that if the standard deviation of a set of data is small, the values are concentrated near the mean, and if the standard deviation is large, the values are scattered widely about the mean. To present this argument on a less intuitive basis (after all, what is small and what is large?), let us refer to an important theorem called Chebyshev's theorem. This theorem states that

<table>
<tr><td>*Chebyshev's theorem*</td><td>*For any set of data (population or sample) and any constant k greater than 1, at least $1 - 1/k^2$ of the data must lie within k standard deviations on either side of the mean.*</td></tr>
</table>

Accordingly, we can be sure that at least $\frac{3}{4}$, or 75 percent, of the values in any set of data must lie between the mean minus 2 standard deviations and the mean plus 2 standard deviations; at least $\frac{8}{9}$, or about 88.9 percent, of the values in any set of data must lie between the mean minus 3 standard deviations and the mean plus 3 standard deviations; and at least $\frac{24}{25}$, or 96 percent, of the values in any set of data must lie between the mean minus 5 standard deviations and the mean plus 5 standard deviations. Here we arbitrarily let $k = 2, 3,$ and 5.

EXAMPLE   If all the 1-pound cans of coffee filled by a food processor have a mean weight of 16.00 ounces with a standard deviation of 0.02 ounce, at least what percentage of the cans must contain between 15.95 and 16.05 ounces of coffee?

SOLUTION   Since k standard deviations, or $k(0.02)$, equals $16.05 - 16.00 = 16.00 - 15.95 = 0.05$, we find that $k = \frac{0.05}{0.02} = 2.5$. Thus, at least $1 - \frac{1}{(2.5)^2} = 1 - \frac{1}{6.25} = 0.84$, or 84 percent, of the cans must contain between 15.95 and 16.05 ounces of coffee.

EXERCISES

3.25 Calculate the standard deviation of the population consisting of the integers 1, 2, 3, 4, 5, and 6.

3.26 Calculate the standard deviation of each of the following samples using both the definition and the computing formulas:
(a) $-1, 5, 0, 4$;   (b) 2, 10, 6, 18, 14.

3.27 A filling machine in a high-production bakery is set to fill open-face pies with 16 fluid ounces of fill. A sample of four pies from a large production lot shows fills of 16.2, 15.9, 15.8, and 16.1 fluid ounces.
(a) Find the range of these fills.
(b) Use the definition formula to calculate the standard deviation of these fills.

3.28 In a national casting championship, the winner of the $\frac{5}{8}$-ounce distance plug event had casts of 410, 418, 426, 422, and 428 feet (to the nearest foot), and in the two-handed distance fly event the same caster won with casts of 208, 190, 195, 196, and 215 feet (to the nearest foot). Find the range and the standard deviation of each of these samples of distances.

3.29 For a recent 10-year period, the production of natural (sundried) raisins in California yielded 715, 825, 640, 900, 790, 965, 895, 700, 915, and 945 trays per acre.

(a) Calculate the range and the standard deviation for the entire period.

(b) Calculate the range and the standard deviation for the most recent five years.

3.30 One of the best measures of overall railroad operating profitability is the percentage of revenues brought down to net railway operating income before federal taxes. The following are the percentages for all the railroads operating in a certain district for a 5-year period: 10.4, 11.6, 12.3, 8.1, and 9.6. Calculate

(a) the mean; (c) the standard deviation.

(b) the range;

3.31 The process of making fish flour as a source of protein, called "fish protein concentrate," consists of grinding whole fish, removing fat chemically, and then drying the resulting flour. Samples from five different experimental batches of such concentrate show the following percentages by weight of the final product of the protein, moisture, ash, and fat contents:

	Batch 1	Batch 2	Batch 3	Batch 4	Batch 5
Protein	70.5	72.4	71.4	73.1	74.0
Moisture	5.1	4.5	6.0	3.0	4.2
Ash	24.0	22.3	22.0	23.1	20.9
Fat	0.4	0.8	0.6	0.8	0.9

Calculate the mean, the range, and the standard deviation for each of these contents.

3.32 With reference to Exercise 3.3 on page 38, find the range and the standard deviation of the ages of the 20 persons.

3.33 With reference to Exercise 3.11 on page 39, find the range and the standard deviation of the lengths of the twenty-five power failures.

3.34 If each item in a set of data has the same constant a added to it, the mean of this new set equals the mean of the original set plus the constant a, but the range and the standard deviation remain unchanged.

(a) Verify that for a sample consisting of the values $-3, 1, 0, 2, -2, 1$, and -6 the mean is -1, the range is 8, and the standard deviation is $\sqrt{8}$, and that after adding 6 to each value the mean becomes $-1 + 6 = 5$, but the range is still 8 and the standard deviation is still $\sqrt{8}$.

(b) The commissions earned by a sample of four automobile salesmen in a given week were $390, $411, $380, and $423. Calculate the standard deviation of these figures after subtracting $400 from each value.

3.35 If each item in a set of data is multiplied by the same positive constant b, the mean, range, and standard deviation of this new set equal the mean, range, and standard deviation of the original set multiplied by b.

(a) With reference to the seven sample values of part (a) of Exercise 3.34, show that if each value is multiplied by 2, the mean becomes -2, the range becomes 16, and the standard deviation becomes $2\sqrt{8}$.

(b) During four pit stops, the front-tire man changed a racing car's right front tire in 10.8, 12.0, 10.5, and 10.7 seconds. Calculate the standard deviation by first multiplying each figure by 10, subtracting 110, determining s for the resulting figures, and then dividing by 10.

3.36 K, a member of a large industrial chemicals sales force whose average sales in one month was $12,000 with standard deviation $2,000, made sales of $14,000 in that month; L, a member of a large hardware supplies sales force whose average sales in the same month was $6,000 with standard deviation $500, made sales of $7,250 in that month. Can we conclude from these data that K's sales performance in that month was almost twice as good as L's?

3.37 An investment service reports for each stock that it lists the price at which it is currently selling, its average price over a certain period of time, and a measure of its variability. Stock A, it reports, has a normal (average) price of $56 with a standard deviation of $11, and is currently selling at $74.50; stock B sells normally for $35, has a standard deviation of $4, and is currently selling at $47. If an investor owns both stocks and wants to dispose of one, which one might he sell and why?

3.38 For a large group of students the mean score on a scholastic aptitude test is 165 points and the standard deviation is 15 points. At least what percentage of these scores must lie between
(a) 135 and 195 points;   (c) 75 and 255 points?
(b) 105 and 225 points;

3.39 Suppose that, for a large group of taxpayers, the elapsed times between the day a notice of deficiency is mailed to an individual taxpayer by the Internal Revenue Service and the day the tax court receives the taxpayer's petition for a hearing have a mean of 64 days and a variance of 16 days^2.
(a) At least what fraction of the elapsed times falls between 52 and 76 days?
(b) Between what values does at least $\frac{24}{25}$ of the elapsed times fall?

3.40 For a large group of persons, the scores on a 30-second tapping test (used to measure speed of movement) have a mean of 180 taps and a standard deviation of 8 taps. Between what two scores must lie
(a) at least $\frac{35}{36}$ of the scores;   (b) at least $\frac{63}{64}$ of the scores?

3.41 One characteristic of the standard deviation as a measure of variation is that it depends on the units of measurement. If, for instance, a set of measurements of the weights of containers of a bleach have a standard deviation of 0.24 ounce, we would look at this variability in one light if the containers are "80-pound" containers, and in quite another light if they are "8-ounce" containers. What we need in a situation like this is a measure of relative variation, such as the coefficient of variation

Coefficient of variation

$$V = \frac{s}{\bar{x}} \cdot 100 \quad or \quad V = \frac{\sigma}{\mu} \cdot 100$$

which expresses the standard deviation as a percentage of the mean.
(a) A sample of 4 specimens of a molded laminated material showed tensile strengths (in thousands of pounds per square inch) of 6.0, 6.9, 6.3, and 6.4. Find the coefficient of variation.

(b) A sample of four specimens of an aluminum-base alloy for die casting showed the following percentages of magnesium: 7.6, 8.1, 8.0, and 7.8. Find the coefficient of variation.

(c) By comparing the coefficients of variation (which are percentages) we can compare the dispersions of two or more sets of data pertaining to different kinds of measurements (say, height, weight, speed, age, and temperature). Compare the coefficients of variation obtained in parts (a) and (b) to judge which set of measurements is relatively more variable.

(d) A sample of five specimens of hard yellow brass showed shear strengths of 48, 51, 50, 52, and 54 thousand psi, and a sample of the gross sales of a boat dock on four summer Sundays showed grosses of 2.1, 1.9, 1.8, and 2.2 thousands of dollars. Which of the two sets of data is relatively more variable?

3.8

The Description of Grouped Data

Published data are often available only in the form of a frequency distribution. For this reason, we shall discuss briefly the calculation of statistical descriptions from grouped data.

As we have already seen, the grouping of data entails some loss of information. Each item loses its identity, so to speak; we only know how many items there are in each class, so we must be satisfied with approximations. In the case of the mean and the standard deviation, we can usually get good approximations by assigning to each item falling into a class the value of the class mark. For instance, to calculate the mean or the standard deviation of the distribution of aptitude scores on page 12, we treat the six values falling into the class 20–29 as if they were all 24.5, the nine values falling into the class 30–39 as if they were all 34.5, ..., and the two values falling into the class 90–99 as if they were both 94.5. This procedure is usually quite satisfactory, since the errors which it introduces into the calculations will more or less "average out."

To write formulas for the mean and the standard deviation of a distribution with k classes, let us designate the successive class marks $x_1, x_2, x_3, \ldots, x_k$, and the corresponding class frequencies $f_1, f_2, f_3, \ldots,$ and f_k. The total that goes into the numerator of the formula for the mean is the sum obtained by adding x_1 times f_1, x_2 times f_2, x_3 times f_3, ..., and x_k times f_k, or $x_1 f_1 + x_2 f_2 + x_3 f_3 + \cdots + x_k f_k$. Using again the $\sum$ notation, we write the formula for the mean of grouped sample data as

Mean of grouped data

$$\bar{x} = \frac{\sum x \cdot f}{n}$$

where $\sum x \cdot f$ represents, in words, the sum of the products obtained by multiplying each class mark by the corresponding class frequency. (If the data constitute a population instead of a sample, we substitute μ for $\bar{x}$ and N for n in this formula.)

Similarly, the total that goes into the numerator of the formulas defining the sample variance and the sample standard deviation is the sum obtained by adding $(x_1 - \bar{x})^2$ times f_1, $(x_2 - \bar{x})^2$ times f_2, $(x_3 - \bar{x})^2$ times $f_3, \ldots$, and $(x_k - \bar{x})^2$ times f_k. Thus, for the sample standard deviation we write

Standard deviation of grouped data

$$s = \sqrt{\frac{\sum (x - \bar{x})^2 \cdot f}{n - 1}}$$

In the computing formulas for s and s^2 we replace $\sum x$ by $\sum x \cdot f$ and $\sum x^2$ by $\sum x^2 \cdot f$, so that the formula for s becomes

Computing formula for the standard deviation of grouped data

$$s = \sqrt{\frac{n(\sum x^2 \cdot f) - (\sum x \cdot f)^2}{n(n - 1)}}$$

EXAMPLE   Calculate the mean and the standard deviation of the distribution of aptitude scores on page 12.

SOLUTION   To get $\sum x \cdot f$ and $\sum x^2 \cdot f$, we perform the calculations shown in the following table, where the first and third columns are copied from the original distribution, the second column contains the class marks, and the fourth and fifth columns contain the products $x \cdot f$ and $x^2 \cdot f$:

Scores	Class marks x	Frequencies f	$x \cdot f$	$x^2 \cdot f$
10–19	14.5	1	14.5	210.25
20–29	24.5	6	147.0	3,601.5
30–39	34.5	9	310.5	10,712.25
40–49	44.5	31	1,379.5	61,387.75
50–59	54.5	42	2,289.0	124,750.5
60–69	64.5	32	2,064.0	133,128.0
70–79	74.5	17	1,266.5	94,354.25
80–89	84.5	10	845.0	71,402.5
90–99	94.5	2	189.0	17,860.5
		150	8,505.0	517,407.5

Then, substitution into the formulas yields

$$\bar{x} = \frac{8,505.0}{150} = 56.7$$

and

$$s = \sqrt{\frac{150(517,407.5) - (8,505.0)^2}{150 \cdot 149}} = 15.4$$

It is apparent from this example that some heavy arithmetic may be required to find the mean or the standard deviation of a distribution. However, we can simplify this work by **coding** the class marks so that we have smaller numbers to work with. Provided the class intervals are all equal, this coding consists of assigning the value 0 to one of the class marks (in manual calculations, preferably at or near the center of the distribution), and representing all the class marks by means of successive integers. For instance, if a distribution has seven classes and the class mark of the middle class is assigned the value 0, the successive class marks of the distribution are assigned the values $-3, -2, -1, 0, 1, 2$, and 3.

Of course, when we code the class marks like this, we must account for it in the formulas for the mean and the standard deviation. Referring to the new (coded) class marks as u's, the formula for the mean of a distribution of sample data becomes

Mean of grouped data (with coding)

$$\bar{x} = x_0 + \frac{\sum u \cdot f}{n} \cdot c$$

where x_0 is the class mark in the original scale to which we assign 0 in the new scale, c is the class interval, n is the number of items grouped, and $\sum u \cdot f$ is the sum of the products obtained by multiplying each of the new class marks by the corresponding class frequency. Similarly, the formula for the standard deviation of a distribution of sample data becomes

Standard deviation of grouped data (with coding)

$$s = c \sqrt{\frac{n(\sum u^2 \cdot f) - (\sum u \cdot f)^2}{n(n-1)}}$$

where $\sum u^2 \cdot f$ is the sum of the products obtained by multiplying the squares of the new class marks by the corresponding class frequencies.

EXAMPLE To demonstrate the simplification brought about by coding, recalculate the mean and the standard deviation of the distribution of the clerical aptitude scores.

Arranging the work, as before, in a table, we get

Class marks x	u	f	$u \cdot f$	$u^2 \cdot f$
14.5	−4	1	−4	16
24.5	−3	6	−18	54
34.5	−2	9	−18	36
44.5	−1	31	−31	31
54.5	0	42	0	0
64.5	1	32	32	32
74.5	2	17	34	68
84.5	3	10	30	90
94.5	4	2	8	32
		150	33	359

where the class mark 54.5 is taken to be 0 in the u scale (shown in the second column of the table). Of course, we could have used any class mark, for instance, 24.5, as the zero of the u scale, but the objective is to make the arithmetic as simple as possible.

Substituting $c = 10$, $x_0 = 54.5$, $n = 150$, $\sum u \cdot f = 33$, and $\sum u^2 f = 359$ into the above formulas for $\bar{x}$ and s, we get

$$\bar{x} = 54.5 + \frac{33}{150} \cdot 10 = 56.7$$

and

$$s = 10 \sqrt{\frac{150(359) - (33)^2}{150 \cdot 149}} = 15.4$$

These results are, as they should be, identical with the ones obtained earlier without coding.

As for the median of a set of grouped data, we cannot calculate the precise value of this measure because of the loss of identity which results from the act of grouping. To approximate it, we shall not assign the value of the class mark to each item falling within any given class; instead, we shall assume that the items lying within each class are spread evenly throughout the class. With this assumption, if the class containing the actual median had, say, 60 values and the median was the 59th largest, the median would be located very close to the upper class boundary instead of at the middle of the class. This entirely reasonable result leads us to define the median of a numerical distribution as the number which is such that half the total area of the rectangles of the histogram of the distribution lies to its left and the other half lies to its right (see Figure 3.1).

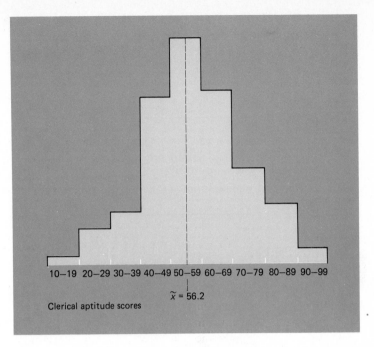

10–19 20–29 30–39 40–49 50–59 60–69 70–79 80–89 90–99

$\tilde{x} = 56.2$

Clerical aptitude scores

To find the dividing line between the two halves of a histogram (each of which represents $n/2$ of the items grouped), we must count $n/2$ of the items starting at one end of the distribution. How this is done is illustrated in the following example:

EXAMPLE   Find the median of the distribution of the clerical aptitude scores.

SOLUTION   Since $\dfrac{n}{2} = \dfrac{150}{2} = 75$, we must count 75 items starting at one end. Counting from the bottom (that is, beginning with the smallest values), we find that $1 + 6 + 9 + 31 = 47$ of the values are less than 50, and $1 + 6 + 9 + 31 + 42 = 89$ are less than 60. Therefore, we must count $75 - 47 = 28$ more items beyond the 47 which are less than 50, and on the assumption that the 42 values of the 50–59 class are spread evenly throughout the class, we can do this by adding $\frac{28}{42}$ of the class interval of 10 to 49.5, the lower boundary of the class. This gives us

$$\tilde{x} = 49.5 + \frac{28}{42} \cdot 10 = 56.2$$

for the median of this distribution.

In general, if L is the lower boundary of the class into which the median must fall, f is its frequency, c is the class interval, and j is the number of items we still lack when we reach L, then the median of the distribution is given by the formula

Median of grouped data

$$\tilde{x} = L + \frac{j}{f} \cdot c$$

If we prefer, we can find the median of a distribution by starting to count at the other end and subtracting an appropriate fraction of the class interval from the upper boundary U of the median class. A general formula for the case where we start counting at the top (beginning with the largest values) is given by

Alternate formula for the median of grouped data

$$\tilde{x} = U - \frac{j'}{f} \cdot c$$

where j' is the number of items we still lack when we reach U.

EXAMPLE Use the alternate formula to find the median of the distribution of aptitude scores.

SOLUTION Since $2 + 10 + 17 + 32 = 61$ of the values fall above 59.5, we need $75 - 61 = 14$ of the 42 values which fall into the next class to reach the median, and we write

$$\tilde{x} = 59.5 - \frac{14}{42} \cdot 10 = 56.2$$

The result is the same, of course.

The procedure we have just described for finding the median of a distribution can also be used to determine more general "positional measures" called **fractiles** or **quantiles**. By definition, a fractile, or quantile, is a value at or below which a given fraction of the data must lie. There are, for instance, the three quartiles Q_1, Q_2, and Q_3, which are such that 25 percent of the data are less than or equal to Q_1, 50 percent are less than or equal to Q_2, and 75 percent are less than or equal to Q_3. Also, there are the nine **deciles**, $D_1, D_2, \ldots,$ and D_9, which are such that 10 percent of the data are less than or equal to D_1, 20 percent are less than or equal to D_2, and so on; and there are the 99 **percentiles**, $P_1, P_2, \ldots,$ and P_{99}, which are such that 1 percent of the data are less than or equal to P_1, 2 percent are less than or equal to P_2, and so on. It should be clear from this that Q_2, D_5, and P_{50} are all equal to the median, and that P_{25} equals Q_1 and P_{75} equals Q_3.

EXAMPLE   Referring again to the distribution of aptitude scores, find Q_1, D_9, and P_{15}.

SOLUTION   Using the formulas for the median and counting in each case the appropriate fraction of the number of items grouped in the distribution, we find that

$$Q_1 = 39.5 + \frac{21.5}{31} \cdot 10 = 46.4$$

$$D_9 = 79.5 - \frac{3}{17} \cdot 10 = 77.7$$

and

$$P_{15} = 39.5 + \frac{6.5}{31} \cdot 10 = 41.6$$

To conclude this discussion of statistical descriptions of grouped data, let us point out that there exist fairly elaborate ways of defining the mode of a distribution. In most cases, however, all we need is the **modal class**, the class with the highest frequency; if a single number is preferred, we can define the mode of a distribution as the midpoint of the modal class.

3.9

Some Further Descriptions

Measures of location and measures of variation are fundamentally important statistical descriptions, but there are many other ways to describe statistical data. In this section we shall consider briefly the problem of describing the overall shape of a distribution.

Distributions of actual data can assume almost any shape or form, but most of those which arise in practice can be described fairly well by one or another of a few standard types. A very important one is the symmetrical **bell-shaped distribution** shown in Figure 3.2. Indeed, there are theoretical reasons why, in many cases, distributions of actual data can be expected to follow this form very closely. The other two distributions of Figure 3.2 can still, by a stretch of the imagination, be called bell-shaped, but they certainly cannot be called symmetrical. Distributions of this sort, having a pronounced "tail" on one side or the other, are said to be **skewed**; those with a tail on the left are **negatively skewed** and those with a tail on the right are **positively skewed**. Distributions of incomes are often positively skewed because of the presence of some relatively high incomes that are not offset by correspondingly low ones. Since these high values tend to affect the mean more than the median, the median is widely used to average incomes. For instance, the Bureau of the

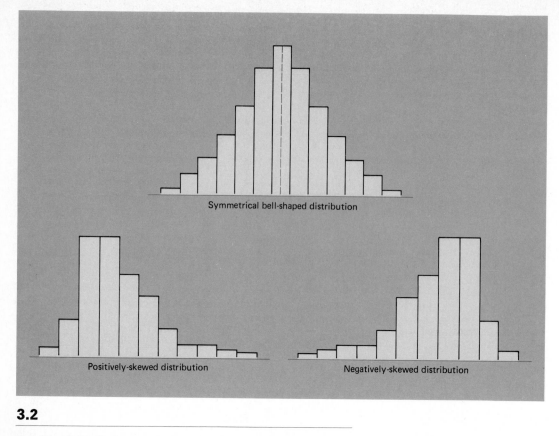

3.2

Bell-shaped distributions.

Census uses the median to measure household incomes (an important indicator of the nation's standard of living).

For a perfectly symmetrical bell-shaped distribution such as the one in Figure 3.2, the values of the mean, median, and mode coincide, and they all lie on the axis of symmetry (the dashed vertical line which divides the histogram of the distribution into equal halves). But, as we have already observed, in a positively skewed distribution the median will generally be exceeded by the mean (see also Figure 3.3), and by the same token in a negatively skewed distribution the median will generally exceed the mean. A simple measure of the extent to which a distribution is skewed is based on this relationship between the median and the mean. Called the **Pearsonian coefficient of skewness**, its formula is

Pearsonian coefficient of skewness

$$SK = \frac{3(mean - median)}{standard\ deviation}$$

3.3

The median and the mean of a positively skewed distribution.

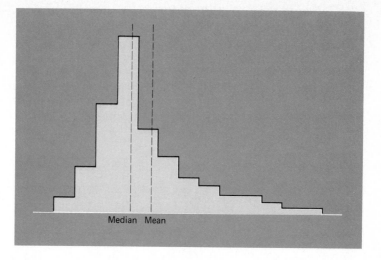

Median   Mean

For a perfectly symmetrical distribution the value of SK is 0, and in general its value must fall between -3 and 3.

EXAMPLE   Find the Pearsonian coefficient of skewness for the distribution of the clerical aptitude scores.

SOLUTION   Substituting into the formula the values of the mean, $\bar{x} = 56.7$, the median, $\tilde{x} = 56.2$, and the standard deviation, $s = 15.4$, we find that

$$SK = \frac{3(56.7 - 56.2)}{15.4} = 0.097$$

On the basis of this result we can say that the distribution is nearly symmetrical.

Two other kinds of distributions which sometimes arise in practice are the **reverse J-shaped** and **U-shaped distributions;** as can be seen from the histograms of Figure 3.4, the names of these distributions quite literally describe

3.4

Histograms of reverse J-shaped and U-shaped distributions.

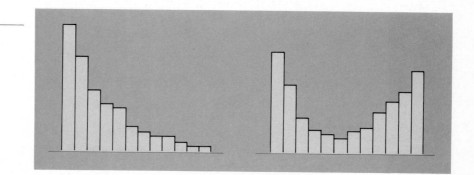

their shape. Examples of such distributions may be found in Exercises 3.54 and 3.55.

EXERCISES

3.42 In a comparative study of bond yields, the following distribution was constructed of the yields to maturity (rounded to the nearest tenth of a percent) of 50 bonds:

Yield (percent)	Number of bonds
1.0– 2.9	5
3.0– 4.9	9
5.0– 6.9	17
7.0– 8.9	12
9.0–10.9	7

Find the mean and the standard deviation of this distribution
(a) without coding;   (b) with coding.

3.43 With reference to the distribution of Exercise 3.42, find
(a) the median;   (b) the mode.

3.44 Use the results of the two preceding exercises to calculate the Pearsonian coefficient of skewness for the distribution of bond yields.

3.45 Find the mean and the standard deviation of the following distribution of the ages of the members of a union:

Age (years)	Frequency
20–24	11
25–29	24
30–34	30
35–39	18
40–44	11
45–49	5
50–54	1

3.46 With reference to the distribution of the preceding exercise, find
(a) the median;   (c) the percentiles P_{90} and P_{95}.
(b) the quartiles Q_1 and Q_3;

3.47 Use the results of the two preceding exercises to calculate the Pearsonian coefficient of skewness for the distribution of the ages of the union members.

3.48 With reference to the distribution of law firm billings on page 8, find
(a) the median;   (b) the quartiles Q_1 and Q_3.
Also explain why it is impossible to find the percentile P_{10} for this distribution.

3.49 With reference to the distribution of Exercise 2.14 on page 20, find
(a) the mean;
(b) the median;
(c) the standard deviation;
(d) the Pearsonian coefficient of skewness.

3.50 With reference to the distribution of clerical aptitude scores on page 12, find
(a) the quartile Q_3;   (c) the percentiles P_5 and P_{95}.
(b) the deciles D_2 and D_8;

3.51 The following is the distribution of the sizes of a sample of 50 orders filled by the special-order department of a large phonograph record wholesaler:

Size of order	Number of orders
$20.00–39.99	3
40.00–59.99	15
60.00–79.99	17
80.00–99.99	13
100.00–119.99	2

Calculate
(a) the mean and the median;   (d) the deciles D_1 and D_9;
(b) the standard deviation;   (e) the percentiles P_{35} and P_{65}.
(c) the quartiles Q_1 and Q_3;

3.52 Use the results of parts (a) and (b) of the preceding exercise to calculate the Pearsonian coefficient of skewness for the distribution of the sizes of the orders.

3.53 Sometimes we use the midquartile $\frac{1}{2}(Q_1 + Q_3)$ as a measure of central location instead of the median or the mean, the semi-interquartile range $\frac{1}{2}(Q_3 - Q_1)$ as a measure of variation instead of the standard deviation or the range, and the coefficient of quartile variation $\dfrac{Q_3 - Q_1}{Q_1 + Q_3} \cdot 100$ as a measure of relative variation instead of the coefficient of variation.
(a) Use the results of part (b) of Exercise 3.46 to calculate all these statistical measures for the age distribution.
(b) Use the results of part (c) of Exercise 3.51 to calculate all these statistical measures for the distribution of the sizes of the orders.
(c) Use the result of part (a) of Exercise 3.50 and the value $Q_1 = 46.4$ to calculate all these statistical measures for the distribution of clerical aptitude scores.

3.54 Roll a pair of dice 120 times and construct a distribution showing how many times there were 0 sixes, how many times there was 1 six, and how many times there were 2 sixes. Draw a histogram of this distribution and describe its shape.

3.55 If a coin is flipped five times, the result may be represented by means of a sequence of H's and T's (for example, HHTTH), where H stands for *heads* and T for *tails*. Having obtained such a sequence of H's and T's, we can then check after each successive flip whether the number of heads exceeds the number of tails. For example, for the sequence HHTTH, heads is ahead after the first flip, after the second flip, after the third flip, not after the fourth flip, but again after the fifth flip; altogether, it is ahead four times. Repeat this experiment 50 times, and construct a histogram showing in how many cases heads was ahead altogether 0 times, 1 time, 2 times, . . . , and 5 times. Explain why the resulting distribution should be U-shaped.

3.10

Technical Note (Summations)

In summation notation, $\sum x$ does not make it clear which, or how many, values of x are to be added. This is taken care of by the more explicit notation

$$\sum_{i=1}^{n} x_i = x_1 + x_2 + \cdots + x_n$$

where it is made clear that we are adding the x's whose subscripts i are 1, 2, . . . , n. We did not use this notation in the text, in order to simplify the overall appearance of the various formulas, assuming that it is clear in each case what x's we are referring to and how many there are.

Using the $\sum$ notation, we shall also have occasion to write such expressions as $\sum x^2$, $\sum xy$, $\sum x^2 f$, . . . , which (more explicitly) represent the sums

$$\sum_{i=1}^{n} x_i^2 = x_1^2 + x_2^2 + x_3^2 + \cdots + x_n^2$$

$$\sum_{j=1}^{m} x_j y_j = x_1 y_1 + x_2 y_2 + \cdots + x_m y_m$$

$$\sum_{i=1}^{n} x_i^2 f_i = x_1^2 f_1 + x_2^2 f_2 + \cdots + x_n^2 f_n$$

Working with two subscripts, we shall also have the occasion to evaluate double summations such as

$$\sum_{j=1}^{3} \sum_{i=1}^{4} x_{ij} = \sum_{j=1}^{3} (x_{1j} + x_{2j} + x_{3j} + x_{4j})$$

$$= x_{11} + x_{21} + x_{31} + x_{41} + x_{12} + x_{22} + x_{32} + x_{42}$$

$$+ x_{13} + x_{23} + x_{33} + x_{43}$$

To verify some of the formulas involving summations that are stated but not proved in the text, it will be convenient to use the following rules:

$$\text{Rule A:} \quad \sum_{i=1}^{n} (x_i \pm y_i) = \sum_{i=1}^{n} x_i \pm \sum_{i=1}^{n} y_i$$

$$\text{Rule B:} \quad \sum_{i=1}^{n} k \cdot x_i = k \cdot \sum_{i=1}^{n} x_i$$

$$\text{Rule C:} \quad \sum_{i=1}^{n} k = n \cdot k$$

The first of these rules states that the summation of the sum (or difference) of two terms equals the sum (or difference) of the individual summations, and it can be extended to the sum or difference of more than two terms. The second rule states that we can, so to speak, factor a constant out of a summation, and the third rule states that the summation of a constant is simply n times that constant. All these rules can be proved by actually writing out in full what each of the summations represents.

EXERCISES

3.56 Write each of the following as summations:
(a) $y_1^2 + y_2^2 + \cdots + y_{30}^2$;
(b) $x_3 y_3 + x_4 y_4 + \cdots + x_9 y_9$;
(c) $x_1 f_1 + x_2 f_2 + \cdots x_n f_n$;
(d) $A_5 + A_6 + \cdots + A_{10}$;
(e) $2x_1 + 2x_2 + \cdots + 2x_{20}$;
(f) $(z_1 - y_1) + (z_2 - y_2) + \cdots + (z_n - y_n)$.

3.57 Write each of the following expressions without summation signs:
(a) $\sum_{i=1}^{9} x_i$; (c) $\sum_{i=3}^{7} x_i y_i$; (e) $\sum_{i=2}^{5} 3x_i^2$;

(b) $\sum_{i=1}^{6} (x_i - k)$; (d) $\sum_{i=4}^{8} y_i^2$; (f) $\sum_{j=1}^{n} (x_j - z_j)$.

3.58 Given $x_1 = 2$, $x_2 = 1$, $x_3 = 2$, $x_4 = 1$, and $x_5 = 3$, find
(a) $\sum x$; (c) $\sum (x - 2)$; (e) $\sum (3x - 2)$;
(b) $\sum x^2$; (d) $\sum (x - 2)^2$; (f) $[\sum (3x - 2)]^2$.

·3.59 Given $x_1 = 1$, $x_2 = 2$, $x_3 = -4$, $f_1 = 3$, $f_2 = -1$, $f_3 = 2$, $y_1 = 3$, $y_2 = 1$, and $y_3 = 5$, find
(a) $\sum x \cdot f$; (c) $\sum x \cdot y$; (e) $\sum (x - y)^2$;
(b) $\sum x^2 \cdot f$; (d) $\sum (x - y)$; (f) $\sum x \cdot y \cdot f$.

3.60 Given that $\sum_{i=1}^{7} x_i = 17$ and $\sum_{i=1}^{7} x_i^2 = 53$, find

(a) $\sum_{i=1}^{7} (x_i - 2)$;   (b) $\sum_{i=1}^{7} (2x_i + 1)$;   (c) $\sum_{i=1}^{7} (x_i + 3)^2$.

3.61 Prove that

(a) $\sum_{i=1}^{n} (x_i - k) = \sum_{i=1}^{n} x_i - nk$;

(b) $\sum_{i=1}^{n} (x_i - \bar{x}) = 0$, where $\bar{x}$ is the mean of the x_i;

(c) $\sum_{i=1}^{n} (x_i - k)^2 = \sum_{i=1}^{n} x_i^2 - 2k \cdot \sum_{i=1}^{n} x_i + nk^2$.

3.62 Is it true in general that $\left(\sum_{i=1}^{n} x_i \right)^2 = \sum_{i=1}^{n} x_i^2$? (*Hint:* Determine whether the equation holds for $n = 2$.)

3.63 Use the sigma notation with subscripts and limits of summation to write the formula for
(a) the sample mean;
(b) the weighted mean;
(c) the grand mean of combined data.

3.11

Technical Note (Unbiased Estimators)

Ordinarily, the purpose of calculating a sample statistic (such as the mean, the standard deviation, or the variance) is to estimate the corresponding population parameter. If we actually took many samples from a population which has the mean μ, calculated the sample means $\bar{x}$, and then averaged all these estimates of μ, we should find that their average is very close to μ. However, if we calculated the variance of each sample by means of the formula $\sum (x - \bar{x})^2/n$, and then averaged all these estimates of σ^2, we would probably find that their average is less (perhaps substantially so) than σ^2. Theoretically, it can be shown that we can compensate for this by dividing by $n - 1$ instead of n in the formula for s^2. Estimators which have the desirable property that their values will on the average equal the quantity they are supposed to estimate are said to be **unbiased**; otherwise, they are said to be **biased**. So, we say that $\bar{x}$ is an unbiased estimator of the population mean μ, and that s^2 is an unbiased estimator of the population variance σ^2. It does not follow from this, however, that s is also an unbiased estimator of σ; but when n is large the bias is small, so we can use s as an estimate of σ.

3.12

A Word of Caution

The fact that there is a certain amount of arbitrariness in the selection of statistical descriptions has led some persons to believe that they can take a set of data, commit some statistics, and prove almost anything they want. To put it more bluntly, a nineteenth-century British statesman once said that there are three kinds of lies: lies, damned lies, and statistics.

To show where such a criticism might be justified, suppose that a paint manufacturer asks his research department to "prove" that on the average a gallon of his paint covers more square feet than those of his two principal competitors. Suppose, furthermore, that the research department tests five cans of each brand, getting the following results (in square feet per gallon can):

Brand A: 505, 516, 478, 513, 503
Brand B: 512, 486, 511, 486, 510
Brand C: 496, 485, 490, 520, 484

If the manufacturer's own brand is brand *A*, the data analyst finds to his delight that the means of the three samples are 503, 501, and 495. He can, thus, claim that in actual tests a can of his employer's product covered on the average more square feet than those of his competitors.

If, however, the manufacturer's own brand is brand *B*, the analyst can no longer base the comparison on the sample means. The sample medians, though, are 505, 510, and 490, and this gives him the results he wants. The median is a perfectly respectable measure of the "average" or "center" of a set of data, and using the medians he can claim that his employer's product came out best in the test.

Finally, suppose that the manufacturer's own brand is brand *C*. After going down the list of various measures of central location, the analyst comes upon one he wants. The **sample midrange** is defined as the mean of the smallest and largest values in a sample, and the midranges of the three samples of this example are 497, 499, and 502. So, he can claim that brand *C*, his employer's product, scored on the average highest in the test.

The moral of this example is that if data are to be compared and special pleading or indoctrination are not to be indulged in, the method of comparison should be decided upon beforehand, or at least without actually looking at the data. All this is aside from the fact that comparisons based on samples are often far from conclusive. It is quite possible that whatever differences there may be among the three means (or three other descriptions) can be attributed entirely to chance.

Another point which must be remembered is that a statistical measure always describes a particular characteristic of a set of data and that it describes this characteristic in a special way. Whether this "special way" is appropriate for a given situation is something which will have to be examined individually in each case. Suppose, for instance, that we want to buy a house and that we are shown one in a neighborhood where, according to the broker, average family income is in excess of $54,000. This gives the impression of a relatively prosperous neighborhood, but it could well be a neighborhood where most families have incomes of less than $9,000 while one very wealthy family has an income of several hundred thousand dollars a year. In this kind of situation the mean is greatly affected by the one extreme value and it would be much more informative to say here that the median family income is less than $9,000, mentioning, perhaps, the special situation created by the one wealthy family. Further examples of this kind may be found in *How to Lie with Statistics*, the book by D. Huff referred to in the Bibliography at the end of this book.

3.13

Check List of Key Terms

Arithmetic mean, 31
Bell-shaped distribution, 57
Biased estimator, 64
Chebyshev's theorem, 48
Coding, 53
Coefficient of quartile variation, 61
Coefficient of variation, 50
Decile, 56
Deviation from mean, 44
Fractile, 56
Geometric mean, 40
Grand mean of combined data, 37
Harmonic mean, 40
Mean, 31, 51
Measures of location, 31
Measures of relative variation, 50
Measures of variation, 42
Median, 34, 54, 56
Midquartile, 61
Midrange, 65
Mode, 36
Parameter, 32

Pearsonian coefficient of skewness, 58
Percentile, 56
Population, 30
Population size, 30
Quantile, 56
Quartile, 56
Range, 43
Reverse J-shaped distribution, 59
Sample, 30
Sample size, 30
Semi-interquartile range, 61
Sigma notation, 32, 62
Skewed distribution, 57
Standard deviation, 44, 45, 52
Standard units, 47
Statistic, 32
U-shaped distribution, 59
Unbiased estimator, 64
Variance, 44, 45
Weighted mean, 36
z-scores, 47

3.14

Review Exercises

3.64 A large manufacturer, seeking target companies for acquisition, analyzes a sample of six possibilities and finds that the companies had 12.8, 10.5, 13.0, 11.6, 12.5, and 14.0 percent after-tax returns on a discounted cash-flow basis. Calculate the mean, median, range, and variance of these returns.

3.65 The lengths of a large shipment of chromium strips have a mean of 44 inches and a variance of 0.01 inch². At least what percent of these lengths must lie between

(a) 43.8 and 44.2 inches;   (c) 43.0 and 45.0 inches?

(b) 43.6 and 44.4 inches;

3.66 If, in a large population consisting of the times required to settle industrial accident claims, the time required to settle a particular claim is 99 days, the standard unit corresponding to this time is -2, and the population coefficient of variation is 5 percent, what are the population mean and standard deviation?

3.67 If a former U.S. president, addressing a luncheon meeting of the Executive's Club, spoke for 42 minutes at an average rate of 6,780 words per hour and received a fee of $14,000, what was the average fee per word of his speech?

3.68 In a certain class, a student scored 80, 85, and 90 on the first quiz, midterm, and final. If the instructor considers the midterm to be twice as important as the first quiz and the final to be three times as important as the midterm in determining the course grade, what was the student's average score on the three tests?

3.69 Decide in each case whether it is possible to find the mean and the median of the given distribution, and explain your answers:

(a) Grade	f	(b) IQ	f	(c) Weight	f
40–49	5	less than 90	3	100 or less	41
50–59	18	90– 99	14	101–110	13
60–69	27	100–109	22	111–120	8
70–79	15	110–119	19	121–130	3
80–89	6	more than 119	7	131–140	1

3.70 In a benefit sale the Friends of the Library raised a total of $3,063.35 on the sales of 1,025 books at an average price of $1.15, 454 books at an average price of $2.40, and some other books at an average price of $3.75. What was the average price of all the books sold?

3.71 In a factory, the time during working hours in which a machine is not operating as a result of breakage or failure is called the "downtime." The following distribution shows a sample of 100 downtimes of a certain machine (rounded to the nearest minute):

Downtime	Frequencies
0–9	3
10–19	13
20–29	30
30–39	25
40–49	14
50–59	8
60–69	4
70–79	2
80–89	1

Find the mean and the standard deviation of this distribution.

3.72 With reference to the preceding exercise, find
(a) the median;   (b) the quartiles Q_1 and Q_3.

3.73 Use the results of the two preceding exercises to find the Pearsonian coefficient of skewness for the distribution of downtimes.

3.74 As part of a study conducted in California, it was found that in a sample of San Francisco households the mean number of different medications on hand was 29.5. If the sample coefficient of variation was 20 percent, find the sample standard deviation.

3.75 A producer of television commercials knows exactly how much money was spent on the production of each of twelve commercials. Give one example each of a problem in which these data would be looked upon as
(a) a population;   (b) a sample.

3.76 A room is $14\frac{1}{16}$ feet wide, 20 feet long, and has a 12-foot ceiling. What is the geometric mean of the dimensions of this room?

3.77 In a comparative study of escalating food prices it was found that the July 2, 1975, price of one pound of sirloin steak was 62 cents in Buenos Aires, 77 cents in Brasilia, $2.12 in Washington, $4.73 in Copenhagen, $3.24 in Paris, and $15.00 in Tokyo.
(a) Calculate the mean of these prices.
(b) What would be the average cost per pound of 20 pounds of sirloin steak, if 5 pounds were bought in Buenos Aires, 4 each in Brasilia and Washington, 3 each in Copenhagen and Paris, and 1 in Tokyo?

3.78 If $\sum_{i=1}^{5} x_i = 15$ and $\sum_{i=1}^{5} x_i^2 = 51$ for a sample of size $n = 5$, find
(a) the mean;   (b) the standard deviation.

3.79 A large manufacturer has decided to enter a market with a new product. Thirty company executives, asked whether the product should be offered in 2, 3, 4, or 5 models, replied as follows: 5, 4, 3, 3, 2, 4, 3, 5, 5, 4, 3, 4, 3, 3, 2, 5, 5, 4, 2, 3, 3, 3, 4, 3, 5, 4, 3, 4, 5, and 2. What is the "consensus" opinion of the group?

3.80 An electronically controlled automatic bulk food filler is set to fill tubs with 60 pounds of cottage cheese. A random sample of five tubs from a large production lot shows filled weights of 60.00, 59.99, 60.05, 60.02, and 60.01 pounds. Find the mean and the median of these fills.

3.81 If a person spends $24 on blank cassette tapes costing $8 per dozen and another $24 on tapes costing $12 per dozen, what is her average cost per tape?

3.82 A general science test is given to a large group of students in a metropolitan school district. If the mean score is 30 points and the standard deviation is 4 points, what is the minimum fraction of the scores which must lie between 10 and 50 points?

3.83 The following is the distribution of the number of raffle tickets sold by the 70 members of a social-service organization:

Number of raffle tickets	Frequency
0–14	5
15–29	19
30–44	23
45–59	10
60–74	8
75–89	5

Find the mean, the median, the mode, and the standard deviation of this distribution.

3.84 With reference to the distribution of the preceding exercise, find
(a) Q_1 and Q_3;   (b) D_1 and D_9;   (c) P_5 and P_{95}.

3.85 On the maximum number of gallons of gasoline allowed for the Indianapolis 500 in one year, a car which averaged 1.6 miles per gallon would fall 52 miles short of the finish line.
(a) What is the maximum gas allowance?
(b) How many miles must a car average per gallon in order to finish the race?

3.86 Four women, working full time in industrial sales, earned the following amounts (in thousands of dollars) in a given year: 16.5, 24.5, 42.2, and 21.8. Calculate the mean, the median, the range, and the standard deviation of these earnings.

3.87 Of the total cash gifts made by seven large corporations in a recent year, 10.5, 12.9, 15.2, 8.8, 11.8, 16.0, and 12.0 percent were for local community-betterment projects. Find the mean, the median, and the range of these percentages.

3.88 If eight persons shopped on the average in 5.6 stores, is it possible that at least six of them shopped in more than seven stores?

3.89 Given a sample for which $\sum_{i=1}^{4} x_i = 3.6$ and $\sum_{i=1}^{4} x_i^2 = 3.44$, find

(a) $\sum_{i=1}^{4} (x_i + 3)$;   (b) $\sum_{i=1}^{4} (5x_i - 1)$;   (c) $\sum_{i=1}^{4} (x_i - 4)^2$.

3.90 A sample of five senior human resources personnel executives (who were responsible for developing and implementing company manpower planning systems) earned base annual salaries of 85, 75, 82, 78, and 80 thousand dollars. Also, a sample of eight senior industrial personnel executives (who did not normally perform company manpower planning functions) earned base annual salaries of 50, 55, 48, 53, 49, 46, 58, and 57 thousand dollars. Calculate the means, medians, modes, and standard deviations of these two samples.

3.91 The mean contributions in a fraternity building fund drive were: "large" donors, $800; "small" donors, $50; and "overall" donors, $300.
(a) What fraction of the donors were "small" donors?
(b) If there were, in fact, 100 "small" donors, what was the total amount contributed in the drive?

Summarizing Data: Index Numbers

Index numbers are numbers used to measure, or indicate, how much something has changed from one time to another, or how something compares with something else. For instance, if we compare the 1980 first-quarter profits of 543 large U.S. corporations with the 1979 first-quarter profits of these same concerns, we find that the 1980 profits were 117 percent of those in 1979. Also, we find that the 1980 first-quarter profits of eight railway equipment companies were 233 percent of their profits in the like quarter of 1979, but that the 1980 first-quarter profits of nine rubber companies were only 29 percent of their profits in the corresponding 1979 quarter. Further comparisons show that machine tool shipments for the year 1979 were 128 percent of such shipments for the year 1978, and that Americans paid 106 percent as much in state and local taxes in 1979 as they did in 1978.

Following some preliminary remarks on index numbers in Section 4.1, we shall discuss in Section 4.2 the basic problems which arise in their construction. In Sections 4.3 and 4.4 we shall present some general index number formulas, then devote Section 4.5 to a discussion of some very important indexes and Section 4.6, the concluding section, to some special applications of index numbers.

4.1

Some Preliminary Remarks

Percentages which compare two things—like the percentage which compares the state and local taxes that Americans paid in 1978 and 1979—are simple examples of index numbers. We call them "simple" because there are also index numbers which are intended to indicate changes in such complicated phenomena as industrial production, business activity, stock prices, transportation costs, consumer confidence, consumer buying plans, and in the prices consumers pay for the "basic necessities" of life (food, housing, energy, and health care), and also for all goods and services. In fact, there are sometimes many different indicators of the same phenomenon. There are, for instance, at least 16 different indicators of the condition of stocks on the New York Stock Exchange, or of the direction the market is heading, and many different indicators of the state of the nation's business in general. Among the latter indexes, some which tend to anticipate, or lead, future movements of business activity are of absolutely basic importance to both government and business. Indeed, the government constructs a single composite "Index of Leading Economic Indicators" which measures the overall change in such individual leading indicators as contracts and orders for new plants and equipment, the average workweek, the layoff rate, building permits, sensitive raw material prices, stock prices, new orders for consumer goods, and the nation's money supply and its total liquid assets.

The principal use of index numbers in business and economics is to make comparisons between two different time periods and many companies construct indexes to measure changes through time in their own raw material, labor, energy, and transportation costs. However, index numbers can also be used to make other kinds of comparisons, such as a comparison of the 1980 warranty costs of automobiles built in one assembly plant with the 1980 warranty costs of automobiles built in another, and a comparison of the 1980 production of cotton in Texas with the 1980 production of cotton in California.

Although index numbers are commonly associated with business and economics, they are also widely used in other fields. Psychologists measure intelligence quotients which are essentially index numbers comparing a person's intelligence with that of an average for the person's age, health authorities prepare indexes to show changes in the adequacy of hospital and other health care facilities, state boards of education construct indexes to measure the effectiveness of school systems, sociologists construct indexes measuring population changes, the National Weather Service has devised a "discomfort index" to measure the combined effect of heat and humidity on individuals, and so on.

In recent years the use of index numbers has been extended to so many fields of human activity that some knowledge of these measures really belongs under the heading of "general education." Index numbers showing changes in consumer prices, and the effect of these changes on purchasing power, are of tremendous importance to all consumers, and the Bureau of Labor Statistics' "Consumer Price Index for Urban Wage Earners and Clerical Workers," especially, is closely watched by many. The movements of this index are vital to millions of workers covered by collective-bargaining agreements containing escalator clauses which provide for automatic wage (and sometimes pension-benefit) increases when the index rises by a specified amount, and to millions of others, too—social security beneficiaries, retired military and government personnel, postal workers, and food stamp recipients, for instance—whose incomes are by statute affected by changes in the level of this index. Index numbers are of great concern to farmers whose subsidies depend on the Parity Index of the federal government, as well as to the parties in alimony agreements and the recipients of trust fund payments, which by "indexing" have been made to vary with an index of the purchasing power of the dollar. Serious suggestions have recently been made that not only wages and salaries, but taxes, interest rates, and various other economic realities be tied to a consumer price index. Aside from major problems connected with the exact nature of the index, or indexes, to be used as the escalator in such a system, many basic questions, all of them outside the scope of our study, arise as to the short- and long-range effect—good or bad—such a move would have on the economy.

4.2

Basic Problems

Like other statistical measures, index numbers are usually constructed to serve definite purposes. Sometimes, the stated purpose of an index is such that the only problem which arises in its construction is that of locating the necessary data. For instance, in a study of housing availability and costs in the United States it became necessary to construct an index showing the change in the sales of new, single-family homes from February 1979 to February 1980. Comparable figures from government sources showed sales of 532,000 units in February 1980 and 715,000 units in February 1979, so

$$\text{index} = \frac{532,000}{715,000} = 0.74 \quad \text{(or 74 percent)}$$

In contrast, there are many situations in which some very complex problems arise as soon as the purpose of an index has been stated. The most critical among these are (1) the availability and comparability of data; (2) the selection of items to be included in the comparison(s); (3) the choice of time periods (localities, and so on) to be compared; (4) the selection of appropriate weights measuring the relative importance of the various items which enter into the index; and (5) the choice of a suitable formula for combining the data into an index number. Some aspects of these basic problems are discussed below.

The availability and comparability of data. It is hardly necessary to point out that comparisons cannot be made and index numbers cannot be constructed unless the required data can be collected. On various occasions research workers have found that sales data needed by brand were available only by type of merchandise, that insurance losses were given per risk and not per claim, and so on. The problem of availability of data also arises if we want to make a comparison of commodity prices in 1981 with those of 1914. Television sets, some major appliances, contact lenses, frozen foods, transistorized devices, and many other items now widely used were not sold commercially in 1914.

The question of comparability of data can also be quite troublesome. In congressional hearings, some labor organizations and others have complained that, to some extent, the Consumer Price Index for Urban Wage Earners and Clerical Workers reflects deterioration in quality rather than an actual change in prices. It does not matter here whether this criticism is valid, but we must always make sure that in price comparisons the prices are actually comparable; that is, they must be the prices of goods and services which, for all practical purposes, are identical in quality. If they are not, this must be taken into account in some way.

The comparability of statistical data may also be questioned when there is a possibility of confusion due to the use of different definitions, or when parts of the data come from different sources. For instance, it is very confusing to read in the same edition of a financial paper that the nation's money supply in one July was up 5 percent from the year earlier figure, and also that it was up 10.2 percent from the year earlier figure. Actually, both statements are correct, but in the first the money supply is understood to be currency in circulation plus money on deposit in checking accounts (what the Federal Reserve then called M-1), whereas in the second the supply includes also savings accounts and other bank time deposits, except large certificates of deposit and nonbank thrift deposits (what the Federal Reserve then called M-3).

So far as data from different sources are concerned, consider the following table, based on figures published by the United Nations, which shows the importers' and exporters' versions of how many metric tons of four commodities the second country received from the first in a given year:

Commodity	Exporters' data		Importers' data	
Butter	France	5,005	United Kingdom	3,787
Eggs	Denmark	63,534	West Germany	51,273
Coffee	El Salvador	38,153	United States	27,693
Wheat	Canada	288,535	Netherlands	164,578

These figures were deliberately selected to illustrate our point, but major discrepancies are often found in employment and production figures quoted by different sources, in sickness and accident data supplied by different agencies, and in other kinds of data. Whether such instances are the exception or the rule, they illustrate the seriousness of the problem of locating relevant and comparable data.

The selection of items to be included in the comparison. If an index is designed for the special purpose of comparing the price of a commodity at two different times, there is no question as to what figures should be included. However, the situation is entirely different in the construction of general-purpose indexes; for instance, those designed to measure general changes in consumer prices. As it is physically impossible to include in such a comparison all commodities and all services, and to include, furthermore, all prices at which these commodities and services are traded in every single transaction throughout the country, the only thing index makers or sponsors can do is to take samples in such a way that, in their professional judgment, the items and transactions included adequately reflect the overall phenomenon being described. For example, the Consumer Price Index we spoke of is based on about 400 items (goods and services) which play a significant role in the typical budget of persons belonging to a certain population group. These index items constitute a sample of the many goods and services people buy each day, and their prices are collected from a sample of the nation's cities and of the sellers in these cities.

The choice of time periods that are to be compared. If an index number is designed for the specific purpose of comparing 1981 figures with those of, say, 1967, it is customary to refer to 1981 as the given year and to 1967 as the base year, indicating the latter by writing 1967 = 100. In general, the year or period whose data are to be compared with those of another period is called the given year or given period, and the year or period with reference to which the comparison is made is called the base year or base period.

The choice of a base year or base period presents no problem if an index is to be constructed to make a specific comparison. So far as general-purpose indexes describing complex phenomena are concerned, it is desirable to base the comparison on a relatively recent period of relative economic stability—if such a thing exists. The reason for choosing a stable period is that during times of abnormal economic conditions (during a war, for example) there may be no free trading of some commodities, there may be black markets, and the buying habits of the public may be irregular due to shortages of products that would otherwise be readily available. A major reason for choosing a relatively recent period is that it facilitates the construction of series for important new products (hand-held electronic calculators, for example) for which price and production data are not available for earlier years.

The federal government has made it a practice to establish a standard reference base for use by federal agencies. About every 10 years this base has been brought forward. In 1940, for example, a reference base of 1935–1939 was established, in 1951 a new base of 1947–1949 was designated, and in 1960 a new three-year base period 1957–1959 was established by the Bureau of the Budget for all government general-purpose indexes. It is worth noting that the 1957–1959 period is probably as nearly stable a period as can be found in the post–World War II era. But, as the government has pointed out, the selection of a base period does not imply "normality" in any real sense; a base period is merely a convenient and necessary reference point if comparisons are to be made. At the time this is written, most government indexes use 1967 = 100, but there are exceptions; for instance, the indexes of prices received and prices paid by farmers are still tied by law to the pre–World War I period 1910–1914.

The choice of appropriate weights.   There are many situations, particularly in index-number construction, in which figures cannot be meaningfully averaged without paying due attention to the relative importance of each item. Suppose, for example, that a manufacturer wants to construct an index comparing the 1981 and 1975 prices of replacement parts required for machines of a certain type, and that he chooses parts AX345 and AX765 to represent the changes in these prices. Suppose, further, that for part AX345 the index with 1975 = 100 is ($5.40/$3.60)·100 = 150 percent, and for part AX765 it is ($9.60/$4.80)· 100 = 200 percent. On the basis of these figures alone, we might assert that the price of replacement parts in 1981 was (150 + 200)/2 = 175 percent of what it was in 1975. This result would be valid if the same numbers of each of the two parts were replaced, but if in 1981 the company actually replaced, say, 40 times as many parts AX765 as parts AX345, then the two price changes are clearly not equally important. Generally speaking, valid comparisons of price changes require that prices be weighted in some way so as to account for their relative importance in practice. As it is virtually impossible to explain weighting schemes without referring to specific index-number formulas, we shall defer discussion of this matter until we study weighted index numbers.

The choice of a suitable formula.   There are various ways in which the average of a set of data can be described, and there are also various ways in which relative changes in prices or quantities can be described. In the next two sections we shall treat some of the simpler index-number formulas, and this discussion should make it clear that any choice among different formulas must ultimately depend on practical considerations and on some of their mathematical niceties. In the notation used in the remainder of this chapter, we refer to index numbers as I, to base-year prices as p_0, to given-year prices as p_n, to base-year quantities as q_0, and to given-year quantities as q_n.

4.3

Unweighted Index Numbers

To illustrate some of the simpler methods used in index-number construction, let us begin with the **simple aggregative method**. The formula for the index it leads to, called a **simple aggregative index**, is

Simple aggregative index

$$I = \frac{\sum p_n}{\sum p_0} \cdot 100$$

where $\sum p_n$ is the sum of the given-year prices, $\sum p_0$ is the sum of the base-year prices, and the ratio of the first to the second is multiplied by 100 to express the index as a percentage.

EXAMPLE The prices shown in the following table are the prices in dollars per pound of four foods sold in cash markets in June 1979 and in June 1980:

	June 1979	*June 1980*
Hams	0.665	0.545
Beef	0.995	1.035
Pork loins	0.930	0.720
Broilers	0.467	0.413

Construct a simple aggregative index comparing the June 1980 prices of these foods with their prices one year earlier.

SOLUTION Dividing the sum of the June 1980 prices by the sum of the June 1979 prices and multiplying by 100, we get

$$I = \frac{0.545 + 1.035 + 0.720 + 0.413}{0.665 + 0.995 + 0.930 + 0.467} \cdot 100 = \frac{2.713}{3.057} \cdot 100$$

$$= 88.7 \text{ percent}$$

This tells us that the combined June 1980 prices of the four foods are 88.7 percent of their combined prices in June 1979, or in other words, that prices dropped by 11.3 percent over the year. The usual way to express this change is to say that with June 1979 = 100, the index stood at 88.7 for June 1980.

A simple aggregative index is easy to construct and easy to understand, but it does not meet a criterion of adequacy called the **units test**. This test requires that the index yield the same results regardless of the units for which the prices of the various items are quoted. For instance, if in the preceding example we compare the combined prices of 1 pound of ham, 20 pounds of beef, 10 pounds of pork loin, and 5 pounds of broilers at the same two time periods as before, we find that the index stands at 94.8 instead of 88.7. Largely for this reason, simple aggregative indexes are not widely used today.

Another way to compare two sets of prices is to first calculate a separate index for each item and then average all these indexes, or **price relatives,** using some measure of central location. If we use the arithmetic mean, we get the **arithmetic mean of price relatives**; symbolically, its formula is

Arithmetic mean of price relatives

$$ I = \frac{\sum \frac{p_n}{p_0} \cdot 100}{k} $$

where k is the number of items whose price relatives are being combined into an index. In principle, price relatives can be averaged with any measure of central location, but in practice the arithmetic mean and the geometric mean are most widely used.

EXAMPLE Based on the data of the preceding example, construct an arithmetic mean of price relatives measuring the overall change in the prices of the four foods from June 1979 to June 1980.

SOLUTION Dividing the 1980 price of each food by its June 1979 price and multiplying by 100, we get

	Price relatives
Hams	$\frac{0.545}{0.665} \cdot 100 = 82.0$
Beef	$\frac{1.035}{0.995} \cdot 100 = 104.0$
Pork loins	$\frac{0.720}{0.930} \cdot 100 = 77.4$
Broilers	$\frac{0.413}{0.467} \cdot 100 = 88.4$

Then, calculating the arithmetic mean of these price relatives, we get

$$I = \frac{82.0 + 104.0 + 77.4 + 88.4}{4} = 88.0 \text{ percent}$$

It is a matter of historical interest that the earliest index number on record is an arithmetic mean of price relatives. In the middle of the eighteenth century, G. R. Carli, an Italian, calculated the effect of the import of silver on the value of money, using a formula like the one given above to compare the 1750 prices of oil, grain, and wine with those of the year 1500.

The formulas we have given in this section are all price index formulas. However, if we replace the p's by q's we get **quantity indexes**—index numbers which compare, for example, quantities put into trade during one period with those put into trade during another period.

4.4

Weighted Index Numbers

Today the need for weighting index items has been almost universally accepted, and few indexes are actually computed without using weights. Prior to 1914, the Wholesale Price Index of the Bureau of Labor Statistics was computed as an arithmetic mean of the price relatives of about 250 commodities. As a result of an important study by W. C. Mitchell in 1915, which has affected index-number construction since that time, the index was changed to a weighted index. Among the important government indexes only the daily Index of Spot Market Prices is still calculated as an unweighted (geometric) mean of price relatives.

In order to construct an index number which reflects differences in importance of the index items, we note first that the importance of changes in the price of a commodity in trade or use is best determined by the quantity of the commodity which is bought or sold, or produced or consumed. Hence, we can construct an index measuring the overall change in the index items by weighting the prices of the items by the corresponding quantities produced in the base year, the given year, or in some other year or period. Such an index, which is 100 times the ratio of the weighted mean of the given year prices to the weighted mean of the base-year prices, is called a **weighted aggregative index**. If we use base-year weights, it is also called a **Laspeyres Index**, named after the statistician who first suggested its use. Canceling the denominators of the two weighted means, $\sum q_0$, we can write the general formula for a Laspeyres Index as

Laspeyres Index

$$I = \frac{\sum p_n q_0}{\sum p_0 q_0} \cdot 100$$

Clearly, this kind of index reflects changes in prices alone—the same quantities of goods (the base-year quantities) are priced at two different times and any difference between the given-year total (the quantity in the numerator of the index) and the base-year total (the quantity in the denominator of the index) must be accounted for by changes in price.

EXAMPLE The following table shows the prices (in dollars per ton) and the quantities produced (in thousands of tons) of grapes for all uses in California in the years 1970, 1974, and 1979:

	PRICES			QUANTITIES		
	1970	1974	1979	1970	1974	1979
Wine varieties	118.00	146.00	202.15	537	1,214	1,692
Table varieties	108.05	135.20	140.10	336	617	413
Raisin varieties	72.40	132.10	155.50	1,890	1,958	2,300

With 1974 as the base year, and using base-year quantities as weights, construct a weighted aggregative (Laspeyres) index which measures the change in the price of California grapes for all uses from 1974 to 1979.

SOLUTION Substituting into the Laspeyres formula, we get

$$I = \frac{(202.15)(1,214) + (140.10)(617) + (155.50)(1,958)}{(146.00)(1,214) + (135.20)(617) + (132.10)(1,958)} \cdot 100$$

$$= 122.5 \text{ percent}$$

We cannot construct a price index by weighting the given-year prices with given-year quantities and the base-year prices with base-year quantities. Since $\sum p_n q_n$ is the total value of the goods in the given year and $\sum p_0 q_0$ is the total value of the goods in the base year, the ratio of $\sum p_n q_n$ to $\sum p_0 q_0$ reflects changes in value rather than changes in price; that ratio is, in fact, a value index. We can, however, use given-year quantities to weight both the base-year prices and the given-year prices (that is, price the given-year quantities at the two different times), and construct a weighted aggregative index with given-year weights. Sometimes called a Paasche Index, the formula is

Paasche Index

$$I = \frac{\sum p_n q_n}{\sum p_0 q_n} \cdot 100$$

Using the 1979 grape production quantities as weights in the preceding example, we find that the Paasche formula with 1974 = 100 leads to a 1979 index number of 124.9 percent.

Most of the important index numbers constructed by the federal government are published in series, that is, regularly every day, every week, every month, or every year. For these it would be highly impractical to use the Paasche formula, because new quantity weights would be required for each new day, week, month, or year. An index that is currently in great favor is the **fixed-weight aggregative index**, whose formula is

Fixed-weight aggregative index

$$I = \frac{\sum p_n q_a}{\sum p_0 q_a} \cdot 100$$

where the weights are quantities for some other period than the base year 0 or the given year n. Although it is actually calculated somewhat differently, one of the most important fixed-weight aggregative indexes is the Wholesale Price Index (now called the Producer Price Index) of the Bureau of Labor Statistics. Its current base period is 1967 and the q_a are quantities marketed in 1963.

In addition to weighted aggregative indexes of the sort we have discussed above, we can also obtain weighted indexes by weighting the individual price relatives. The formula for a **weighted arithmetic mean of price relatives** is

Weighted arithmetic mean of price relatives

$$I = \frac{\sum \frac{p_n}{p_0} \cdot w}{\sum w} \cdot 100$$

where the w's are suitable weights assigned to the individual price relatives of the index items (written as proportions, not as percentages).

Since the importance of the relative change in the price of a commodity is most reasonably reflected by the total amount of money spent on it, it is customary to use **value weights** for the w's of the last formula. This raises the question of whether to use the values (prices times quantities) of the base year, those of the given year, or perhaps some other fixed-value weights. Actually, if base-year value weights $p_0 q_0$ are used, we do not get a new index, for with these weights, the formula directly above reduces to that of a weighted aggregative index with base-year weights, the Laspeyres Index.

EXAMPLE With reference to the example on page 80, construct a weighted arithmetic mean of price relatives with 1970 value weights to measure the change in the price of California grapes for all uses from 1974 to 1979.

Calculating first the price relatives $p_n/p_0 = p_{79}/p_{74}$ and the value weights $w = p_a q_a = p_{70} q_{70}$ (in thousands of dollars), we get

	Price relatives p_{79}/p_{74}	Values $p_{70} q_{70}$
Wine varieties	1.385	63,366
Table varieties	1.036	36,305
Raisin varieties	1.177	136,836

Then, substituting into the index-number formula, we find that

$$I = \frac{(1.385)(63,366) + (1.036)(36,305) + (1.177)(136,836)}{63,366 + 36,305 + 136,836} \cdot 100$$

$$= 121.1 \text{ percent}$$

We now have three measures, or indicators, of the change in price of California grapes for all uses from 1974 to 1979: 122.5 (the Laspeyres Index), 124.9 (the Paasche Index), and 121.1 (the weighted arithmetic mean of price relatives with 1970 value weights). This illustrates the fact that different methods applied to the measurement of the same phenomenon can lead to different results. The differences here do not seem large, but in practice, where often millions of dollars ride on a change of less than one point (as in labor–management agreements containing escalator clauses), the question of choosing an appropriate index is a serious one indeed.

EXERCISES

4.1 The following are the estimated average retail prices (to the nearest cent) of selected dairy products in the years 1972, 1973, 1974, and 1980:

	1972	1973	1974	1980
Milk, fresh (grocery), $\frac{1}{2}$ gallon	60	65	78	92
Ice cream, $\frac{1}{2}$ gallon	86	91	108	229
Butter, pound	87	92	95	199
Cheese, American process, $\frac{1}{2}$ pound	54	60	73	129

Calculate the 1972, 1973, 1974, and 1980 values of a simple aggregative index with 1972 = 100 for the overall price change of these items.

4.2 With reference to the preceding exercise, find the arithmetic mean of the price relatives comparing the 1980 prices with those of 1972.

4.3 The total 1978 and 1979 factory shipments of selected appliances (in thousands of units) were

	1978	1979
Refrigerators, electric	5,890	5,701
Freezers	1,521	1,859
Ranges, electric	3,217	3,003
Water heaters, gas	2,921	2,887
Dishwashers	3,558	3,488
Disposers, food waste	3,312	3,316
Air conditioners, room	4,037	3,749

(a) Calculate a simple aggregative index comparing the 1979 shipments of these goods with those of 1978.

(b) Find the arithmetic mean of the seven relatives comparing the 1979 shipments of these goods with those of 1978.

4.4 The following are the annual average prices (in dollars per pound) of three selected imported metals:

	1970	1975	1976	1977	1978
Copper	0.63	0.53	0.58	0.58	0.58
Nickel	1.30	1.88	2.07	2.15	1.93
Tin	1.66	3.35	3.31	4.37	5.52

(a) Find the 1975, 1976, 1977, and 1978 values of a simple aggregative index with 1970 = 100.

(b) Find a simple aggregative index comparing the 1978 prices with those of 1976.

(c) Find for 1977 and 1978 the arithmetic mean of the price relatives using 1970 = 100.

4.5 With reference to the preceding exercise, compare the 1978 prices of copper and nickel with those of 1976 by finding
(a) the arithmetic mean of the price relatives;
(b) the geometric mean of the price relatives.

4.6 Inventories of table wines in California for the years 1975 through 1979 were 321, 347, 383, 412, and 446 millions of gallons. Construct an index series showing changes of these inventories during the years 1975–1979 with 1977 = 100.

4.7 The following figures, in millions of dollars, are the cash receipts of farms in the East North Central United States and in the state of Kansas in 1978:

	Crops	Livestock	Government payments
Ohio	1,730	1,273	24.3
Indiana	1,921	1,557	53.4
Illinois	3,985	2,139	102.8
Michigan	1,129	998	45.5
Wisconsin	674	2,971	44.1
Kansas	1,490	2,956	300.9

(a) Construct a simple aggregative index for each of the East North Central states (Ohio, Indiana, Illinois, Michigan, Wisconsin), comparing its total cash receipts of farms with that of Kansas.

(b) Find the arithmetic mean of the relatives comparing the cash receipts of farms in Indiana with those in Illinois.

4.8 Calculate the geometric mean of the relatives of the four foods given on page 78.

4.9 With reference to the example on page 77,

(a) verify the value of 94.8 given for an index based on the prices of 1 pound of ham, 20 pounds of beef, 10 pounds of pork loins, and 5 pounds of broilers;

(b) what value would we get if we compare the combined prices of 10 pounds each of ham, beef, pork loins, and broilers;

(c) what value would we get if we compare the combined prices of 1 pound each of ham, pork loins, and broilers, and 100 pounds of beef?

4.10 Verify that the Paasche Index number referred to on page 81 is 124.9.

4.11 Show that if we substitute base-year value weights into the formula for a weighted arithmetic mean of price relatives, we obtain the formula for the Laspeyres Index.

4.12 It is interesting to note that the Laspeyres formula can generally be expected to overestimate price changes, while the Paasche formula will generally do just the opposite. Explain why this is so, using as an illustration a consumer price index intended to measure changes in the prices of a market basket consisting of several hundred consumer goods and services.

4.13 In the Ideal Index, which has never been widely used for practical reasons, the upward and downward biases of the Laspeyres Index and the Paasche Index computed from the same data (see Exercise 4.12) will be more or less averaged out by taking the geometric mean of the two indexes. If a Laspeyres Index stands at 134.0 and a Paasche Index calculated from the same data stands at 132.5, what is the Ideal Index for this specific comparison?

4.14 Write a general formula for the Ideal Index (see Exercise 4.13).

4.15 The following table contains prices received by fishermen (in cents per pound) and the quantities of catch (in millions of pounds) of four selected species of fish:

| | PRICES | | | QUANTITIES | | |
	1974	1975	1978	1974	1975	1978
Cod, Atlantic	18.8	23.3	22.9	59	56	87
Flounder	26.8	33.5	44.8	156	162	181
Tuna	41.0	33.8	59.8	386	393	409
Haddock	28.8	26.8	27.3	8	16	39

(a) Use the 1974 quantities as weights and 1974 = 100 to find weighted aggregative indexes for the 1975 and 1978 prices.

(b) Calculate a weighted aggregative index comparing the 1978 prices of the fish with those of 1974, using the 1978 quantities as weights.

(c) Calculate a weighted aggregative index comparing the 1978 prices of the fish with those of 1974, using the averages of the 1975 and 1978 quantities as weights.

(d) With 1974 = 100, calculate for 1978 the weighted arithmetic mean of price relatives, using the base-year values as weights.

(e) With 1974 = 100, calculate for 1978 the weighted arithmetic mean of price relatives, using the given-year values as weights.

(f) Interchanging the p's and q's in the formula used in part (e), construct an index comparing the 1978 catch of the four species with that of 1974.

4.16 The following are the 1976, 1977, and 1978 prices (in cents per can) and production (in millions of standard cases of 24 No. 303 cans) of selected canned fruits and vegetables:

| | PRICES | | | QUANTITIES | | |
	1976	1977	1978	1976	1977	1978
Fruit cocktail	46.0	47.8	48.8	13.6	13.0	11.1
Peas, green	38.6	38.3	37.7	31.9	30.2	25.3
Tomatoes	35.1	37.6	38.0	42.8	54.1	49.2

(a) Use 1976 quantities as weights to construct aggregative indexes comparing the 1977 and 1978 prices with those of 1976.

(b) Use 1977 quantities as weights to construct aggregative indexes comparing the 1977 and 1978 prices with those of 1976.

(c) Use 1978 quantities as weights to construct an aggregative index comparing the 1978 prices with those of 1976.

(d) Use the means of the 1976 and 1978 quantities as weights to construct an aggregative index comparing the 1978 prices with those of 1976.

(e) With 1976 = 100, calculate for 1978 the weighted arithmetic mean of price relatives, using base-year values as weights. Compare this result with the 1978 index number calculated in part (a).

(f) With 1976 = 100, calculate for 1978 the weighted arithmetic mean of the price relatives, using given-year values as weights.

4.17 Compare the 1978 prices of the three foods in Exercise 4.16 with those of 1976 by means of the Ideal Index (see Exercise 4.13).

4.18 Show that the formula for the weighted arithmetic mean of price relatives with value weights of the form $w = p_0 q_a$ reduces to the formula for a fixed-weight aggregative index.

4.5

Some Important Indexes

Among the many important indexes intended to describe assorted phenomena, some are prepared by private organizations. Financial institutions, utility companies, and university bureaus of business research, for example, often prepare indexes of such things as employment, factory hours and wages, and retail sales for the regions they serve; trade associations prepare indexes of price and quantity changes vital to their particular interests; and so on. Many of these indexes are widely used and highly respected indicators of the phenomena they describe. However, by far the most widely circulated and widely used indexes are those prepared by the federal government. Of the many important government indexes, we shall describe briefly the Consumer Price Index, the Producer Price Index, and a quantity index called the Index of Industrial Production.

What may be considered by most people to be "the" Consumer Price Index is called officially the **Consumer Price Index for Urban Wage Earners and Clerical Workers.** It is constructed by the Bureau of Labor Statistics and it has been published by the Bureau since 1921. Now, as then, the index is intended to measure the effect of price changes of a collection of goods and services called a "market basket" on the living costs of urban wage earners and clerical workers, both families and single persons living alone. This market basket now consists of fixed quantities of some 400 consumption items, described by detailed specifications to assure—insofar as possible—comparability in successive periods. Included among these items are meats, dairy products, residential rents, clothing, appliances, new and used cars, gasoline and oil and parking fees, physicians' and dentists' services, drugs, haircuts, toothpaste, television sets and replacement tubes, newspapers, cigarettes, beer, and the like, which are of major importance in consumer purchases and which, in fact, account for the greater part of consumer spending.

The items which comprise the market basket are periodically priced in a sample of 85 Standard Metropolitan Statistical Areas (SMSA's) and cities chosen to represent all urban places in the United States whose populations are over 2,500. The prices that enter the index are for items classified into six major groups and subgroups: (1) food (at home and away from home); (2) housing (shelter, fuel and utilities, household furnishings and operation); (3) apparel and upkeep (men's and boys', women's and girls', footwear, miscellaneous apparel, and apparel services); (4) transportation (private and public); (5) health and recreation (medical care, personal care, reading and recreation); and (6) other goods and services (tobacco products, alcoholic beverages, financial and miscellaneous personal expenses). Group indexes are regularly calculated for 26 cities (or areas), and they are then combined into overall city indexes, group indexes for all cities combined, and one index covering all groups and all cities. When the data for the individual groups are combined into a city index, each group is assigned a weight (differing from city to city) which is intended to represent the "relative importance" of the group in the average budget, or family expenditure, of families covered by the index. The combined price index for the whole country is calculated from the data obtained for the various cities, giving those for each city a weight in proportion to the part of the total wage-earner and clerical-worker population it represents in the overall index. The current base period for the index is 1967.

In calculating this Consumer Price Index and its various group and subindexes, the Bureau of Labor Statistics uses two formulas, which differ in the mechanics of construction but which are mathematically the same. Both formulas represent fixed-weight aggregative indexes of the sort now commonly in use. Looking at the index as a weighted average of price relatives, the value (price times quantity) weights used vary from time to time, but this is due entirely to changes in prices, since the quantities making up the market basket are held fixed. In this way, the index measures the change in the prices of fixed quantities of goods and services.

The index we have been speaking of above is the result of several revisions of the index first published in 1921. The (January) 1964 revision, or "series," represents a major revision of the preceding series, which began in January 1953. Intended to account for various important social and economic developments which had taken place in this country since the earlier series began, the 1964 revision included, among other things, increases in the number of sample cities and in the number of prices collected; an increase in the number of discount houses, suburban stores, and physicians included in the sample; and an increase in the size of the market basket from 325 to 400 items, giving broader coverage to such categories as apparel, home furnishings, automobiles, and restaurant meals, and pricing a number of items, including precooked foods and "miracle" fabrics, for the first time.

One crucial problem in constructing this Consumer Price Index is that of determining the specific goods and services to be included in the market basket and establishing as part of the value weights the relative importance of the different groups of goods and services in the expenditures of the index families and individuals. The information which leads to these determinations is gathered in periodic surveys of consumer expenditures, savings, and incomes. The Consumer Expenditure Survey which provided the information for the 1964 revision was based on a large sample of wage-earner and clerical-worker families and individuals carefully selected in 1960–1961. In calculating the relative importance weights, the survey data (compiled from about 10,000 usable schedules) were adjusted to 1963 prices and consumption levels.

Under normal conditions the relative importance of the various categories of goods and services in a person's expenditures (food 22.94 percent, housing 32.89 percent, and so on) changes fairly slowly, but at any time other than at the point when revised weights are introduced into the index, the relative importance weights are only approximations to the actual expenditure distribution. Consequently, a time always comes when changes in consumer buying patterns require another revision of the revised Consumer Price Index. Accordingly, in 1972 the Bureau of Labor Statistics began a major revision of the then-current 1964 series and in 1978 published the first figure in the new series. The purpose of this improved version of the original index (now designated CPI-W) was to meet the needs of collective bargaining. At the same time in 1978 the Bureau of Labor Statistics began publication of a second index, this one intended to provide a more comprehensive measure of price changes in the economy than the original index. Called the "Consumer Price Index for All Urban Households" (CPI-U), this new index covers about 80 percent of the nation's civilian noninstitutional population.

In constructing these two indexes, the determination of where people buy things, what they buy, and the relative importance of what they buy in their overall expenditures was based on a series of sample surveys. The most important of these was another large-scale Consumer Expenditure Survey (the eighth of its kind the federal government has conducted), which furnished information on what people buy. This latest survey covered the years 1972 and 1973 and consisted of two separate surveys using different questionnaires and different samples. The first consisted of a series of quarterly interviews with a sample of about 20,000 families spread over 216 areas of the country; the second was a diary survey in which a separate sample of about 20,000 families kept records over two one-week periods on such items usually bought on a daily or weekly basis as food and beverage and personal care products.

Since the federal government published the first figure in its first index series, it has been concerned over the accuracy of its indexes. No further lesson is needed to appreciate the tremendous importance of the accuracy of the Consumer Price Indexes than to realize that, according to the Bureau of Labor Statistics, an increase of 1 percent in the revised index causes an increase of at least $1 billion in income payments under escalation agreements; hence, a measurement error of only 0.1 percent leads to the misdirection of over $100 million. When in 1978 the Bureau of Labor Statistics began publishing the numbers in the revised and the new indexes, it planned to publish both indexes for at least three years, and at some time decide what was really needed to meet the demands of the nation for consumer price information: only one index, both indexes plus possibly a third index to measure the difference between the two, or perhaps an entire family of indexes. As of May 1981 both indexes were still being published, although both had come under increasingly heavy criticism recently, largely because of the way the Bureau handles the housing component of the indexes. We shall not discuss this matter in any detail, but we should note that, by treating homes as consumption items, like gasoline, for example, the estimated housing costs are probably higher than the true housing costs. The Bureau itself has conceded this and a Congressional Budget analyst has said that if housing costs were computed on the basis of what it would cost homeowners to rent comparable accommodations, the revised index would have been 2 points lower in 1979; this, in turn, would have resulted in a total saving to the federal government of $4 billion.

Nevertheless, the Bureau of Labor Statistics has said that it plans no hasty revisions of the indexes, and no further revisions seem possible before 1983 at the earliest.

Also constructed by the Bureau of Labor Statistics, the **Producer Price Index** (formerly called the Wholesale Price Index) is intended to measure changes in the prices of commodities at their first important commercial transaction. Thus, the word "wholesale" does not refer to prices received by wholesalers, jobbers, and distributors, but to prices of large lots in primary markets. Most of the prices used are those quoted on organized exchanges or markets, or received by manufacturers and other producers. The price changes measured are of individual commodities, groups of commodities, and the components of these, and they are designed to be "real" price changes, that is, price changes unaffected by changes in quality, quantity, terms of sale, level of distribution, the unit priced, or the source of the price.

Like other comprehensive indexes, the Producer Price Index is based on a sample. Because of the importance of wholesale price movements in the many subdivisions of the economy, about 2,800 items are presently being priced (usually on Tuesday of the week containing the 15th of the month) to get the prices used in the index. These index commodities are not randomly selected from the many commodities traded at the wholesale level; they are carefully chosen by professionals in whose judgment they appear to be the most important or representative ones in their categories.

At present, separate price indexes are calculated for commodities classified in various ways. One major classification is by stages of processing (crude materials for further handling; intermediate materials, supplies, and components; and finished goods). In another classification, all farm products are subclassified as fresh and dried fruits and vegetables, grains, live poultry, and livestock. Industrial commodities are grouped in 13 subclassifications, including chemical and allied products, furniture and household durables, lumber and wood products, machinery and equipment, rubber and rubber products, textile products and apparel, and transportation equipment. Some of these are, in turn, further subdivided; the furniture and household durables group, for instance, includes household appliances, household furniture, and home electronic equipment, all of which are themselves groupings of other commodities.

Basically, although not strictly, the Producer Price Index is calculated as a fixed-weight aggregative index. However, this type of index is identical with one in which price relatives p_n/p_0 are weighted with value weights $p_0 q_a$. Commencing with the January 1967 figures, the weights used in calculating the index represent the total net selling value of commodities produced in, processed in, or imported into the United States and flowing into primary markets in the year 1963. For technical reasons, the formula actually used to calculate the index is a variation of the second one on page 81 with $w = p_0 q_a$, and currently 1967 = 100.

Unlike indexes measuring changes in prices, the Federal Reserve Board's **Index of Industrial Production** measures changes in the physical volume or quantity of output of the nation's factories, mines, and electric and gas utilities. Forerunners of the current index appeared in 1919 and 1922 and were succeeded in 1927 by the "New Index of Industrial Production." Based on 60 series for manufactured goods and mineral products, this index represented, either directly or indirectly, about 80 percent of the nation's production of these goods and products. In an effort to keep pace with the output of a rapidly expanding economy, major revisions of the index were made in 1940, 1953, and 1959. In the latter revision, aimed at giving broader coverage and also permitting better comparisons with production indexes of other countries, the output of electric and gas utilities was added. The most recent revision, again extending the index's coverage, was made in 1971. The fabrication of materials into final products is now represented directly, but the distribution of industrial products and their use in construction is still only indirectly covered. Production in the construction industry itself, on farms, in transportation, and in various trade and service industries is not covered by the index.

The monthly composite index showing changes in the nation's total industrial production is arrived at by combining 227 individual monthly series into a number of different groups and then combining the various group indicators into one overall figure. One such grouping of individual series is the "industry grouping," whose main components are manufacturing, mining, and utilities. In this classification, manufacturing is separated into durable and nondurable manufactures. The durable-goods group includes primary metals and fabricated metal products; machinery; transportation equipment; lumber and related products; clay, glass, and stone products; furniture and fixtures; and miscellaneous manufactures. For their part, the primary metals group, for instance, includes iron and steel, and nonferrous metals; and the transportation group includes motor vehicles and parts, and aerospace and miscellaneous equipment. The nondurable-manufactures group consists of 10 subgroups, including textile mill products, apparel products, leather and leather products, paper and paper products, printing and publishing, chemicals and chemical products, petroleum products, rubber and plastic products, foods, and tobacco products. For their part, mining and utilities are both groupings (utilities of electricity and gas, and mining of metal; stone and related minerals; and coal, oil, and gas).

In addition to the industry groupings, the 1959 revision added another combination of production series called "market groupings." In the present arrangement, the 227 component series of the total industrial production index are grouped broadly into final products and intermediate products, with further divisions and subdivisions as in the industry groupings. One major advantage of the market groupings is that they make possible careful studies of various marketing developments, such as the relationships between changes in production and dollar expenditures. Somewhat more broadly, it is hoped that the improved physical volume measures in the current index will permit penetrating analyses of economic developments, and a wider understanding of these developments, in the dynamic and complex American economy.

We shall not describe in detail the fairly complicated way in which the individual series are combined into group indexes and eventually into the overall production index. We observe, however, that for the total index the calculations actually performed lead to a fixed-weight aggregative index, which is the ratio of total value added in a given month to the corresponding value added in 1967 (using in both cases 1967 value-added-per-unit prices); that is,

$$I = \frac{\sum q_n p_{67}}{\sum q_0 p_{67}} \cdot 100$$

Currently, the published index relates to the base 1967.

Unlike the Consumer and Producer Price Indexes, the Index of Industrial production is intended to be a measure of current business conditions, and this sensitive, comprehensive business indicator furnishes fundamentally important direction to many persons and institutions whose current activities and future plans are in some way affected by the ever-changing overall business conditions.

Some Special Applications

Many studies made, and polls taken, in the last few years have shown that the consensus feeling of the American people is that price inflation is the most crucial problem facing the nation today. In times of inflation prices rise faster, often much faster, than personal incomes, with the result that a dollar loses some of its value and will buy less than it would have bought at an earlier time. It is this "shrinking value of the dollar" which so troubles producers and consumers and governments alike. Of course, there are all sorts of "dollars"— dollars spent for energy, for food, for housing, for medical care, and so on— and their values do not necessarily all move in the same direction at the same time; the value of the hand-held-calculator dollar, for example, has increased dramatically in recent years while the value of the housing dollar has dropped sharply.

It is easy to show how the value of some single-commodity dollar is measured. If a baker paid $5.20 in 1975 for a fixed quantity of honey and paid $10.40 for the same quantity of the same honey in 1980, the 1980 value of the "honey dollar" has been cut in half and is only 50 cents compared to its 1975 value. This follows from the fact that the honey price has doubled and that it takes two dollars in 1980 to buy what one dollar bought in 1975. Generally speaking, the value, or the purchasing power, of a dollar relative to some period in time is the reciprocal of an appropriate price index (written as a proportion). If prices have doubled since some reference period, the price index on that base is 2.00 (or 200 percent) and the dollar will buy only $\frac{1}{2.00} = \frac{1}{2}$ of what it would have bought in the earlier period. In other words, the purchasing power of the dollar is $\frac{1}{2}$ of what it was, or 50 cents. Similarly, if prices have risen by 50 percent, the price index stands at 150 and the purchasing power of the dollar is $\frac{1}{1.50} = \frac{2}{3}$ of what it was, or about 67 cents.

The same argument applies also when we speak of the purchasing power of, say, a construction dollar, a food dollar, a rent dollar, or a medical care dollar, none of which refers to a single commodity. Comprehensive indexes of construction, food, rent, and medical care prices are available, so we just take their reciprocals to arrive at the purchasing powers of the different dollars. To get an estimate of the purchasing power of what is the nearest thing to an omnibus dollar used to buy the full range of goods and services available in the economy, called "the" dollar, the reciprocals of such comprehensive indexes as the Consumer Price Index CPI-U or the Producer Price Index is usually used. Figures giving the purchasing power of the dollar, based on these indexes, are regularly published by the government.

Another important application of price indexes is in the calculation of "real wages" as distinct from "money wages." Since money is not an end in itself, wage earners are usually interested in what their wages will buy, and this depends on two things: how much they earn and the prices of the things they want to buy. Clearly, persons would be worse off this year than last year if their money wages have doubled over this period but the prices of the things they want to buy have tripled.

To calculate real wages, we can either multiply actual money wages by a quantity measuring the purchasing power of the dollar, or better, divide money wages by an appropriate price index. This process is called *deflating*, and the price index used as a divisor is called a *deflator*.

EXAMPLE From 1978 to 1980 the average weekly wages of one class of city maintenance workers in a large city increased from $298.60 to $327.15. Over the same period an index of consumer prices in that city increased 22.0 percent. Calculate the real average weekly wages of these employees for 1980.

SOLUTION The first column of the table below shows the 1978 and 1980 money wages and the second column shows the consumer price index with 1978 = 100:

	Average weekly wages	*Consumer prices 1978 = 100*	*Real wages*
1978	298.60	100.0	298.60
1980	327.15	122.0	268.16

Based on these figures, we get the values in the third column, the real wages, by dividing the index numbers (expressed as proportions) into the corresponding actual wages.

Over the period covered in this example, the money wages increased by 9.6 percent, and if prices had remained unchanged, real wages would have increased by precisely the same amount. However, prices increased by 22.0 percent, and since $\frac{1}{1.220} = 0.820$, there was a decrease of 18.0 cents in the purchasing power of the 1980 dollar. As a result, since $\frac{268.16}{298.60} = 0.898$, real wages actually decreased by 10.2 percent. Expressed in another way, a market basket of goods and services which could have been bought for $268.16 in 1978 cost $327.15 in 1980.

The procedure we have illustrated here is frequently used to deflate individual values, value series, or value indexes. It is used, for example, in analyzing the movements through time of dollar sales or dollar inventories (which vary in response to both price and quantity changes) of manufacturers, wholesalers, and retailers; the total values of construction contracts or construction put in place; money incomes; and money wages. The only real problem in deflating series such as these is that of finding appropriate deflators.

EXERCISES

4.19 Would it be possible for the revised Consumer Price Index to reach an all-time high in a given month when at the same time prices on the New York Stock Exchange and industrial production were falling sharply and unemployment had reached a record high and was still increasing?

4.20 An elderly couple, planning to move from Pittsburgh to either Houston or Atlanta, finds from a government publication that the 1978 value of the revised Consumer Price Index for Atlanta stands at 192.6, while that for Houston is 208.2. Comment on their conclusion that it would cost them 15.6 percent more to live in Houston than in Atlanta.

4.21 Comment on the following statements, both wrong, which appeared in a "popular" article on index numbers:
(a) "Probably the most important use of the Wholesale Price Index is in forecasting later movements in the Consumer Price Index."
(b) "A direct comparison of the Wholesale Price Index and the Consumer Price Index gives a very close estimate of the profit margins between primary markets and other distributive levels."

4.22 It has been said that in a dynamic economy production indexes, like the Federal Reserve Board's Index of Industrial Production, always understate production. Why is a downward bias inherent in production indexes?

4.23 For the year 1980 relative to the year 1979 the number of personal bankruptcies decreased by 14 percent in city A and increased by 5 percent in city B. Is it reasonable to conclude that for 1980 the number of bankruptcies in A was 19 percent lower than in B?

4.24 It is often desirable, or necessary, to change the point of reference, or shift the base, of an index number series from one period to another. Ordinarily, this is done simply by dividing each value in the series by the original index number for the period which is to be the new base, then multiplying by 100. For instance, to shift the revised Consumer Price Index for the years 1975 through 1980 from 1967 = 100 to 1975 = 100, we divide each of the six yearly values by 161.2, the original value of the index for 1975.

The following are the average weekly wages of part-time legal office employees in a large city for the years 1975 through 1980: 187.55, 196.92, 203.82, 217.88, 239.67, and 252.85 dollars.

(a) Construct an index showing the changes in these wages from the base year 1975.

(b) Shift the base of the index in part (a) to the year 1978.

(c) If a consumer price index for this city showed an increase of 20 percent from 1978 to 1980, how much did these employees earn in 1980 in real wages (constant 1978 dollars)?

4.25 In the months of July for the years 1975 through 1979, 4.09, 3.93, 3.79, 4.00, and 3.88 million people were in farm employment in the United States.

(a) Construct an index of July farm employment for this 5-year period with July 1975 = 100.

(b) Shift the base of the index of part (a) to July 1979.

4.26 For the months of January through December of 1979, family clothing stores in the United States had estimated sales of 179, 166, 231, 244, 244, 243, 222, 302, 263, 273, 321, and 536 millions of dollars.

(a) Construct a monthly index of these sales using January 1979 as the base month.

(b) Shift the base of the index of part (a) to July 1979.

4.27 In a large metropolitan area the 1975 through 1979 gross average weekly earnings of workers in one type of employment were 310.25, 337.82, 363.40, 394.18, and 440.57 dollars. For the same years, 1975 through 1979, the values of a "cost of living" index for the area were 159.8, 168.9, 182.1, 195.0, and 216.7 with 1967 = 100.

(a) Shift the base of the index to 1975, then use these index values to express the actual earnings in constant 1975 dollars (that is, deflate the actual earnings).

(b) Construct an index of the purchasing power of the dollar for this period with 1975 = 100.

(c) What were the percentage changes from 1975 to 1979 in money earnings, real earnings, living costs, and purchasing power?

4.28 Suppose that in the area referred to in the preceding exercise, the gross average weekly earnings for 1975–1979 had been 310.25, 327.93, 353.69, 378.51, and 420.70 dollars. How well would their actual earnings gains from year to year over the period have protected the workers from the inflation in living costs?

4.29 In 1978 the average weekly wages of one class of laborers in a certain area were \$173.10 and in 1980 the corresponding earnings were \$198.34. A regional cost of living index stood at 179.5 for 1978 and 218.9 for 1980 with 1967 = 100. Express the 1980 dollar wages of these laborers in terms of constant 1978 dollars.

4.30 When we deflate the 1980 value of a single commodity to, say, 1975 prices, we divide its value by an index expressing the 1980 price of the commodity as a relative of the 1975 price. Show symbolically that this process leads to the value of the commodity in 1980 at 1975 prices. (Although this argument does not apply strictly when we deflate an aggregate of the values of several commodities, we are in a sense estimating the total value of the same goods at base-year prices.)

4.31 Since index numbers are designed to compare two sets of figures, it seems reasonable that if an index for 1981 with the base year 1967 stands at 200, the same index for 1967 with the base year 1981 should be equal to 50. (If one thing is twice as big as another, the second should be half as big as the first.) To test whether an index meets this criterion, called the time-reversal test, we need only interchange the subscripts 0 and n wherever they appear in the formula and then see whether the resulting index (written as a proportion) is the reciprocal of the first. Determine which indexes among the simple aggregative index, the weighted aggregative index, the arithmetic mean of price relatives, the geometric mean of price relatives, and the Ideal Index (see Exercise 4.13 on page 84) satisfy this criterion.

4.32 As has been suggested in the text, price index formulas can be changed into quantity index formulas simply by replacing the p's with q's and the q's with p's. Using this relationship between the formula for a price index and the corresponding formula for a quantity index, the factor-reversal test requires that the product of the two (written as proportions) equal the value index $\sum p_n q_n / \sum p_0 q_0$. Show that this criterion is satisfied if we compare the prices, quantities, and values of a single commodity and for the Ideal Index (see Exercise 4.13 on page 84), but not for any of the other index number formulas given in this chapter.

4.7

A Word of Caution

We now add to our discussion of index numbers a word of caution about their use and interpretation. Trouble always arises when attempts are made to generalize beyond the stated purpose of an index to phenomena it was never intended to describe. The word "general" serves well enough to distinguish more-or-less comprehensive general-purpose indexes from those that are narrowly limited, or "special," in scope, but it is quite misleading in another sense: most "general-purpose" indexes are themselves strictly limited in purpose and scope.

Perhaps the most widely misunderstood index of all is the revised Consumer Price Index of the Bureau of Labor Statistics. The index is widely thought to measure not only the "cost of living" for everybody everywhere, but also to measure current business conditions—neither of which it does. In view of the government's many careful explanations of just what this Consumer Price Index is and is not intended to measure, this is hard to understand. Whatever remote or indirect connection may exist between the phenomenon described by the index and business conditions in general is unintended. Moreover, there is little basis for thinking of the index as a measure of everyone's cost of living, even though for years it was officially called a cost-of-living index. Actually, as the government is now careful to point out, the index measures the effect of price changes of a (fixed) market basket of goods and services on the cost of living of the families and individuals to which it applies. But a person's cost of living also depends to some extent on his level of living, and changes in the level of living are not reflected in the index because purchases are held constant. Also, the index does not take into account, among other things, federal and state income taxes, social security taxes, and such noncash consumption items as food grown at home. Unfortunately, no true cost of living index—one which, for example, would measure changes while holding satisfaction or utility, rather than purchases, constant—exists for this country. Nevertheless, some professionals have asserted that, under "normal" conditions, whatever they may be, the revised Consumer Price Index can be considered to be a good approximation to changes in the cost of living; just how good no one knows.

There are some persons who would like to see the government develop bigger, better, and more general indexes, say, a truly general "all-consumer" price index (or even a cost-of-living index) covering all families and all goods and services. Others feel that the worth of an index decreases more or less in proportion to the increase in its scope. From the latter point of view, such phenomena as changing retail prices, wholesale prices, industrial production, and so on, are far too broad ever to be described in terms of a single number. No matter how one feels about this problem, it is true that the reduction of a large set of data to a single number often entails the loss of such a tremendous amount of information that the whole procedure may have little practical value, if any. There are, indeed, some formidable problems connected both with the construction of index numbers by the professional and their use and interpretation by the layman. As one economist has pointed out:

It ought to be conceded that index numbers are essentially arbitrary. Being at best rearrangements of data wrenched out of original market and technological contexts, they strictly have no economic meaning. Changes in tastes technology, population composition, etc., over time increase their arbitrariness. But, of course, there is no bar to the use of indexes 'as if' they did have some unequivocal meaning provided that users remember that they themselves made up the game and do not threaten to "kill the umpire" when the figures contradict expectations.†

In any case, professionals must continue to construct indexes, and yesterday's platoon of layman index watchers has now reached battalion strength and is increasing at such a rate that it is sure to become an army soon.

†I. H. Siegel, in a letter to the editor of *The American Statistician*, February 1952.

4.8

Check List of Key Terms

Arithmetic mean of price relatives, 78
Base year (or period), 75
Consumer Price Index, 86
Deflating, 93
Deflator, 93
Factor-reversal test, 96
Fixed-weight aggregative index, 81
Given year (or period), 75
Ideal index, 84
Index of Industrial Production, 90
Laspeyres Index, 79
Paasche's Index, 80
Price relatives, 78

Producer Price Index, 89
Quantity index, 79
Shift the base, 94
Simple aggregative index, 77
Time-reversal test, 96
Units test, 78
Unweighted index number, 77
Value index, 80
Value weights, 81
Weighted aggregative index, 79
Weighted arithmetic mean of price
  relatives, 81
Weighted index number, 79

4.9

Review Exercises

4.33 For the months of January through April of 1980, the average weekly hours per worker in durable-goods manufacturing in the United States were 40.8, 40.6, 40.4, and 40.1. Construct an index with January 1980 = 100 to measure changes in the hours worked. Then shift the base of this index to April 1980.

4.34 A manufacturing company buys replacement parts C305, RM22, and 5C-1 for use in automatic machines. The following are the unit prices in dollars the company paid for the parts in 1967, 1975, and 1980, and the quantities of each it used in those years:

	PRICES			QUANTITIES		
	1967	1975	1980	1967	1975	1980
C305	0.50	0.95	1.48	260	245	256
RM22	1.10	2.00	3.05	110	124	118
5C-1	0.80	0.92	1.22	48	45	52

(a) Using the 1967 quantities as weights, construct aggregative indexes comparing the 1975 and 1980 prices of the parts with those of 1967.

(b) Using the 1980 quantities as weights, construct aggregative indexes comparing the 1975 and 1980 prices of the parts with those of 1967.

(c) Construct an aggregative index comparing the 1980 prices of the parts with the 1967 prices, in which the weights are the quantities of the parts used in 1975.

(d) With 1967 = 100, calculate for 1980 the weighted arithmetic mean of price relatives using base-year values as weights.

(e) With 1967 = 100, calculate for 1980 the weighted arithmetic mean of price relatives using given-year values as weights.

(f) Interchanging the p's and q's in the formula used in part (a), construct an index comparing the 1980 use of the three parts with that of 1967.

4.35 If in 1980 average factory wages in one region were 118 percent of what they were in 1978 and a consumer price index for that region stood at 80 in 1978 with 1980 = 100, did wages keep up with inflation?

4.36 In the years 1972 through 1977 there were 76, 75, 71, 70, 67, and 66 million industrial life insurance policies in force in the United States.

(a) Construct an index measuring the year-to-year change in the number of these policies in force using 1976 as the base year.

(b) Shift the base of the index of part (a) to 1972.

4.37 The following are the prices (in dollars) of one pound each of five oils in cash markets on a day in July 1979 and the prices in the same markets one year later:

	1979	1980
Coconut oil	0.575	0.310
Cottonseed oil	0.345	0.275
Linseed oil	0.320	0.280
Peanut oil	0.390	0.270
Palm oil	0.330	0.285

(a) Construct a simple aggregative price index comparing the prices of the oils in 1980 with their prices in 1979.

(b) Find the arithmetic mean of the relatives comparing the 1980 prices with those of 1979.

4.38 The following shows the production (in thousands of tons) and the farm value (in millions of dollars) of pears and cherries in 1976 and 1978:

	PRODUCTION		VALUE	
	1976	1978	1976	1978
Pears	821	727	102	158
Cherries, sweet	168	155	64	108

Construct a simple aggregative index measuring the change in the overall prices of these two items from 1976 to 1978.

4.39 In 1979 the overall inflation rate in the United States was 13.3 percent, virtually all of which was accounted for by the rising prices of energy, food, housing, and medical care. Why should the prices of these four items be of such great concern to all consumers?

4.40 In 1977 the average weekly earnings of department store employees in a certain area were $169.50 and in 1980 the corresponding earnings were $202.35. A regional cost of living index stood at 172.9 for 1977 and 216.5 for 1980 with 1967 = 100. Express the 1980 dollar wages of these department store employees in terms of constant 1977 dollars.

5

Possibilities, Probabilities, and Expectations

e can hardly predict the outcome of a presidential election unless we know what candidates are running for office, and we cannot very well predict what records will be among the "top 10" unless we know at least which ones are on the market. More generally, we cannot make intelligent predictions or decisions unless we know at least what is possible, or to put it differently, we must know what is possible before we can judge what is probable. Thus, Sections 5.1 and 5.2 will be devoted to the problem of determining what is possible in given situations. Then, in Section 5.3 we shall learn how to judge also what is probable, and in Section 5.4 we shall introduce the related concept of a mathematical expectation.

5.1

Counting

The simple process of counting still plays an important role in business and economics. One still has to count 1, 2, 3, 4, . . . , for example, when taking inventory, when determining the number of damaged cases in a shipment of wines from France, or when preparing a report showing how many times certain stock-market indexes went up during a given month. Sometimes, the process of counting can be simplified by using mechanical devices (for instance, when counting spectators passing through turnstiles), or by performing counts indirectly (for instance, by subtracting the serial numbers of invoices to determine the total number of sales). At other times, the process of counting can be simplified greatly by means of special mathematical techniques, such as the ones given below.

In the study of "what is possible," there are essentially two kinds of problems. First there is the problem of listing everything that can happen in a given situation, and then there is the problem of determining how many different things can happen (without actually constructing a complete list). The second kind of problem is especially important, because in many cases we really do not need a complete list, and hence, can save ourselves a great deal of work. Although the first kind of problem may seem straightforward and easy, this is not always the case.

EXAMPLE Three applicants for real estate licenses in the state of Arizona are planning to take the required examination in October, and repeat it, if necessary, in November and December. If we are interested only in how many of the applicants pass the examination in each of the three months, how many different possibilities are there?

SOLUTION Clearly, there are many. For instance, all three applicants might pass the examination in October; one might pass in October, another in December, and the third fail all three times; one might pass in November and the other two in December; and all three of the applicants might fail each month. Continuing this way carefully, we may determine that there are altogether 20 possibilities.

To handle problems like this systematically, it helps to construct a tree diagram such as that of Figure 5.1. This diagram shows that for October there are four possibilities (four branches), corresponding to 0, 1, 2, or 3 of the applicants passing the examination; for November

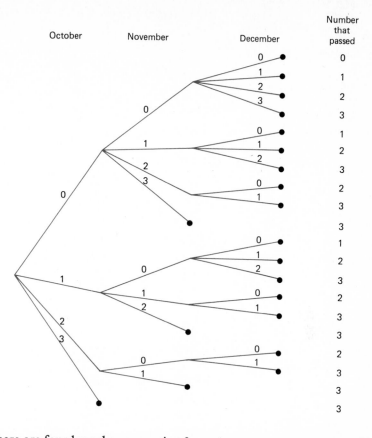

October November December

Number that passed

5.1

Tree diagram for real-estate-license example.

there are four branches emanating from the top branch, three from the second branch, two from the third branch, and none from the bottom branch. Evidently, there are still four possibilities (0, 1, 2, or 3) when no one passes in October, but only three possibilities (0, 1, or 2) when one passes in October, two possibilities (0 or 1) when two pass in October, and there is no need to go on when all three of the applicants pass in October. The same sort of reasoning applies also to December, and (going from left to right) we find that there are altogether 20 different paths along the "branches" of the tree. In other words, there are 20 distinct possibilities in this situation. It can also be seen from this diagram that in ten of the cases all three of the applicants pass the examination (sooner or later) during the three months, in six of the cases two of them pass, in three cases only one of them passes, and in one case none of the applicants passes.

EXAMPLE A helicopter service connecting two airports has four pilots and three helicopters. In how many different ways can one pilot and one helicopter be assigned to a job?

If we label the four pilots A, B, C, and D, the three helicopters I, II, and III, and draw the tree diagram of Figure 5.2, we find that there are 12 different ways in all. The first path along the branches of the tree corresponds to the choice of pilot A and helicopter I, the second path corresponds to the choice of pilot A and helicopter II, . . . , and the 12th path corresponds to the choice of pilot D and helicopter III.

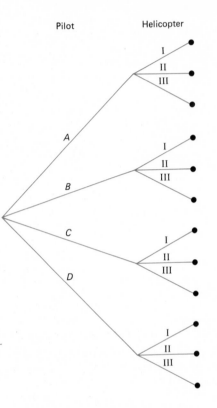

Pilot   Helicopter

5.2

Tree diagram for
helicopter-service
example.

The answer we got in the second example is $4 \cdot 3 = 12$, the product of the number of ways in which one pilot can be selected and the number of ways in which one helicopter can be selected. Generalizing from this example, let us state the following rule:

*Multiplication
of choices*

> *If a choice consists of two steps, the first of which can be made in m ways and for each of these the second can be made in n ways, then the whole choice can be made in m·n ways.*

To prove this, we have only to draw a tree diagram similar to that of Figure 5.2. First there are m branches corresponding to the possibilities in the first step, and then there are n branches emanating from each of these branches to represent the possibilities in the second step. This leads to $m \cdot n$ paths along the branches of the tree diagram, and hence $m \cdot n$ possibilities.

EXAMPLE   If a firm has 4 warehouses and 12 retail outlets, in how many different ways can it ship an item from one of the warehouses to one of the stores?

SOLUTION   Since $m = 4$ and $n = 12$, there are $4 \cdot 12 = 48$ ways.

EXAMPLE   If a travel agency offers trips to 15 different cities, either by air, rail, or bus, in how many different ways can such a trip be arranged?

SOLUTION   Since $m = 15$ and $n = 3$, there are $15 \cdot 3 = 45$ ways.

By use of appropriate tree diagrams, it is easy to generalize the foregoing rule so that it will apply to choices involving more than two steps. For k steps, where k is a positive integer, we arrive at the following rule:

Multiplication of choices (generalized)

> *If a choice consists of k steps, the first of which can be made in n_1 ways, for each of these the second can be made in n_2 ways, . . . , and for each of these the kth can be made in n_k ways, then the whole choice can be made in $n_1 \cdot n_2 \cdot \ldots \cdot n_k$ ways.*

EXAMPLE   If a new-car buyer is faced with a choice of five body styles, three engines, and 10 colors, in how many different ways can he choose a body style, an engine, and a color for his car? Also, if the buyer can choose a car with or without automatic transmission, with or without air conditioning, and with or without bucket seats, how many different choices does he have?

SOLUTION   For the first question, $n_1 = 5$, $n_2 = 3$, and $n_3 = 10$, so the buyer can choose his car in $5 \cdot 3 \cdot 10 = 150$ different ways. For the second question, $n_1 = 5$, $n_2 = 3$, $n_3 = 10$, $n_4 = 2$, $n_5 = 2$, and $n_6 = 2$, so there are altogether $5 \cdot 3 \cdot 10 \cdot 2 \cdot 2 \cdot 2 = 1,200$ different choices open to him.

EXAMPLE   If a test consists of ten multiple-choice questions, each permitting three possible answers, in how many different ways can a student mark her paper with one answer to each question?

SOLUTION   Since $n_1 = 3$, $n_2 = 3$, . . . , and $n_{10} = 3$, there are $3 \cdot 3 \cdot 3 \cdot 3 \cdot 3 \cdot 3 \cdot 3 \cdot 3 \cdot 3 \cdot 3 = 59,049$ ways. In one of the 59,049 cases the answers will all be correct, and in $2 \cdot 2 \cdot 2 \cdot 2 \cdot 2 \cdot 2 \cdot 2 \cdot 2 \cdot 2 \cdot 2 = 1,024$ of them they will all be wrong.

5.2

Permutations and Combinations

The rule for the multiplication of choices and its generalization are often applied when several choices are made from one set and we are concerned with the order in which they are made.

EXAMPLE In how many different ways can the judges choose the winner and the first runner-up from among the 10 finalists in a student essay contest?

SOLUTION Since the winner can be chosen in $m = 10$ ways and the first runner-up must be one of the other $n = 9$ finalists, there are $10 \cdot 9 = 90$ ways.

EXAMPLE In how many different ways can the 48 members of a college fraternity choose a president, a vice-president, a secretary, and a treasurer?

SOLUTION Since $n_1 = 48$, $n_2 = 47$, $n_3 = 46$, and $n_4 = 45$ (regardless of which officer is elected first, second, third, and fourth), there are $48 \cdot 47 \cdot 46 \cdot 45 = 4,669,920$ ways.

In general, if r objects are selected from a set of n objects, any particular arrangement (order) of these objects is called a **permutation**. For instance, 3 2 1 4 is a permutation of the first four positive integers; Vermont, Massachusetts, Connecticut is a permutation (a particular ordered arrangement) of three of the six New England states; and

Yankees, Orioles, Brewers, Red Sox
Tigers, Indians, Yankees, Brewers

are two different permutations (ordered arrangements) of four of the six baseball teams in the Eastern Division of the American League.

EXAMPLE Determine the number of possible permutations of two of the five vowels a, e, i, o, and u, and list them all.

SOLUTION Since $m = 5$ and $n = 4$, there are $5 \cdot 4 = 20$ permutations; they are

ae	ai	ao	au	ei	eo	eu	io	iu	ou
ea	ia	oa	ua	ie	oe	ue	oi	ui	uo

To find a formula for the total number of permutations of r objects selected from n distinct objects, such as the six baseball teams or the five vowels, we observe that the first selection is made from the whole set of n objects, the second selection is made from the $n - 1$ objects which remain after the first selection has been made, the third selection is made from the $n - 2$ objects which remain after the first two selections have been made, ..., and the rth and final selection is made from the $n - (r - 1) = n - r + 1$ objects which remain after the first $r - 1$ selections have been made. Now, direct application of the generalized rule for the multiplication of choices shows that the total number of permutations of r objects selected from n distinct objects, which we shall denote $_nP_r$, is $n(n - 1)(n - 2) \cdot \ldots \cdot (n - r + 1)$.

Since products of consecutive integers arise in many problems relating to permutations and other kinds of special arrangements or selections, it is convenient to introduce here the **factorial notation**. In this notation, the product of all positive integers less than or equal to the positive integer n is called "n factorial" and denoted by $n!$. Thus,

$$1! = 1$$
$$2! = 2 \cdot 1 = 2$$
$$3! = 3 \cdot 2 \cdot 1 = 6$$
$$4! = 4 \cdot 3 \cdot 2 \cdot 1 = 24$$
$$5! = 5 \cdot 4 \cdot 3 \cdot 2 \cdot 1 = 120$$
$$6! = 6 \cdot 5 \cdot 4 \cdot 3 \cdot 2 \cdot 1 = 720$$

and in general $n! = n(n - 1)(n - 2) \cdot \ldots \cdot 3 \cdot 2 \cdot 1$. Also, to make various formulas more generally applicable, we let $0! = 1$ by definition.

To express the formula for $_nP_r$ in terms of factorials, we note, for instance, that $15 \cdot 14 \cdot 13! = 15!$, $8 \cdot 7 \cdot 6 \cdot 5! = 8!$, $36 \cdot 35 \cdot 34 \cdot 33 \cdot 32 \cdot 31! = 36!$, and similarly,

$$_nP_r \cdot (n - r)! = n(n - 1)(n - 2) \cdot \ldots \cdot (n - r + 1) \cdot (n - r)!$$
$$= n!$$

so that $_nP_r = \dfrac{n!}{(n - r)!}$. To summarize

Number of permutations of n objects taken r at a time

The number of permutations of r objects selected from a set of n distinct objects is

$$_nP_r = n(n - 1)(n - 2) \cdot \ldots \cdot (n - r + 1)$$

or, in factorial notation,

$$_nP_r = \frac{n!}{(n - r)!}$$

---

and we now have two formulas for $_nP_r$.

EXAMPLE   Find the number of ways in which three of ten real estate salespersons can be ranked first, second, and third according to market knowledgeability.

SOLUTION   For $n = 10$ and $r = 3$ the first formula yields

$$_{10}P_3 = 10 \cdot 9 \cdot 8 = 720$$

and the second formula yields

$$_{10}P_3 = \frac{10!}{7!} = \frac{10 \cdot 9 \cdot 8 \cdot 7!}{7!} = 720$$

EXAMPLE   Find the number of permutations of zero objects selected from a set of 25 distinct objects.

SOLUTION   We cannot use the first formula here, but substituting $n = 25$ and $r = 0$ into the second formula, we get

$$_{25}P_0 = \frac{25!}{25!} = 1$$

This result may be trivial, but it shows that the factorial notation makes the formula for the number of permutations more generally applicable.

To find the formula for the number of permutations of n distinct objects taken all together, we substitute $n = r$ into the second formula for $_nP_r$, getting $\frac{n!}{(n-n)!} = \frac{n!}{0!} = n!$ (since $0! = 1$ by definition). Hence,

Number of permutations of n objects taken all together

$$_nP_n = n!$$

EXAMPLE   Find the number of ways in which nine teaching assistants can be assigned to nine sections of a course, and the number of ways in which 12 different package designs for a new product can be ranked in order of preference.

SOLUTION   For the nine teaching assistants we get $9! = 362,880$, and for the 12 package designs we get $12! = 479,001,600$.

There are many problems in which we want to know the number of ways in which r objects can be selected from a set of n objects, but we do not want to include in our count all the different orders in which the selection can be made. For instance, three persons, P, Q, and R, can be assigned to a three-person committee in $3! = 6$ different orders (PQR, PRQ, QPR, QRP, RPQ, and RQP), but there is only one committee, not six.

To obtain a formula which applies to problems like this, let us consider the following 24 permutations of three of the first four letters of the alphabet:

$$
\begin{array}{cccccc}
abc & acb & bac & bca & cab & cba \\
abd & adb & bad & bda & dab & dba \\
acd & adc & cad & cda & dac & dca \\
bcd & bdc & cbd & cdb & dbc & dcb
\end{array}
$$

Inspection of this table shows that if we do not count the different orders in which three letters are chosen from the four letters a, b, c, and d, there are only four ways in which the selection can be made. These are shown in the first column—abc, abd, acd, and bcd. Each row of the table merely contains the $3! = 6$ different permutations of the letters shown in the first column.

In general, there are $r!$ permutations of any r objects selected from a set of n distinct objects, so that the $_nP_r$ permutations of r objects selected from a set of n distinct objects contain each set of r objects $r!$ times. Therefore, to find the number of ways in which r objects can be selected from a set of n distinct objects, also called the number of **combinations** of n objects taken r at a time and denoted by $\left(\begin{array}{c} n \\ r \end{array} \right)$, we divide $_nP_r$ by $r!$, and we get

---

Number of combinations of n objects taken r at a time

> The number of ways in which r objects can be selected from a set of n distinct objects is
>
> $$\left(\begin{array}{c} n \\ r \end{array} \right) = \frac{n(n-1)(n-2) \cdot \ldots \cdot (n-r+1)}{r!}$$
>
> or, in factorial notation,
>
> $$\left(\begin{array}{c} n \\ r \end{array} \right) = \frac{n!}{r!(n-r)!}$$

For $n = 0$ to $n = 20$, the values of $\left(\begin{array}{c} n \\ r \end{array} \right)$ may be read from Table IX at the end of the book, where these quantities are called **binomial coefficients** (see Exercise 5.30 on page 114).

EXAMPLE Find the number of ways in which a person can select four stocks from a list of eight stocks (the number of combinations of eight things taken four at a time).

SOLUTION   For $n = 8$ and $r = 4$, the first formula yields

$$\binom{8}{4} = \frac{8 \cdot 7 \cdot 6 \cdot 5}{4!} = 70$$

and the second formula yields

$$\binom{8}{4} = \frac{8!}{4!4!} = \frac{8 \cdot 7 \cdot 6 \cdot 5}{4 \cdot 3 \cdot 2 \cdot 1} = 70$$

EXAMPLE   In how many ways can a dean choose 2 of 50 faculty members to review a student grade appeal?

SOLUTION   For $n = 50$ and $r = 2$, the first formula yields

$$\binom{50}{2} = \frac{50 \cdot 49}{2!} = 1{,}225$$

The result of the first example, but not that of the second, can be read from Table IX.

EXAMPLE   In how many ways can 4 good switches and 2 defective switches be chosen from a lot containing 20 good and 5 defective switches?

SOLUTION   The 4 good switches can be selected in $\binom{20}{4}$ ways, the 2 defective switches in $\binom{5}{2}$ ways, and by the multiplication of choices we have

$$\binom{20}{4} \cdot \binom{5}{2} = 4{,}845 \cdot 10 = 48{,}450$$

In this case we looked up the binomial coefficients in Table IX.

When r objects are selected from a set of n distinct objects, $n - r$ of the objects are left, and consequently there are as many ways of leaving (or selecting) $n - r$ objects from a set of n distinct objects as there are ways of selecting r objects. Symbolically, we write

Rule for binomial coefficients

$$\binom{n}{r} = \binom{n}{n-r} \quad \textit{for } r = 0, 1, 2, \ldots, n$$

Sometimes this rule serves to simplify details and sometimes it is needed in connection with the use of Table IX.

Determine the value of $\binom{85}{82}$.

To avoid having to write down the product $85 \cdot 84 \cdot 83 \cdot \ldots \cdot 4$ and cancel $82 \cdot 81 \cdot \ldots \cdot 4$, we write directly

$$\binom{85}{82} = \binom{85}{3} = \frac{85 \cdot 84 \cdot 83}{3!} = 98{,}770$$

Find the value of $\binom{17}{13}$.

$\binom{17}{13}$ cannot be looked up directly in Table IX, but $\binom{17}{17-13} = \binom{17}{4} = 2{,}380$ can.

EXERCISES

5.1 A person with $2 in his pocket bets $1, even money, on the flip of a coin, and he continues to bet $1 so long as he has any money. Draw a tree diagram to show the various things that can happen during the first three flips of the coin. In how many of the cases will he be
   (a) exactly $1 ahead; (b) exactly $1 behind?

5.2 Suppose that in a baseball World Series (in which the winner is the first team to win four games) the National League champion leads the American League champion three games to two. Construct a tree diagram to show the number of ways in which these teams may win or lose the remaining game or games.

5.3 A student can study 0, 1, or 2 hours for an accounting test on any given night. Construct a tree diagram to show that there are six different ways in which she can study altogether 4 hours for the test on three consecutive nights.

5.4 There are four routes, A, B, C, and D, between a person's home and his place of work, but route A is one-way so that he cannot take it on the way to work, and route D is one-way so that he cannot take it on the way home.
   (a) Draw a tree diagram showing the various ways he can go to and from work.
   (b) Draw a tree diagram showing the various ways he can go to and from work, but does not go by the same route both ways.

5.5 A purchasing agent places his orders by telephone, by telegram, or by mail, requesting in each case that his order be confirmed by telegram or by mail. Draw a tree diagram to show the various ways in which one of his orders can be placed and confirmed.

5.6 A woman can buy a vacuum cleaner in any one of three sizes and in any one of five colors. How many choices does she have?

5.7 In a real estate development new houses are offered with 2, 3, or 4 bedrooms, with or without air conditioning, with carport or garage, and in several different exterior finishes. If there are 72 possible choices open to a buyer, how many different exterior finishes are available?

5.8 If the NCAA has applications from four universities for hosting its inter-collegiate swimming championships in 1984 and 1985, in how many ways can it select the sites for these championship meets
(a) if they are not to be held at the same university;
(b) if they may be held at the same university?

5.9 The Standard and Poor's Corporation regularly rates common stocks, assign-ing them the ratings A+, A, A−, B+, B, B−, and C.
(a) In how many ways can it assign ratings to three different stocks?
(b) In how many ways can it assign at least A− ratings to two different stocks?
(c) In how many ways can it assign ratings to two different stocks, if one of them (but not both) is to have a rating of A or A+?

5.10 A customer can buy from a department store either the standard or the deluxe model of an item, in any one of four colors, and on any one of four payment plans. How many different buy options are open to a customer?

5.11 In a market study, heads of households are classified into six categories accord-ing to income, into five categories according to the extent of their education, and into four categories according to their place of residence. In how many different ways can the head of a household be classified?

5.12 A psychologist preparing three-letter nonsense words for use in a memory test chooses the first letter from among the consonants q, w, x, and z; the second letter from among the vowels e, i, and u; and the third letter from among the consonants c, f, p, and v.
(a) How many different three-letter nonsense words can she construct?
(b) How many of these nonsense words will begin either with a w or an x?
(c) How many of these nonsense words will end with a c?
(d) How many of these nonsense words ending in a c begin with an x?

5.13 A true–false test consists of 15 questions. In how many ways can a student mark her answers to these questions?

5.14 Trailer license plates in one state consist of three digits, the first of which cannot be 0, followed by two letters of the alphabet, the first of which cannot be I, O, or Q. How many different plates are possible using this scheme?

5.15 How many different permutations are there of three of the six New England states?

5.16 Determine whether each of the following is true or false:
(a) $11! = 11 \cdot 10 \cdot 9 \cdot 8!$;   (c) $3! + 4! = 7!$;
(b) $4! \cdot 3! = 12!$;   (d) $13! = \dfrac{14!}{14}$.

5.17 On a trip to Pennsylvania, a person wants to visit four of 16 historical sites. If the order of the visits matters, in how many ways can this person plan the trip?

5.18 If there are seven cars in a race, in how many different ways can they place first, second, and third?

5.19 In how many different ways can a window dresser arrange four shirts in a horizontal row in a store window?

5.20 In how many ways can a television director schedule six different commercials for the six time slots allocated to commercials during the telecast of the final period of a hockey game?

5.21 The number of ways in which n distinct objects can be arranged in a circle is $(n - 1)!$.
(a) Present an argument to justify this formula.
(b) In how many ways can six persons be seated at a round table (if it matters only who sits on whose left and right)?
(c) In how many ways can a window dresser display four shirts in a circular arrangement?

5.22 If among n objects r are alike, and the others are all distinct, the number of permutations of these n objects taken all together is $n!/r!$.
(a) How many permutations are there of the letters in the word "cool"?
(b) In how many ways (according only to manufacturer) can five cars place in a stock-car race, if three of the cars are Fords, one is a Chevrolet, and one is a Dodge?
(c) Justify the formula given in this exercise.

5.23 If among n objects r_1 are identical, another r_2 are identical, and the rest are all distinct, the number of permutations of these n objects taken all together is $n!/(r_1! \cdot r_2!)$.
(a) How many permutations are there of the letters in the word "greater"?
(b) In how many ways can the television director of Exercise 5.20 schedule the commercials, if there are two commercials, each of which is to be shown three times?
(c) Generalize the formula so that it applies if among n objects r_1 are identical, another r_2 are identical, another r_3 are identical, and the rest are all distinct. In how many ways can the television director of part (b) schedule the commercials, if there are three commercials, each of which is to be shown twice?
(d) In its cookbook section, a bookstore has four copies of *The New York Times Cookbook*, two copies of *The Joy of Cooking*, five copies of the *Better Homes and Gardens Cookbook*, and one copy of *The Secret of Cooking for Dogs*. If these books are sold one at a time, in how many different sequences can they be sold?

5.24 Calculate the number of ways in which a discount chain can choose 2 of 12 locations for the construction of new stores.

5.25 Calculate the number of ways in which the Internal Revenue Service can choose 4 of 14 income tax returns for a special audit.

5.26 Among the 12 nominees for the Board of Directors of a farm cooperative there are 8 men and 4 women. In how many ways can the members elect as directors
(a) any two of the nominees;
(b) two of the male nominees;
(c) one of the male nominees and one of the female nominees?

5.27 The personnel manager of a store wants to fill five openings in its training program with three college graduates and two persons who are not college graduates. In how many ways can these openings be filled if among 21 applicants 12 are college graduates?

5.28 A shipment of 15 alarm clocks contains one that is defective. In how many ways can an inspector choose three of the clocks for inspection so that
(a) the defective clock is not included;
(b) the defective clock is included?

5.29 Suppose that among the 15 alarm clocks of the preceding exercise there are two defectives. In how many ways can the inspector choose three of the clocks for inspection so that
(a) neither of the defective clocks is included;
(b) one of the defective clocks is included;
(c) both of the defective clocks are included?

5.30 The quantity $\binom{n}{r}$ is called a binomial coefficient because it is, in fact, the coefficient of $a^{n-r}b^r$ in the binomial expansion of $(a + b)^n$. Verify that this is true for $n = 2, 3,$ and 4, by expanding $(a + b)^2, (a + b)^3,$ and $(a + b)^4$ and comparing the coefficients with the corresponding values of $\binom{n}{r}$ given in Table IX.

5.31 A table of binomial coefficients is easy to construct by following the pattern shown below, which is called **Pascal's triangle.**

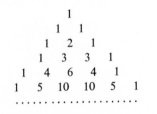

In this arrangement, each row begins with a 1, ends with a 1, and each other entry is given by the sum of the nearest two entries in the row immediately above.
(a) Use Table IX to verify that the third row of the triangle contains the values of $\binom{2}{r}$ for $r = 0, 1,$ and 2, the fourth row contains the values of $\binom{3}{r}$ for $r = 0, 1, 2,$ and 3, and the fifth row contains the values of $\binom{4}{r}$ for $r = 0, 1, 2, 3,$ and 4.
(b) Construct the next two rows of the triangle and use Table IX to verify the results.

5.32 Verify the identity $\binom{n+1}{r} = \binom{n}{r} + \binom{n}{r-1}$ by expressing each of the binomial coefficients in terms of factorials. Explain why this identity justifies the method used in the construction of Pascal's triangle in Exercise 5.31.

5.3

Probability

So far we have studied only what is possible in a given situation. In some instances we listed all possibilities and in others we merely determined how many different possibilities there are. Now we shall go one step further and judge also what is probable and what is improbable.

The most common way of measuring the uncertainties connected with events (say, the success of a new product, the effectiveness of an advertising campaign, or the return on an investment) is to asisgn them **probabilities**; alternatively, we may specify the **odds** at which it would be fair to bet that the events will occur. In this section we shall see how probabilities are interpreted and their numerical values are determined; then in Section 5.4 and in Chapter 6 we shall see how they can be used to make choices, among different courses of action, which promise to be the most profitable, or otherwise most desirable. "Odds" will be defined later, in Section 6.3, where we shall also study the relationship between probabilities and odds.

Historically, the oldest way of measuring uncertainties is the **classical probability concept.** It was developed originally in connection with games of chance, and it lends itself most readily to bridging the gap between possibilities and probabilities. This concept applies only when all possible outcomes are equally likely, in which case we can say that

The classical probability concept

> *If there are n equally likely possibilities, one of which must occur and s are regarded as favorable, or as a "success," then the probability of a "success" is given by the ratio $\frac{s}{n}$.*

In the application of this rule, the terms "favorable" and "success" are used rather loosely—what is favorable to one player is unfavorable to his opponent, and what is a success from one point of view is a failure from another. Thus, the terms "favorable" and "success" can be applied to any particular kind of outcome, even if "favorable" means that a television set does not work, or a "success" means that someone caught the flu. This usage dates back to the days when probabilities were studied only in connection with games of chance.

EXAMPLE What is the probability of drawing an ace at random from a well-shuffled deck of 52 playing cards?

SOLUTION There are $s = 4$ aces among the $n = 52$ cards, so we get

$$\frac{s}{n} = \frac{4}{52} = \frac{1}{13}$$

EXAMPLE What is the probability of rolling a 5 or a 6 with a balanced die?

SOLUTION Since $s = 2$ and $n = 6$, we get

$$\frac{s}{n} = \frac{2}{6} = \frac{1}{3}$$

Although equally likely possibilities are found mostly in games of chance, this probability concept applies also in a great variety of situations where gambling devices are used to make **random selections**—say, when offices are assigned to sales agents by lot, when machine parts are chosen for inspection so that each part produced has the same chance of being selected, or when each family in a certain market area has the same chance of being included in a sample survey.

EXAMPLE If 2 of 20 tires are defective and 4 of the 20 are randomly chosen for inspection, what is the probability that none of the defective tires will be chosen?

SOLUTION There are $\binom{20}{4} = 4,845$ equally likely ways of choosing 4 of the 20 tires by reason of the random selection. The number of favorable outcomes is the number of ways in which none of the defective tires and 4 of the nondefective tires can be selected, or $\binom{2}{0}\binom{18}{4} = 1 \cdot 3,060 = 3,060$, and it follows that the probability is

$$\frac{s}{n} = \frac{3,060}{4,845} = \frac{12}{19}$$

The values of the binomial coefficients $\binom{20}{4}$, $\binom{2}{0}$, and $\binom{18}{4}$ were read directly from Table IX.

The major shortcoming of the classical probability concept (where the possibilities must all be equally likely) is that there are many situations in which the possibilities that arise cannot be regarded as equally likely. This might be the case, for example, if we are concerned with the question whether there will be rain, sunshine, snow, or hail; when we wonder whether or not a person will receive a promotion; or when we want to predict the success of a new business or the behavior of the stock market.

Among the various other probability concepts, most widely held is the frequency interpretation, according to which

The frequency interpretation of probability

> *The probability of an event (happening or outcome) is the proportion of the time that events of the same kind will occur in the long run.*

If we say that the probability is 0.78 that a jet from San Francisco to Phoenix will arrive on time, we mean that such flights arrive on time 78 percent of the time. Also, if the Weather Service predicts that there is a 40 percent chance of rain (that the probability is 0.40 that it will rain), it means that under the same weather conditions it will rain 40 percent of the time. More generally, we say that an event has a probability of, say, 0.90, in the same sense in which we might say that our car will start in cold weather 90 percent of the time. We cannot guarantee what will happen on any particular occasion—the car may start and then it may not—but if we kept records over a long period of time, we should find that the proportion of "successes" is very close to 0.90.

In accordance with the frequency concept of probability, we estimate the probability of an event by observing what fraction of the time similar events have occurred in the past.

EXAMPLE   If records show that (over a period of time) 516 of 600 jets from Denver to Chicago arrived on time, what is the probability that any one jet from Denver to Chicago will arrive on time?

SOLUTION   Since in the past $\frac{516}{600} = 0.86$ of the flights arrived on time, we use this fraction as an estimate of the probability.

EXAMPLE   If 687 of 1,854 freshmen who entered a men's college (over a number of years) dropped out before the end of their freshman year, what is the probability that a freshman entering this college will drop out before the end of his freshman year?

SOLUTION   Since in the past $\frac{687}{1,854} = 0.37$ of the freshmen dropped out before the end of their freshman year, we use this figure as an estimate of the probability.

When probabilities are estimated in this way, it is only reasonable to ask just how good the estimates are. Later we shall answer this question in some detail, but for now let us refer to an important theorem called the Law of Large Numbers. Informally, this theorem may be stated as follows:

> *If a situation, trial, or experiment is repeated again and again, the proportion of successes will tend to approach the probability that any one outcome will be a success.*

To illustrate this law, we repeatedly flipped a balanced coin and recorded the accumulated proportion of heads after every fifth flip. The results are shown in Figure 5.3, where the proportion of heads can be seen to fluctuate, but come closer and closer to $\frac{1}{2}$, the probability of heads for each flip of the coin.

In the frequency interpretation, the probability of an event is defined in terms of what happens to similar events in the long run, so let us consider briefly whether it is at all meaningful to talk about the probability of an event which can occur only once. For instance, can we assign a probability to the event that Ms. Barbara Smith's broken arm, broken for the first time, will heal within a month? If we put ourselves in the position of Ms. Smith's doctor, we could check medical records, discover that such fractures have healed within one month in (say) 39 percent of the thousands of reported cases, and apply this figure to Ms. Smith's arm. This may not be of much comfort to Ms. Smith, but it does provide a meaning for a probability statement concerning her arm: the probability that it will heal within a month is 0.39.

This illustrates that when we make a probability statement about a specific (nonrepeatable) event, the frequency concept of probability leaves us no choice but to refer to a set of similar events. As can well be imagined, however, this can easily lead to complications, since the choice of "similar" events is often

5.3

Graph illustrating the Law of Large Numbers.

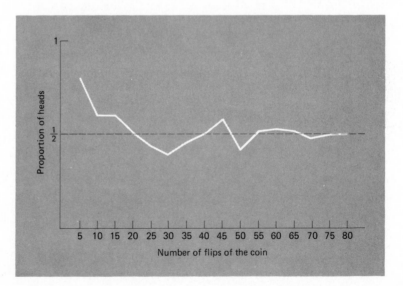

neither obvious nor straightforward. For instance, with reference to Ms. Smith's arm, we might consider as "similar" only those cases where the fracture was in the same arm, or only those in which the patients were the same age as Ms. Smith, or only those in which the patients were also the same height and weight as Ms. Smith.

This shows that the choice of "similar" events is ultimately a matter of personal judgment, and it is by no means contradictory that we can arrive at different probabilities concerning the same event. It should be observed, however, that the more we narrow things down, the less information we have to estimate probabilities.

An alternative point of view, which is currently gaining in favor, is to interpret probabilities as **personal** or **subjective** evaluations. Such probabilities express the strength of one's belief with regard to the uncertainties that are involved, and they apply especially when there is little or no direct evidence, so that there really is no choice but to consider collateral (indirect) information, "educated guesses," and perhaps intuition and other subjective factors. Subjective probabilities are sometimes determined by putting the issues in question on a "put up or shut up" basis, as will be explained in Sections 5.4 and 6.3.

EXERCISES

5.33 When one card is drawn from a well-shuffled deck of 52 playing cards, what are the probabilities of getting
(a) a red king;
(b) a queen, king, or ace of any suit;
(c) a red card;
(d) a 3, 4, 5, or 6 of any suit?

5.34 If H stands for heads and T for tails, the four possible outcomes in two flips of a coin are HH, HT, TH, and TT. If it can be assumed that these four possibilities are equally likely, what are the probabilities of getting 0, 1, or 2 heads?

5.35 If H stands for heads and T for tails, the eight possible outcomes in three successive flips of a coin are HHH, HHT, HTH, THH, HTT, THT, TTH, and TTT. Assuming that these eight possibilities are equally likely, what are the probabilities of getting 0, 1, 2, or 3 heads?

5.36 A bowl contains 17 red beads, 10 white beads, 20 blue beads, and 3 black beads. If one of these beads is drawn at random, what are the probabilities that it will be
(a) red;
(b) blue or white;
(c) black;
(d) neither white nor black?

5.37 If we roll a balanced die, what are the probabilities of getting
(a) a 3 or a 4;
(b) an even number?

5.38 If two different cards are drawn from a well-shuffled deck of 52 playing cards, what are the probabilities of getting
(a) two hearts;
(b) two aces;
(c) a king and a queen?

5.39 If 2 of 20 tires are defective and 4 of them are randomly chosen for inspection, what is the probability that both of the defective tires will be chosen?

5.40 If one of eight applications for liquor licenses contains fraudulent information, and the three to be granted are chosen in such a way that each possible choice has the same probability, find
   (a) the total number of ways in which any three of the eight applications can be selected;
   (b) the number of ways in which three of the eight applications can be selected so that the one with fraudulent information will be included;
   (c) the probability that the application containing fraudulent information will be granted.

5.41 New York City, Los Angeles, and Philadelphia are among the twelve largest cities in the United States. Assuming that the selection is random (that each set of three of the twelve cities has the same chance of being selected), what are the probabilities that a survey conducted in three of the twelve largest cities in the United States will include
   (a) New York City;   (b) Los Angeles and Philadelphia?

5.42 Among 842 armed robberies in a certain city, 143 were never solved. Estimate the probability that an armed robbery in this city will never be solved.

5.43 If a department store's records show that 1,564 of 1,840 women who entered the store on a Saturday afternoon made at least one purchase, estimate the probability that a woman who enters the store on a Saturday afternoon will make at least one purchase.

5.44 In a sample of 400 cans of mixed nuts (taken from a very large shipment), 124 contained no pecans. Estimate the probability that there will be no pecans in a can of mixed nuts randomly selected from this shipment.

5.45 In an early morning radar check on a Los Angeles freeway, 214 of 856 cars were found to exceed the legal speed limit of 55 mph. Estimate the probability that a car traveling on that freeway at that time of the day will exceed the 55 mph limit.

5.46 If 1,558 of 2,050 persons visiting a national park said that they would like to return, estimate the probability that any randomly chosen visitor to this park would like to return.

5.4

Mathematical Expectation

If an insurance agent tells us that a 45-year-old woman can expect to live 33 more years, this does not mean that anyone really expects a 45-year-old woman to live until her 78th birthday and then die the next day. Similarly, if we read that a person living in the United States can expect to eat 10.4 pounds of cheese and 324.7 eggs a year, it must be obvious that the word "expect" is not being used in its colloquial sense. Most persons do not eat 0.7 egg, and it would be surprising, indeed, if we found somebody who has actually eaten 10.4 pounds of cheese in a given year. So far as the first statement is concerned, some 45-year-old women will live another 12 years, some will live another 25 years, some will live another 38 years, . . . , and the life expectancy of "33 more years" must be interpreted as a particular kind of average called an expected value, or a mathematical expectation.

Originally, the concept of a mathematical expectation arose in connection with games of chance, and in its simplest form it is given by the product of the amount a player stands to win and the probability that he or she will win this amount.

EXAMPLE   What is our mathematical expectation if we stand to win $5 if and only if a balanced coin falls heads?

SOLUTION   The coin is balanced, so the probability of heads is $\frac{1}{2}$ and our mathematical expectation is $5 \cdot \frac{1}{2} = \$2.50$.

EXAMPLE   What is our mathematical expectation if we buy one of 1,000 raffle tickets issued for a prize, a color television set, worth $480?

SOLUTION   Since the probability that we will win is $\frac{1}{1,000}$, our mathematical expectation is $480 \cdot \frac{1}{1,000} = \$0.48$ or 48 cents. Thus, in a strict monetary sense, it would be foolish to pay more than 48 cents for the ticket.

In both our examples there was a single prize, but in each case there were two possible payoffs—$5 or $0 in the first example and $480 or $0 in the other. Indeed, in the second example we can argue that 999 of the tickets will not pay anything at all, one ticket will pay $480 (or the equivalent in merchandise), so that altogether the 1,000 tickets pay $480, or on the average 48 cents per ticket, which is the mathematical expectation.

To generalize the concept of a mathematical expectation, let us consider the following modification of the raffle of the preceding example:

EXAMPLE   What is our mathematical expectation if we buy one of 1,000 raffle tickets for a first prize of a color television set worth $480, a second prize of a record player worth $120, and a third prize of a radio worth $40?

SOLUTION Now we can argue that 997 of the tickets will not pay anything at all, one ticket will pay the equivalent of $480, another will pay the equivalent of $120, while a third will pay the equivalent of $40; altogether, the 1,000 raffle tickets will pay $480 + $120 + $40 = $640, or on the average 64 cents per ticket—this is the mathematical expectation of each ticket. Looking at the problem in a different way, we could argue that if the raffle were repeated many times, we would win nothing $\frac{997}{1,000} \cdot 100 = 99.7$ percent of the time and win each of the three prizes $\frac{1}{1,000} \cdot 100 = 0.1$ percent of the time. On the average we would win

$$0(0.997) + 480(0.001) + 120(0.001) + 40(0.001) = \$0.64$$

or 64 cents, which is the sum of the products obtained by multiplying each amount by the corresponding proportion or probability.

Generalizing from this example, let us now give the following definition:

Mathematical expectation

> *If the probabilities of obtaining the amounts $a_1, a_2, \ldots,$ or a_k are $p_1, p_2, \ldots,$ and p_k, then the mathematical expectation is*
>
> $$E = a_1 p_1 + a_2 p_2 + \ldots + a_k p_k$$

Each amount is multiplied by the corresponding probability, and the mathematical expectation, E, is given by the sum of all these products. It is important to keep in mind that the a's are positive when they represent profits, winnings, or gains (amounts which we receive) and that they are negative when they represent losses, penalties, or deficits (amounts which we must pay).

EXAMPLE What is our mathematical expectation if we win $5 if a balanced coin falls heads and lose $5 if it falls tails?

SOLUTION The amounts are $a_1 = 5$ and $a_2 = -5$, the probabilities are $p_1 = \frac{1}{2}$ and $p_2 = \frac{1}{2}$, and the mathematical expectation is

$$E = 5 \cdot \frac{1}{2} + (-5) \cdot \frac{1}{2} = 0$$

Games, like this one, in which the mathematical expectation is zero and neither player is favored, are said to be fair, or equitable.

EXAMPLE   The probabilities are 0.24, 0.35, 0.29, and 0.12 that a speculator will be able to sell a subdivision lot within a year at a profit of $12,500, at a profit of $8,000, at a profit of $1,000, or at a loss of $2,500. What is his expected profit?

SOLUTION   Substituting $a_1 = 12,500$, $a_2 = 8,000$, $a_3 = 1,000$, $a_4 = -2,500$, $p_1 = 0.24$, $p_2 = 0.35$, $p_3 = 0.29$, and $p_4 = 0.12$ into the formula for E, we get

$$E = 12,500(0.24) + 8,000(0.35) + 1,000(0.29)$$
$$- 2,500(0.12)$$
$$= \$5,790$$

In all our examples, the a's were dollar amounts or the cash equivalent of merchandise, and it is customary in that case to refer to the mathematical expectation as the **expected monetary value,** or EMV. However, the a's need not be monetary values at all. For instance, if we say that a child in the age group from 6 to 16 can expect to go to the dentist 1.9 times a year, we are actually referring to the result obtained by multiplying 0, 1, 2, 3, 4, . . . , by the probabilities that a child in this age group will visit a dentist that many times a year, and then adding all these products.

EXAMPLE   If the probabilities are 0.05, 0.17, 0.24, 0.19, 0.18, 0.09, 0.05, 0.02, and 0.01 that an airline office at a certain airport will receive 0, 1, 2, 3, 4, 5, 6, 7, or 8 complaints about its luggage handling on any one day, how many such complaints can be expected per day?

SOLUTION   The expected number is

$$E = 0(0.05) + 1(0.17) + 2(0.24) + 3(0.19) + 4(0.18)$$
$$+ 5(0.09) + 6(0.05) + 7(0.02) + 8(0.01)$$
$$= 2.91$$

It has been suggested that a person's behavior is rational if, when choosing between alternatives in situations involving uncertainties and risks, the person chooses the alternative having the highest mathematical expectation. This may seem to be a reasonable criterion for rational behavior and in many cases it is, but there are exceptions and they involve a number of difficulties which we shall discuss in Chapter 7 (see also Exercise 5.61 on page 125). For the moment, let us merely show how this decision-making criterion can be used to determine subjective probabilities.

Defending a liability suit against a client, a lawyer must decide whether to charge a straight fee of $1,500 or a contingent fee of $4,500, which she will get only if her client wins the case. How does she feel about her client's chances if she prefers the straight $1,500 fee?

If she feels that the probability is p that her client will win and she accepts the contingent fee, her mathematical expectation is $4,500p + 0(1 - p) = 4,500p$. Since she feels that the certainty of getting $1,500 is preferable to a mathematical expectation of $4,500p$, we write

$$1,500 > 4,500p$$

which yields $p < \dfrac{1,500}{4,500}$ and, hence, $p < \frac{1}{3}$. To narrow things down further, we might ask the lawyer if she would still prefer the straight $1,500 fee if the contingent fee were, say, $6,000 (see Exercise 5.58 on page 125).

EXERCISES

5.47 If a service club sells 500 raffle tickets for a cash prize of $100, what is the mathematical expectation of a person who buys one of the tickets?

5.48 A charitable organization raises funds by selling 2,000 raffle tickets for a $400 first prize and a $100 second prize. What is the mathematical expectation of a person who buys one of the tickets?

5.49 If someone offers to give us $5.00 each time we roll a 5 or a 6 with a balanced die, how much should we give him each time we roll a 1, 2, 3, or 4 to make the game fair?

5.50 To introduce his new cars to the public, a dealer offers a first prize of $2,500 and a second prize of $1,000 to some lucky persons who come to his showroom and submit contest entry cards. The winning cards are to be drawn at random. What is each entrant's mathematical expectation if 8,750 persons submit entry cards? Does this make it worthwhile to spend 50 cents on gasoline to drive to the dealer's showroom?

5.51 The two finalists in a golf tournament play 18 holes, with the winner getting $20,000 and the runner-up getting $12,000. What are the two players' mathematical expectations if
(a) they are evenly matched;
(b) their probabilities of winning are $\frac{3}{4}$ and $\frac{1}{4}$?

5.52 If the two league champions are evenly matched, the probabilities that a "best of seven" basketball play-off will take 4, 5, 6, or 7 games are $\frac{1}{8}$, $\frac{1}{4}$, $\frac{5}{16}$, and $\frac{5}{16}$. Under these conditions, how many games can we expect such a play-off to last?

5.53 A union negotiator feels that the probabilities are 0.25, 0.60, 0.10, and 0.05 that the union members will get a $1.20-an-hour raise, an 80-cent raise, a 40-cent raise, or no raise at all. What is their expected raise?

5.54 An importer is offered a shipment of pearls for $20,000, and the probabilities that she will be able to sell it for $24,000, $22,000, $20,000, or $18,000 are 0.22, 0.47, 0.26, and 0.05. If she buys the pearls, what is her expected gross profit?

5.55 A police chief knows that the probabilities of 0, 1, 2, 3, 4, or 5 car thefts on any given day are 0.23, 0.34, 0.26, 0.12, 0.04, and 0.01. How many car thefts can he expect per day? (It is assumed here that the probability of more than five car thefts is negligible.)

5.56 The following table gives the probabilities that a woman who enters "The Dress Shop" will buy 0, 1, 2, 3, or 4 dresses:

Number of dresses	0	1	2	3	4
Probability	0.11	0.37	0.35	0.12	0.05

How many dresses can a woman entering this shop be expected to buy?

5.57 A grab-bag contains 5 packages worth $1 apiece, 5 packages worth $3 apiece, and 10 packages worth $5 apiece. Is it rational to pay $4 for the privilege of selecting one of these packages at random?

5.58 With reference to the example on page 124, how does the lawyer feel about her client's chances if she prefers a contingent fee of $6,000 to the straight fee of $1,500?

5.59 A salesperson must choose between a straight salary of $28,800 and a salary of $24,000 plus a bonus of $9,600 if her sales exceed a certain quota. How does she assess her chances of exceeding the quota if she chooses the lower salary with the possibility of a bonus?

5.60 The manufacturer of a new battery additive must decide whether to sell his product for $1.00 a can, or for $1.25 with a "double-your-money-back-if-not-satisfied guarantee." How does he feel about the chances that a person will actually ask for double his money back if
(a) he decides to sell the product for $1.00;
(b) he decides to sell the product for $1.25 with the guarantee;
(c) he cannot make up his mind?

5.61 A contractor must choose between two jobs. The first job promises a profit of $120,000 with a probability of $\frac{3}{4}$ or a loss of $30,000 (due to strikes and other delays) with a probability of $\frac{1}{4}$; the second job promises a profit of $180,000 with a probability of $\frac{1}{2}$ or a loss of $45,000 with a probability of $\frac{1}{2}$.
(a) Which job should the contractor choose if he wants to maximize his expected profit?
(b) Which job would the contractor probably choose if his business is in fairly bad shape and he will go broke unless he can make a profit of at least $150,000 on his next job?

5.5

A Word of Caution

Many fallacies involving probabilities are due to inappropriate assumptions concerning the equal likelihood of events. Consider, for example, the following situation:

Among three identical file trays one contains two current records, one contains one current and one dead record, and the other contains two dead records. After taking one of these trays at random, a clerk randomly takes one record from it. If this record is a current one, what is the probability that the other record on this tray is also a current one?

Without giving the matter too much thought, it may seem reasonable to say that this probability is $\frac{1}{2}$. After all, the current record must have come from the first or second tray. For the first tray the other record is a current one, for the second tray the other record is a dead one, and it would seem reasonable to say that these two possibilities are equally likely. Actually, this is not the case: The correct value of the probability is $\frac{2}{3}$, and the reader can verify this by drawing an appropriate tree diagram. (When drawing such a tree diagram showing the six possible outcomes corresponding to which of the six records is actually chosen, it will be convenient to label the two current records on the first tray C_1 and C_2 and the two dead records on the third tray D_1 and D_2.)

5.6

Check List of Key Terms

5.7

Review Exercises

5.62 The employees of a company are classified into five categories according to age and into four categories according to marital status. In how many ways can an employee thus be classified?

5.63 If a college drama club has to choose three of nine half-hour skits to present on one evening, in how many different ways can it arrange the evening's schedule?

5.64 The probabilities that a person shopping at "The Bookstore" will buy 0, 1, 2, 3, or 4 books are 0.31, 0.45, 0.17, 0.06, and 0.01. How many books can a person shopping at this bookstore be expected to buy?

5.65 In how many different orders can a salesperson telephone six customers?

5.66 Determine the number of ways in which a person can buy a pound each of three of the 12 kinds of cheese carried by a gourmet food shop.

5.67 In a union election, Mr. Brown, Ms. Green, and Ms. Jones are running for president, while Mr. Adams, Ms. Roberts, and Mr. Smith are running for vice-president. Construct a tree diagram showing the nine possible outcomes, and use it to determine the number of ways in which the two union officials elected will not be of the same sex.

5.68 In how many ways can ten accounts be assigned to three stockbrokers, A, B, and C, so that A gets two accounts, B gets five accounts, and C gets three accounts?

5.69 An insurance company agrees to pay the promoter of a drag race $15,000 if the race is rained out. If the company's actuary feels that $2,400 is a fair net premium for this risk, what does this tell us about his assessment of the probability that the race will be rained out?

5.70 If 441 of 700 television viewers interviewed in a certain area feel that local news coverage is inadequate, estimate the probability that a television viewer randomly selected in that area will feel this way.

5.71 In how many different ways can 12 suggested company logos be ranked first, second, third, and fourth best by a panel of judges?

5.72 If a statistics department schedules six lecture sections and twelve discussion groups for a course in business statistics, in how many different ways can a student choose a lecture section and a discussion group?

5.73 The probabilities are 0.12, 0.40, 0.36, and 0.12 that Ms. Green will get $119,000, $124,000, $129,000, or $134,000 for her house. What is her mathematical expectation?

5.74 Determine whether each of the following is true or false:

(a) $6! = \dfrac{8!}{56}$;

(b) $\dfrac{1}{2!} + \dfrac{1}{2!} = 1$;

(c) $2! + 2! = 4!$;

(d) $7! = 7 \cdot 5!$.

5.75 What is the probability of rolling a 10 with a pair of balanced dice?

5.76 A, B, C, D, and E are the five chapter finalists competing for a national fraternity's public service award. Draw a tree diagram showing the different ways in which the judges can choose the winner and the first runner-up.

5.77 How many different sums of money can be formed with one or more of the following coins: a penny, a nickel, a dime, and a quarter?

5.78 If we receive 45 cents each time we draw a spade at random from an ordinary deck of 52 playing cards, how much should we pay when we draw a heart, a diamond, or a club so as to make the game fair?

5.79 In how many different ways can a person arrange eight books on a shelf?

5.80 A questionnaire sent through the mail as part of a market research study consists of eight questions, each with five different answers. In how many different ways can a person answer the eight questions?

5.81 Among the classified ads of a newspaper are listings for eight two-bedroom homes, six three-bedroom homes, and two four-bedroom homes. In how many different ways can a person choose two of the two-bedroom homes, four of the three-bedroom homes, and one of the four-bedroom homes to inspect?

5.82 If 804 of 1,200 letters mailed by a firm were delivered within 48 hours, estimate the probability that any one letter mailed by the firm will be delivered within 48 hours.

In the study of probability there are basically three kinds of questions: (1) What do we mean when we say, for example, that the probability of rain tomorrow is 0.80, that the probability that a new record shop will succeed is 0.35, or that the probability a candidate will be elected mayor is 0.60? (2) How are the numbers we call probabilities determined, or measured in actual practice? (3) What are the mathematical rules which probabilities must obey?

We have already studied the first two of these questions in Chapter 5. In the classical probability concept we are concerned with equally likely possibilities and count "favorable" outcomes; in the frequency interpretation we are concerned with proportions of "successes" in the long run and base our estimates on what happened in the past; and in the subjective probability concept we are concerned with a measure of a person's belief and we observe how the person will react in risk-taking situations.

In this chapter, after some preliminaries in Section 6.1, we shall study the question of what basic rules probabilities must obey, or how they must "behave." As we shall see in Section 6.2, there are essentially three of these rules, called the postulates of probability, and they must be obeyed regardless of whether we interpret probabilities in terms of equally likely possibilities, as proportions in the long run, or as subjective evaluations.

Then, in Section 6.3 we shall see how probabilities are related to odds, and in Section 6.4 we shall study the important concepts of conditional probability and independence.

Some Rules of Probability

6.1

Sample Spaces and Events

In statistics, a set of all possible outcomes of an experiment is called a **sample space** and it is usually denoted by the letter S. For instance, if a broker must choose three of 24 stocks to suggest to a client, the sample space consists of the $\binom{24}{3} = 2,024$ ways in which this choice can be made; if the dean of a college must assign two of her 84 faculty members as advisors to a journalism club, the sample space consists of the $\binom{84}{2} = 3,486$ ways in which this can be done.

Also, if we are concerned with the number of days it rains in Pittsburgh during the month of March, the sample space is the set

$$S = \{0, 1, 2, 3, 4, \ldots, 30, 31\}$$

To avoid misunderstandings about the terms "outcome" and "experiment," let us make it clear that they are used here in a very wide sense. For lack of a better term, "experiment" refers to any process of observation or measurement. Thus, an **experiment** may consist of determining the number of injury accidents in a large motor freight terminal in one year; it may consist of the simple process of noting whether a light is on or off; or it may consist of the complicated process of obtaining and evaluating data to predict gross national product. The results one obtains from an experiment, whether they are instrument readings, counts, "yes" or "no" answers, or values obtained through extensive calculations, are called the **outcomes** of the experiment.

When we study the outcomes of an experiment, we usually identify the various possibilities with numbers, points, or other kinds of symbols, so that we can treat all questions concerning the outcomes mathematically, without having to go through long verbal descriptions of what has taken place, is taking place, or will take place. For instance, if there are eight applicants for a job and we let $a, b, c, d, e, f, g,$ and h denote the events that it is offered to Arnold, Betty, Clark, and so on, then the sample space for this experiment is the set

$$S = \{a, b, c, d, e, f, g, h\}$$

The use of points rather than letters or numbers has the added advantage that it is easier to visualize the various possibilities, and perhaps discover some special features which several of the outcomes may have in common. For instance, if two contractors, among others, bid on two construction jobs and we are interested in how many jobs each of the two will get, we could write the six possible outcomes as (0, 0), (1, 0), (0, 1), (2, 0), (1, 1), and (0, 2). Here, (0, 1) represents the outcome that the first contractor gets neither job and the second gets one, and (1, 1) represents the outcome that each contractor gets one job. Geometrically, this situation may be pictured as in Figure 6.1, from which it is apparent, for instance, that they get the same number of jobs in two of the six possibilities, and that, between them, they get both jobs in three of the six possibilities.

Usually, we classify sample spaces according to the number of elements, or points, which they contain. The ones we have studied so far in this section contained 2,024, 3,486, 32, 8, and 6 elements, and we call them all **finite**, since the number of possibilities is in each case finite, or fixed. In this chapter we shall consider only finite sample spaces, but in later chapters we shall consider also **infinite sample spaces.**

In statistics, any subset of a sample space is called an **event**, and usually designated by a capital letter. By subset we mean any part of a set, including the set as a whole, and trivially, a set called the **empty set** and denoted by $\varnothing$, which has no elements at all. For instance, for the sample space of the number of days that it rains in Pittsburgh during the month of March,

$$A = \{15, 16, 17, 18, 19, 20\}$$

6.1

Outcomes of the two-contractor example.

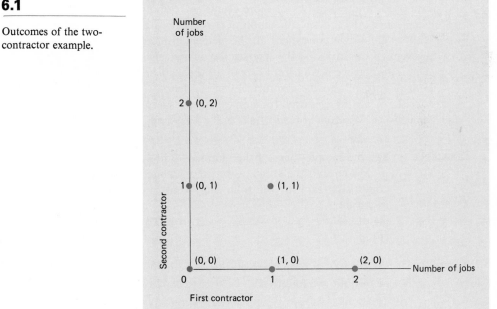

131   Sec. 6.1 : Sample Spaces and Events

is the event that there will be from 15 to 20 rainy days, and

$$B = \{18, 19, 20, \ldots, 30, 31\}$$

is the event that there will be at least 18 rainy days. Also, with reference to Figure 6.1,

$$C = \{(1, 0), (0, 1)\}$$

is the event that, between them, the two contractors get only one job,

$$D = \{(0, 0), (1, 0), (0, 1), (1, 1)\}$$

is the event that neither contractor gets both jobs, and

$$E = \{(1, 1)\}$$

is the event that each contractor gets one job.

Also, in the two-contractor example events C and E have no elements in common. Such events are called **mutually exclusive,** which means that they cannot both occur at the same time. Evidently, if, between them, the two contractors get only one job, each one cannot get one job. On the other hand, events D and E are not mutually exclusive since they both contain the outcome $(1, 1)$, where each contractor gets one job.

In many probability problems we are interested in events that can be expressed in terms of two or more events by forming **unions, intersections,** and **complements.** In general, the union of two events X and Y, denoted by $X \cup Y$, is the event which consists of all the elements (outcomes) either in event X or in event Y, or in both; the intersection of two events X and Y, denoted by $X \cap Y$, is the event which consists of all the elements (outcomes) contained in both X and Y; and the complement of X, denoted by X', is the event which consists of all the elements (outcomes) of the sample space that are not contained in X. We usually read $\cup$ as "or," $\cap$ as "and," and X' as "not X."

EXAMPLE For the sample space of the number of days that it rains in Pittsburgh in March and the events A and B as defined above, list the outcomes comprising each of the following events and also express the events in words:

(a) $A \cup B$: (c) B';
(b) $A \cap B$; (d) $A' \cap B'$.

SOLUTION (a) Since $A \cup B$ contains all the elements that are either in A or in B, or in both, we find that

$$A \cup B = \{15, 16, 17, \ldots, 30, 31\}$$

and this is the event that there will be at least 15 rainy days; (b) since $A \cap B$ contains all the elements that are in both A and B, we find that

$$A \cap B = \{18, 19, 20\}$$

and this is the event that there will be from 18 to 20 rainy days; (c) since B' contains all the elements of the sample space that are not in B, we find that

$$B' = \{0, 1, 2, \ldots, 16, 17\}$$

and this is the event that there will be fewer than 18 rainy days; (d) since $A' \cap B'$ contains all the elements of the sample space that are neither in A nor in B, we find that

$$A' \cap B' = \{0, 1, 2, \ldots, 13, 14\}$$

and this is the event that there will be fewer than 15 rainy days.

Sample spaces and events, particularly relationships among events, are often pictured by means of **Venn diagrams** such as those of Figures 6.2 and 6.3. In each case, the sample space is represented by a rectangle, and events by circles or parts of circles within the rectangle. The tinted regions of the four Venn diagrams of Figure 6.2 represent the event X, the complement of event X, the union of events X and Y, and the intersection of events X and Y.

6.2

Venn diagrams.

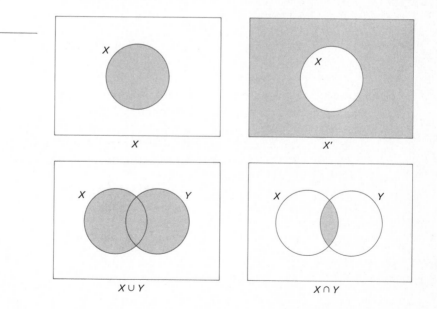

EXAMPLE   If X is the event that stock prices will go up and Y is the event that interest rates will go up, what events are represented by the tinted regions of the four Venn diagrams of Figure 6.2?

SOLUTION   The tinted region of the first diagram represents the event that stock prices will go up; the tinted region of the second diagram represents the event that stock prices will not go up; the tinted region of the third diagram represents the event that either stock prices or interest rates, or both, will go up; and the tinted region of the fourth diagram represents the event that stock prices and interest rates will both go up.

When we deal with three events, we draw the circles as in Figure 6.3. In this diagram, the circles divide the sample space into eight regions, numbered 1 through 8, and it is easy to determine whether the corresponding events are in X or in X', in Y or in Y', and in Z or in Z'.

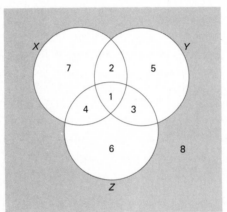

6.3

Venn diagram.

EXAMPLE   With reference to a newly designed engine, X represents the event that its gasoline consumption will be low, Y represents the event that its maintenance cost will be low, and Z represents the event that it can be sold at a profit. Express in words what events are represented by the following regions of the Venn diagram of Figure 6.3:

(a) region 4;
(b) regions 1 and 3 together;
(c) regions 3, 5, 6, and 8 together.

SOLUTION   (a) Since this region is contained in X and in Z but not in Y, it represents the event that the engine's gasoline consumption will be low, that it can be sold at a profit, but that its maintenance cost will not be low; (b) since this is the region common to Y and Z, it represents the event that the engine's maintenance cost will be low and that it can be sold at a profit; (c) since this is the entire region outside X, it represents the event that the engine's gasoline consumption will not be low.

EXERCISES

6.1 In an experiment, persons are asked to pick a number from 1 to 10, so that for each person the sample space is the set $S = \{1, 2, \ldots, 9, 10\}$. If $A = \{3, 4, 5, 6, 7\}$, $B = \{1, 2, 3, 4\}$, and $C = \{6, 7, 8, 9\}$, list the elements of the sample space comprising each of the following events, and also express the events in words:
(a) B';
(c) $A \cap B$;
(b) $A \cup C$;
(d) $B' \cap C'$.

6.2 With reference to the illustration on page 130, suppose that a, b, c, d, e, f, g, and h denote the events that Arnold, Betty, Clark, David, Eric, Francis, George, or Hilda will be offered the job, and that $D = \{a, c, d, f, g\}$, $E = \{d, g, h\}$, and $F = \{a, b, e\}$. List the elements of the sample space comprising each of the following events, and also express the events in words:
(a) D';
(c) $E \cap F$;
(b) $D \cup E$;
(d) $D \cap F'$.

6.3 To construct sample spaces for experiments in which we deal with categorical data, we often code the various alternatives by assigning them numbers. For instance, an airline passenger's complaint might be coded 1, 2, 3, 4, or 5, depending on whether it is about baggage handling, ticketing and boarding, seats or leg room, food service, or carry-on facilities. Express each of the following events in words:
(a) $K = \{1, 5\}$;
(c) $M = \{1, 2, 3\}$.
(b) $L = \{3, 5\}$;

6.4 With reference to the preceding exercise, list the elements of the sample space comprising each of the following events, and also express the events in words:
(a) K';
(c) $K \cap L$;
(b) $K \cup L$;
(d) $K \cap M$.

6.5 With reference to the two-contractor illustration on page 131 and Figure 6.1, describe each of the following events in words:
(a) $F = \{(0, 0), (1, 1)\}$;
(c) $H = \{(0, 0), (0, 1), (0, 2)\}$.
(b) $G = \{(2, 0), (1, 1), (0, 2)\}$;

6.6 With reference to the two-contractor illustration on page 131 and Figure 6.1, list the points of the sample space which comprise the following events:
(a) One of the contractors gets both jobs.
(b) The second contractor gets one job.
(c) The second contractor does not get either job.

6.7 A company providing shuttle service between two nearby airports has two helicopters which leave the airports every hour on the hour; the larger of the two can carry four passengers, the smaller one can carry only three passengers.
(a) Use two coordinates so that $(1, 3)$, for example, represents the event that when the helicopters take off at a given hour the larger helicopter has one passenger while the smaller helicopter has three, and $(2, 0)$ represents the event that the larger helicopter has two passengers while the smaller helicopter is empty. Draw a diagram (similar to that of Figure 6.1) showing the 20 points of the corresponding sample space.

(b) Describe in words the event which is represented by each of the following sets of points of the sample space: the event Q, which consists of the points (2, 3), (3, 2), (3, 3), (4, 1), (4, 2), and (4, 3); the event R, which consists of the points (0, 0), (1, 1), (2, 2), and (3, 3); the event T, which consists of the points (0, 1), (0, 2), (0, 3), (1, 2), (1, 3), and (2, 3); and the event U, which consists of the points (0, 3), (1, 2), (2, 1), and (3, 0).

6.8 With reference to the preceding exercise, which of the following are mutually exclusive events:
(a) R and T;
(b) R and U';
(c) Q and T;
(d) Q and U?

6.9 A small real-estate office has only three part-time salespersons.
(a) Using two coordinates so that (2, 1), for example, represents the event that two of the salespersons are at work but only one is busy with a customer, and (3, 0) represents the event that all three salespersons are at work but none of them is busy with a customer, draw a diagram (similar to that of Figure 6.1) showing the 10 points of the corresponding sample space.
(b) List the points which comprise the event K that at least two of the salespersons are busy with customers, the event L that only one of the salespersons is at work, and the event M that all the salespersons who are at work are busy with customers.
(c) With reference to part (b), list the points of the sample space which represent M' and $M \cap L$, and describe in words the corresponding events.
(d) With reference to part (b), which of the pairs of events, K and L, K and M, and L and M are mutually exclusive?

6.10 A wholesale appliance salesman has four customers in San Jose, whom he may or may not have time to call during a two-day visit to this city. He will not call any one of these customers more than once.
(a) Using two coordinates so that (2, 1), for example, represents the event that he will call two of the customers on the first day and one on the second day, and (0, 2) represents the event that he will not call any of these customers on the first day and two on the second day, draw a diagram similar to that of Figure 6.1, which shows the 15 possibilities.
(b) List the points of the sample space of part (a) which constitute the following events: event R that he will call all four of the customers during his visit, event M that he will call more of them on the first day than on the second day, event T that he will call at least three of the customers on the second day, and event U that he will call only one of the customers during his visit.
(c) With reference to part (b), list the points of the sample space which comprise the events R', $R \cup T$, $R \cap M$, and $R' \cap T$.

6.11 With reference to the preceding exercise, which of the following are mutually exclusive events:
(a) R and T;
(b) M and T;
(c) U and M;
(d) U and T?

6.12 Which of the following pairs of events are mutually exclusive? Explain your answers.
(a) Having rain and sunshine on July 4, 1981.

(b) Being under 25 years of age and being president of the United States.

(c) One person wearing yellow socks and brown shoes at the same time.

(d) A driver getting a ticket for speeding and a ticket for going through a red light at the same time.

(e) A person leaving San Francisco by jet at 11:45 P.M. and arriving in Washington, D.C., on the same day.

(f) Drawing a king and an ace on a single draw from an ordinary deck of 52 playing cards.

(g) Drawing a king and a black card on a single draw from an ordinary deck of 52 playing cards.

(h) A baseball player getting a walk and hitting a home run in the same game.

(i) A baseball player getting a walk and hitting a home run in the same time at bat.

6.13 In Figure 6.4, U is the event that the unemployment rate will go down and I is the event that the inflation rate will go up. Explain in words what events are represented by regions 1, 2, 3, and 4.

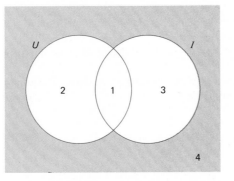

6.4

Venn diagram for Exercise 6.13.

6.14 With reference to the preceding exercise, what events are represented by

(a) regions 3 and 4 together;

(b) regions 2 and 3 together;

(c) regions 1, 2, and 3 together?

6.15 In Figure 6.5, L is the event that a person arrested for car theft can afford to pay a lawyer and G is the event that the person is found guilty of the crime. Explain in words what events are represented by regions 1, 2, 3, and 4.

6.5

Venn diagram for Exercise 6.15.

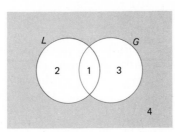

6.16 With reference to the preceding exercise, what events are represented by
  (a) regions 1 and 3 together;
  (b) regions 2 and 4 together;
  (c) regions 1, 3, and 4 together?

6.17 With reference to the illustration on page 134 dealing with the newly designed engine and Figure 6.3, express in words what events are represented by the following regions of the Venn diagram:
  (a) region 2;
  (b) regions 1 and 4 together;
  (c) regions 2 and 5 together;
  (d) regions 4, 6, and 7 together;
  (e) regions 2, 5, 7, and 8 together.

6.18 Suppose that a group of students is planning to visit a factory and that P is the event that the students will get to see the public relations director, R is the event that they will get to see the research and development department, and T is the event that they will get very tired. With reference to the Venn diagram of Figure 6.6, list (by number) the regions or combinations of regions which represent the events that the students
  (a) will get to see the public relations director and the research and development department, but not get very tired;
  (b) will see neither the public relations director nor the research and development department, yet get very tired;
  (c) will get to see the research and development department and get very tired;
  (d) will get to see the public relations director or the research and development department (or both), but not get very tired.

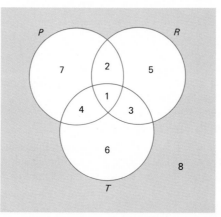

6.6

Venn diagram for Exercise 6.18.

6.19 With reference to the preceding exercise and the Venn diagram of Figure 6.6, express in words what events are represented by the following regions:
  (a) region 1;
  (b) region 5;
  (c) region 8;
  (d) regions 3 and 5 together;
  (e) regions 1 and 4 together;
  (f) regions 6 and 8 together.

6.20 Venn diagrams are often used to verify relationships among sets, subsets, or events, without requiring rigorous proofs based on a formal algebra of sets. We simply show that the expressions which are supposed to be equal are represented by the same region of a Venn diagram. Use Venn diagrams to show that

(a) $A \cup (A \cap B) = A$;

(b) $(A \cap B) \cup (A \cap B') = A$;

(c) $(A \cap B)' = A' \cup B'$ and $(A \cup B)' = A' \cap B'$;

(d) $(A \cup B) = (A \cap B) \cup (A \cap B') \cup (A' \cap B)$;

(e) $A \cap (B \cup C) = (A \cap B) \cup (A \cap C)$.

6.2

Some Basic Rules

We turn now to the question of how probabilities must "behave," and we begin by stating the three basic postulates. To formulate these postulates and some of their immediate consequences, we shall continue the practice of denoting events by capital letters, and we shall write the probability of event A as $P(A)$, the probability of event B as $P(B)$, and so forth. As before, we shall denote the set of all possible outcomes, the sample space, by the letter S. As we shall formulate them here, the three postulates of probability apply only when the sample space S is finite.

First two postulates of probability

> **1** *The probability of any event is a positive real number or zero; symbolically,* $P(A) \geq 0$ *for any event* A.
>
> **2** *The probability of any sample space is equal to 1; symbolically,* $P(S) = 1$ *for any sample space* S.

Let us justify these two postulates, as well as the one which follows, with reference to the classical probability concept and the frequency interpretation; in Section 6.3 we shall see to what extent the postulates are compatible also with subjective probabilities.

So far as the first postulate is concerned, the fraction s/n is always positive or zero, and so are percentages or proportions. The second postulate states indirectly that certainty is identified with a probability of 1; after all, one of the possibilities included in S must occur, and it is to this certain event that we assign a probability of 1. For equally likely outcomes, $s = n$ for the whole sample space and $s/n = n/n = 1$; and in the frequency interpretation, a probability of 1 implies that the event will occur 100 percent of the time, or in other words, that it is certain to occur.

In actual practice, we also assign a probability of 1 to events which are "practically certain" to occur. For instance, we would assign a probability of 1 to the event that at least one person will vote in the next presidential election, and that among all new cars sold during any one model year at least one will be involved in an accident before it has been driven 12,000 miles.

The third postulate of probability is especially important, but it is not quite so obvious as the other two.

Third postulate of probability

> **3** *If two events are mutually exclusive, the probability that one or the other will occur equals the sum of their probabilities. Symbolically,*
>
> $$P(A \cup B) = P(A) + P(B)$$
>
> *for any two mutually exclusive events A and B.*

For instance, if the probability that a manufacturers' raw material price index will go up during a certain month is 0.82 and the probability that it will remain unchanged is 0.13, then the probability that it will either go up or remain unchanged is $0.82 + 0.13 = 0.95$. Similarly, if the probabilities that a student will get an A or a B in a course are 0.17 and 0.35, then the probability that she will get either an A or a B is $0.17 + 0.35 = 0.52$.

This postulate is also compatible with the classical probability concept and the frequency interpretation. In the classical concept, if s_1 of n equally likely possibilities constitute event A and s_2 others constitute event B, then these $s_1 + s_2$ equally likely possibilities constitute event $A \cup B$, and we have

$$P(A) = \frac{s_1}{n}, \quad P(B) = \frac{s_2}{n}, \quad P(A \cup B) = \frac{s_1 + s_2}{n},$$

and

$$P(A) + P(B) = P(A \cup B)$$

In accordance with the frequency interpretation, if one event occurs, say, 36 percent of the time, another event occurs 41 percent of the time, and they cannot both occur at the same time (they are mutually exclusive), then one or the other will occur $36 + 41 = 77$ percent of the time; this satisfies the third postulate.

By using the three postulates of probability, we can derive many further rules according to which probabilities must "behave"—some of them are easy to prove and some are not, but they all have important applications. Among the immediate consequences of the three postulates, we find that probabilities can never be greater than 1, that an event which cannot occur has the probability 0, and that the probabilities that an event will occur and that it will not occur always add up to 1. Symbolically,

The first of these results simply expresses the fact that there cannot be more favorable outcomes than there are outcomes, and that an event cannot occur more than 100 percent of the time. The second result expresses the fact that when an event cannot occur there are $s = 0$ favorable outcomes, and that such an event occurs zero percent of the time. In actual practice, we also assign 0 probabilities to events which are so unlikely that we are "practically certain" they will not occur. For instance, we would assign a probability of 0 to the event that a monkey set loose on a typewriter will by chance type Plato's *Republic* word for word without a single mistake.

The third result can also be derived from the postulates of probability, but it can easily be seen that it is compatible with the classical probability concept and the frequency interpretation. In the classical concept, if there are s "successes" there are $n - s$ "failures," the corresponding probabilities are $\dfrac{s}{n}$ and $\dfrac{n-s}{n}$, and their sum is $\dfrac{s}{n} + \dfrac{n-s}{n} = 1$. In accordance with the frequency interpretation, if shipments arrive late 16 percent of the time, then they do not arrive late 84 percent of the time, the corresponding probabilities are 0.16 and 0.84, and their sum is 1.

The examples which follow illustrate how the postulates and the further rules are put to use in actual practice:

EXAMPLE   If A is the event that the price of a certain stock will remain unchanged on a given trading day and B is the event that its price will go up, $P(A) = 0.64$ and $P(B) = 0.21$, find

     (a) $P(A')$;
     (b) $P(A \cup B)$;
     (c) $P(A \cap B)$.

SOLUTION   (a) From the third of the further rules, we find that $P(A') = 1 - P(A)$ $= 1 - 0.64 = 0.36$; (b) since A and B are mutually exclusive, it follows from the third postulate that $P(A \cup B) = P(A) + P(B) = 0.64 + 0.21 = 0.85$; (c) since A and B are mutually exclusive, $A \cap B = \emptyset$ and it follows that $P(A \cap B) = 0$ in accordance with the second of the further rules.

In problems like this, it often helps to draw a Venn diagram, fill in the probabilities associated with the various regions, and then read the answers directly off the diagram.

EXAMPLE   If C is the event that a certain lawyer will be in her office on a given afternoon and D is the event that she will be in court, $P(C) = 0.48$ and $P(D) = 0.27$, find the value of $P(C' \cap D')$, the probability that she will be neither in her office nor in court.

SOLUTION   Drawing the Venn diagram as in Figure 6.7, we first put a 0 into region 1 because C and D are mutually exculsive events. It follows that the 0.48 probability of event C must go into region 2, the 0.27 probability of event D must go into region 3, and since the probability of the entire sample space must equal 1, we put $1 - (0.48 + 0.27) = 0.25$ into region 4. Since the event $C' \cap D'$ is represented by region 4, the region outside both circles, we find that $P(C' \cap D') = 0.25$.

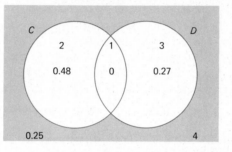

6.7

Venn diagram.

The third postulate applies only to two mutually exclusive events, but it can easily be generalized; repeatedly using this postulate, it can be shown that

Generalization of Postulate 3

> *If k events are mutually exclusive, the probability that one of them will occur equals the sum of their individual probabilities; symbolically*
>
> $$P(A_1 \cup A_2 \cup \cdots \cup A_k) = P(A_1) + P(A_2) + \cdots + P(A_k)$$
>
> *for any mutually exclusive events $A_1, A_2, \ldots,$ and A_k.*

where, again, $\cup$ is usually read "or."

EXAMPLE   The probabilities that a woman will buy a new party dress at Bullock's, the Broadway Southwest department store, or the May Co. are 0.35, 0.22, and 0.18. What is the probability that she will buy the dress at one of these stores?

SOLUTION   Since the three possibilities are mutually exclusive, direct substitution into the formula yields

$$0.35 + 0.22 + 0.18 = 0.75$$

If the probabilities that the Standard and Poor's Corporation will rate a company's stock A+, A, A−, or B+ are 0.10, 0.16, 0.31, and 0.36, what is the probability that the stock will get one of these ratings?

SOLUTION Since the four possibilities are mutually exclusive, direct substitution into the formula yields

$$0.10 + 0.16 + 0.31 + 0.36 = 0.93$$

The job of assigning probabilities to all possible events connected with a given situation can be very tedious, to say the least. If there are only five outcomes (or points) in a sample space S, there is $\binom{5}{0} = 1$ subset (the empty set) which contains no outcomes at all, there are $\binom{5}{1} = 5$ subsets which contain one outcome, $\binom{5}{2} = 10$ subsets which contain two outcomes, $\binom{5}{3} = 10$ subsets which contain three outcomes, $\binom{5}{4} = 5$ subsets which contain four outcomes, and there is $\binom{5}{5} = 1$ subset (the sample space itself) which contains all five outcomes. Thus, there are $1 + 5 + 10 + 10 + 5 + 1 = 32$ different subsets of a sample space with five outcomes, and as the number of outcomes increases slowly the number of subsets increases rapidly.

In general, if we let $a = 1$ and $b = 1$ in the binomial expansion of $(a + b)^n$, it follows directly that a sample space with n outcomes has 2^n different subsets, namely, that

$$\binom{n}{0} + \binom{n}{1} + \binom{n}{2} + \cdots + \binom{n}{n-1} + \binom{n}{n} = 2^n$$

where $\binom{n}{0}, \binom{n}{1}, \binom{n}{2}, \ldots, \binom{n}{n-1}$, and $\binom{n}{n}$ are the numbers of subsets containing $0, 1, 2, \ldots, n - 1$, and n outcomes. As we saw in the preceding paragraph, a sample space with $n = 5$ outcomes has $2^5 = 32$ different subsets; also, a sample space with $n = 20$ outcomes has $2^{20} = 1,048,576$ different subsets.

Fortunately, it is seldom necessary to assign probabilities to all possible events, and the following rule (which is a direct application of the foregoing generalization of the third postulate) makes it relatively easy to determine the probability of any event on the basis of the probabilities assigned to the individual outcomes (points) of the corresponding sample space:

Rule for calculating the probability of an event

> *The probability of any event A is given by the sum of the probabilities of the individual outcomes comprising A.*

This rule is illustrated in Figure 6.8, where the dots represent the individual (mutually exclusive) outcomes.

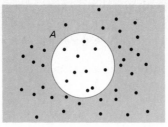

6.8

Sample space.

EXAMPLE   If the probabilities are 0.05, 0.14, 0.17, 0.33, 0.20, and 0.11 that a consumer testing service will rate a new lawn mower very poor, poor, fair, good, very good, or excellent, what are the probabilities that it will rate the lawn mower

    (a) very poor or poor;

    (b) good, very good, or excellent;

    (c) poor, fair, good, or very good?

SOLUTION   Adding the probabilities of the respective outcomes, we get (a) $0.05 + 0.14 = 0.19$, (b) $0.33 + 0.20 + 0.11 = 0.64$, and (c) $0.14 + 0.17 + 0.33 + 0.20 = 0.84$.

EXAMPLE   Referring again to the two-contractor illustration on page 131, suppose that the six points of the sample space have the probabilities shown in Figure 6.9. Find the probabilities that

6.9

Sample space with probabilities.

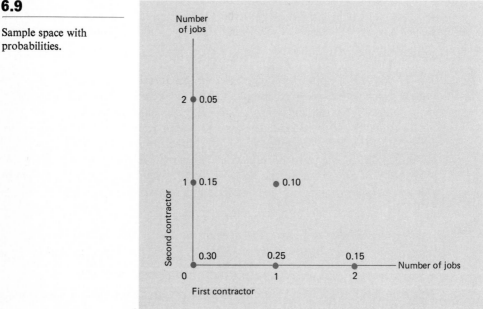

(a) the first contractor will not get either job;

(b) between them, the two contractors will get both jobs;

(c) the second contractor will get at most one job.

SOLUTION (a) Adding the probabilities associated with the points $(0, 0)$, $(0, 1)$, and $(0, 2)$, we get $0.30 + 0.15 + 0.05 = 0.50$; (b) adding the probabilities associated with the points $(2, 0)$, $(1, 1)$, and $(0, 2)$, we get $0.15 + 0.10 + 0.05 = 0.30$; (c) adding the probabilities associated with the points $(0, 0)$, $(1, 0)$, $(2, 0)$, $(0, 1)$, and $(1, 1)$, we get $0.30 + 0.25 + 0.15 + 0.15 + 0.10 = 0.95$.

If the individual outcomes are all equiprobable, the calculations are even simpler, for the rule directly above leads to the formula $\frac{s}{n}$, which we introduced in Chapter 5 in connection with the classical probability concept.

Since the third postulate applies only to mutually exclusive events, it cannot be used, for example, to find the probability that at least one of two roommates will pass a final exam in economics, the probability that a person will break an arm or a rib in an automobile accident, or the probability that a customer will buy a shirt or a tie while shopping at Macy's. Both roommates can pass the exam; a person can break an arm and a rib; and a customer can buy both a shirt and a tie.

To find a formula for $P(A \cup B)$ which holds whether the events A and B are mutually exclusive or not, let us consider the Venn diagram of Figure 6.10, which concerns the appointment of a college president. The letter G stands for the event that the appointee will be a graduate of the given college, and the letter W stands for the event that the appointee will be a woman. It follows from the figures in the Venn diagram that

$$P(G) = 0.59 + 0.08 = 0.67$$
$$P(W) = 0.08 + 0.04 = 0.12$$

and

$$P(G \cup W) = 0.59 + 0.08 + 0.04 = 0.71$$

6.10

Venn diagram.

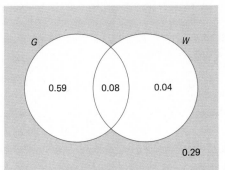

Here we added the probabilities because they represent mutually exclusive events (nonoverlapping regions of the Venn diagram).

Erroneously using the third postulate of probability to calculate $P(G \cup W)$, we get $P(G) + P(W) = 0.67 + 0.12 = 0.79$, which exceeds the correct value by 0.08. The error here results from adding $P(G \cap W)$ in twice, once in $P(G) = 0.67$ and once in $P(W) = 0.12$. However, we can correct for this by subtracting $P(G \cap W) = 0.08$ from $P(G) + P(W) = 0.79$, writing

$$P(G \cup W) = P(G) + P(W) - P(G \cap W)$$
$$= 0.67 + 0.12 - 0.08$$
$$= 0.71$$

This agrees, as it should, with the result which we obtained before.

Since the argument we used in this example holds for any two events A and B, we can now state the following **general addition rule**, which applies whether A and B are mutually exclusive or not:

General addition rule

$$P(A \cup B) = P(A) + P(B) - P(A \cap B)$$

When A and B are mutually exclusive, $P(A \cap B) = 0$ (since by definition the two events cannot both occur at the same time), and this formula reduces to that of the third postulate of probability. In this connection, the third postulate is also called the **special addition rule**.

EXAMPLE   If the probabilities are $0.20, 0.15$, and 0.03 that a student will get a failing grade either in accounting or in marketing, or in both, what is the probability that he will fail at least one of these subjects?

SOLUTION   Direct substitution into the formula above yields

$$0.20 + 0.15 - 0.03 = 0.32$$

EXAMPLE   If the probabilities are $0.87, 0.36$, and 0.29 that a family, randomly chosen as part of a sample survey in a large metropolitan area, owns a color television set, or a black-and-white set, or both, what is the probability that a family in this area will own one, or the other, or both kinds of sets?

SOLUTION   Substituting these values into the formula for the general addition rule, we get

$$0.87 + 0.36 - 0.29 = 0.94$$

The general addition rule can be generalized further so that it applies to more than two events (see, for example, Exercise 6.37 on page 149).

6.21 In a study of the adequacy of fuel supplies, C stands for the event that a power plant uses coal and E is the event that it is able to provide enough electricity. State in words what probabilities are expressed by

(a) $P(C')$;

(b) $P(E')$;

(c) $P(C \cup E)$;

(d) $P(C \cap E)$;

(e) $P(C' \cup E)$;

(f) $P(C' \cap E')$.

6.22 If F is the event that a dishonest land developer is in financial difficulties, T is the event that he has tax problems, and Q is the event that he uses questionable sales practices, write in symbolic form the probabilities that a dishonest land developer

(a) has tax problems and uses questionable sales practices;

(b) is not in financial difficulties but has tax problems;

(c) uses questionable sales practices or has tax problems;

(d) has neither financial difficulties nor tax problems.

6.23 Explain why there must be a mistake in each of the following statements:

(a) The probability that a corporation will pay its regular quarterly dividend is 0.83, and the probability that it will not pay its regular quarterly dividend is 0.27.

(b) The probability that a new service station will lose money during its first year of operation is 0.33, and the probability that it will break even or make a profit is 0.57.

(c) The probability that on any given working day an office worker in a certain city drives to work is 0.62, and the probability that he either drives to work or takes a bus is 0.54.

(d) The probability that a married student prefers living on campus is 0.29, and the probability that he and his wife both prefer living on campus is 0.43.

(e) The probability that an insurance salesman will sell a life insurance policy to a friend is 0.48, the probability that he will sell him automobile insurance but no life insurance is 0.36, and the probability that he will sell him neither kind of insurance is 0.12.

6.24 The probabilities that a typist will make at most five mistakes when typing a report, or make from 6 to 10 mistakes, are 0.64 and 0.21. Find the probabilities that the typist will make

(a) at least six mistakes;

(b) at most ten mistakes;

(c) more than ten mistakes.

6.25 If F and T are the events that a one car owner in a certain income bracket will drive a Ford or a Toyota, and $P(F) = 0.34$ and $P(T) = 0.08$, find the probabilities that such a person will

(a) not drive a Ford;

(b) drive a Ford or a Toyota;

(c) drive neither a Ford nor a Toyota.

6.26 Given the mutually exclusive events A and B for which $P(A) = 0.37$ and $P(B) = 0.41$, find

(a) $P(A')$;

(b) $P(B')$;

(c) $P(A \cap B)$;

(d) $P(A \cup B)$;

(e) $P(A' \cup B')$;

(f) $P(A' \cap B')$.

6.27 The probabilities that a review board will rate a given movie X, R, or PG are 0.43, 0.28, and 0.12. What is the probability that the movie will get one or another of these three ratings?

6.28 The probabilities that a student will get an A, a B, or a C in an accounting course are 0.08, 0.14, and 0.45. What is the probability that the student will get a grade lower than a C?

6.29 With reference to Figure 6.9, find the probabilities that
(a) the first contractor will get only one job;
(b) between them, the two contractors will get only one job;
(c) neither contractor will get both jobs.

6.30 The probabilities that a consumer testing service will rate a new antipollution device for cars poor, fair, adequate, very good, or excellent are 0.10, 0.19, 0.33, 0.23, and 0.15. Find the probabilities that the service will rate the device
(a) poor or fair;
(b) at least adequate;
(c) at best adequate;
(d) neither poor nor excellent.

6.31 The probabilities that 0, 1, 2, 3, 4, 5, 6, or at least 7 persons will inquire about a piece of industrial property on the first day that it is advertised for sale are 0.002, 0.013, 0.039, 0.081, 0.125, 0.155, 0.160, and 0.425. What are the probabilities that
(a) at most 4 persons will inquire about the property on that day;
(b) at least 2 persons will inquire about the property on that day;
(c) from 3 to 5 persons will inquire about the property on that day?

6.32 Given two events A and B for which $P(A) = 0.56$, $P(B) = 0.43$, and $P(A \cap B) = 0.18$, find
(a) $P(A')$;
(b) $P(B')$;
(c) $P(A \cup B)$;
(d) $P(A' \cap B)$;
(e) $P(A' \cup B)$;
(f) $P(A \cap B')$.

6.33 The probabilities that a convicted reckless driver will be fined, have his license revoked, or both are 0.88, 0.62, and 0.55. What is the probability that he will be fined or have his license revoked?

6.34 The probability that a person stopping at a gas station will ask to have his tires checked is 0.12, the probability that he will ask to have his oil checked is 0.29, and the probability that he will ask to have them both checked is 0.07. What is the probability that a person stopping at this station will ask to have
(a) either his tires or his oil checked;
(b) neither his tires nor his oil checked?

6.35 A businessman has two secretaries. The probability that the one he hired most recently will be absent on any given day is 0.08, the probability that the other secretary will be absent on any given day is 0.07, and the probability that they will both be absent on any given day is 0.02. What is the probability that
(a) either or both secretaries will be absent on any given day;
(b) at least one secretary comes to work on any given day;
(c) only one secretary comes to work on any given day?

6.36 A student artist who has entered an oil painting and a watercolor in a show feels that the probabilities are, respectively, 0.19, 0.13, and 0.11 that she will sell the oil painting, the watercolor, or both. What is the probability that she will sell

(a) either of these works;

(b) neither of these works;

(c) the oil painting but not the watercolor?

6.37 It can be shown that for any three events, A, B, and C, the probability that at least one of them will occur is given by $P(A \cup B \cup C) = P(A) + P(B) + P(C) - P(A \cap B) - P(A \cap C) - P(B \cap C) + P(A \cap B \cap C)$. Use this formula in the following problems:

(a) The probabilities that a person visiting a certain dentist will have his teeth cleaned, a cavity filled, a tooth extracted, his teeth cleaned and a cavity filled, his teeth cleaned and a tooth extracted, a cavity filled and a tooth extracted, or his teeth cleaned, a cavity filled, and a tooth extracted are 0.47, 0.29, 0.22, 0.08, 0.06, 0.07, and 0.03. What is the probability that a person visiting this dentist will have at least one of these three things done?

(b) Suppose that if a person visits Disneyland, the probabilities that he will go on the Jungle Cruise, the Monorail, the Matterhorn ride, the Jungle Cruise and the Monorail, the Jungle Cruise and the Matterhorn ride, the Monorail and the Matterhorn ride, or the Jungle Cruise, the Monorail, and the Matterhorn ride are 0.74, 0.70, 0.62, 0.52, 0.46, 0.44, and 0.34. What is the probability that a person visiting Disneyland will go on at least one of these three rides?

6.38 The following is a proof of the fact that $P(A) \leq 1$ for any event A: By definition A and A' represent mutually exclusive events and $A \cup A' = S$ (since A and A' together comprise all the points of the sample space S). So, we can write $P(A \cup A') = P(S)$, and it follows that

$$P(A) + P(A') = P(S) \qquad \text{step 1}$$
$$P(A) + P(A') = 1 \qquad \text{step 2}$$
$$P(A) = 1 - P(A') \qquad \text{step 3}$$
$$P(A) \leq 1 \qquad \text{step 4}$$

State which of the three postulates of probability justify the first, second, and fourth steps of this proof; the third step is simple arithmetic. Note also that in step 2 we actually proved that $P(A') = 1 - P(A)$.

6.3

Probabilities and Odds

If an event is twice as likely to occur as not to occur, we say that the **odds** are 2 to 1 that it will occur; if an event is three times as likely to occur as not to occur, we say that the odds are 3 to 1; if an event is ten times as likely to occur as not to occur, we say that the odds are 10 to 1; and so forth. In general, the odds that an event will occur are given by the ratio of the probability that it will occur to the probability that it will not occur. Symbolically,

*Formula relating
odds to probabilities*

> *If the probability of an event is p, the odds for its occurrence are a to b, where a and b are positive values such that*
>
> $$\frac{a}{b} = \frac{p}{1-p}$$

It is customary to express odds as a ratio of two positive integers having no common factor, and if an event is more likely not to occur as to occur, to give the odds that it will not occur rather than the odds that it will occur.

EXAMPLE What are the odds for the occurrence of an event if its probability is
(a) $\frac{5}{9}$;
(b) 0.85;
(c) 0.20?

SOLUTION (a) By definition, the odds are $\frac{5}{9}$ to $1 - \frac{5}{9} = \frac{4}{9}$, or 5 to 4; (b) by definition, the odds are 0.85 to $1 - 0.85 = 0.15$, 85 to 15, or 17 to 3; (c) by definition, the odds are 0.20 to $1 - 0.20 = 0.80$, 20 to 80, or 1 to 4, but we say instead that the odds against the occurrence of the event are 4 to 1.

In gambling, the word "odds" is also used to denote the ratio of the wager of one party to that of another. For instance, if a gambler says that he will give 3 to 1 odds on the occurrence of an event, he means that he is willing to bet $3 against $1 (or perhaps $30 against $10 or $1,200 against $400) that the event will occur. If such betting odds equal the odds that the event will occur, we say that the betting odds are fair.

EXAMPLE Records show that $\frac{1}{12}$ of the trucks weighed at a certain check point in Nevada are overloaded. If someone offers to bet $40 against $4 that the next truck weighed at this check point is not overloaded, are these odds fair?

SOLUTION Since the probability is $1 - \frac{1}{12} = \frac{11}{12}$ that the truck is not overloaded, the odds are 11 to 1, and the bet would be fair if the person offered to bet $44 against $4 that the next truck weighed at the check point is not overloaded. Thus, the original bet of $40 against $4 favors the person offering the bet; it is not fair.

The preceding discussion provides the groundwork for a way of measuring subjective probabilities. If a businessman feels that the odds on the success of a new clothing store are 3 to 2, this means that he is willing to bet (or considers it fair to bet), say, $300 against $200 that the new store will be a success. In this way he is expressing his belief regarding the uncertainties connected with the success of the store. To convert the odds into a probability we solve the equation $\frac{a}{b} = \frac{p}{1-p}$ for p, and get the following result:

Formula relating probabilities to odds

> *If the odds are a to b that an event will occur, the probability of its occurrence is*
>
> $$p = \frac{a}{a+b}$$

EXAMPLE Convert the businessman's 3 to 2 odds on the success of the new clothing store into a probability.

SOLUTION Substituting $a = 3$ and $b = 2$ into the formula for p, we get $p = \frac{3}{3+2} = \frac{3}{5}$.

EXAMPLE If an applicant for a managerial position feels that the odds are 7 to 4 that she will get the job, what probability does she assign to her getting the job?

SOLUTION Substituting $a = 7$ and $b = 4$ into the formula for p, we get $p = \frac{7}{7+4} = \frac{7}{11}$.

Let us now see whether subjective probabilities, determined in this way, "behave" in accordance with the postulates of probability. Since a and b are positive quantities, $\frac{a}{a+b}$ cannot be negative and this satisfies the first postulate. In connection with the second postulate, we note that the surer we are that an event will occur, the "better" odds we should be willing to give—say, 100 to 1, 1,000 to 1, or even 1,000,000 to 1. The corresponding probabilities are $\frac{100}{100+1} = 0.99$, $\frac{1,000}{1,000+1} = 0.999$, and $\frac{1,000,000}{1,000,000+1} = 0.999999$, and it can be seen that the surer we are that an event will occur, the closer its probability will be to 1.

The third postulate of probability—$P(A \cup B) = P(A) + P(B)$ for any two mutually exclusive events A and B—does not necessarily apply to subjective probabilities, but proponents of the subjectivist point of view impose it as a **consistency criterion**. In other words, if a person's subjective probabilities "behave" in accordance with the third postulate, he is said to be **consistent**; otherwise, he is said to be **inconsistent** and his probability judgments cannot be taken seriously.

EXAMPLE   If an economist feels that the odds are 2 to 1 that the price of beef will go up during the next month, 1 to 5 that it will remain unchanged, and 8 to 3 that it will go up or remain unchanged, are the corresponding probabilities consistent?

SOLUTION   The probabilities that the price of beef will go up during the next month, that it will remain unchanged, and that it will go up or remain unchanged are $\frac{2}{2+1} = \frac{2}{3}, \frac{1}{1+5} = \frac{1}{6}$, and $\frac{8}{8+3} = \frac{8}{11}$. Since $\frac{2}{3} + \frac{1}{6} = \frac{5}{6} \neq \frac{8}{11}$, the probabilities are not consistent.

EXERCISES   6.39 Convert each of the following probabilities to odds:
(a) The probability that the last digit of a car's license plate is 2, 3, 4, 5, 6, or 7 is $\frac{6}{10}$.
(b) The probability of getting at most one head in five flips of a balanced coin is $\frac{6}{32}$.
(c) The probability of getting at least three heads in six flips of a balanced coin is $\frac{42}{64}$.

6.40 If the probability is 0.64 that an executive trainee will stay with a banking firm for at least two years, what are the odds that a randomly chosen trainee will stay with the firm for at least two years?

6.41 If the probability that a certain shipment will arrive on time is $\frac{5}{16}$, what are the odds that it will not arrive on time?

6.42 Convert each of the following odds to probabilities:
(a) The odds are 5 to 3 that a union membership will vote down an employer's proposed strike settlement package.
(b) The odds against rolling a 7 or an 11 with a pair of balanced dice are 7 to 2.

6.43 A sportswriter feels that the odds are 5 to 1 that the home team will lose its next basketball game. What subjective probability expresses his feelings about the home team's winning the game?

6.44 A labor union wage negotiator feels that the odds are 3 to 1 that the union members will get a raise of 80 cents in their hourly wage, the odds are 17 to 3 against their getting a raise of 40 cents in their hourly wage, and the odds are 9 to 1 against their getting no raise at all.

(a) Find the corresponding probabilities that they will get an 80-cent raise, a 40-cent raise, or no raise in their hourly wage.

(b) What is the expected raise in their hourly wage?

6.45 If a stockbroker is unwilling to bet $60 against $180 that the price of a certain stock will go up within a week, what is his subjective probability that the price of the stock will go up within a week? (*Hint:* The answer should read "less than. . . .")

6.46 Suppose that a student is willing to bet $5 against $1, but not $7 against $1, that he will get a passing grade in a certain course. What personal probability does he assign to his getting a passing grade in the course? (*Hint:* The answer should read "at least . . . but less than. . . .")

6.47 A television executive is willing to bet $500 against $4,500, but not $600 against $4,400, that a new game show will be a success. What is the executive's subjective probability that the new show will not be a success? (*Hint:* The answer should read "greater than . . . but at most. . . .")

6.48 A branch bank manager feels that the odds are 7 to 5 against her getting a $1,000 raise and 11 to 1 against her getting a $2,000 raise. Furthermore, she feels that the odds are even (1 to 1) that she will get either one or the other raise. Discuss the consistency of these subjective probabilities.

6.49 There are two Porsches in a race, and a driver feels that the odds against their winning are, respectively, 3 to 1 and 4 to 1. To be consistent, what odds should he assign to the event that neither car will win?

6.50 A stockbroker feels that the odds are 2 to 1 that the price of Xerox stock will close up the following day and 3 to 1 that it will close down. Can these odds be right? Explain.

6.4

Conditional Probabilities

It is often meaningless (or at least very confusing) to speak of the probability of an event without specifying the sample space on which the event is defined. For instance, if we ask for the probability that a lawyer makes more than $40,000 per year, we may well get many different answers, and they can all be correct. One of these might apply to all lawyers in the United States, another to lawyers handling only divorce cases, a third to corporation lawyers, another to lawyers handling only tax cases, and so on. Since the choice of the sample space (that is, the set of all possibilities under consideration) is by no means always self-evident, it is helpful to use the symbol $P(A|S)$ to denote the **conditional probability** of event A relative to the sample space S, or as we often call it "the probability of A given S." The symbol $P(A|S)$ makes it explicit that we are referring to a particular sample space S, and it is generally preferable to the abbreviated notation $P(A)$ unless the tacit choice of S is clearly understood.

It is also preferable when we have to refer to different sample spaces in the same problem.

To elaborate on the idea of a conditional probability, suppose that a consumer research organization has studied the service under warranty provided by the 200 tire dealers in a large city, and that its findings are summarized in the following table:

	Good service under warranty	Poor service under warranty
Name-brand tire dealers	84	36
Off-brand tire dealers	38	42

Suppose, too, that a person randomly selects the name of one of these tire dealers. Since each of the dealers has the same chance, a probability of $\frac{1}{200}$, of being selected, the probability of choosing a name-brand dealer who provides good service under warranty is $\frac{84}{200} = 0.42$, and if we let N denote the selection of a name-brand tire dealer and G the selection of a tire dealer who provides good service under warranty, this probability can be written as

$$P(N \cap G) = 0.42$$

Also, it can be seen that the probability of choosing a name-brand tire dealer is

$$P(N) = \frac{84 + 36}{200} = 0.60$$

and that the probability of choosing a tire dealer who provides good service under warranty is

$$P(G) = \frac{84 + 38}{200} = 0.61$$

where all these probabilities were calculated by means of the formula $\frac{s}{n}$ for equally likely possibilities.

Suppose now that the person wants to limit the selection to name-brand dealers. This reduces the number of equally likely choices to $84 + 36 = 120$. Hence, the probability of choosing a dealer who provides good service under warranty given that he is a name-brand dealer is

$$P(G|N) = \frac{84}{120} = 0.70$$

Note that this conditional probability, 0.70, can also be written as

$$P(G \mid N) = \frac{84/200}{120/200} = \frac{P(N \cap G)}{P(N)}$$

which is the ratio of the probability of choosing a name-brand dealer who provides good service under warranty to the probability of choosing a name-brand dealer.

Generalizing from this example, let us now make the following definition of conditional probability, which applies to any two events A and B belonging to a given sample space S:

Definition of conditional probability

> *If $P(B)$ is not equal to zero, then the conditional probability of A relative to B, namely, the probability of A given B, is*
>
> $$P(A \mid B) = \frac{P(A \cap B)}{P(B)}$$

EXAMPLE With reference to the tire dealers of the above illustration, what is the probability that an off-brand tire dealer will give good service under warranty?

SOLUTION As can be seen from the table, $P(G \cap N') = \frac{38}{200}$ and $P(N') = \frac{38 + 42}{200}$, so that substitution into the formula yields

$$P(G \mid N') = \frac{\dfrac{38}{200}}{\dfrac{38 + 42}{200}} = \frac{38}{38 + 42} = 0.475$$

Of course, the fraction $\frac{38}{38 + 42}$ could have been obtained directly from the second row of the table.

Although we introduced the formula $P(A \mid B)$ by means of an example in which the possibilities were all equally likely, this is not a requirement for its use. The only restriction is that $P(B)$ must not equal zero.

EXAMPLE A paint manufacturer feels that the probability is 0.72 that the raw materials needed to fill an order will arrive on time, and the probability is 0.54 that the raw materials will arrive on time and the order will be filled on time. What is the probability that the order will be filled on time given that the raw materials arrived on time?

If R and F denote the events that the raw materials will arrive on time and that the order will be filled on time, then $P(R) = 0.72$ and $P(F \cap R) = 0.54$. Substituting into the formula for conditional probabilities, we get

$$P(F \mid R) = \frac{P(F \cap R)}{P(R)} = \frac{0.54}{0.72} = 0.75$$

To introduce another concept which is important in the study of probability, let us consider the following problem:

EXAMPLE The probabilities that a student will fail accounting, art history, or both are $P(A) = 0.20$, $P(H) = 0.15$, and $P(A \cap H) = 0.03$. What is the probability that he will fail accounting given that he will fail art history?

SOLUTION Substituting into the formula for conditional probabilities, we get

$$P(A \mid H) = \frac{P(A \cap H)}{P(H)} = \frac{0.03}{0.15} = 0.20$$

What is special, and interesting, about this result is that $P(A \mid H) = P(A) = 0.20$; that is, the probability of event A is the same regardless of whether event H has occurred (occurs, or will occur).

In general, if $P(A \mid B) = P(A)$, we say that event A is **independent** of event B and since it can be shown that event B is independent of event A whenever event A is independent of event B, we say simply that A and B are independent whenever one is independent of the other. Intuitively, we might say that two events are independent if the probability of the occurrence of either is in no way affected by the occurrence or nonoccurrence of the other (see also Exercise 6.55 on page 159). If two events A and B are not independent, we say that they are **dependent**.

So far we have used the formula $P(A \mid B) = \dfrac{P(A \cap B)}{P(B)}$ only to calculate conditional probabilities, but if we multiply both sides of the equation by $P(B)$, we get the following formula, called the **general multiplication rule**, which enables us to calculate the probability that two events will both occur:

General multiplication rule

$$P(A \cap B) = P(B) \cdot P(A \mid B)$$

In words, this formula states that the probability that two events will both occur is the product of the probability that one of the events will occur and the conditional probability that the other event will occur given that the first event has occurred (occurs, or will occur). It does not matter which event is referred to as A and which is referred to as B, so the formula above can also be written as

$$P(A \cap B) = P(A) \cdot P(B \mid A)$$

EXAMPLE If we randomly select two hair dryers, one after the other, from a carton containing 12 hair dryers, three of which are defective, what is the probability that both of them will be defective?

SOLUTION Since the selections are random, the probability that the first one we pick will be defective is $\frac{3}{12}$, and the probability that the second one we pick will be defective given that the first one was defective is $\frac{2}{11}$. Clearly, there are only 2 defective dryers among the 11 which remain after one defective dryer has been picked. Hence, the probability of getting 2 defective dryers is

$$\frac{3}{12} \cdot \frac{2}{11} = \frac{1}{22}$$

Using the same kind of argument we find that the probability of getting 2 good dryers is

$$\frac{9}{12} \cdot \frac{8}{11} = \frac{12}{22}$$

and it follows, by subtraction, that the probability of getting one good dryer and one defective dryer is $1 - \frac{1}{22} - \frac{12}{22} = \frac{9}{22}$.

When A and B are independent events, we can substitute $P(A)$ for $P(A \mid B)$ in the first of the two formulas for $P(A \cap B)$, or $P(B)$ for $P(B \mid A)$ in the second, and we obtain

Special multiplication rule

$$P(A \cap B) = P(A) \cdot P(B)$$

In words, the probability that two independent events will both occur is simply the product of their probabilities. This rule is sometimes used as the definition of independence; in any case, it may be used to determine whether two given events are independent.

EXAMPLE   What is the probability of getting two heads in two flips of a balanced coin?

SOLUTION   Since the probability of heads is $\frac{1}{2}$ for each flip and the two flips are independent, the probability is $\frac{1}{2} \cdot \frac{1}{2} = \frac{1}{4}$.

EXAMPLE   If $P(C) = 0.65$, $P(D) = 0.40$, and $P(C \cap D) = 0.24$, are the events C and D independent?

SOLUTION   Since $P(C) \cdot P(D) = (0.65)(0.40) = 0.26 \neq 0.24$, the two events are not independent.

EXAMPLE   What is the probability of getting two consecutive aces in two cards drawn at random from an ordinary deck of 52 playing cards if
(a) the first card is replaced before the second card is drawn;
(b) the first card is not replaced before the second card is drawn?

SOLUTION   (a) Since there are four aces among the 52 cards, we get

$$\frac{4}{52} \cdot \frac{4}{52} = \frac{1}{169}$$

(b) since there are only three aces among the 51 cards which remain after one ace has been removed from the deck, we get

$$\frac{4}{52} \cdot \frac{3}{51} = \frac{1}{221}$$

The distinction between the two parts of the preceding example is important in statistics, where we sometimes **sample with replacement** and sometimes **sample without replacement**.

The special multiplication rule can easily be generalized so that it applies to the occurrence of three or more independent events—again, we simply multiply all the individual probabilities together.

EXAMPLE   What is the probability of getting three heads in three flips of a balanced coin?

SOLUTION   The flips of the coin are independent and we get

$$\frac{1}{2} \cdot \frac{1}{2} \cdot \frac{1}{2} = \frac{1}{8}$$

EXAMPLE   What is the probability of first rolling four 3's and then another number in five rolls of a balanced die?

SOLUTION   Multiplying the five probabilities, we get

$$\frac{1}{6} \cdot \frac{1}{6} \cdot \frac{1}{6} \cdot \frac{1}{6} \cdot \frac{5}{6} = \frac{5}{7,776}$$

For three or more dependent events the multiplication rule becomes somewhat more complicated, as is illustrated in Exercise 6.70 on page 161.

EXERCISES

6.51 If E is the event that an applicant for a home mortgage is employed, M is the event that he is married, and A is the event that the application is approved, state in words what probabilities are expressed by
(a) $P(A|E)$;
(b) $P(M|A)$;
(c) $P(A|M')$;
(d) $P(A'|E')$;
(e) $P(A|E \cap M)$;
(f) $P(A \cap E|M')$.

6.52 If H is the event that a job has a high starting salary and F is the event that it has a good future, express in symbolic form the probabilities that
(a) a job with a high starting salary will also have a good future;
(b) a job with a good future will also have a high starting salary;
(c) a job which does not have a high starting salary will have a good future;
(d) a job which does not have a good future will not have a high starting salary either.

6.53 If A and B are independent events and $P(A) = 0.25$ and $P(B) = 0.40$, find
(a) $P(A|B)$;
(b) $P(A \cap B)$;
(c) $P(A \cup B)$;
(d) $P(A' \cap B')$.

6.54 Given $P(A) = 0.4$, $P(B|A) = 0.3$, and $P(B'|A') = 0.2$, find
(a) $P(A')$;
(b) $P(B|A')$;
(c) $P(B)$;
(d) $P(A \cap B)$;
(e) $P(A|B)$.

6.55 Given $P(A) = 0.30$, $P(B) = 0.50$, and $P(A \cap B) = 0.15$, verify that
(a) $P(A|B) = P(A)$;
(b) $P(A|B') = P(A)$;
(c) $P(B|A) = P(B)$;
(d) $P(B|A') = P(B)$.
(It is interesting to note that if any one of these four relations is satisfied, the other three must also be satisfied and the events A and B are independent.)

6.56 As part of a promotional scheme in California and Nevada, a company distributing frozen foods will award a grand prize of $100,000 to some person sending in his name on an entry blank, with the option of including a label from one of the company's products. A breakdown of the 150,000 entries received is shown in the following table:

	With label	Without label
California	80,000	28,000
Nevada	20,000	22,000

If the winner of the grand prize is chosen by lot, C represents the event that it will be won by an entry from California, and L represents the event that it will be won by an entry which included a label, find each of the following probabilities:

(a) $P(C)$;

(b) $P(L')$;

(c) $P(C' \cap L')$;

(d) $P(C \cap L')$;

(e) $P(C \cup L)$;

(f) $P(C'|L')$;

(g) $P(C|L)$;

(h) $P(L|C')$;

(i) $P(L'|C)$.

6.57 With reference to the preceding exercise, suppose that the drawing is rigged so that by including a label each entry's probability of winning the grand prize is doubled. Recalculate the probabilities (a) through (i) of that exercise.

6.58 The probability that a bus from Buffalo to Rochester will leave on time is 0.70, and the probability that it will leave on time and also arrive on time is 0.56. What is the probability that if such a bus leaves on time it will also arrive on time?

6.59 The probability that a certain concert will be well advertised is 0.82, and the probability that it will be well advertised and also well attended is 0.75. What is the probability that if the concert is well advertised it will be well attended?

6.60 If the probabilities are 0.28, 0.15, and 0.09 that a person traveling through Arizona will visit the Grand Canyon, the Petrified Forest, or both, find the probability that

(a) a person traveling through Arizona who visits the Grand Canyon will also visit the Petrified Forest;

(b) a person traveling through Arizona who visits the Petrified Forest will also visit the Grand Canyon.

6.61 With reference to the example on page 156, show that event H is also independent of event A, namely, that $P(H|A) = P(H) = 0.15$.

6.62 If $P(A) = 0.60$, $P(B) = 0.45$, and $P(A \cap B) = 0.21$, are events A and B independent?

6.63 If the probability that the manager of bank branch A can decipher a garbled teletype message is $\frac{1}{4}$ and the probability that the manager of bank branch B can decipher it independently of the first manager is $\frac{2}{3}$, what is the probability that the message will be deciphered if the two managers do, in fact, work on it independently?

6.64 A bank manager has six male tellers and nine female tellers. If she selects two of them at random for special training, what are the probabilities that

(a) both will be males;   (b) both will be females?

6.65 Among the 80 pieces of luggage loaded on a plane in San Francisco, 55 are destined for Seattle and 25 for Vancouver. If two of the pieces of luggage are sent to Portland by mistake and the "selection" is random, what are the probabilities that

(a) both should have gone to Seattle;

(b) both should have gone to Vancouver;

(c) one should have gone to Seattle and one to Vancouver?

6.66 What is the probability of getting two 6's in two rolls of a balanced die?

6.67 What is the probability of getting six heads in six tosses of a balanced coin?

6.68 What is the probability of getting four spades in four random draws from an ordinary deck of 52 playing cards, if each card is replaced before the next card is drawn?

6.69 One critical operation in assembling a delicate electronic device requires that a skilled operator fit one part to another precisely. If the operator succeeds in matching the parts on his first attempt, he moves on to the next assembly; otherwise, he repeats his (independent) attempts until he gets a match. What is the probability that an operator with a constant match probability of $\frac{2}{3}$ will succeed in matching the parts in a given assembly (a) on the fourth attempt and (b) within four attempts? (In this example the set of all possible outcomes is not finite, and when this is the case we must modify the third postulate of probability so that it applies to the union of any number of mutually exclusive events; nevertheless, it is possible to solve this problem with the methods discussed in this chapter.)

6.70 The problem of determining the probability that any number of events will occur becomes more complicated when the events are not independent. For three events A, B, and C, for example, the probability that they will all occur is obtained by multiplying the probability of A by the probability of A given B, and then multiplying the result by the probability of C given $A \cap B$. For instance, the probability of drawing (without replacement) three aces in a row from an ordinary deck of 52 playing cards is

$$\frac{4}{52} \cdot \frac{3}{51} \cdot \frac{2}{50} = \frac{1}{5,525}$$

Clearly, there are only three aces among the 51 cards which remain after the first ace has been drawn, and only two aces among the 50 cards which remain after the first two aces have been drawn.
(a) A carton contains 12 hair dryers, 3 of which are defective and the rest are good. What is the probability that if 3 of the dryers are randomly selected from the carton without replacement, they will all be defective?
(b) If a person randomly picks 3 of 15 coins without replacement, and 6 of the coins are counterfeits, what is the probability that the coins picked are all counterfeits?
(c) If 5 of a company's 12 delivery trucks do not meet emission standards and 4 of the 12 are randomly picked for inspection, what is the probability that none of them meets emission standards?
(d) If four patch cords are randomly picked from a tray containing six blue and five yellow cords, what is the probability that the cords selected are all of the same color?
(e) The only supermarket in a small town offers two brands of frozen orange juice, brand A and brand B. Among its customers who buy brand A one week, 80 percent will buy brand A and 20 percent will buy brand B the next week, and among its customers who buy brand B one week, 40 percent will buy brand B and 60 percent will buy brand A the next week. To sim-

plify matters, it will be assumed that each customer buys frozen orange juice once a week.

(1) What is the probability that a customer who buys brand *A* one week will buy brand *B* the next week, brand *B* the week after that, and brand *A* the week after that?

(2) What is the probability that a customer who buys brand *B* one week will buy brand *B* the next two weeks, and brand *A* the week after that?

(3) What is the probability that a customer who buys brand *A* in the first week of a month will also buy brand *A* in the third week of that month? (*Hint:* Add the probabilities associated with the two mutually exclusive possibilities corresponding to his buying brand *A* or brand *B* in the second week.)

(f) A department store which bills its charge-account customers once a month has found that if a customer pays promptly one month, the probability is 0.90 that he will also pay promptly the next month; however, if a customer does not pay promptly one month, the probability that he will pay promptly the next month is only 0.50.

(1) What is the probability that a customer who pays promptly one month will also pay promptly the next three months?

(2) What is the probability that a customer who does not pay promptly one month will also not pay promptly the next three months and then make a prompt payment the month after that?

(3) What is the probability that a customer who pays promptly one month will also pay promptly the third month after that?

Parts (e) and (f) of this exercise deal with sequences of experiments, called **Markov chains**, in which the outcome of each experiment depends only on what happened in the preceding experiment.

6.5

Check List of Key Terms

Addition rules, 146
Betting odds, 150
Complement, 132
Conditional probability, 153, 155
Consistency criterion, 152
Dependent events, 156
Empty set, 131
Event, 131
Experiment, 130
General addition rule, 146
General multiplication rule, 156
Generalized addition rule, 142
Independent events, 156
Intersection, 132

Markov chain, 162
Multiplication rules, 156, 157
Mutually exclusive events, 132
Odds, 149
Outcome, 130
Postulates of probability, 139, 140
Sample space, 130
Sampling with replacement, 158
Sampling without replacement, 158
Special addition rule, 146
Special multiplication rule, 157
Subjective probability, 119
Union, 132
Venn diagram, 133

6.6

Review Exercises

6.71 Convert each of the following probabilities to odds:
  (a) If a person has eight $1 bills, five $5 bills, and one $20 bill in her purse and randomly pulls out three of the bills, the probability that they will not all be $1 bills is $\frac{11}{31}$.
  (b) If a management teacher randomly selects two of eight students to discuss a proposed corporate strategy, the probability that any one particular student will be chosen is $\frac{2}{8}$.

6.72 Shoppers in a market are asked whether they prefer brand U coffee to brand V, whether they prefer brand V to brand U, or whether they have no preference, and these three alternatives are assigned the codes 1, 2, and 3.
  (a) Use two coordinates to represent, in order, the responses of two shoppers, and draw a diagram (similar to that of Figure 6.1) which shows the nine points of the corresponding sample space.
  (b) Describe in words the event which is represented by each of the following sets of points of the sample space of part (a): the event D, which consists of the points $(1, 1), (1, 2)$, and $(1, 3)$; the event E, which consists of the points $(1, 2), (2, 2)$, and $(3, 2)$; the event F, which consists of the points $(1, 1), (1, 2), (2, 1)$, and $(2, 2)$; and the event G, which consists of the points $(1, 3)$ and $(3, 1)$.
  (c) With reference to part (b), describe in words the events which are denoted by F', $D \cup E$, and $F \cap E$. Also list the points which comprise each of these three events.

6.73 With reference to part (b) of the preceding exercise, determine whether events D and F are mutually exclusive, and also whether events E and G are mutually exclusive.

6.74 Of the many dwellings in a large district of a major city 70 percent are single- and the rest multiple-family dwellings, 60 percent were built prior to 1939 and the rest since then, and of the pre-1939 dwellings three-fourths are single and the rest multiple units. If one dwelling is selected at random from all the dwellings in this district, what are the probabilities that this dwelling is
  (a) a pre-1939 single one;
  (b) neither a multiple nor a pre-1939 one;
  (c) either a multiple or a pre-1939 one;
  (d) a multiple one given that it is not pre-1939.

6.75 If the probability is 0.18 that any one woman will name blue as her favorite color, what is the probability that four women, selected at random, will all name blue as their favorite color?

6.76 Ms. Jones is looking for a job. If H is the event that she will find a job near her home and G is the event that she will find a job with a good pension plan, state in words what probabilities are expressed by
  (a) $P(G|H)$;
  (b) $P(H'|G)$;
  (c) $P(H|G')$;
  (d) $P(G'|H')$.

6.77 If $P(M) = 0.55$, $P(N) = 0.18$, and $P(M \cap N) = 0.099$, are the events M and N dependent or independent?

6.78 Convert each of the following odds to probabilities:

(a) If three eggs are randomly chosen from a carton of 12 eggs of which three are cracked, the odds are 34 to 21 that at least one of them will be cracked.

(b) The odds are 27 to 5 against getting four heads and one tail in five flips of a balanced coin.

6.79 Suppose that the numbers 1, 2, 3, 4, 5, and 6 are used to denote that a reviewer feels that a new book is terrible, poor, fair, good, very good, or excellent. If $K = \{1, 2, 3, 4\}$ and $L = \{2, 3, 4, 5\}$, list the elements of the sample space comprising each of the following events, and also express the events in words:

(a) K'; (c) $K \cap L$;

(b) $K \cup L$; (d) $K \cap L'$.

6.80 If someone feels that 17 to 8 are fair odds that a construction job will be finished on time, what subjective probability does he assign to this event?

6.81 The probabilities are 0.16, 0.24, and 0.03 that a police department will buy Uniroyal, Goodyear, or Michelin tires for its patrol cars. What is the probability that it will buy one of these brands of tires?

6.82 The probabilities that a television station will receive 0, 1, 2, 3, ..., 8, or at least 9 complaints after showing a controversial program are 0.01, 0.03, 0.07, 0.15, 0.19, 0.18, 0.14, 0.12, 0.09, and 0.02. What are the probabilities that after showing such a program the station will receive

(a) at most 4 complaints;

(b) at least 6 complaints;

(c) from 5 to 8 complaints?

6.83 The probability that a play will get a good review in the *Republic* is 0.32, and the probability that it will get a good review in the *Republic* as well as the *Progress* is 0.12. What is the probability that it will get a good review in the *Progress* given that it will get a good review in the *Republic*?

6.84 If the dots of Figure 6.8 on page 144 all represent equally likely outcomes, what is the probability of event A?

6.85 There are 60 applicants for a job in the news department of a television station. Some are college graduates and some are not, some have at least three years' experience and some have not, with the exact breakdown being

	College graduates	*Not college graduates*
At least three years' experience	12	6
Less than three years' experience	24	18

If the order in which the applicants are interviewed by the station manager is random, G is the event that the first applicant interviewed is a college graduate, and T is the event that the first applicant interviewed has at least three years' experience, determine each of the following probabilities directly from the entries and the row and column totals of the table:

(a) $P(G)$;
(b) $P(T')$;
(c) $P(G \cap T)$;

(d) $P(G' \cap T')$;
(e) $P(T \mid G)$;
(f) $P(G' \mid T')$.

6.86 With reference to the preceding exercise, verify that

(a) $P(T \mid G) = \dfrac{P(G \cap T)}{P(G)}$;

(b) $P(G' \mid T') = \dfrac{P(G' \cap T')}{P(T')}$.

6.87 Given $P(U) = 0.35$, $P(V) = 0.61$, and $P(U \cap V) = 0$, find

(a) $P(U')$;
(b) $P(V')$;

(c) $P(U \cup V)$;
(d) $P(U' \cap V')$.

6.88 A game of chance is played as follows: A player draws a bead at random from an urn containing 6 red and 4 blue beads. If the bead drawn is red, the game ends right there. However, if the bead drawn is blue, the player draws one more bead without replacing the first bead, after which the game positively ends. If the player wins $5 for each blue bead he draws and loses $4 for each red bead he draws, what is his expectation?

6.89 If A is the event that a university's football team is rated among the top twenty by AP and U is the event that it is rated among the top twenty by UPI, what events are represented by the four regions of the Venn diagram of Figure 6.11?

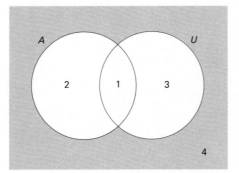

6.11

Venn diagram for Exercise 6.89.

6.90 A movie producer feels that the odds are 7 to 1 that his new movie will not be rated G, 13 to 3 that it will not be rated PG, and 11 to 5 that it will not get either of these two ratings. Are the corresponding probabilities consistent?

6.91 The probability that John will get an M.B.A. degree is 0.40, and the probability that with an M.B.A. degree he will get a well-paying job is 0.85. What is the probability that he will get an M.B.A. degree and a well-paying job?

6.92 Explain why there must be a mistake in each of the following statements:
   (a) The probability that a certain missile will explode on lift-off is -0.002.
   (b) The probability that a student will get an A in a course is 0.12, but he is ten times as likely to get a C.
   (c) The probability that everyone will attend a certain conference is 0.69, and the probability that this will not be the case is 0.21.

6.93 If a student answers the 12 questions on a true–false test by flipping a balanced coin, what is the probability that she will answer all 12 questions correctly?

6.94 If the probabilities are 0.20 that a shipment will be late, 0.16 that it will arrive in poor condition, and 0.11 that it will arrive late and in poor condition, what is the probability that the shipment will be late or arrive in poor condition?

6.95 In Figure 6.12, B is the event that a salesperson will visit customers in Blythe, P is the event that she will visit customers in Palm Springs, and E is the event that she will visit customers in El Centro. Explain in words what events are represented by the following regions or combinations of regions:
  (a) region 7;
  (b) regions 1 and 2 together;
  (c) regions 2 and 7 together;
  (d) regions 4, 6, and 7 together;
  (e) regions 7 and 8 together.

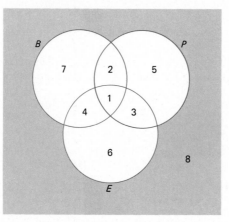

6.12

Venn diagram for Exercise 6.95.

6.96 From an urn containing 7 red beads and 3 white beads, player A is to draw 2 beads at random without replacement. Show formally that if A wins $3 for each red bead he draws, he should pay $7 for each white bead he draws in order to make the game fair.

The approach to statistics as the art, or science, of decision making in the face of uncertainty is called **decision theory**. The initial impact of this very general approach to statistics, which dates back to the middle of this century, was largely on the theoretical level, but in the last few years it has made itself felt more and more in practical situations, especially in business applications.

Since the study of statistical decision theory is quite complicated mathematically, we shall introduce here only some of the most basic ideas. However, even these are extremely important, because the decision-theory approach has the positive advantage of forcing one to formulate problems clearly, to anticipate the various consequences of one's actions, to retain the relevant and eliminate the irrelevant, to place cash values on the consequences of one's actions, and so on. No matter how little or how far it is pursued, a systematic approach like this is bound to help.

After a brief introduction to basic concepts in Sections 7.1 and 7.2, Sections 7.3 through 7.5 are devoted to problems where the decision maker has **no information about relevant events over which he has no control.** In Sections 7.6 and 7.7 we learn what a decision maker might do when such events are controlled by a **competitor,** and in Sections 7.8 through 7.10 we assume that the decision maker has **some knowledge about the probabilities** that these events will occur. Finally, in Section 7.11 we introduce the concept of **utility,** which often plays an important role in decision making.

7

Decision Analysis

Payoff Tables and Decision Trees

The first step in the mathematical analysis of any problem is to translate the problem into the language of mathematics, that is, express the given information in terms of equations, tables, graphs, charts, or other kinds of mathematical objects. In the analysis of decision problems, the first step usually consists of summarizing the given information in the form of a **payoff table** or in the form of a **decision tree**. To illustrate the former, suppose that a manufacturer of office equipment must decide whether to expand his plant capacity now or wait another year. His advisors tell him that if he expands now and economic conditions remain good, there will be a profit of $369,000; if he expands now and there is a recession, there will be a loss (negative profit) of $90,000; if he waits another year to expand and economic conditions remain good, there will be a profit of $180,000; and if he waits another year and there is a recession, there will be a small profit of $18,000. Schematically, this information can be presented in the following table:

	Expand now	*Delay expansion*
Economic conditions remain good	$369,000	$180,000
There is a recession	−$90,000	$18,000

The columns "Expand now" and "Delay expansion" represent the two actions the manufacturer can take, the two rows represent the economic conditions over which he has no control, and for each combination of actions and economic conditions the table shows the corresponding profit.

Borrowing from the language of **game theory**, we refer to the entries in the table, the various profits, as the **payoffs**, and to the table itself as a **payoff table**, or a **payoff matrix**. The advantage of the table should be obvious—it gives a much easier-to-grasp picture of the whole situation than the original lengthy verbal formulation. As in many other problems of decision making, the payoffs in our example are based on accounting data, but there are also situations where they follow directly from the formulation of the problem (see, for example, Exercise 7.3 on page 170.)

In general, if a decision maker has the choice of k actions, $A_1, A_2, \ldots,$ and A_k, and their consequences depend on r events (possibilities, alternatives, or conditions), $E_1, E_2, \ldots,$ and E_r, over which he has no direct control, the corresponding payoffs are denoted $p_{11}, p_{12}, \ldots,$ and p_{rk}, with the first subscript being like that of event E and the second like that of action A. Thus, for $r = 3$ and $k = 4$ the payoff table would be

	A_1	A_2	A_3	A_4
E_1	p_{11}	p_{12}	p_{13}	p_{14}
E_2	p_{21}	p_{22}	p_{23}	p_{24}
E_3	p_{31}	p_{32}	p_{33}	p_{34}

The events E, over which the decision maker has no direct control, may reflect facts unavailable at the time the decision must be made; they may reflect the decision maker's ignorance; they may reflect unknown decisions made by a competitor; or they may reflect situations brought about by fate or chance. Because of the latter, the E's are often called **states of nature**.

To simplify our example, we based the payoffs on the opinions of advisors; in actual practice, if the values of the payoffs are not given, they must be determined from pertinent information (see, for example, Exercises 7.3, 7.4, and 7.6 on pages 170 and 171). Also, the payoffs are dollar profits in our example, but this need not be the case. They may also be losses, production costs, sales revenues, quantities produced or consumed, the IQ's of job applicants, mileages, and so forth.

In the beginning of this section we said that there are essentially two ways of expressing decision problems mathematically, using either payoff tables or decision trees. A decision tree is simply a tree diagram whose branches represent the A's and the E's. Furthermore, to make it clear whether a branch represents an A or an E, branches representing A's emanate from small squares and branches representing E's emanate from small circles (see Figure 7.1). Also,

7.1

Decision tree.

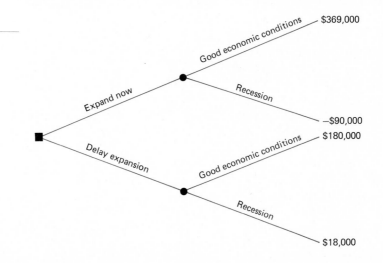

the payoff associated with a given path along the branches of a decision tree is usually shown on the right, at the end of the path.

For the problem in which the manufacturer must decide whether to expand his plant capacity now or later, we drew the decision tree in Figure 7.1. However, there is really no advantage to constructing a decision tree in a simple problem like this; the payoff table is more compact and gives an easier-to-grasp picture of the whole situation. On the other hand, decision trees can be of great help in more complicated kinds of decision problems; for instance, in problems where it is hard even to enumerate all the decision maker's courses of action.

EXERCISES

7.1 In handling a liability suit against one of her clients, a lawyer must decide whether to charge a straight fee of $4,000 or a contingent fee of $20,000 which she will get only if she wins the case.
(a) Construct a payoff table in which the payoffs are the amounts the lawyer receives.
(b) Draw a decision tree.

7.2 The management of an oil company must decide whether to continue drilling at a certain location. If it continues to drill and there is oil, this will be worth $2,000,000 to the company; if it continues to drill and there is no oil, this will entail a loss of $1,200,000; if it stops drilling and there is oil (for a competitor to use), this will entail a loss of $800,000; and if it stops drilling and there is no oil, this will be worth $200,000 to the company (because funds allocated to the project will remain unspent).
(a) Construct a payoff table.
(b) Draw a decision tree.

7.3 A truck driver has to deliver a load of lumber to one of three construction sites which are 8, 12, and 15 miles from the lumberyard. The distance between the first two sites is 6 miles, that between the first and third is 11 miles, and that between the second and third is 9 miles. The telephone at the lumberyard is out of order, but if the driver arrives at a wrong site, he can find out where the lumber should go by calling one of the other sites.
(a) Construct a payoff table showing the total mileages he may have to drive to deliver the lumber, depending on where he decides to go first and where the lumber should go.
(b) Draw a decision tree.

7.4 A retailer has shelf space for four highly perishable items which are destroyed at the end of the day if they are not sold. The unit cost of the item is $2, the selling price is $5, so the profit is $3 per item sold. Each day the retailer must decide whether to stock 1, 2, or 3 of the items, and there may be a demand for 0, 1, 2, 3, or 4 or more of the items.
(a) Construct a payoff table in which the payoffs are the retailer's profits.
(b) Draw a decision tree.

7.5 A person who is very gasoline-economy-minded plans to buy one of three cars, A, B, or C. Car A yields 21 miles per gallon (mpg) in city driving, 28 mpg in freeway driving, and 24 mpg driving on rural roads; car *B* yields 17 mpg in city driving, 25 mpg in freeway driving, and 19 mpg driving on rural roads; and car *C* yields 23 mpg in city driving, 25 mpg in freeway driving, and 23 mpg driving on rural roads.

(a) Using these mpg's as payoffs, construct a payoff table for this decision problem.

(b) Draw a decision tree.

7.6 A manufacturer produces an item consisting of two components, which must both work if the item is to function properly. The cost to the manufacturer of having one of the items returned for repairs is $12, the cost of inspecting one of the components is $10, and the cost of repairing a faulty component is $25. He can ship each item without inspection with the guarantee that each component will be put in perfect working condition at his factory in case it does not work; he can inspect both components and repair them if necessary; or he can randomly select one of the two components for inspection, repair it if necessary, and ship the item with the same guarantee as in the first case. Using the manufacturer's expected cost of inspecting, returning, and repairing an item as the payoffs, construct a payoff table for his choice among the inspection procedures.

7.2

Decision Making Under Uncertainty

Although we shall be concerned here only with decision making under uncertainty, there also exist problems of **decision making under certainty**. Suppose, for instance, that a student wants to sell her car and she has two offers. The first offer consists of a payment of $3,550 now, and the second consists of three payments, $1,200 now, $1,200 six months from now, and $1,200 one year from now, and we assume that money earns interest at 6 percent compounded quarterly. Even though the three payments total $3,600, their present value (calculated with the use of a compound discount table) is only $3,495.42, which is less than $3,550. So, it would be to the student's advantage to accept the first offer, and that is all there is to this problem of decision making under certainty. All the pertinent facts are known in a problem like this, and a decision is reached after appropriate calculations.

In all the decision problems we shall study in this chapter, both the actions, the *A*'s, open to the decision maker and the events, the *E*'s, which are beyond the decision maker's control will be known and stated, and the payoffs, the *p*'s, will either be given or can be calculated (as in the exercises preceding this section). So far as our work here is concerned, the uncertainties which the decision maker faces arise in connection with the *E*'s; that is, we shall study problems in which we do not know how, by whom, under what circumstances, or with what probabilities the occurrences of the *E*'s are controlled.

To cover these various possibilities, we shall divide our study of decision making into the following three parts:

1. We shall study problems in which the decision maker has no knowledge whatever about the E's, other than what they are.

2. We shall study problems in which the decision maker knows that the choice of the E's is controlled by a competitor whose interests are diametrically opposed to those of the decision maker.

3. We shall study problems in which the decision maker has some idea about the probabilities of the occurrences of the different E's.

The next three sections will be devoted to the first kind of problem; the other two kinds will be discussed after that.

7.3

Dominance and Admissibility

When a decision maker has no information about the E's except that he knows what they are, it may be possible, nevertheless, to make a choice among the A's, or at least eliminate some of the possibilities. Suppose, for instance, that we want to invest in one of four mutual funds, basing our choice on the ratings (from A high to F low) given to the performance of these funds in up markets and in down markets by *Forbes* magazine. The ratings shown in the following table are taken from a recent issue of that magazine:

	Fund I	Fund II	Fund III	Fund IV
Performance in up markets	A	B	C	D
Performance in down markets	F	F	B	C

Comparing the ratings of fund IV with those of the other three, we find that compared to funds I and II, fund IV is rated lower for up markets and higher for down markets (D is lower than A or B and C is higher than F), but compared to fund III, fund IV is rated lower for both markets (D is lower than C and C is lower than B). Thus, we say that the ratings of fund III **dominate** those of fund IV, and that the selection of fund IV is **not admissible**; that is, fund IV should not be selected so long as fund III is available for investment. Similarly, comparing the ratings of fund II with those of funds I and III, we find that compared to fund III, fund II is rated higher for up markets and lower for down markets (B is higher than C and F is lower than B), but compared to fund I, fund II is never rated higher but it is rated lower for up markets (B is lower than A and one F is as low as the other). Thus, the ratings of fund I dominate those of fund II, and the selection of fund II is not admissible. As a result of this analysis, the decision problem has been simplified quite a bit—we have eliminated two funds from consideration and, instead of having to choose one of four mutual funds, we now have to choose only one of two, fund I or fund III. Evidently, this is as far as we can go in this way, since fund I is rated higher for up markets and fund III is rated higher for down markets (A is higher than C and B is higher than F).

In general, **one action of a decision maker is said to dominate a second action, and the second is said to be nonadmissible, if none of the payoffs of the second action is preferable to the corresponding payoff of the first, and at least one payoff of the first action is preferable to the corresponding payoff of the second.** If the columns representing the payoffs of two different actions are identical, there is no way to make a logical choice between the two actions and they may, in fact, be regarded as one so far as the decision analysis is concerned.

For instance, if after eliminating the nonadmissible funds II and IV in the preceding example we consider another fund, fund V rated C for up markets and B for down markets like fund III, we group funds III and V together as one action and arrive at the following payoff table:

	Fund I	Fund III or Fund V
Performance in up markets	A	C
Performance in down markets	F	B

The process of eliminating nonadmissible actions or discovering actions that are equivalent can be of great help, and it is usually the first step in the analysis of a decision-making problem.

7.4

The Maximin and Maximax Criteria

The basic problem of decision making is to find decision-making procedures, or **decision rules**, which meet certain criteria of desirability. If we knew or could calculate the mathematical expectations associated with the various actions open to us, it would seem reasonable to follow the rule of choosing whichever action is best for us, but for the time being we are assuming that we have no knowledge of the probabilities of the E's (the "states of nature" over which the decision maker has no control), and, hence, cannot calculate expectations of any kind. However, there are criteria which apply even in such cases, but they are based on the decision maker's attitudes rather than on straight economic considerations.

To illustrate one of these, let us return to the plant-expansion example, where the payoff table is

	Expand now	Delay expansion
Economic conditions remain good	$369,000	$180,000
There is a recession	−$90,000	$18,000

Suppose now that the office equipment manufacturer is a confirmed pessimist, who always expects the worst to happen. Looking at the situation through dark-colored glasses, he finds that if he expands his plant capacity now, the worst that can happen is a loss of $90,000, and if he decides to delay the expansion, the worst that can happen is a profit of $18,000. A profit of $18,000 is definitely preferable to a loss of $90,000, so the pessimistic manufacturer would decide to delay expanding his plant capacity.

The criterion we have introduced in this example is called the **maximin criterion**, since it suggests that one maximize the minimum payoffs corresponding to the various actions. To apply this criterion, we look for the worst that can happen, namely, the smallest value of each column of the payoff table. Then we choose the column (action) which maximizes this minimum payoff.

EXAMPLE Referring to the example concerning the four mutual funds, which one will we choose if we base our choice on the maximin criterion?

SOLUTION The payoff table with the nonadmissible choices deleted is

	Fund I	Fund III
Performance in up markets	A	C
Performance in down markets	F	B

and we find that the lowest ratings are F for fund I and C for fund III. Since C exceeds F, it follows that the maximin criterion leads to the choice of fund III. By making this choice, the pessimistic decision maker protects himself against a fund rated lower than C in either an up or down market.

The maximin criterion applies only when the payoffs are quantities such as profits or benefits, which we want to make large. If the payoffs are quantities such as losses or costs, which we want to make small, we use the minimax criterion instead. To apply this criterion, we again look for the worst that can happen, in this case the largest value of each column of the payoff table. Then we choose the column (action) which minimizes this maximum payoff. For all practical purposes, though, the maximin and minimax criteria are equivalent, since minimizing maximum losses is the same as maximizing minimum profits.

Besides the maximin and minimax criteria, there are several other criteria which are based on the attitudes of the decision maker. Referring again to the office equipment manufacturer, let us suppose, for instance, that he is a confirmed optimist, who always expects the best to happen. Looking at his payoff table through rose-colored glasses, he finds that if he expands his plant capacity now, he might make a profit of $369,000, but if he decides to delay the expansion, the profit cannot be more than $180,000. A profit of $369,000 is preferable to a profit of $180,000, so the optimistic manufacturer would decide to expand his plant capacity right away.

The criterion illustrated in the preceding example is called the maximax criterion, since it requires that one maximize the maximum payoffs corresponding to the different actions. To apply this criterion, we look for the best that can happen, namely, the largest value of each column of the payoff table. Then we choose the column (action) which maximizes this maximum payoff.

EXAMPLE Referring again to the mutual funds example, which fund will we choose if we base our decision on the maximax criterion?

SOLUTION The payoff table with the nonadmissible choices deleted is

	Fund I	Fund III
Performance in up markets	A	C
Performance in down markets	F	B

and we find that the highest ratings are A for fund I and B for fund III. Since A exceeds B, it follows that the maximax criterion leads to the choice of fund I.

Finally, there is also the minimin criterion, which an optimistic decision maker would use instead of the maximax criterion when the payoffs are losses, costs, or other quantities which he wants to make small.

7.5

Opportunity Losses

Decisions can also be based on a person's fear of not taking the best possible action open to him and missing, or losing, a good opportunity. For instance, if the office equipment manufacturer we have been discussing is this kind of person, he might argue that if he decided to expand his plant capacity now and there is a recession, he would have been better off by $18,000 − (−$90,000) = $108,000 (the $18,000 profit he could have made plus the $90,000 he lost) if he had decided to delay. On the other hand, if he decided to delay the expansion of his plant capacity and economic conditions remain good, he would have been better off by $369,000 − $180,000 = $189,000 (the $369,000 profit he could have made minus the $180,000 profit he made) if he had decided to expand right away. These quantities, usually called opportunity losses or regrets, are shown in the following opportunity-loss table:

	Expand now	*Delay expansion*
Economic conditions remain good	0	$189,000
There is a recession	$108,000	0

To explain the two zeros, note that when the manufacturer decides to expand his plant capacity right away and economic conditions remain good, he has made the best possible decision and there is no loss of opportunity. This is also true if he decides to delay expanding his plant capacity and there is a recession.

In general, opportunity losses are calculated for each row of an opportunity-loss table as follows. If the payoffs are profits or other quantities which we want to make large, we subtract each value in a row from the largest value in the row; if the payoffs are losses or other quantities which we want to make small, we subtract the smallest value in a row from each value in the row.

EXAMPLE Mr. H is giving Mrs. H the choice of one of two kinds of sofas for her birthday. Three furniture stores carry the sofas and he is preparing to take her to one of these stores and buy the sofa right there, on the spot. Since she has not yet made up her mind, though, which sofa she wants, he does not know which store they should go to. Construct the opportunity-loss table which corresponds to the following payoff table, where the payoffs are the prices of the two sofas at the three stores; A_1, A_2, and A_3 represent Mr. H's decisions to go to the first, second, or third store; and the E's represent his wife's ultimate choice of the first kind or second kind of sofa:

	A_1	A_2	A_3
E_1	$285	$253	$250
E_2	$345	$352	$360

SOLUTION — Since the first kind of sofa is cheapest at the third store, the opportunity losses in the first row are $285 − $250 = $35, $253 − $250 = $3, and $250 − $250 = 0, and since the second kind of sofa is cheapest at the first store, the opportunity losses in the second row are $345 − $345 = 0, $352 − $345 = $7, and $360 − $345 = $15. All this is summarized in the following opportunity-loss table:

	A_1	A_2	A_3
E_1	$35	$3	0
E_2	0	$7	$15

Having learned how opportunity losses are calculated, let us now see how they may be used in making decisions. Since opportunity losses are quantities we want to make small, it seems that we could select either the minimax or minimin criterion and apply it to opportunity-loss tables. However, since opportunity losses are never negative and there is a zero in each row, the minimin criterion will generally not lead to a unique course of action. This leaves the minimax criterion for making decisions based on opportunity-loss tables, and according to this criterion we choose whichever action minimizes the maximum opportunity loss.

EXAMPLE — Continuing the preceding example, determine where Mr. H should go so as to minimize his maximum loss of opportunity.

SOLUTION — We see from the opportunity-loss table that the maximum opportunity loss is $35 for A_1, $7 for A_2, and $15 for A_3. Since $7 is the smallest of the three, the minimax criterion leads Mr. H to take his wife shopping for the sofa at the second store.

Determine what the office equipment manufacturer should do so as to minimize the maximum loss of opportunity.

From the opportunity-loss table on page 176, we see that the maximum opportunity loss is $108,000 if he expands his plant capacity now and $189,000 if he delays the expansion. Since the first of these figures is smaller than the second, the minimax criterion leads the manufacturer to expand his plant capacity now.

EXERCISES

7.7 The following payoff table shows two surveys' data on the numbers of one-family homes within a 2-mile radius of three different commercial building sites:

	Building site 1	Building site 2	Building site 3
Survey I	275	280	264
Survey II	243	291	275

If a restaurant chain is interested in acquiring one of these building sites, and it prefers one in an area of high population density, which choice, or choices, are not admissible? With no information about the credibility of these surveys, is it possible to make a first choice and a second choice?

7.8 With reference to Exercise 7.5 on page 170, which choice of cars is not admissible regardless of whether the car is to be driven on freeways, in the city, or on rural roads?

7.9 The values in the following table are the prices charged for a pound of certain kinds of candies, cookies, and peanuts in three different supermarkets:

	Supermarket A	*Supermarket B*	*Supermarket C*
Candies	$1.41	$1.25	$1.23
Cookies	$0.81	$0.85	$0.82
Peanuts	$1.27	$1.29	$0.99

Mrs. K has to go to one of these supermarkets to buy 10 pounds of the food item she promised to get for a church picnic, but she cannot remember whether she is supposed to bring the candies, cookies, or peanuts. Hoping that she will remember once she gets to the market, find

(a) which choice of supermarkets is not admissible;

(b) the supermarket where she would go if she applied the minimax criterion to the amount of money she will have to pay.

7.10 With reference to the preceding exercise, construct an opportunity-loss table and find the supermarket where Mrs. K should go so as to minimize her maximum loss of opportunity.

7.11 With reference to Exercise 7.1 on page 170, what should the lawyer do if she wants to

(a) maximize her minimum fee;

(b) maximize her maximum fee?

7.12 With reference to Exercise 7.3 on page 170, where should the truck driver go first if he wants to

(a) minimize the maximum distance he will have to drive;

(b) minimize the minimum distance he will have to drive;

(c) minimize his maximum loss of opportunity?

7.13 With reference to Exercise 7.4 on page 170, how many of the perishable items should the retailer stock if he wants to

(a) maximize his minimum profit;

(b) maximize his maximum profit;

(c) minimize his maximum loss of opportunity?

7.14 A dinner guest wants to show his appreciation to his hostess by taking her a box of candy, a bottle of wine, or flowers. He remembers, though, that either she is on a strict diet or she is a teetotaler, but he cannot remember which. In any case, he feels that the reaction to his gift will be as shown in the following table, where the payoffs are in units of "appreciation":

	Candy	Wine	Flowers
Hostess is dieting	−2	3	4
Hostess is teetotaler	6	−9	4

What should he take if he wants to

(a) maximize the minimum appreciation of his gift;

(b) maximize the maximum appreciation of his gift;

(c) minimize his maximum loss of opportunity?

7.15 With reference to Exercise 7.6 on page 171, what inspection procedure should the manufacturer choose if he wants to

(a) minimize the maximum expected cost;

(b) minimize the maximum expected loss of opportunity?

7.16 Because of various difficulties, the supplier of glue used in the manufacture of a laminated fiberboard product can guarantee the manufacturer only that it will deliver on schedule the required quantity of either glue Q or glue R (but not some of both). Because of time requirements, however, the manufacturer must set up his production process prior to knowledge of which glue will be available with no later change possible if he is to meet contractual obligations. Both glues can be used with any one of six production methods open to the company, but for technical reasons the profit per piece differs substantially from one method to another for the same glue. The estimated unit profits (in cents) for methods 1 through 6 using glue Q are, respectively, 108, 158, 147, 172, 137, and 156, while the corresponding figures for glue R are 267, 128, 187, 207, 247, and 214.

(a) Construct a payoff table and eliminate the nonadmissible production methods.

(b) Which production method should the manufacturer use if he wants to maximize his minimum unit profit?

(c) Which production method should the manufacturer use if he wants to minimize his maximum loss of opportunity?

7.6

Decision Making Under Competition†

Let us now consider the case where the choice among the E's (the events over which the decision maker has no control) is made by a competitor. It is customary in this kind of problem to let the payoffs be the decision maker's losses, so that his gains will be represented by negative numbers. Furthermore, it will be assumed that whatever the decision maker gains, his competitor loses, and vice versa. In mathematics, this is called a zero-sum two-person game, where "game" is just a word meaning "competitive situation," the two persons are the decision maker and his competitor, and the "zero-sum" means that whatever one person loses the other person gains. In other words, in a zero-sum game there is no "cut for the house" as in professional gambling, and no capital is created or destroyed during the course of play.

To illustrate these ideas, suppose that a small town has two service stations which share the town's market for gasoline. The owner of station I is debating whether or not to give away table glasses to his customers as part of a promotional scheme, and the owner of station II is debating whether or not to give away kitchen knives. Their decisions will be based on the information (from similar situations elsewhere) that if station I gives away glasses and station II does not give away knives, station I's share of the market will increase by 12 percent; if station II gives away knives and station I does not give away glasses, station II's share of the market will increase by 16 percent; and if both stations give away the respective items, station II's share of the market will increase by

†This section and Section 7.7 may be omitted without loss of continuity.

5 percent. Schematically, this information may be represented in the following payoff table:

	Station I	
	A_1 (no glasses)	A_2 (glasses)
Station II — E_1 (no knives)	0	−12
Station II — E_2 (knives)	16	5

The zero represents the case where both owners decide not to give away these items, so that there is no change in their shares of the market. Also, station I's gain of 12 percent is denoted −12, for as we said, the payoffs are the decision maker's losses, and a loss of −12 percent is the same as a gain of 12 percent. In the scheme above, we let the owner of station I play the role of the decision maker and the owner of station II the role of the competitor, but this is an arbitrary choice; their roles may be reversed and the signs of all the numbers changed accordingly.

In decision making under competition, as in the cases which we studied earlier, the basic problem is to determine **optimum choices** that is, choices which are in some respect the most desirable. This applies not only to the decision maker who must choose one of the A's, but also to the competitor who must choose one of the E's. Clearly, the decision maker must judge what the competitor might consider most profitable, and then account for this in making his decision. Also, it will be assumed that the decision maker as well as the competitor must make his choice without knowledge of what the other one has done or is planning to do, and once he has made his choice, it cannot be changed.

Having begun the analysis of a decision-making problem under competition by constructing a payoff table, we continue by looking for dominances and nonadmissible choices. We do this for the A's as well as the E's, for surely no thinking person would want to make a choice which is worse than another regardless of what the other person decides to do.

EXAMPLE With reference to the illustration directly above, find the best choices for the two station owners.

SOLUTION We see from the payoff table that it would be foolish for the owner of station II to choose E_1, since E_2 is preferable to E_1 regardless of the choice made by the owner of station I. Clearly, a 16 percent increase in the market share is preferable to no increase, and a 5 percent increase is preferable to a 12 percent decrease. Thus, E_1 is not admissible, and if we eliminate it from the payoff table, we find that a 5 percent decrease in the market is preferable to a 16 percent decrease, and, hence, that

A_2 dominates A_1. This leaves A_2 and E_2 as the optimum choices for the two owners. Evidently, the situation favors the owner of station II, and this suggests that the owner of station I might well consider some other promotional scheme; however, this is not part of the problem as formulated here.

The process of discarding dominated alternatives can be of great help in solving problems of decision making under competition (that is, in finding optimum choices for the decision maker and his competitor), but what do we do when no dominances exist?

To illustrate this situation, let us consider a problem of decision making under competition which has the following payoff table:

| | | Decision maker | | |
		A_1	A_2	A_3
	E_1	-2	5	-3
Competitor	E_2	1	3	5
	E_3	-3	-7	11

The payoffs are the decision maker's losses, say, in dollars, which he wants to make small. Inspection shows that there are no dominances among the A's or the E's, but if we look at the problem from the decision maker's point of view, we might argue as follows. If he chooses A_1, the worst that can happen is that he loses \$1; if he chooses A_2, the worst that can happen is that he loses \$5; and if he chooses A_3, the worst that can happen is that he loses \$11. Looking at the problem from this rather pessimistic point of view, it would seem advantageous for the decision maker to minimize his maximum loss by choosing A_1; that is, it would be advantageous for him to apply the minimax criterion to the loss he might incur.

If we apply the same kind of argument to the competitor's choice, we find that if he chooses E_1, the most he can lose is \$3; if he chooses E_2, the worst that can happen is that he wins \$1; and if he chooses E_3, the most he can lose is \$7. Thus, the competitor would minimize his maximum losses (or maximize his minimum gain, which is the same thing) by choosing E_2.

The use of the minimax criterion in decision making under competition is really quite reasonable. By choosing A_1 in our example, the decision maker is assured that his competitor can win at most \$1, and by choosing E_2, the competitor makes sure that he actually does win this amount. A very important aspect of the results obtained in our example is that the choices are completely "spyproof" in the sense that neither the decision maker nor the competitor can profit from knowledge of the other's choice. Even if the decision maker announces publicly that he will choose A_1, it is still best for the competitor to choose E_2, and if the competitor announces publicly that he will choose E_2, it is still best for the decision maker to choose A_1.

The method we used here works nicely in our example—the minimax choices are spyproof—but this will not always be the case. Consider the following example.

EXAMPLE   Show that if the payoff corresponding to A_1 and E_1 is 3 instead of -2 in the preceding example, the minimax choices are not spyproof.

SOLUTION   With this modification, the payoff table becomes

		Decision maker		
		A_1	A_2	A_3
	E_1	3	5	-3
Competitor	E_2	1	3	5
	E_3	-3	-7	11

and it can be seen that the minimax criterion leads to the same choices as before, A_1 and E_2. However, they are no longer spyproof; if the competitor knows that the decision maker will use the minimax criterion and choose A_1, he can switch to E_1 and thus assure for himself a gain of $3 instead of a gain of $1.

Fortunately, there is a fairly easy way of deciding for any given problem whether minimax choices are spyproof. What we do is look for pairs of choices, called saddle points, for which the payoff entry is the smallest value in its row and also the largest value in its column. We cannot prove it here, but it can be shown that the choices which correspond to a saddle point are optimum choices, and if there is more than one saddle point in a given problem (see Exercise 7.19 on page 187), the corresponding payoffs will be the same and it does not matter which saddle point is used for making optimum choices among the A's and E's. When there is a saddle point, we say that the decision problem is strictly determined.

EXAMPLE   With reference to the payoff table on page 182, verify that the optimum choices constitute a saddle point.

SOLUTION   Since the entry corresponding to A_1 and E_2 is the smallest value in its row (1 is less than 3 or 5) and also the largest value in its column (1 is greater than -2 or -3), it follows from the definition that A_1 and E_2

constitute a saddle point. Note, however, that after we modify the payoff table as on page 183, the entry which corresponds to A_1 and E_2 is no longer the largest value in its column (1 is greater than -3 but less than 3), and A_1 and E_2 do not constitute a saddle point.

7.7

Randomized Decisions†

If there are no saddle points in a zero-sum two-person game, game theory has some other ideas on how the decision maker and the competitor should proceed in determining their optimum choices. To illustrate, let us consider the following problem of decision making under competition. A country has two missile bases, one with installations worth $20,000,000 and the other with installations worth $100,000,000. It can defend only one of these bases against an attack by its enemy. The enemy, on the other hand, can attack only one of the bases, and can capture it only if the base is left undefended. In trying to decide which base to defend in the face of the uncertainty as to which base the enemy might attack, the defender considers its payoff in this (deadly) game to be the dollar worth of an installation lost to the enemy. If A_1 represents the decision to defend the smaller base and A_2 the decision to defend the larger base, and E_1 represents the decision to attack the smaller base and E_2 the decision to attack the larger base, the payoff table is

		Defending country	
		A_1	A_2
Attacking country	E_1	0	20
	E_2	100	0

where the units are in millions of dollars. Since the smallest value in each row is 0 and neither 0 is the largest value in its column, there is no saddle point. One might argue, though, that to the defending country a maximum loss of $20 million is preferable to a maximum loss of $100 million and, hence, that A_2 is preferable to A_1. However, if the attacking country knows that the defending country always uses minimax decision procedures, in this case A_2, it can take advantage of this by choosing E_1 and capturing the smaller base. This sounds fine, unless the defending country reasons that this is precisely what the attacking country intends to do and switches to A_1 and prepares to defend the smaller base. This argument can be continued ad infinitum. If the attacking country

†This section, based on the material in Section 7.6, may be omitted without loss of continuity.

thinks that the defending country will try to outwit it by choosing A_1, it can, in turn, try to outwit the defending country by choosing E_2; if the defending country thinks that this is precisely what the attacking country will do, it can switch to A_2 and prepare to defend the larger base; and so on, and so on.

To avoid the possibility of being outguessed or outsmarted, it seems reasonable for each decision maker to mix up his decisions deliberately in some way or other, and the best way to do this is to introduce an element of chance into the final choice.

To illustrate how this can be done, suppose that the defending country in the preceding example uses some kind of gambling device which leads it to choose A_1 with probability p, and to choose A_2 with probability $1 - p$. The defenders can then argue as follows: If the enemy chooses E_1, the expected loss will be

$$E = 0 \cdot p + 20 \cdot (1 - p) = 20 - 20p$$

and if the enemy chooses E_2, the expected loss will be

$$E = 100 \cdot p + 0 \cdot (1 - p) = 100p$$

This situation is described graphically in Figure 7.2, where we have plotted the two lines whose equations are $E = 20 - 20p$ and $E = 100p$ for values of p

7.2

Expected loss of decision maker.

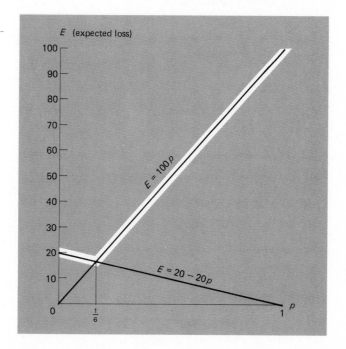

from 0 to 1. (Actually, we drew the lines by connecting in both cases the two values of E which correspond to $p = 0$ and $p = 1$.)

To apply the minimax criterion to the expected loss of the defending country, observe from Figure 7.2 that the most the defenders can expect to lose (the larger of the two values of E for any given value of p, as indicated by the shading) is least where the two lines intersect, and to find the value of p which leads to this minimax expected loss, we must set $20 - 20p$ equal to $100p$, and solve for p. We have

$$20 - 20p = 100p$$

$$20 = 120p$$

and, finally,

$$p = \frac{20}{120} = \frac{1}{6}$$

Therefore, in order to minimize the maximum expected loss, the defenders could do the following. Label one tag A_1 and five tags A_2, put the tags in an urn, and draw one of them from the urn at random; then if the tag drawn is A_1, take action A_1 (defend the smaller base); otherwise, take action A_2 (defend the larger base). This minimax criterion assures the defending country of holding its maximum expected loss to

$$E = 100 \cdot \frac{1}{6} = 16\frac{2}{3} \text{ million dollars}$$

and there is nothing the enemy can do about it. Observe that this expected loss is less than the possible loss of $100 million the defenders are exposed to by the direct (nonchance) choice of A_1 and the possible loss of $20 million they are exposed to by the direct choice of A_2.

If a decision maker's ultimate choice is left to chance, his overall decision procedure is referred to as **randomized** or **mixed**. If a decision maker chooses directly an action open to him without introducing an element of chance, his decision procedure is said to be **pure**. Of course, it may be hard to convince anyone that it makes any sense at all, much less is "best," to gamble with one's country's security, but if one really wants a decision procedure, or **strategy**, that is absolutely spyproof, there is no alternative. Although game theory has been used in analyzing complicated military problems and in playing war games on computers, it should be observed that the illustration we gave here pictures an "ivory tower" kind of situation. In warfare there seems always to be some intelligence and counterintelligence, no matter how little or how poor it is, and it is usually possible to change one's decision (adjust one's strategy) after it becomes known what one's opponent is doing or intends to do. This is certainly not the case, however, in all competitive business situations.

EXERCISES

7.17 Each of the following is the payoff table for a problem of decision making under competition, with the payoffs being the losses of the decision maker who has to choose among the columns. Eliminate all dominated choices and thus determine the best choice for each "player":

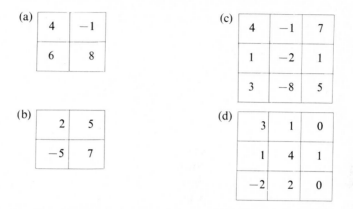

(a)

4	−1
6	8

(b)

2	5
−5	7

(c)

4	−1	7
1	−2	1
3	−8	5

(d)

3	1	0
1	4	1
−2	2	0

7.18 Find the saddle point for each part of the preceding exercise.

7.19 Each of the following is the payoff table for a problem of decision making under competition, with the payoffs being the losses of the decision maker who has to choose among the columns. Find the saddle point or saddle points, and show that if there is more than one, it does not matter which one is selected.

(a)

−1	5	−2
0	3	1
−2	−4	5

(b)

3	2	4	9
4	4	4	3
5	6	5	6
5	7	5	9

7.20 A certain type of computer is made by only two companies who share the market for the machine equally. Both would prefer not to introduce a new model at this time, but both suspect that the other is readying a new model and that if it is introduced some sales will be lost to the competitor. If neither brings out a new model or if both bring out new models, the status quo will be maintained and both will continue to get their same relative share of the market. If, however, one brings out a new model and the other does not, there will be a loss of 10 percent in share of the market to the competitor with the new machine. What is the best strategy for the two companies to use with respect to the introduction of a new model?

7.21 Show that in the problem of decision making under competition given by the following payoff table, the maximin criterion leads the competitor to choose the second alternative:

Decision maker

−2	200	200	200	200
0	0	0	0	0

Competitor

It has been suggested that in a situation like this it would be wholly irrational for the competitor to choose the second alternative. Give examples where
(a) this suggestion would be reasonable;
(b) this suggestion would not be reasonable.

7.22 Only two cities are being considered as sites for a national political convention. Each city can send a lobbyist, a delegation, or neither, to the meeting of the "convention site" committee, and the corresponding decreases in city *A*'s probability of getting the convention are as shown in the following payoff table:

		City A Lobbyist	Delegation	Neither
	Lobbyist	−0.01	0.04	0.03
City B	Delegation	0.00	0.02	0.10
	Neither	−0.05	−0.08	0.00

If they base their decisions on the minimax and maximin criteria, what should the two cities decide to do?

7.23 The following is the payoff table for a zero-sum two-person game, with the payoffs being the amounts player *A* loses to player *B*:

Player A

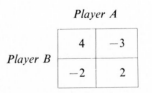

Player B	4	−3
	−2	2

(a) What randomized decision procedure should player *A* use so as to minimize his maximum expected loss?
(b) What randomized decision procedure should player *B* use so as to maximize his minimum expected gain?

7.24 Two persons agree to play the following game: The first writes either 1 or 4 on a slip of paper and at the same time the second writes either 0 or 3 on another slip of paper. If the sum of the two numbers is odd, the first wins this amount in dollars; otherwise, the second wins $2.

(a) Construct a payoff table in which the payoffs are the first person's losses.

(b) What randomized decision procedure should the first person use so as to minimize his maximum expected loss?

(c) What randomized decision procedure should the second person use so as to maximize his minimum expected gain?

7.25 Suppose that the manufacturer of Exercise 7.16 on page 179 has the choice of only three production methods for which the unit profits (in cents) using glue Q are, respectively, 120, 90, and 200. If the corresponding figures for glue R are 240, 175, and 180, what randomized decision procedure should the manufacturer use so as to maximize his minimum expected unit profit?

7.26 With reference to Exercise 7.2 on page 170, what randomized decision procedure should the management of the oil company use if it wants to minimize the maximum expected loss? Explain under what conditions this decision procedure might be regarded as rational.

7.27 There are two gas stations in a certain block, and the owner of the first station knows that if neither station lowers its prices, he can expect a net profit of $100 on any given day. If he lowers his prices while the other station does not, he can expect a net profit of $140; if he does not lower his prices but the other station does, he can expect a net profit of $70; and if both stations participate in this "price war," he can expect a net profit of $80. The owners of the two stations decide independently what prices to charge on any given day, and it is assumed that they cannot change their prices after they discover those charged by the other.

(a) Should the owner of the first station charge his regular prices or should he lower them, if he wants to maximize his minimum net profit?

(b) Assuming that the foregoing profit figures apply also to the second station, how might the two owners collude so that each could expect a net profit of $105? (Note that this "game" is not zero-sum, so that the prospect of collusion opens entirely new possibilities.)

7.8

The Bayes Decision Rule

Until now we have considered only problems in which the decision maker has no information about the probabilities of the events E over which he has no control. In practical business applications, ideally, at any rate, this is the exception rather than the rule. Decisions are ordinarily delegated to knowledgeable, trained, and experienced persons, whose expertise should enable them to assign valid (not to say correct) probabilities to the occurrences of the different events. In many instances, these probability estimates are based on subjective evaluations, collateral (indirect) information, intuition, and other factors, that are all acquired prior to the time that the decision-making situation arises. For this reason, we refer to such probabilities as **prior probabilities**, and to a decision analysis based on such probabilities as a **prior analysis**; another term is **decision making under risk**.

Once we have assigned probabilities to the E's, we can calculate mathematical expectations, and base decisions on whichever action promises the maximum expected profit, the minimum expected cost, the maximum expected sales, the minimum expected spoilage, and so on. When we do this we are said to be using the **Bayes decision rule**, named after the Reverend Thomas Bayes (1702–1761), and what we are doing is called **Bayesian decision making**. It is called this because in most real applications we base our decisions on prior information as well as direct sample evidence collected specially to aid in making the decisions, and combining these two kinds of evidence requires the use of a formula attributed to Bayes which we shall present in Chapter 9.

The following is an example of Bayesian decision making based on the values of the payoffs and prior probabilities of the events E over which the decision maker has no control.

EXAMPLE   Referring again to the example on page 168, where the payoff table was

	Expand now	Delay expansion
Economic conditions remain good	$369,000	$180,000
There is a recession	−$90,000	$18,000

find the decision which will maximize the manufacturer's expected profit if he feels (on the basis of relevant information available to him) that the odds on a recession are

   (a) 2 to 1;
   (b) 3 to 2.

SOLUTION   (a) If the odds on a recession are 2 to 1, the probability of a recession is $\frac{2}{2+1} = \frac{2}{3}$ and the probability that economic conditions will remain good is $\frac{1}{1+2} = \frac{1}{3}$. Thus, if he expands his plant capacity right away, the expected profit is

$$369,000 \cdot \frac{1}{3} + (-90,000) \cdot \frac{2}{3} = \$63,000$$

and if the expansion is delayed, the expected profit is

$$180{,}000 \cdot \frac{1}{3} + 18{,}000 \cdot \frac{2}{3} = \$72{,}000$$

Since an expected profit of \$72,000 is obviously preferable to an expected profit of \$63,000, the Bayes decision rule leads the manufacturer to delay expanding his plant capacity. (b) If the odds on a recession are 3 to 2, the probability of a recession is $\frac{3}{3+2} = \frac{3}{5}$ and the probability that economic conditions will remain good is $\frac{2}{2+3} = \frac{2}{5}$. Thus, if he expands his plant capacity right away, the expected profit is

$$369{,}000 \cdot \frac{2}{5} + (-90{,}000) \cdot \frac{3}{5} = \$93{,}600$$

and if the expansion is delayed, the expected profit is

$$180{,}000 \cdot \frac{2}{5} + 18{,}000 \cdot \frac{3}{5} = \$82{,}800$$

Here the Bayes decision rule leads the manufacturer to expand his plant capacity right away, which is the reverse of the decision arrived at in part (a).

The preceding example illustrates that in Bayesian decision making we must be fairly sure that the prior probabilities are "correct" (or at least reasonably close). Just how sensitive our decisions are to changes (errors?) in the prior probabilities is a matter of sensitivity analysis, which we shall illustrate here by finding the "changeover point," namely, the value of the prior probability (of economic conditions remaining good) at which the manufacturer's choice of action would change from one to the other.

If we let p denote the probability that economic conditions will remain good, so that the probability is $1 - p$ that there will be a recession, the manufacturer's expected profit is

$$369{,}000p + (-90{,}000)(1 - p) = 459{,}000p - 90{,}000$$

if he expands his plant capacity right away, and it is

$$180{,}000p + (18{,}000)(1 - p) = 162{,}000p + 18{,}000$$

if the expansion is delayed. These two expected profits are represented by the two lines of Figure 7.3, and it can be seen that the second alternative is preferable

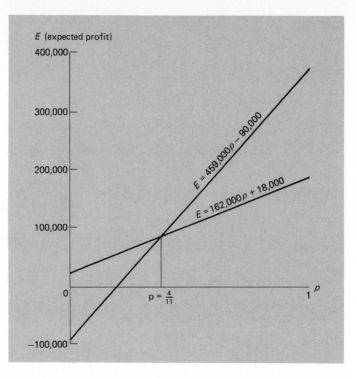

7.3

Diagram for
sensitivity analysis.

(has the higher expected profit) up to the value of p where the two lines intersect, and that the first alternative is preferable for values of p greater than that. To find the value of p where the lines intersect, we equate the two expected profits, getting

$$459{,}000p - 90{,}000 = 162{,}000p + 18{,}000$$

and, hence, $297{,}000p = 108{,}000$ and $p = \dfrac{108{,}000}{297{,}000} = \dfrac{4}{11}$, or approximately 0.364. So, for $p = \frac{4}{11}$ the expected profit is the same regardless of whether or not the manufacturer decides to expand his plant capacity right away.

As the problem was originally formulated in part (a) of the example on page 190, the manufacturer felt that the odds on a recession were 2 to 1 and, hence, assigned the probability $p = \frac{1}{3}$ to economic conditions remaining good. However, if this probability had been slightly higher (higher by more than $\frac{4}{11} - \frac{1}{3} = \frac{1}{33}$ or approximately 0.03), the manufacturer's decision would have gone the other way. Surely, this is not much of a margin for error.

Another question that comes to one's mind in Bayesian decision making is how sensitive it may be to changes in the payoffs. Indeed, this is a crucial factor in any kind of decision analysis (see, for example, Exercise 7.37 on page 198).

7.9

Expected Opportunity Losses

Earlier in this chapter, we suggested that decisions could be based on opportunity losses instead of the payoffs themselves and, in the absence of any prior information, used the minimax criterion to choose a course of action. Once again, opportunity losses are calculated for each row of an opportunity loss table as follows: If the payoffs are profits or other quantities which we want to make large, we subtract each value in a row from the largest value in the row; if the payoffs are losses or other quantities which we want to make small, we subtract the smallest value in a row from each value in the row. Now, if we can assign prior probabilities to the "states of nature," we can proceed as in the preceding section and apply the Bayesian decision rule; that is, we can minimize the **expected opportunity loss**, also referred to as the **EOL**.

EXAMPLE   In the example on page 176, the opportunity losses which Mr. H faced in deciding where to go to buy his wife a sofa are

	Furniture Stores		
	A_1	A_2	A_3
E_1	$35	$3	0
E_2	0	$7	$15

Sofas

What store should he go to in order to minimize the EOL if he feels that the odds are 4 to 1 that his wife will decide on the first kind of sofa?

SOLUTION   If he chooses the first store, the expected opportunity loss is

$$35(0.80) + 0(0.20) = \$28.00$$

if he chooses the second store, the EOL is

$$3(0.80) + 7(0.20) = \$3.80$$

and if he chooses the third store, the EOL is

$$0(0.80) + 15(0.20) = \$3.00$$

Thus, the EOL is least if he decides to go to the third store, and this should not come as a surprise—the prior probabilities strongly favor E_1, and for E_1 and A_3 (the third store) the opportunity loss is zero.

In the office equipment manufacturer problem on page 176, the opportunity-loss table is

	A_1	A_2
E_1	0	$189,000
E_2	$108,000	0

Which decision will minimize the EOL if the prior probabilities of E_1 and E_2 are

(a) $\frac{1}{3}$ and $\frac{2}{3}$;

(b) $\frac{2}{5}$ and $\frac{3}{5}$?

SOLUTION (a) If the manufacturer decides to expand his plant capacity right away, the EOL is

$$0 \cdot \frac{1}{3} + 108,000 \cdot \frac{2}{3} = \$72,000$$

and if he delays, the EOL is

$$189,000 \cdot \frac{1}{3} + 0 \cdot \frac{2}{3} = \$63,000$$

and he will minimize his expected opportunity loss by deciding to delay. (b) If he decides to expand his plant capacity right away, the EOL is

$$0 \cdot \frac{2}{5} + 108,000 \cdot \frac{3}{5} = \$64,800$$

and if he delays, the EOL is

$$189,000 \cdot \frac{2}{5} + 0 \cdot \frac{3}{5} = \$75,600$$

and he will minimize his expected opportunity loss by deciding to expand right away.

In all these examples the decisions are identical with those arrived at earlier by applying the Bayes decision rule directly to the expected profits. This is by no means a coincidence, as it can be shown mathematically that minimizing the expected loss of opportunity will always lead to the same decision as maximizing the expected profit or minimizing the expected loss, as the case may be (see Exercise 7.38 on page 198). Therefore, the method of this section does not really present anything new, and we have given it here mainly because it is essential to understanding another important concept in decision analysis, which we shall introduce in the next section.

7.10

Expected Value of Perfect Information

In the two preceding sections we studied problems in which decisions were based on prior information, without considering the possibility of acquiring further information before making a decision. We shall see later how decisions may be based on prior information as well as direct sample evidence, but for the moment let us merely consider whether it is really worthwhile to delay a decision until we collect further information; more specifically, how much should we be willing to pay for such information?

Let us introduce first the concept of an **expected payoff with perfect information**, that is, the payoff we could expect, with reference to a given set of prior probabilities, if we were called upon to make such a decision an indefinitely large number of times and we always made the best possible decision.

EXAMPLE In the office equipment manufacturer problem on page 168, the payoffs are

	Expand now	*Delay expansion*
Economic conditions remain good	$369,000	$180,000
There is a recession	−$90,000	$18,000

Find the expected payoff with perfect information when the prior probabilities of economic conditions remaining good and of there being a recession are $\frac{1}{3}$ and $\frac{2}{3}$.

SOLUTION If economic conditions remain good, the optimum choice would be to expand the plant capacity right away, and if there is a recession, the optimum choice would be to delay the expansion. The corresponding payoffs are $369,000 and $18,000, so that for prior probabilities of $\frac{1}{3}$ and $\frac{2}{3}$ the expected payoff with perfect information is

$$369,000 \cdot \frac{1}{3} + 18,000 \cdot \frac{2}{3} = \$135,000$$

The importance of knowing an expected payoff with perfect information is that it enables us to determine how much of an improvement perfect information is over merely knowing the prior probabilities of the E's. For instance, in the office equipment manufacturer problem we saw on page 191 that, based only on the prior probabilities $\frac{1}{3}$ and $\frac{2}{3}$, the maximum expected profit is $72,000. Since this is $135,000 - $72,000 = $63,000 less than the expected payoff with perfect information calculated above, it follows that the manufacturer should be willing to spend not more than $63,000 for additional information, and spend that much only for perfect information.

As suggested by this discussion, we are really interested mostly in the difference between the expected payoff with perfect information and the optimum expected payoff based only on the prior probabilities. We call this difference the **expected value of perfect information**, or simply the **EVPI**. In practice, the EVPI can be determined without first calculating the expected payoff with perfect information. As the reader will be asked to demonstrate in Exercise 7.39 on page 199 (for the special case where there are only two choices A_1 and A_2 and two events E_1 and E_2), **the expected value of perfect information is always equal to the minimum expected opportunity loss**; that is, the EVPI is always equal to the smallest of the EOL's.

EXAMPLE   In the office equipment manufacturer problem on page 168, verify that the EVPI equals the smallest of the EOL's.

SOLUTION   In the preceding discussion we showed that the EVPI = $63,000 and on page 194 we showed that for the same prior probabilities, $\frac{1}{3}$ and $\frac{2}{3}$, the two EOL's are $72,000 and $63,000. Thus, the EVPI is, indeed, equal to the smallest of the EOL's.

The expected value of perfect information is an important concept in decision analysis, and it is sometimes referred to as the **cost of uncertainty**. It must be understood, though, that it depends on the choice of the prior probabilities, where even a small inaccuracy or misjudgment can have a pronounced effect (see Exercise 7.35 on page 197).

EXERCISES   7.28 Mr. Cooper is planning to attend a convention in Washington, D.C., and he must send in his room reservation immediately. The convention is so large that the activities are held partly in hotel I and partly in hotel II, and Mr. Cooper does not know where the particular session he wants to attend will be held. He is planning to stay only one day, which would cost him $36.00 at hotel I and $32.40 at hotel II, but it will cost him an extra $6.00 for cab fare if he stays at the wrong hotel.
(a) Construct a payoff table for this decision problem.
(b) If Mr. Cooper feels that the odds are 3 to 1 that the session he wants to attend will be held at hotel I and he wants to minimize his expected cost, where should he make his reservation?

(c) Would it be worthwhile to spend $1.20 on a long-distance call to find out in which hotel the session will be held?

7.29 With reference to part (b) of the preceding exercise, how far off could the prior probabilities be without affecting Mr. Cooper's decision?

7.30 With reference to Exercise 7.2 on page 170, find
(a) what decision would maximize the oil company's expected gain if it is felt that the probability of finding oil is 0.25;
(b) at most how much the company should be willing to spend to find out for certain whether there is any oil.

7.31 With reference to the preceding exercise, how far off could the 0.25 probability be without affecting the decision?

7.32 With reference to Exercise 7.3 on page 170, which construction site should the truck driver go to first if he wants to minimize the expected mileage and feels that the prior probabilities for sites 1, 2, and 3 are 0.10, 0.70, and 0.20?

7.33 With reference to Exercise 7.4 on page 170, how many of the perishable items should the retailer stock so as to maximize his expected profit if he feels that the prior probabilities of a demand for 0, 1, 2, 3, or 4 or more of the items are
(a) 0.10, 0.40, 0.30, 0.10, and 0.10;
(b) 0.10, 0.10, 0.30, 0.30, and 0.20?

7.34 Mrs. Jones, who lives in a suburb, plans to spend an afternoon shopping in downtown Manhattan, and she must decide whether or not to take along her raincoat. If it rains, she will be inconvenienced if she does not take it, and if it does not rain, she will be inconvenienced if she does. On the other hand, it will be convenient to have the raincoat if it rains, and she will be neither convenienced nor inconvenienced if she does not bring the coat and it does not rain. To express all this numerically, suppose that the payoffs in the following table are in units of inconvenience, so that the negative value reflects convenience:

	She takes the raincoat	She does not take the raincoat
It rains	—75	150
It does not rain	45	0

(a) What should Mrs. Jones do to minimize her expected inconvenience if she feels that the probabilities of rain and no rain are 0.15 and 0.85?
(b) If the cash equivalent of a unit of inconvenience is 30 cents, at most how much should she be willing to spend to find out whether or not it is going to rain?
(c) How much could the prior probabilities of part (a) be off without affecting her decision?

7.35 With reference to the problem dealing with the two kinds of sofas and the three stores on page 176, find the EVPI when the prior probabilities are 0.80 and 0.20 by

(a) subtracting the expected payoff with perfect information from the smallest of the expected prices for the three stores;

(b) determining the smallest of the EOL's.

Also, repeat part (b) when the prior probabilities are 0.70 and 0.30, and discuss the effect of this change in the prior probabilities.

7.36 At a time when the Republican party has not decided where to hold its national convention but has narrowed its choice to either San Francisco or Miami, a women's specialty store in San Francisco is offered either 1,000 or 2,000 novelty bottles of perfume in the form of the party elephant at a cost of $2 each. If the store buys 1,000 bottles and the convention goes to Miami, it expects to sell all of them at $5 each; if buys 2,000 bottles, however, it expects to be able to sell only 1,000 bottles at $5 each and be forced to sell the remainder at 50 cents each. If the convention comes to San Francisco, the store expects to sell 2,000 bottles at $5 each, but if it buys only 1,000, it will have to rebuy and pay $4 each for the additional 1,000 bottles.

(a) If there is a fifty–fifty chance that the convention will come to San Francisco, what should the initial order be so as to maximize the expected profit?

(b) Show that if we use the same probability as in part (a), minimizing the expected opportunity loss would lead to the same decision as maximizing the expected profit.

(c) Calculate first the expected profit with perfect information and then the expected value of perfect information, and verify that the EVPI equals the expected opportunity loss corresponding to the optimum choice.

7.37 With reference to the example on page 190, would the Bayes decision rule lead to different decisions if

(a) the $369,000 profit is replaced by a $450,000 profit and the prior probabilities of economic conditions remaining good or there being a recession are $\frac{1}{3}$ and $\frac{2}{3}$;

(b) the $90,000 loss is replaced by a $135,000 loss and the prior probabilities of economic conditions remaining good or there being a recession are $\frac{2}{5}$ and $\frac{3}{5}$?

7.38 The following is the payoff table of a decision problem in which the payoffs are profits, $b > a$, $c > d$, and the prior probabilities of E_1 and E_2 are p and $1 - p$:

	A_1	A_2
E_1	a	b
E_2	c	d

(a) Find an expression for the difference between the expected profits corresponding to A_1 and A_2.

(b) Construct an opportunity-loss table and find an expression for the difference between the expected opportunity losses corresponding to A_1 and A_2.

(c) Combining the results of parts (a) and (b), explain why maximizing the expected profit leads to the same decision as minimizing the expected opportunity loss.

7.39 To show that the expected value of perfect information equals the minimum expected loss of opportunity, refer to the decision problem of the preceding exercise.

(a) Find an expression for the expected profit with perfect information.

(b) Subtract the expected profits corresponding to A_1 and A_2 from the expression obtained in part (a) to show that the expected value of perfect information is either $(b - a)p$ or $(c - d)(1 - p)$, depending on whether A_1 or A_2 is the optimum selection.

(c) Based on the opportunity-loss table of part (b) of the preceding exercise, find expressions for the expected opportunity losses corresponding to A_1 and A_2.

(d) Use the results of parts (b) and (c) to verify that the EVPI equals the minimum EOL.

7.11

Further Considerations: Utility

In all the problems in which we applied the Bayes decision rule, we assumed that it is rational to take whichever action maximizes expected profit, minimizes expected cost, minimizes expected opportunity loss, and so on. Of course, there are many situations where this kind of decision making is justified, but there are exceptions. Suppose, for example, that the office equipment manufacturer (see page 168) is in bad financial condition and feels that, unless he makes a profit in the next year, he will be forced to file for bankruptcy. This would make the $90,000 loss in the payoff table very heavy, and he might well decide to delay expanding his plant capacity in spite of the prior probabilities of $\frac{2}{5}$ and $\frac{3}{5}$ and expectations which tell him to do otherwise.

This illustrates the fact that the value, usefulness, or **utility** of a payoff may be greater or less than its cash equivalent, and it raises the question of how such intangibles might actually be measured. Of course, we could always ask a person directly how much something is worth to him, but "talk is cheap" and unless there is something at stake, it is hard to get meaningful answers.

To illustrate another way of measuring utilities, suppose that a football fan tells us that he would "give his right arm" for a ticket to a bowl game that has been sold out for weeks. Naturally, this is only a figure of speech, but it is meant to imply that the value which he attaches to a game ticket is very high. If we asked him to be more specific, he might say that he would be willing to pay $50 or perhaps even $100, but unless we can put this on a "put-up or shut-up" basis, it really does not have much significance. So, suppose that we make him the following proposition. If he will pay us $5.00, we will let him draw at random one of 10 sealed envelopes, nine of which contain a dollar bill and the tenth a ticket to the game. If he accepts this deal and we let U denote the value, or utility, which he assigns to the ticket, we can argue that his expected utility, $\frac{1}{10} \cdot U + \frac{9}{10} \cdot 1$, must be worth at least \$5.00 to him. Symbolically,

$$\frac{1}{10} \cdot U + \frac{9}{10} \cdot 1 \geq 5$$

and this leads to $U + 9 \geq 50$, and hence to $U \geq 41$. We have arrived here at the result that the fan feels that the ticket to the game is worth at least \$41 to him, and if we varied the odds, the contents of the envelopes, or the amount he is asked to pay, we could be more precise about it than that. For instance, if he is willing to pay \$6 but not \$7.50 in this situation, this would lead to \$51 $\leq$ $U <$ \$66 (see Exercise 7.40 on page 202).

In this example we assumed that \$5 is worth five times as much as \$1 and \$50 is worth ten times as much as \$5. Assumptions like this are often reasonable when we are dealing with fairly small amounts. It is a well-known fact, however, that the utility which a person assigns to a sum of money does not always equal its monetary value. Very often, the reason for this is that "the second \$100 gain is worth less than the first," "the second \$1,000 gain is worth less than the first," . . . , or that "the second \$1,000 loss hurts more than the first." Of course, \$100 is \$100 and \$1,000 is \$1,000, but the value a person attaches to a sum of money will depend on how much a person already has or owes; in technical language, this is a matter of **marginal utility**

To give an example, suppose that we offer someone the choice between an outright gift of \$120 or a gamble on the flip of a fair coin which pays him \$250 if it falls heads and nothing if it falls tails. Even though the mathematical expectation of the gamble is $250 \cdot \frac{1}{2} + 0 \cdot \frac{1}{2} =$ \$125, which exceeds the dollar value of the outright gift, it would be surprising if a person actually chose the gamble. In some cases this reluctance to gamble may be based on moral or religious grounds, or on a person's general dislike of taking chances, but in others it simply reflects the fact that the first \$120 is worth more to a person than the next \$130.

To continue this argument, suppose that the gamble is modified so that the person wins the \$250 if he draws at random a white bead from an urn containing 100 beads, some black and some white. If he feels that the value of the gamble equals the value of the outright gift of \$120 when there are 60 white beads and 40 black beads in the urn, we can write

$$\frac{60}{60 + 40} \cdot U = 120$$

where U is the value, or utility, which the person attaches to \$250. This leads to $U = \frac{5}{3} \cdot$ \$120, and it tells us that the utility which the person assigns to \$250 is only five thirds the utility he assigns to \$120. It would be misleading, though, to multiply out $\frac{5}{3} \cdot$ \$120 and get \$200, since the utility which a person attaches to \$200 may not be five thirds the utility he assigns to \$120.

To avoid this difficulty, we introduce an artificial unit of utility and call it a **utile**. For instance, in our last example we could let \$0 be 0 utiles and \$120 be 1 utile, and we could then say that the person assigns $250 a utility of $\frac{5}{3} \cdot 1 = 1\frac{2}{3}$ utiles.

Proceeding with the example, suppose that we make the same person another offer; he will toss a fair coin and if it falls heads he wins and we will pay him $120, otherwise he loses and he will pay us $120. This is certainly a fair offer in a monetary sense, but suppose the person declines it because, to him, the possibility of losing $120 outweighs the possibility of winning $120. Instead, he makes us this counter offer: he will match with us but if he loses he will pay us only $100, not $120. We would not accept this offer, but it does tell us something about the utility U which the person attaches to a loss of $100. Since he considers a bet of $100 against $120 fair and since his probabilities of winning and losing are both $\frac{1}{2}$, we can assign his winning $120 a utility of 1 utile, as before, and write

$$\frac{1}{2} \cdot U + \frac{1}{2} \cdot 1 = 0$$

Solving this equation for U, we get $U = -1$.

Continuing in this way, we could find the utilities, the numbers of utiles, which the person attaches to various other amounts, and thus arrive at a **utility curve** like the one shown in Figure 7.4.

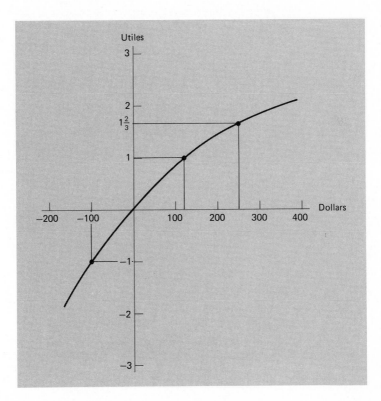

When utilities are measured in this way, the concept of utility is closely interwoven with a person's eagerness or reluctance to take chances. For instance, if we are willing to bet $120 against someone's $100 on the flip of a balanced coin, we are the kind of person who is eager to bet even when the odds are against him (a compulsive gambler, perhaps), and our utility curve is like the one pictured in Figure 7.5. If we are willing to bet only when the situation is fair to both parties in the monetary sense, we are said to be **neutral to risk**, and our utility curve is a straight line.

We have introduced the concept of utility here mainly to make Bayesian decision making more generally applicable. Let us add, though, that the construction of meaningful utility curves is usually a formidable task. A Bayesian decision analysis based on maximizing one's expected utility is illustrated in Exercise 7.49 on page 203.

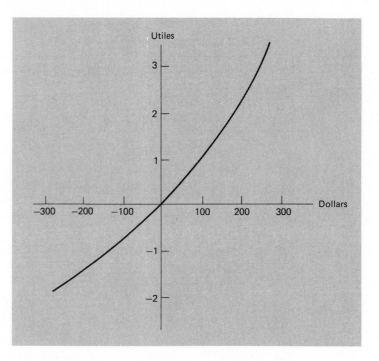

7.5

Utility curve for person eager to take chances.

EXERCISES

7.40 With reference to the example on page 199, verify that if the fan is willing to pay $6 but not $7.50 to play the game, the value he assigns to the ticket is greater than or equal to $51 but less than $66.

7.41 Ms. Smith feels that it is a toss-up whether to accept a $13.00 cash gift or to gamble on drawing at random a bead from an urn containing 25 white beads and 75 red beads, where she is to receive $2.50 if she draws a white bead or a digital clock if she draws a red bead. What value, or utility, does she assign to the clock?

7.42 Mr. Jones would love to beat Mr. Brown in the club tennis tournament, but his chances are nil unless he takes $200.00 worth of extra lessons, which (according to the club tennis pro) will give him a fifty-fifty chance. If Mr. Jones assigns the utility U to his beating Mr. Brown and the utility $-\frac{1}{3}U$ to his losing to Mr. Brown, find U if Mr. Jones decides that it is just about worthwhile to spend the $200.00 on extra lessons.

7.43 Suppose that the person referred to on page 200, who attaches a utility of 1 utile to $120, is willing to pay $120, but not more than $120, for a gamble which will pay him $600 if he draws a spade from an ordinary deck of 52 playing cards, but nothing if he draws a heart, diamond, or club. What utility does he assign to $600?

7.44 Suppose that the person referred to on page 201, who attaches a utility of −1 utile to a loss of $100, actually owes a business associate $100. To settle the debt, he proposes the following scheme, which he considers fair. He will toss a balanced die and if it falls 1, 2, 3, 4, or 5, the debt is canceled; but if it falls 6, he will pay the associate $300. What utility does he assign to this $300 loss?

7.45 The person whose utility curve is given in Figure 7.4 takes a job which promises to pay him a fee of $100, $200, $300, or $400 with probabilities of 0.24, 0.32, 0.28, and 0.16. Read the utilities corresponding to these amounts off the graph of Figure 7.4 and calculate the expected utility which the person assigns to this job.

7.46 Discuss under what conditions it would be wise for a person to have
(a) complete health insurance;
(b) only major medical insurance with a $500 deductible;
(c) no health insurance at all.

7.47 Suppose that the utility which Miss Brown assigns to the amount A (in dollars) is given by

$$U = A - 0.001A^2$$

for $-\$200 \leq A \leq \$200$. Calculate U for $A = -\$200, -\$100, -\$50, \$0, \$50, \$100$, and 200, and sketch the graph of this utility curve.

7.48 Would Miss Brown of the preceding exercise prefer to pay $100 outright for an oil painting or match the artist "double or nothing" for the painting using a balanced coin?

7.49 With reference to the office equipment manufacturer on page 168, suppose that he will be forced into bankruptcy if he cannot show a positive profit during the next year, and that he translates the payoffs into the utiles shown in the following table:

	Expand now	Delay expansion
Economic conditions remain good	30	20
There is a recession	−485	5

Which decision will maximize the manufacturer's expected utility, if he feels that the prior probabilities for economic conditions remaining good and there being a recession are $\frac{1}{3}$ and $\frac{2}{3}$?

7.50 With reference to the preceding exercise, show that expanding his plant capacity right away will maximize the manufacturer's expected utility so long as the prior probability of a recession is less than or equal to 0.02. Does it matter what he does when this prior probability is 0.02?

7.12

A Word of Caution

One of the greatest difficulties in applying the methods of this chapter (and more general methods) to realistically complex problems in statistics, business management, and economics, is that we seldom know the exact values of all the risks that are involved. That is, we seldom know the exact values of the payoffs corresponding to the various eventualities, and we seldom have sufficient information about the values of all relevant probabilities. For instance, if a manufacturer must decide whether to market a new drug right away, how can he put a cash value on the damage that might be caused by not waiting for a more thorough evaluation of the side effects of the drug, or on the lives that might be lost by not marketing the drug? Similarly, if a management consultant must decide whether to recommend that a new detergent be marketed or not marketed, how can he possibly take into account all the effects which his advice (good or bad) might have on himself, on the prospective marketer, and ultimately the consumer, of the product?

So far as the prior probabilities of the events over which the decision maker has no control are concerned, small changes (perhaps errors of judgment or differences of opinion) can lead to different decisions. Indeed, the problem of "policing" subjective probabilities—of deciding which ones are to be trusted and which ones are not and which ones are to be used and which ones are not—can pose serious difficulties in practical applications.

Nevertheless, there is much good to be said for decision theory and we repeat here what we said in the introduction to this chapter: "The decision-theory approach has the positive advantage of forcing one to formulate problems clearly, to anticipate the various consequences of one's actions, to retain the relevant and eliminate the irrelevant, to place cash values on the consequences of one's actions, and so on. No matter how little or how far it is pursued, such a systematic approach is bound to help."

7.13

Check List of Key Terms

Bayesian decision making, 190
Bayes decision rule, 190
Cost of uncertainty, 196
Decision making under certainty, 171
Decision making under competition, 180
Decision making under risk, 190
Decision making under uncertainty, 171
Decision theory, 167
Decision tree, 168
Dominance, 172
Expected opportunity loss, EOL, 193
Expected payoff with perfect information, 195
Expected value of perfect information, EVPI, 196
Game theory, 168
Maximax criterion, 175
Maximin criterion, 174
Minimax criterion, 175

Minimin criterion, 175
Mixed decision procedure, 186
Opportunity loss, 176
Opportunity-loss table, 176
Optimum choices, 181
Payoff, 168
Payoff matrix, 168
Payoff table, 168
Prior analysis, 190
Pure decision procedure, 186
Randomized decision procedure, 186
Regret, 176
Saddle point, 183
Sensitivity analysis, 191
State of nature, 169
Strictly determined game, 183
Utile, 200
Utility, 199
Zero-sum two-person game, 180

7.14

Review Exercises

7.51 Tom is starting out to meet his friends at the beach, but he cannot remember whether he is supposed to meet them in La Jolla or in Mission Beach, which are 4 miles apart. He lives 11 miles from the spot where he would meet them in La Jolla and 9 miles from the spot where he would meet them in Mission Beach. Construct a payoff table which shows the number of miles Tom has to drive to meet his friends depending on where he goes first and where they are actually supposed to meet.

7.52 With reference to the preceding exercise, where should Tom go first if he wants to
(a) minimize the maximum distance he has to drive;
(b) minimize the minimum distance he has to drive?

7.53 With reference to Exercise 7.51, where should Tom go first if he wants to minimize the distance he can expect to drive to meet his friends and he feels that
(a) the odds are 5 to 1 that they are to meet in La Jolla;
(b) the odds are 2 to 1 that they are to meet in La Jolla;
(c) the odds are 3 to 1 that they are to meet in La Jolla?

7.54 With reference to Exercise 7.51, what is the EVPI (in miles) when the odds are 5 to 1 that they are to meet in La Jolla?

7.55 The regents of a university must decide whether to authorize funds for the construction of a new stadium. They know that if the new stadium is built and the university has a good football team there will be a profit of $240,000; if the new stadium is built and the university has a poor football team there will be a deficit of $40,000; if the old stadium is used and the university has a good football team there will be a profit of $120,000; and if the old stadium is used and the university has a poor football team there will be a profit of $20,000.
(a) Construct a payoff table.
(b) Construct a decision tree.

7.56 With reference to the preceding exercise, construct an opportunity-loss table and determine which decision of the regents will
(a) minimize the maximum loss of opportunity;
(b) minimize the minimum loss of opportunity.

7.57 Based on the opportunity loss table of Exercise 7.56, what should the regents do if they feel that the prior probabilities for a good or a poor football team are 0.30 and 0.70, and they want to minimize the EOL?

7.58 With reference to the preceding exercise, how much could the 0.30 prior probability be increased without changing the regents' decision?

7.59 Mr. Green has the choice of staying home and reading a good book or going to a party. If he goes to the party he might have a terrible time (to which he assigns a utility of 0), or he might have a wonderful time (to which he assigns a utility of 30 units). If he feels that the odds against his having a good time are 4 to 1 and he decides not to go, what can we say about the utility which he assigns to staying home and reading a good book?

7.60 The following are the frequency-of-repairs records for some 1976 models of imported cars:

	Audi Fox	Fiat 131	Honda Civic
Brakes	very poor	very poor	poor
Engine	poor	average	average
Exhaust system	very poor	very poor	very poor
Fuel system	very good	poor	good

Based on these considerations only, which choice of cars is not admissible?

7.61 The credit manager of a mortgage company figures that if an applicant for a certain-size loan is a good risk and the company accepts him the company's profit will be $4,112, and if he is a bad risk and the company accepts him the company will lose $655. If the credit manager turns down the loan applicant, there will be no direct profit or loss either way.

(a) Construct a payoff table for this decision problem.

(b) If the credit manager feels that the probabilities are 0.10 and 0.90 that a certain loan applicant is a good risk or a bad risk, what should he decide to do so as to maximize the company's expected profit?

(c) Construct an opportunity-loss table and use it to determine at most how much he should be willing to spend to find out for certain whether the loan applicant is a good risk or a bad risk?

(d) Suppose that the credit manager's superior feels that the credit manager rates the applicant too low, and that the probabilities of his being a good risk or a bad risk should be 0.15 and 0.85. Would it affect the credit manager's decision if he assigned these probabilities to the two events?

7.62 In a football game it is third down and 4 yards to go. The offensive team must decide whether to call a running play or a forward pass, and the defensive team must decide whether to expect a running play or a forward pass and set up the defense accordingly. Depending on these choices, the probabilities that the offensive team will make the first down are as shown in the following table:

| | | Offensive team | |
		Running play	Forward pass
Defensive team	Running play	0.30	0.70
	Forward pass	0.50	0.40

(a) What randomized decision procedure will maximize the offensive team's minimum expected probability of making the first down?

(b) If the offensive team uses the procedure of part (a), what is the probability that it will make the first down?

7.63 The following is the payoff table for a zero-sum two-person game, with the payoffs being the amounts player A loses to player B:

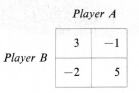

Player A

3	−1
−2	5

Player B

(a) What randomized decision procedure should player A use so as to minimize the maximum expected loss?

(b) What randomized decision procedure should player B use so as to maximize the minimum expected gain?

7.64 The following is the payoff table for a problem of decision making under competition, with the payoffs being the losses (in thousands of dollars) of the decision maker who must choose a column:

−2	6	1
3	5	2
4	0	−1

Find the saddle point.

7.65 Ms. Adams feels that it is a toss-up whether to accept a cash gift of $10 or to gamble on drawing a card at random from an ordinary deck of playing cards, where she is to receive $2 if she draws a club, a diamond, or a heart, or a transistor radio if she draws a spade. What utility does she attach to the radio?

In most statistical problems we are interested in only one aspect, or at most in a few aspects, of the outcomes of experiments. For example, a student who takes a true–false test may be interested only in how many questions he misses, since his grade depends on just this. An annuitant considering a stock for investment may be interested only in the yield on the stock and care nothing about the earnings per share of the issuing company or the stock's current price–earnings ratio, but a stock analyst in a brokerage firm must be concerned with all three of these things. Also, a grape hybridizer may be interested not only in the yield per acre of a new grape variety but also in the total acidity and the sugar content of the grapes; and an automotive engineer may be interested in both the durability and brightness of the headlights proposed for a new model car.

In these five cases, the student, the annuitant, the stock analyst, the hybridizer, and the engineer are all interested in numbers that are associated with the outcomes of situations involving an element of chance, or more specifically, in the values taken on by **random variables**.

In the study of random variables, we are usually interested in the probabilities with which they take on the various values within their range, namely, in their **probability distributions**. The study of random variables and probability distributions in Sections 8.1 and 8.2 will be followed by the discussion of various special probability distributions in Sections 8.3 through 8.6, and the description of their most important features in Sections 8.7, 8.8, and 8.9.

Probability Distributions

Random Variables

To be more explicit about the concept of a random variable, let us consider Figure 8.1, which, like Figure 6.9, pictures the sample space for the example dealing with the two contractors who have bid on two jobs. Here, however, we have added another number to each point—the number 0 to the point (0, 0); the number 1 to the points (1, 0) and (0, 1); and the number 2 to the points (2, 0), (1, 1), and (0, 2). In this way, we have associated with each point of the sample space the total number of jobs which, between them, the two contractors will get.

Since associating numbers with the points of a sample space is just a way of defining a function over the points of the sample space, random variables are, strictly speaking, functions. Conceptually, though, most beginners find it easier to think of random variables simply as quantities which can take on different values depending on chance. For instance, the number of speeding tickets issued each day on the freeway from Phoenix to Tucson is a random variable, and so is the annual production of soybeans in the United States, the number of persons visiting Disneyland each week, the number of defectives produced each

8.1

Sample space with values of random variable.

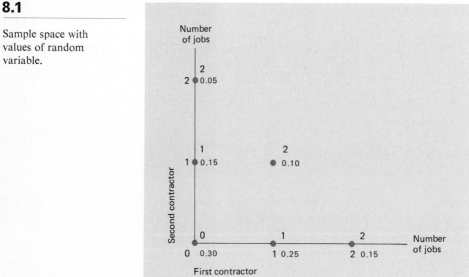

day by a machine, and the number of classified ads in the Sunday edition of a metropolitan newspaper.

Random variables are usually classified according to the number of values which they can assume. In this chapter we shall limit our discussion to random variables called **discrete random variables** which can take on only a finite number of values, or a countable infinity of values (as many as there are whole numbers). Infinite random variables will be taken up in Chapter 10.

8.2

Probability Distributions

The tables in the illustrations which follow serve to show what we mean by a probability distribution. With reference to Figure 8.1, if we add the probabilities associated with the different points, we find that the random variable "the number of jobs which, between them, the two contractors will get" takes on the value 0 with probability 0.30, the value 1 with probability $0.25 + 0.15 = 0.40$, and the value 2 with probability $0.15 + 0.10 + 0.05 = 0.30$. All this is summarized in the following table:

Number of jobs	Probability
0	0.30
1	0.40
2	0.30

As this table shows, a probability distribution is a correspondence which assigns probabilities to the values of a random variable.

For the probability distribution of the number of points which show in one roll of a fair die, we have the correspondence shown in the following table:

Number of points rolled with a die	Probability
1	$\frac{1}{6}$
2	$\frac{1}{6}$
3	$\frac{1}{6}$
4	$\frac{1}{6}$
5	$\frac{1}{6}$
6	$\frac{1}{6}$

Also, for four flips of a balanced coin there are the 16 equally likely possibilities, HHHH, HHHT, HHTH, HTHH, THHH, HHTT, HTHT, HTTH, THHT, THTH, TTHH, HTTT, THTT, TTHT, TTTH, and TTTT, and, counting the number of heads in each case and using the formula $\frac{s}{n}$ for equally likely possibilities, we get

Number of heads	Probability
0	$\frac{1}{16}$
1	$\frac{4}{16}$
2	$\frac{6}{16}$
3	$\frac{4}{16}$
4	$\frac{1}{16}$

Whenever possible, we try to express probability distributions by means of mathematical formulas which enable us to calculate the probabilities associated with the various values of a random variable. For instance, for the number of points which we roll with a fair die, we can write

$$f(x) = \frac{1}{6} \quad \text{for } x = 1, 2, 3, 4, 5, \text{ and } 6$$

where $f(1)$ represents the probability of rolling a 1, $f(2)$ represents the probability of rolling a 2, and so on, in the usual functional notation.†

In the sections which follow, we shall see how other probability distributions can be expressed mathematically as formulas.

To conclude this introduction to probability distributions, let us state the following two general rules which the values of all probability distributions must obey:

> Since the values of a probability distribution are probabilities, they must be numbers on the interval from 0 to 1.

> Since a random variable has to take on one of its values, the sum of all the values of a probability distribution must be equal to 1.

†We shall write the probability that a random variable takes on the value x as $f(x)$, but we could just as well write it as $g(x)$, $h(x)$, $b(x)$, and so on.

8.3

The Binomial Distribution

There are many applied problems in which we are interested in the probability that an event will occur "x times in n trials." For instance, we may be interested in the probability of getting 34 responses to 300 mail questionnaires, the probability that 8 of 24 newly franchised travel agencies will go bankrupt within two years, or the probability that 132 of 400 television viewers (interviewed by a rating service) will recall what products were advertised on a given program. To borrow again from the language of games of chance, we say that in each of these examples we are interested in the probability of getting x successes and $n - x$ failures in n trials.

In the problems which we shall study in this section, it will always be assumed that the number of trials is fixed, that the probability of a success is the same for each trial, and that the trials are all independent. This means that the theory we shall develop will not apply, for example, if we are interested in the number of pairs of shoes a woman will try on before she finally buys a pair (where the number of trials is not fixed), if we check hourly whether traffic is congested at an important intersection (where the probability of congestion is not the same for each trial), or if we are interested in a student's passing in order French I, French II, French III, and French IV (where the trials are not independent).

To solve problems which do meet the conditions listed in the preceding paragraph, we use a formula obtained in the following way. If p and $1 - p$ are the probabilities of a success and a failure on any given trial, then the probability of getting x successes and $n - x$ failures in some specific order is $p^x(1 - p)^{n-x}$; clearly, in this product of p's and $(1 - p)$'s there is one factor p for each success, one factor $1 - p$ for each failure, and the x factors p and $n - x$ factors $1 - p$ are all multiplied together by virtue of the generalized multiplication rule for more than two independent events. Since this probability applies to all points of the sample space which represent x successes and $n - x$ failures (in any specific order), we have only to count how many points of this kind there are, and then multiply $p^x(1 - p)^{n-x}$ by this number. Now, the number of ways in which we can select the x trials on which there is to be a success is $\binom{n}{x}$, the number of combinations of x objects selected from a set of n objects, and we have arrived at the following result:

Binomial distribution

> The probability of getting x successes in n independent trials is
>
> $$f(x) = \binom{n}{x} p^x (1 - p)^{n-x} \quad \text{for } x = 0, 1, 2, \ldots, \text{ or } n$$
>
> where p is the constant probability of a success for each trial.

It is customary to say here that the number of successes in n trials is a random variable having the **binomial probability distribution**, or simply the **binomial distribution**. The binomial distribution is called by this name because for $x = 0, 1, 2, \ldots,$ and n, the values of the probabilities are the successive terms of the binomial expansion of $[(1 - p) + p]^n$.

EXAMPLE Write the formula for the binomial distribution of the number of heads obtained in four flips of a balanced coin.

SOLUTION Substituting $n = 4$ and $p = \frac{1}{2}$ into the formula, we get

$$f(x) = \binom{4}{x}\left(\frac{1}{2}\right)^x \left(1 - \frac{1}{2}\right)^{4-x} = \binom{4}{x}\left(\frac{1}{2}\right)^4 = \frac{\binom{4}{x}}{16}$$

for $x = 0, 1, 2, 3,$ and 4. For instance, for $x = 2$ we get $f(2) = \binom{4}{2}\Big/16$ $= \frac{6}{16}$, and for $x = 4$ we get $\binom{4}{4}\Big/16 = \frac{1}{16}$.

EXAMPLE If the probability is 0.20 that any one shoplifter will get caught, what is the probability that in a random sample of eight shoplifters three will get caught?

SOLUTION Substituting $x = 3$, $n = 8$, $p = 0.20$, and $\binom{8}{3} = 56$ into the formula, we get

$$f(3) = \binom{8}{3}(0.20)^3(1 - 0.20)^{8-3}$$
$$= 56(0.20)^3(0.80)^5$$
$$= 0.147$$

The following is an example in which we calculate all the probabilities of a binomial distribution:

EXAMPLE The probability is 0.30 that a person making under \$20,000 per year owns common stocks. Find the probabilities that among six randomly selected persons making under \$20,000 per year there are 0, 1, 2, 3, 4, 5, or 6 who own common stocks.

SOLUTION Substituting $n = 6$, $p = 0.30$, and $x = 0, 1, 2, 3, 4, 5,$ and 6 into the formula for the binomial distribution, we get

$$f(0) = \binom{6}{0}(0.30)^0(0.70)^6 = 0.118$$

$$f(1) = \binom{6}{1}(0.30)^1(0.70)^5 = 0.303$$

$$f(2) = \binom{6}{2}(0.30)^2(0.70)^4 = 0.324$$

$$f(3) = \binom{6}{3}(0.30)^3(0.70)^3 = 0.185$$

$$f(4) = \binom{6}{4}(0.30)^4(0.70)^2 = 0.060$$

$$f(5) = \binom{6}{5}(0.30)^5(0.70)^1 = 0.010$$

$$f(6) = \binom{6}{6}(0.30)^6(0.70)^0 = 0.001$$

A histogram of this binomial distribution is shown in Figure 8.2.

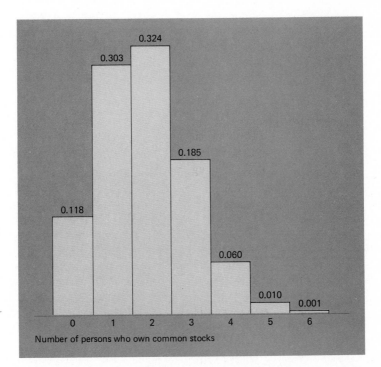

8.2

Histogram of binomial distribution with $n = 6$ and $p = 0.30$.

In actual practice, binomial probabilities are seldom calculated by direct substitution into the formula. Sometimes we use approximations such as those discussed later, but more often we refer to special tables such as Table V at the end of the book or the more detailed tables listed in the Bibliography. Table V is limited to the binomial probabilities for $n = 2$ to $n = 15$, and $p = 0.05$, 0.1, 0.2, 0.3, 0.4, 0.5, 0.6, 0.7, 0.8, 0.9, and 0.95. Where values are omitted in the table, they are 0.0005 or less.

EXAMPLE Suppose that the probability is 0.40 that a car stolen in a given city will be recovered. Use Table V to find the probabilities that
(a) at most three of 10 cars stolen in this city will be recovered;
(b) at least seven of 10 cars stolen in this city will be recovered.

(a) For $n = 10$ and $p = 0.40$ the entries in Table V corresponding to $x = 0, 1, 2,$ and 3 are 0.006, 0.040, 0.121, and 0.215, and the probability that at most three of 10 cars will be recovered is

$$0.006 + 0.040 + 0.121 + 0.215 = 0.382$$

(b) For $n = 10$ and $p = 0.40$ the entries in Table V corresponding to $x = 7, 8,$ and 9 are 0.042, 0.011, and 0.002; the probability that $x = 10$ is at most 0.0005, so the probability that at least seven of 10 cars will be recovered is

$$0.042 + 0.011 + 0.002 = 0.055$$

EXAMPLE If the probability is 0.05 that any one person will dislike the taste of a new mouth wash, what is the probability that at least two of 15 randomly selected persons will dislike it?

SOLUTION Adding the tabled probabilities for $n = 15$, $p = 0.05$, and $x = 2, 3, 4,$ and 5, we get

$$0.135 + 0.031 + 0.005 + 0.001 = 0.172$$

The probabilities which have not been added here, those for $x = 6, 7, 8, \ldots,$ and 15, are all no greater than 0.0005 and they are not shown in Table V.

EXERCISES

8.1 In each case determine whether the given values can be looked upon as the values of a probability distribution of a random variable which can take on only the values 1, 2, 3, and 4, and explain your answers:
(a) $f(1) = 0.24$, $f(2) = 0.24$, $f(3) = 0.24$, $f(4) = 0.24$;
(b) $f(1) = \frac{1}{6}$, $f(2) = \frac{2}{6}$, $f(3) = \frac{3}{6}$, $f(4) = \frac{4}{6}$;
(c) $f(1) = 0.15$, $f(2) = 0.38$, $f(3) = 0.24$, $f(4) = 0.23$;
(d) $f(1) = 0.23$, $f(2) = 0.42$, $f(3) = -0.03$, $f(4) = 0.38$;
(e) $f(1) = \frac{1}{2}$, $f(2) = \frac{1}{4}$, $f(3) = \frac{1}{8}$, $f(4) = \frac{1}{16}$.

8.2 Determine whether the following can be probability distributions, defined in each case for the given values of x, and explain your answers:

(a) $f(x) = \dfrac{1}{3}$ for $x = 0, 1, 2, 3$;

(b) $f(x) = \dfrac{x}{15}$ for $x = 0, 1, 2, 3, 4, 5$;

(c) $f(x) = \dfrac{x - 2}{5}$ for $x = 0, 1, 2, 3, 4, 5$;

(d) $f(x) = \dfrac{x^2}{30}$ for $x = 0, 1, 2, 3, 4$.

8.3 In a large government agency, illness is given as the reason for 90 percent of all absences from work. Find the probability that three of four absences from work (randomly selected from the agency's records) were claimed to be due to illness, by using

(a) the formula for the binomial distribution;
(b) Table V.

8.4 A multiple-choice test consists of eight questions and three answers to each question (only one of which is correct). If a student answers each question by rolling a balanced die and checking the first answer if he gets a 1 or a 2, the second answer if he gets a 3 or a 4, and the third answer if he gets a 5 or a 6, find the probability of getting
(a) exactly three correct answers;
(b) no correct answers;
(c) at least six correct answers.

8.5 Assuming that 1 in 5 industrial accidents are due to fatigue, find the probability that 2 of 12 randomly selected industrial accidents will be due to fatigue, using
(a) the formula for the binomial distribution;
(b) Table V.

8.6 If it is known from experience that 50 percent of all persons who get a mail-order catalog will order something from it, find the probability that three of nine randomly selected persons who get this catalog will order something from it by using
(a) the formula for the binomial distribution;
(b) Table V.

8.7 In the game of "chuck-a-luck" three dice are thrown and a player bets on the occurrence of a number which he can choose. If he wins $1 if his number appears on only one die, $2 if his number appears on two dice, $3 if his number appears on all three dice, and he loses $1 if his number appears on none of the dice, find the probabilities of his winning $1, $2, $3, and the probability of his losing $1. Also determine the player's mathematical expectation.

8.8 A bank knows that 60 percent of its checking-account customers will pay extra for personalized checks. Find the probability that at most two of eight randomly selected checking-account customers will pay extra for personalized checks.

8.9 Find the constant probability that a census enumerator will find any one family on his list at home in the afternoon, if the probability that all four of the families he plans to visit on an afternoon are out equals $\frac{16}{81}$.

8.10 An agricultural cooperative claims that 95 percent of the watermelons that are shipped out are ripe and ready to eat. Find the probabilities that among ten watermelons randomly selected to be shipped out
(a) at least eight are ripe and ready to eat;
(b) from seven to nine are ripe and ready to eat;
(c) at most eight are ripe and ready to eat.

8.11 A study has shown that 50 percent of the families in a certain large area have at least two cars. Find the probabilities that among 15 families randomly selected in this area for a market research study
(a) 8 have at least two cars;
(b) more than 11 have at least two cars;
(c) at most 5 have at least two cars;
(d) from 9 to 12 have at least two cars.

8.12 Suppose that a civil service examination is designed so that 70 percent of all persons with an IQ of 90 can pass it. Find the probabilities that among 15 persons with an IQ of 90 who take the test
(a) at most six will pass;
(b) at least 12 will pass;
(c) from 9 to 12 will pass.

8.13 The quality-control engineer of an electronics firm claims that 95 percent of the components that are shipped out are in good working condition. Find the probabilities that among 14 components which are shipped out, 0, 1, 2, 3, . . . , 13, or 14 will be in good working condition, and draw a histogram of this probability distribution.

8.14 A food distributor claims that 80 percent of his 6-ounce cans of mixed nuts contain at least three pecans. To check this, a consumer testing service decides to take 10 of the 6-ounce cans at random from a very large production lot, reject the claim if less than 7 of them contain at least three pecans, and otherwise accept it. What are the probabilities that the testing service will
(a) reject the claim even though it is true;
(b) accept the claim when in reality only 50 percent of the cans contain at least three pecans;
(c) accept the claim when in reality only 30 percent of the cans contain at least three pecans.

8.15 In some situations where the binomial distribution applies, we are interested in the probabilities that the first success will occur on any given trial. For this to happen on the xth trial, it must be preceded by $x - 1$ failures for which the probability is $(1 - p)^{x-1}$, and it follows that the probability that the first success will occur on the xth trial is

Geometric distribution

$$f(x) = p(1 - p)^{x-1} \quad for \ x = 1, 2, 3, 4, \ldots$$

This distribution is called the geometric distribution (because its successive values constitute a geometric progression) and it should be observed that there is a countable infinity of possibilities.† Using the formula for the geometric distribution, we find, for example, that in repeated rolls of a balanced die, the probability that the first 6 will occur on the fifth roll is $\frac{1}{6}\left(\frac{5}{6}\right)^{5-1} = \frac{625}{7,776}$.

(a) When taping a television commercial, the probability that a certain actor will get his lines straight on any one take is 0.40. What is the probability that this actor will get his lines straight for the first time on the fourth take?
(b) Suppose the probability is 0.25 that any given person will believe a rumor about the private life of a certain politician. What is the probability that the fifth person to hear the rumor will be the first one to believe it?

---

†As formulated in Chapter 6, the postulates of probability apply only when the sample space is finite. When the sample space is countably infinite (that is, when there are as many outcomes as there are whole numbers), as is the case here, the third postulate has to be modified so that for any sequence of mutually exclusive events $A_1, A_2, A_3, \ldots$,

$$P(A_1 \cup A_2 \cup A_3 \cup \cdots) = P(A_1) + P(A_2) + P(A_3) + \cdots$$

8.4

The Hypergeometric Distribution

In Chapter 6 we spoke of sampling "with replacement" to illustrate the multiplication rule for independent events, and of sampling "without replacement" to illustrate the rule for dependent events. Now, to introduce a probability distribution which applies when we sample without replacement, let us consider the following problem. A company ships automatic dishwashers in lots of 24. Before they are shipped, though, an inspector randomly selects 4 dishwashers from each lot, and the lot passes this inspection only if all 4 are in perfect condition; otherwise, each dishwasher is checked out individually and repaired, if necessary, at a considerable cost. Clearly, this kind of sampling inspection involves certain risks—it is possible for a lot to pass this inspection even though 12, or even 20, of the 24 dishwashers have serious defects, and it is possible for a lot to fail this inspection even though only one of the dishwashers has a slight defect. This raises many questions. For instance, it may be of special interest to know the probability that a lot will pass the inspection when, say, 4 of the dishwashers are not in perfect condition. This means that we shall have to find the probability of 4 successes (perfect dishwashers) in 4 trials, and we might be tempted to argue that since 20 of the dishwashers are in perfect condition, the probability of getting such a dishwasher is $\frac{20}{24} = \frac{5}{6}$, and hence the probability of "4 successes in 4 trials" is

$$f(4) = \binom{4}{4}\left(\frac{5}{6}\right)^4\left(1 - \frac{5}{6}\right)^{4-4} = 1\left(\frac{5}{6}\right)^4 = \frac{625}{1,296} = 0.482$$

This use of the binomial distribution is correct if sampling is with replacement and each dishwasher is replaced before the next one is selected; otherwise, the assumption of independence is violated. Clearly, if a dishwasher is randomly selected and not replaced, the probability that a second randomly selected dishwasher will be in perfect condition depends on whether or not the first one was in perfect condition.

However, in sampling inspection we seldom, if ever, sample with replacement. To get the correct answer for our problem when sampling is without replacement, we can argue as follows: There are altogether $\binom{24}{4} = 10,626$ ways of selecting four of the 24 dishwashers, and they are all equiprobable since the selection is random. Among these, there are $\binom{20}{4} = 4,845$ ways of selecting four of the 20 dishwashers in perfect condition, and it follows by the rule for equiprobable outcomes that the desired probability is $\frac{4,845}{10,626} = 0.456$.

To generalize the method used in the preceding example, suppose that n objects are to be chosen from a set of a objects of one kind (successes) and b objects of another kind (failures), and that we are interested in the probability of getting "x successes and $n - x$ failures." Arguing as before, we can say that the x successes can be chosen in $\binom{a}{x}$ ways, the $n - x$ failures can be chosen in $\binom{b}{n-x}$ ways, and, hence, x successes and $n - x$ failures can be chosen in $\binom{a}{x} \cdot \binom{b}{n-x}$ ways. Also, n objects can be chosen from the whole set of $a + b$ objects in $\binom{a+b}{n}$ ways, and if we regard all these possibilities as equally likely, it follows that the probability of getting "x successes and $n - x$ failures" is

<table>
<tr><td>Hypergeometric
distribution</td><td>

$$f(x) = \frac{\binom{a}{x} \cdot \binom{b}{n-x}}{\binom{a+b}{n}} \qquad for \ x = 0, 1, 2, \ldots, \ or \ n$$

</td></tr>
</table>

This is the formula for the **hypergeometric distribution**, and it applies only when x does not exceed a and $n - x$ does not exceed b, since we cannot very well get more successes (or failures) than there are in the whole set.

EXAMPLE Among a department store's 16 delivery trucks, five have worn brakes. If eight trucks are randomly picked for inspection, what is the probability that this sample will include at least three trucks with worn brakes?

SOLUTION The probability we want to find is $f(3) + f(4) + f(5)$, where each term in this sum is to be calculated by means of the formula for the hypergeometric distribution with $a = 5$, $b = 11$, and $n = 8$. Substituting these values together with $x = 3$, then $x = 4$, and $x = 5$, we get

$$f(3) = \frac{\binom{5}{3} \cdot \binom{11}{5}}{\binom{16}{8}} = \frac{10 \cdot 462}{12,870} = 0.359$$

$$f(4) = \frac{\binom{5}{4} \cdot \binom{11}{4}}{\binom{16}{8}} = \frac{5 \cdot 330}{12,870} = 0.128$$

$$f(5) = \frac{\binom{5}{5} \cdot \binom{11}{3}}{\binom{16}{8}} = \frac{1 \cdot 165}{12,870} = 0.013$$

and the probability that the sample will include at least three trucks with worn brakes is

$$0.359 + 0.128 + 0.013 = 0.500$$

In the beginning of this section we introduced the hypergeometric distribution with an example in which we first erroneously used the binomial distribution. The error was not large though—we got 0.482 instead of 0.456—and in actual practice the binomial distribution is often used to approximate the hypergeometric distribution. It is generally agreed that this approximation is satisfactory if n constitutes less than 5 percent of $a + b$. The main advantages of the approximation are that the binomial distribution has been tabulated much more extensively than the hypergeometric distribution, and that it is generally easier to use.

8.5

The Poisson Distribution

If n is large and p is small, binomial probabilities are often approximated by means of the formula

Poisson distribution

$$f(x) = \frac{(np)^x \cdot e^{-np}}{x!} \qquad x = 0, 1, 2, 3, \ldots$$

which is that for the **Poisson distribution**. Here the irrational number $e = 2.71828\ldots$ is the base of the system of natural logarithms, and the values of e^{-np} may be read from Table X at the end of the book. For the Poisson distribution, the random variable x can take on the infinite set of values $x = 0, 1, 2, 3, \ldots$; practically speaking, though, this poses no problems, since the probabilities become negligible (very close to zero) after the first few values of x.

EXAMPLE Records show that the probability is 0.00005 that a car will have a flat tire while driving through a certain tunnel. Use the Poisson approximation to the binomial distribution to find the probability that among 10,000 cars passing through this tunnel at least two will have a flat tire.

SOLUTION Subtract from 1 the probabilities that 0 or 1 of the cars will have a flat tire. Since $np = 10,000(0.00005) = 0.5$ and $e^{-0.5} = 0.607$ (see Table X), we find that

$$f(0) = \frac{(0.5)^0(0.607)}{0!} = 0.607$$

$$f(1) = \frac{(0.5)^1(0.607)}{1!} = 0.304$$

and the probability that at least two cars will have a flat tire is $1 - (0.607 + 0.304) = 0.089$.

The Poisson distribution has many important applications which have no direct connection with the binomial distribution. In these cases np is replaced by the parameter λ (Greek lowercase lambda) and we calculate the probability of getting x "successes" by means of the formula

Poisson distribution (with parameter λ)

$$f(x) = \frac{\lambda^x \cdot e^{-\lambda}}{x!} \quad \textit{for } x = 0, 1, 2, 3, \ldots$$

where the parameter λ is interpreted as the expected, or average, number of successes.

The formula above applies in many situations where we can expect a fixed number of "successes" per unit time (or for some other kind of unit), say, when a bank can expect to receive 6 bad checks per day, when 1.6 accidents can be expected per day at a busy intersection, when 12 small pieces of meat can be expected in a frozen meat pie, when 5.2 imperfections can be expected per roll of cloth, when 0.3 complaint per visitor can be expected by the manager of a resort, and so on.

EXAMPLE   If a bank receives on the average $\lambda = 6$ bad checks per day, what is the probability that it will receive four bad checks on a given day?

SOLUTION   Substituting $\lambda = 6$ and $x = 4$ into the formula, we get

$$f(4) = \frac{6^4 \cdot e^{-6}}{4!} = \frac{(1,296)(0.0025)}{24} = 0.135$$

EXAMPLE   If $\lambda = 1.6$ accidents can be expected at a certain intersection per day, what is the probability that there will be three accidents at the intersection on a given day?

SOLUTION   Substituting $\lambda = 1.6$ and $x = 3$ into the formula, we get

$$f(3) = \frac{1.6^3 \cdot e^{-1.6}}{3!} = \frac{(4.096)(0.202)}{6} = 0.138$$

8.6

The Multinomial Distribution

An important generalization of the binomial distribution arises when there are more than two possible outcomes for each trial, the probabilities of the various outcomes remain the same for each trial, and the trials are all independent.

This is the case, for example, in repeated rolls of a die where each trial has six possible outcomes, when students are asked whether they like a certain record, dislike it, or are indifferent toward it, or when a USDA inspector grades beef as prime, choice, good, commercial, or utility.

If there are k possible outcomes for each trial and their probabilities are $p_1, p_2, \ldots,$ and p_k, it can be shown that the probability of x_1 outcomes of the first kind, x_2 outcomes of the second kind, $\ldots,$ and x_k outcomes of the kth kind in n trials is given by

Multinomial distribution

$$\frac{n!}{x_1! x_2! \cdot \ldots \cdot x_k!} p_1^{x_1} \cdot p_2^{x_2} \cdot \ldots \cdot p_k^{x_k}$$

This distribution is called the **multinomial distribution**.

EXAMPLE If four clerks prepare all the billings in a company office and it has been determined that 40 percent of all erroneous billings are prepared by clerk A, 20 percent by clerk B, 10 percent by clerk C, and the rest by clerk D, what is the probability that among seven randomly selected erroneous billings two were prepared by A, one by B, one by C, and three by D?

SOLUTION Substituting $n = 7$, $x_1 = 2$, $x_2 = 1$, $x_3 = 1$, $x_4 = 3$, $p_1 = 0.4$, $p_2 = 0.2$, $p_3 = 0.1$, and $p_4 = 0.3$ into the formula, we get

$$\frac{7!}{2! \cdot 1! \cdot 1! \cdot 3!} (0.4)^2 (0.2)^1 (0.1)^1 (0.3)^3 = 0.036$$

EXERCISES

8.16 Among 15 job evaluation reports there are nine which have not been cleared for release and should be marked "Hold." If someone who is unfamiliar with the reports randomly selects two of them and marks them "Hold," what are the probabilities that
(a) neither should be marked "Hold";
(b) one should be marked "Hold";
(c) both should be marked "Hold?"

8.17 Find the probability that an IRS auditor will get three tax returns with unallowable deductions, if he randomly selects five returns from among 12 returns, six of which contain unallowable deductions?

8.18 Twenty-five members, 10 women and 15 men, of a neighborhood improvement association are in attendance at a large city zoning commission hearing. Three of the association members are selected by lot and allowed to address the meeting formally. Find the probability distribution of the number of women selected and draw a histogram of this distribution.

8.19 If, among 16 delivery trucks, five have worn brakes and ten are chosen at random for inspection, what is the probability that at least three trucks with worn brakes are chosen?

8.20 Among the 12 male applicants for a job with the postal service, 8 have working wives. If two of the applicants are randomly selected for further consideration, find the probabilities that
(a) neither has a working wife;
(b) one has a working wife;
(c) both have working wives.

8.21 Among 15 color television sets which have had tube changes, 4 have had their tube and circuitry lives shortened by improper voltage regulation after the change. Find the probability distribution of the number of sets so affected in a random sample of 3 sets taken without replacement from these 15 sets.

8.22 To pass a quality control inspection, two batteries are chosen from each lot of 12 car batteries, and the lot is passed only if neither battery has any defects; otherwise, each of the batteries in the lot is checked. If the selection of the batteries is random, find the probabilities that a lot will
(a) pass the inspection when one of the 12 batteries is defective;
(b) fail the inspection when three of the batteries are defective;
(c) fail the inspection when six of the batteries are defective.

8.23 A shipment of 120 burglar alarms contains 5 defectives. If 3 of these burglar alarms are randomly selected and shipped to a customer, find the probability that he will get exactly one bad unit using
(a) the formula for the hypergeometric distribution;
(b) the binomial distribution as an approximation.
What is the error of the approximation of part (b)?

8.24 Among the 200 employees of a company, 160 are union members and the others are nonunion. If four of the employees are chosen by lot to serve on the pension fund committee, find the probability that two of them will be union members and the others nonunion, using
(a) the formula for the hypergeometric distribution;
(b) the binomial distribution as an approximation.
What is the error of the approximation of part (b)?

8.25 It is presumed that 1.5 percent of the inhabitants of a border city are illegal immigrants. Use the Poisson approximation to the binomial distribution to determine the probability that in a random sample of 200 inhabitants of the city, two are illegal immigrants.

8.26 It is known from experience that 2.8 percent of the calls received at a company switchboard are wrong numbers. Use the Poisson approximation to the binomial distribution to determine the probability that among 250 calls received at the switchboard, 6 are wrong numbers.

8.27 If 1.8 percent of the fuses delivered to an arsenal are defective, what is the approximate probability that in a random sample of 400 fuses, 8 are defective?

8.28 If the number of complaints which a laundry receives per day is a random variable having the Poisson distribution with $\lambda = 3.5$, find the probabilities that on any given day the laundry will receive
(a) no complaints;   (c) two complaints;
(b) at least one complaint;   (d) four complaints.

8.29 The number of arrivals for service at a tool crib in a plant per quarter hour of first-shift time is a random variable having the Poisson distribution with $\lambda = 2$.

What are the probabilities of 0, 1, and 2 arrivals for service in a randomly chosen quarter hour?

8.30 Suppose that in the inspection of metal produced in continuous rolls the number of imperfections spotted by an inspector during a 10-minute period is a random variable having the Poisson distribution with $\lambda = 1.7$. Find the probabilities that during a 10-minute period an inspector will find

(a) no imperfections;   (c) two imperfections;

(b) one imperfection;   (d) at least three imperfections.

8.31 The manager of television station X tells a prospective advertiser that on Saturday nights his station has 60 percent of the family viewing audience, while his two competitors, stations Y and Z, have 30 percent and 10 percent. If this is so, what is the probability that among eight randomly selected families watching television on a Saturday night, four will be watching station X, two will be watching station Y, and two will be watching station Z?

8.32 Suppose that the probabilities are 0.50, 0.20, 0.20, and 0.10 that a state income tax form will be filled out correctly, that it will contain only errors favoring the taxpayer, that it will contain only errors favoring the government, or that it will contain both kinds of errors. What is the probability that among 12 such tax forms (randomly chosen for audit) five will be filled out correctly, three will contain only errors favoring the taxpayer, three will contain only errors favoring the government, and one will contain both kinds of errors?

8.7

The Mean of a Probability Distribution

When we say that an airline office at an airport can expect 2.91 complaints per day about its luggage handling, we are referring to the sum of the products obtained by multiplying 0, 1, 2, 3, 4, . . . , by the probabilities that it will receive 0, 1, 2, 3, 4, . . . , complaints on any one day. If we apply the same argument to the illustrations of Section 8.2, we find that, between them, the two contractors can expect to get

$$0(0.30) + 1(0.40) + 2(0.30) = 1$$

of the two jobs, the number of points we can expect in the roll of a die is

$$1 \cdot \frac{1}{6} + 2 \cdot \frac{1}{6} + 3 \cdot \frac{1}{6} + 4 \cdot \frac{1}{6} + 5 \cdot \frac{1}{6} + 6 \cdot \frac{1}{6} = 3\frac{1}{2}$$

and the number of heads we can expect in four flips of a balanced coin is

$$0 \cdot \frac{1}{16} + 1 \cdot \frac{4}{16} + 2 \cdot \frac{6}{16} + 3 \cdot \frac{4}{16} + 4 \cdot \frac{1}{16} = 2$$

Of course, we cannot actually roll a $3\frac{1}{2}$ with a die; like all mathematical expectations, this figure must be looked upon as an average.

In general, if a random variable takes on the values $x_1, x_2, x_3, \ldots$, and x_k, with the probabilities $f(x_1), f(x_2), f(x_3), \ldots$, and $f(x_k)$, its expected value (or its mathematical expectation) is given by the quantity

$$x_1 \cdot f(x_1) + x_2 \cdot f(x_2) + x_3 \cdot f(x_3) + \cdots + x_k \cdot f(x_k)$$

called the **mean of the random variable** or the **mean of its probability distribution**. The mean of a random variable is usually denoted by the Greek letter μ (lower-case mu), and using the $\sum$ notation, we can write

Mean of probability distribution

$$\mu = \sum x \cdot f(x)$$

EXAMPLE   With reference to the example on page 214, how many of six randomly selected persons making under \$20,000 per year can be expected to own stocks?

SOLUTION   Substituting $x = 0, 1, 2, 3, 4, 5$, and 6, and the probabilities on pages 214 and 215 into the formula for μ, we get

$$\mu = 0(0.118) + 1(0.303) + 2(0.324) + 3(0.185) + 4(0.060)$$
$$+ 5(0.010) + 6(0.001)$$
$$= 1.802$$

When a random variable can take on many different values, the calculation of μ usually becomes quite laborious. For instance, if we want to know how many among 800 customers entering a store can be expected to make a purchase, and the probability that any one of them will make a purchase is 0.40, we would first have to calculate the 801 probabilities corresponding to 0, 1, 2, $\ldots$, and 800 of them making a purchase. However, if we think for a moment, we might argue that in the long run 40 percent of the customers make a purchase, 40 percent of 800 is 320, and, hence, we can expect 320 of the 800 customers to make a purchase. Similarly, if a balanced coin is flipped 1,000 times, we can argue that in the long run heads will come up 50 percent of the time and, hence, that we can expect $(1,000)(0.50) = 500$ heads. These two values are, indeed, correct, and it can be shown that, in general, there exists the special formula

Mean of binomial distribution

$$\mu = n \cdot p$$

for the mean of a binomial distribution. In words, the mean of a binomial distribution is simply the product of the number of trials and the probability of success on an individual trial.

EXAMPLE   With reference to the example on page 214, find the mean of the probability distribution of the number of persons who own common stocks.

SOLUTION Since we are dealing with a binomial distribution with $n = 6$ and $p = 0.30$, we have $\mu = 6(0.30) = 1.80$. The small difference of 0.002 between this exact value and the value obtained before is due to rounding the probabilities to three decimals in the earlier calculations.

EXAMPLE Find the mean of the probability distribution of the number of heads obtained in four flips of a balanced coin.

SOLUTION For a binomial distribution with $n = 4$ and $p = \frac{1}{2}$, we get $\mu = 4 \cdot \frac{1}{2} = 2$, and this agrees with the result obtained in the beginning of this section.

It is important to remember, of course, that the formula $\mu = n \cdot p$ applies only to binomial distributions. Fortunately, there are other formulas for other distributions; for the hypergeometric distribution, for example, the formula for the mean is

Mean of hypergeometric distribution

$$\mu = \frac{n \cdot a}{a + b}$$

EXAMPLE Among 16 delivery trucks, five have worn brakes. If eight trucks are randomly picked for inspection, how many of them can be expected to have worn brakes?

SOLUTION We have here a hypergeometric distribution with $a = 5$, $b = 11$, and $n = 8$, and substitution into the formula above yields

$$\mu = \frac{8 \cdot 5}{5 + 11} = 2.5$$

This should not come as a surprise; half of the trucks are selected and half of the ones with worn brakes are expected to be included in the sample.

Also, the mean of the Poisson distribution is simply $\mu = \lambda$. Formal proofs of all these special formulas may be found in any textbook on mathematical statistics.

8.8

The Standard Deviation of a Probability Distribution

We saw in Chapter 3 that there are many cases in which we must describe, in addition to the mean or some other measure of location, the variability (spread, or dispersion) of a set of data. The most widely used statistical measures of variation are the variance and its square root, the standard deviation, which both measure variability by averaging the squared deviations from the mean. For probability distributions, we measure variability in almost the same way, but instead of averaging the squared deviations from the mean, we calculate their expected value. If x is a value of some random variable whose probability distribution has the mean μ, the deviation from the mean is $x - \mu$ and we define the **variance of the probability distribution** as the expected value of the squared deviation from the mean, that is, as

Variance
of probability
distribution

$$\sigma^2 = \sum (x - \mu)^2 \cdot f(x)$$

where the summation extends over all values assumed by the random variable. The square root of the variance defines the standard deviation, σ, of a probability distribution, and we write

Standard deviation
of probability
distribution

$$\sigma = \sqrt{\sum (x - \mu)^2 \cdot f(x)}$$

EXAMPLE Use the probabilities obtained in the example on pages 214 and 215, to determine the standard deviation of the probability distribution of the numbers of persons, among six with incomes under \$20,000, who own common stocks.

SOLUTION Here $\mu = 6(0.30) = 1.80$ and we arrange the calculations as follows:

Number of persons	Prob- ability	Deviation from mean	Squared deviation from mean	$(x - \mu)^2 f(x)$
0	0.118	−1.8	3.24	0.38232
1	0.303	−0.8	0.64	0.19392
2	0.324	0.2	0.04	0.01296
3	0.185	1.2	1.44	0.26640
4	0.060	2.2	4.84	0.29040
5	0.010	3.2	10.24	0.10240
6	0.001	4.2	17.64	0.01764

$$\sigma^2 = 1.26604$$

The values in the column on the right were obtained by multiplying each squared deviation from the mean by its probability, and their sum is the variance of the distribution. Also $\sigma = \sqrt{1.26604} = 1.13$.

The calculations in the preceding example were quite easy since the deviations from the mean were small numbers given to one decimal. If the deviations from the mean are large numbers, or if they are given to several decimals, it is usually worthwhile to simplify the calculations by using the computing formula for σ^2 given in Exercise 8.35 on page 231.

As in the case of the mean, the calculation of the variance or the standard deviation of a probability distribution can often be simplified when dealing with special kinds of probability distributions. For instance, for the binomial distribution we have the formula

Standard deviation of binomial distribution

$$\sigma = \sqrt{np(1-p)}$$

EXAMPLE   Use this formula to verify the result obtained in the preceding example.

SOLUTION   For a binomial distribution with $n = 6$ and $p = 0.30$, the formula yields

$$\sigma^2 = 6(0.30)(0.70) = 1.26$$

and this exact value differs from the result obtained before by the small rounding error of $1.26604 - 1.26 = 0.00604$.

EXAMPLE   Find the variance of the probability distribution of the number of heads obtained in four flips of a balanced coin.

SOLUTION   The variance of the binomial distribution with $n = 4$ and $p = \frac{1}{2}$ is

$$\sigma^2 = 4 \cdot \frac{1}{2} \cdot \frac{1}{2} = 1$$

There also exist special formulas for the standard deviation of other special distributions, and they may be found in more advanced texts.

8.9

Chebyshev's Theorem

Intuitively speaking, the variance and the standard deviation of a probability distribution measure its spread or its dispersion: When σ is small, the probability is high that we will get a value close to the mean, and when σ is large, we are more likely to get a value far away from the mean. This important idea is expressed rigorously in a theorem called **Chebyshev's theorem**, which we introduced in Chapter 3. For probability distributions, this theorem can be stated as follows:

Chebyshev's theorem

> The probability that a random variable will take on a value within k standard deviations of the mean is at least
>
> $$1 - \frac{1}{k^2}$$

For instance, the probability of getting a value within two standard deviations of the mean (a value between $\mu - 2\sigma$ and $\mu + 2\sigma$) is at least $1 - \frac{1}{2^2} = \frac{3}{4}$, the probability of getting a value within five standard deviations of the mean (a value between $\mu - 5\sigma$ and $\mu + 5\sigma$) is at least $1 - \frac{1}{5^2} = \frac{24}{25}$, and so forth. When Chebyshev's theorem is used to illustrate the relationship between the standard deviation of a probability distribution and its spread or dispersion, k is often chosen more or less arbitrarily; k can be any positive number, although the theorem becomes trivial when k is 1 or less.

EXAMPLE The number of telephone calls which an answering service receives between 10 A.M. and 11 A.M. is a random variable whose distribution has the mean $\mu = 26$ and the standard deviation $\sigma = 3\frac{1}{3}$. What does Chebyshev's theorem with $k = 3$, for instance, tell us about the number of telephone calls which the answering service may receive between 10 A.M. and 11 A.M.?

SOLUTION Since $\mu - 3\sigma = 26 - 3(3\frac{1}{3}) = 16$ and $\mu + 3\sigma = 26 + 3(3\frac{1}{3}) = 36$, we can assert with a probability of at least $1 - \frac{1}{3^2} = \frac{8}{9}$, or approximately 0.89, that the answering service will receive between 16 and 36 calls.

EXAMPLE What does Chebyshev's theorem with $k = 6$, for instance, tell us about the number of heads we may get in 400 flips of a balanced coin?

SOLUTION For the binomial distribution with $n = 400$ and $p = \frac{1}{2}$, the mean and the standard deviation are $\mu = n \cdot p = 400 \cdot \frac{1}{2} = 200$ and $\sigma = \sqrt{np(1 - p)} = \sqrt{400 \cdot \frac{1}{2} \cdot \frac{1}{2}} = 10$, so that $\mu - 6\sigma = 200 - 6 \cdot 10 = 140$ and $\mu + 6\sigma = 200 + 6 \cdot 10 = 260$. Thus, we can assert with a probability of at least $1 - \frac{1}{6^2} = \frac{35}{36}$, or approximately 0.97, that we will get between 140 and 260 heads.

If we convert the numbers of heads into proportions, we can assert with a probability of at least $\frac{35}{36}$ that the proportion of heads we get in

400 flips of a balanced coin will lie between $\frac{140}{400} = 0.35$ and $\frac{260}{400} = 0.65$. To continue this argument, the reader will be asked to show in Exercise 8.51 on page 233 that the probability is at least $\frac{35}{36}$ that for 10,000 flips of a balanced coin the proportion of heads will lie between 0.47 and 0.53, and that for 1,000,000 flips of a balanced coin it will lie between 0.497 and 0.503. This provides support for the **Law of Large Numbers**, which we mentioned in Section 5.3 in connection with the frequency interpretation of probability.

The probability statement which Chebyshev's theorem enables us to make, though mathematically correct, is often unnecessarily weak—that is, the probability is often unnecessarily small. For instance, in the preceding example we showed that the probability of getting a value within six standard deviations of the mean is at least 0.97, whereas the actual probability that this will happen for a random variable having the binomial distribution with $n = 400$ and $p = \frac{1}{2}$ is about 0.999999998.

EXERCISES

8.33 The probabilities of 0, 1, 2, or 3 armed robberies in a Western city in any given month are 0.4, 0.3, 0.2, and 0.1. Find the mean and the variance of this probability distribution.

8.34 The following table gives the probabilities that a computer will malfunction 0, 1, 2, 3, 4, 5, or 6 times on any given day:

Number of malfunctions	0	1	2	3	4	5	6
Probability	0.15	0.22	0.31	0.18	0.09	0.04	0.01

Calculate the mean and the standard deviation of this probability distribution.

8.35 Using the rules for summations given in Section 3.10, we can derive the following shortcut formula for the variance of a probability distribution:

$$\sigma^2 = \sum x^2 \cdot f(x) - \mu^2$$

The advantage of this formula is that we do not have to work with the deviations from the mean. Instead, we subtract μ^2 from the sum of the products obtained by multiplying the square of each value of the random variable by the corresponding probability. Use this formula to find
(a) the variance of the probability distribution of Exercise 8.33;
(b) the standard deviation of the probability distribution of Exercise 8.34.

8.36 Use the formula of the preceding exercise to recalculate σ^2 for the common-stock example on page 228, where we obtained $\sigma^2 = 1.26604$.

8.37 Find σ^2 for the distribution of the number of times a fair coin falls heads in four flips, using the probabilities on page 212 and
(a) the definition formula on page 228;
(b) the shortcut formula of Exercise 8.35.
Also compare the results with that obtained on page 229 with the special formula $\sigma^2 = np(1 - p)$.

8.38 Find the mean and the standard deviation of the number of points rolled with a balanced die.

8.39 Find the mean and the standard deviation of the binomial distribution with $n = 5$ and $p = 0.10$, using
(a) the definition formulas for the mean and the standard deviation of a probability distribution;
(b) the special formulas for the mean and the standard deviation of a binomial distribution.

8.40 A study shows that 80 percent of all patients coming to a certain medical clinic have to wait at least 20 minutes to see their doctor. Find the mean and the variance of the distribution of the number of randomly selected patients among ten who have to wait at least 20 minutes to see their doctor, using
(a) the definition formula for the mean of a probability distribution and the shortcut formula for σ^2 of Exercise 8.35;
(b) the special formulas for the mean and the variance of a binomial distribution.

8.41 Find the mean and the standard deviation of the distribution of each of the following random variables having binomial distributions:
(a) The number of heads obtained in 576 flips of a balanced coin.
(b) The number of 5's obtained in 405 rolls of a balanced die.
(c) The number of persons (among 400 invited) who will attend the opening of a new branch bank, when the probability is 0.85 that any one of them will attend.
(d) The number of defectives in a sample of 2,400 parts made by a machine, when the probability is 0.04 that any one of the parts is defective.
(e) The number of students (among 800 interviewed) who do not like the food served at the university cafeteria, when the probability that any one of them does not like the food is 0.30.

8.42 If five of a department store's 16 delivery trucks have worn brakes and eight trucks are randomly picked for inspection, the probabilities are 0.359, 0.128, and 0.013 that 3, 4, or 5 of the eight will have worn brakes.
(a) Show that the probabilities are 0.013, 0.128, and 0.359 that 0, 1, or 2 of the eight will have worn brakes.
(b) Using all these probabilities, calculate the mean of the distribution of the number of trucks with worn brakes that will be included among the eight.
(c) Compare the value obtained for μ in part (b) with the exact value $\mu = 2.5$.

8.43 With reference to the preceding exercise, find the variance of the distribution of the number of trucks with worn brakes that will be included among the eight.

8.44 Two of the 12 cans of tomatoes in a carton have seam leaks and the rest do not.

(a) Find the probability distribution of the number of cans with seam leaks in a random sample of two cans drawn without replacement from this carton.

(b) Use the probabilities obtained in part (a) to calculate the mean of this distribution, and compare it with the value yielded by the special formula for the mean of a hypergeometric distribution.

8.45 The probabilities that there will be 0, 1, 2, 3, 4, or 5 fires caused by lightning during a summer storm are 0.449, 0.360, 0.144, 0.038, 0.008, and 0.001. Calculate the mean of this Poisson distribution with $\lambda = 0.8$, and use the result to confirm (subject to an error due to rounding) the special formula $\mu = \lambda$ for the mean of a Poisson distribution.

8.46 Use the probabilities of the preceding exercise and the shortcut formula of Exercise 8.35 to calculate the variance of the given distribution, and use the result to confirm (subject to an error due to rounding) the special formula $\sigma^2 = \lambda$ for the variance of a Poisson distribution.

8.47 If the number of imperfections in 10 yards of a certain kind of cloth is a random variable having the Poisson distribution with $\lambda = 4$, the probabilities of 0, 1, 2, 3, . . . , or 12 imperfections are 0.018, 0.073, 0.147, 0.195, 0.195, 0.156, 0.104, 0.060, 0.030, 0.013, 0.005, 0.002, and 0.001.

(a) Calculate the mean of this distribution, and use the result to verify (subject to an error due to rounding) the special formula $\mu = \lambda$ for the mean of a Poisson distribution.

(b) Calculate the standard deviation of this distribution, and use the result to verify (subject to an error due to rounding) the special formula $\sigma = \sqrt{\lambda}$ for the standard deviation of a Poisson distribution.

8.48 If a student answers the 100 questions on a true–false test by flipping a fair coin, marking "true" when the coin falls heads and "false" when it falls tails, what does Chebyshev's theorem with $k = 4$ tell us about the number of correct answers he will get?

8.49 The daily number of customers served breakfast on a weekday in a certain restaurant is a random variable with $\mu = 132$ and $\sigma = 12$. According to Chebyshev's theorem, with what probability can we assert that between 72 and 192 customers will be served breakfast in the restaurant on any given weekday?

8.50 The weekly number of special orders processed by a wholesale drug house is a random variable with $\mu = 126$ and $\sigma = 9$.

(a) What does Chebyshev's theorem with $k = 12$ tell us about the number of such orders processed in any given week?

(b) According to Chebyshev's theorem, with what probability can we assert that between 96 and 156 special orders will be processed in a week?

8.51 Use Chebyshev's theorem to show that the probability is at least $\frac{35}{36}$ that

(a) in 10,000 flips of a balanced coin there will be between 4,700 and 5,300 heads, and hence the proportion of heads will be between 0.47 and 0.53;

(b) in 1,000,000 flips of a balanced coin there will be between 497,000 and 503,000 heads, and hence the proportion of heads will be between 0.497 and 0.503.

8.10

A Word of Caution

When we use a specific probability distribution to describe a given situation, we must be sure that we are using the right model. For instance, we must make sure that we do not use the binomial distribution when we should be using the hypergeometric distribution, or that we do not use the Poisson distribution when we should be using some other distribution, unless, of course, we are intentionally making approximations. Specifically, we must always make sure that the assumptions underlying the distribution which we choose are actually met. Thus, it would be a mistake to use the binomial distribution to determine, say, the probability that there will be five rainy days at a given resort during the first two weeks in August, or the probability that 8 of 100 persons (whose age ranges from 18 to 79) will be hospitalized at least once during the coming year. In the first case the successive trials are clearly not independent, and in the second case the probability of being hospitalized is not the same for each trial (person).

8.11

Check List of Key Terms

8.12

Review Exercises

8.52 If only 4 percent of all medical students plan to specialize in psychiatry, what is the probability that in a random sample of three medical students
(a) one plans on this specialization;
(b) at least one plans on this specialization?

8.53 In attempting to simplify purchase control, a retail hardware store uses a color-code system, putting green tags on stock items bought the first half of a year and yellow tags on items bought the second half of the year. If there are

10 cake pans on a table, 6 tagged green and the rest yellow, and a customer buys 5 of them, selecting them at random, what are the probabilities that she selects

(a) 4 bought the first half of the year;

(b) at least 4 bought the first half of the year?

8.54 In each case determine whether the given values can be looked upon as the values of a probability distribution of a random variable which can take on the values 1, 2, and 3, and explain your answers:

(a) $f(1) = 0.32, f(2) = 0.35, f(3) = 0.35$;

(b) $f(1) = 0.47, f(2) = 0.69, f(3) = -0.16$;

(c) $f(1) = \frac{1}{3}, f(2) = \frac{5}{9}, f(3) = \frac{1}{9}$.

8.55 The probabilities that a building inspector will find 0, 1, 2, 3, 4, or 5 violations of the building code in a home built in a large development are 0.41, 0.22, 0.17, 0.13, 0.05, and 0.02. Find the mean of this probability distribution.

8.56 Find the variance of the probability distribution of the preceding exercise.

8.57 If the probability is 0.70 that a child exposed to a certain contagious disease will catch it, what is the probability that the third child exposed to the disease will be the first one to catch it?

8.58 A market bin contains 12 bottles of household bleach, 7 of brand Q and 5 of brand R. If a woman randomly selects two bottles from this bin and puts them into her cart, what is the probability that both bottles are of the same brand?

8.59 Among 20 persons empaneled for jury duty, six are over 65 years old. If four of the 20 are called at random, find

(a) the probabilities that 0, 1, 2, 3, or 4 are over 65 years old;

(b) the mean of this probability distribution.

Also, verify the result of part (b) by using the special formula for the mean of such a distribution.

8.60 The owner of a bookstore knows from experience that the probabilities are 0.50, 0.40, and 0.10 that a person browsing at the store will buy no books, one book, or two or more books. What is the probability that among nine persons browsing at the store, six will buy no books, two will buy one book, and one will buy two or more?

8.61 Find the mean and the variance of the binomial distribution with $n = 6$ and $p = 0.50$, using

(a) the formulas for the mean and the variance of a probability distribution;

(b) the special formulas for the mean and the variance of a binomial distribution.

8.62 The daily number of calls for a certain kind of service in a branch bank has a Poisson distribution with $\lambda = 2.4$. Find the probability that on one randomly selected day there are at least two calls for the service.

8.63 The number of marriage licenses issued in a certain city during the month of June averages $\mu = 134$ with a standard deviation of $\sigma = 7.5$.

(a) What does Chebyshev's theorem with $k = 9$ tell us about the number of marriage licenses issued there during a month of June?

(b) According to Chebyshev's theorem, with what probability can we assert that between 74 and 194 marriage licenses will be issued there during a month of June?

8.64 Suppose that of all franchise operations of a certain kind 20 percent are profitable. What is the probability that two of six randomly chosen franchise operations of this kind are profitable?

8.65 Eighty percent of all insurance policies of a certain kind remain in force for at least four years. What are the probabilities that in a random sample of 13 policies of this kind
(a) at most six remain in force for at least four years;
(b) from 9 to 11 remain in force for at least four years;
(c) at least 10 remain in force for at least four years?

8.66 Suppose that 4 percent of all the employees who request a car from a company's motor pool are not properly licensed. Use the Poisson approximation to the binomial distribution to determine the probability that among 60 employees who request a car, three are not properly licensed.

8.67 If a balanced die is rolled 180 times, with what probability can we assert according to Chebyshev's theorem that a 5 will turn up between 10 and 50 times?

8.68 Determine whether the following can be probability distributions (defined in each case only for the given values of x) and explain your answers:

(a) $f(x) = \dfrac{x+1}{14}$ for $x = 1, 2, 3, 4$;

(b) $f(x) = \dfrac{\binom{2}{x}}{4}$ for $x = 0, 1, 2$;

(c) $f(x) = \dfrac{x^2 + 1}{18}$ for $x = 1, 2, 3$.

8.69 If 4 of the 24 air-conditioning compressors in a lot are defective, find the probabilities that among 4 compressors randomly selected from the lot for inspection
(a) 2 will be defective;   (c) all 4 will be defective;
(b) 3 will be defective;   (d) at least 2 will be defective.

8.70 The owner of a used record store offers to sell an album that cost him $3 to a buyer for two bills chosen at random from the buyer's wallet, which contains one $5 bill and five $1 bills. Show that the odds are 2 to 1 that the owner will lose money on the sale but that, despite this, the owner expects to make $\frac{1}{3}$. How can we explain this seeming contradiction?

8.71 A department store has seven Brand C AM-FM cassette players in stock; two of them have faulty tape counters and will require service during the warranty period.
(a) If a person buys two of these players, selecting them at random from the lot, what is the probability that he gets 0, 1, or 2 players with faulty tape counters?
(b) How many players with faulty counters can we expect the buyer to get?
(c) What are the odds against the buyer's getting two players with faulty counters?

Although the symbols $P(A|B)$ and $P(B|A)$ may look alike, conceptually there is a great difference between the probabilities which they represent. For instance, if A is the event that Christmas retail sales are high in a given metropolitan area in a given year and B is the event that the weather is good, then $P(A|B)$ is the probability that retail sales are high given that the weather is good, but $P(B|A)$ is the probability that the weather is good given that sales are high. Also, if A is the event that a man committed a certain crime of embezzlement and B is the event that he is convicted of the crime, then $P(A|B)$ is the probability that the man actually committed the crime given that he is convicted of it, but $P(B|A)$ is the probability that he is convicted of the crime given that he committed it.

There are many problems in statistics which involve such pairs of conditional probabilities, and in Section 9.1 we shall introduce a very important formula, that of Bayes' theorem, which enables us to express the mathematical relationship between probabilities of the form $P(A|B)$ and $P(B|A)$. Then, in Sections 9.2 and 9.3 we shall see how Bayes' theorem may be applied in statistical inference and decision making.

Bayes' Theorem and the Revision of Probabilities

9

Bayes' Theorem

In order to find a formula which expresses $P(B|A)$ in terms of $P(A|B)$ for any two events A and B, let us equate the two expressions for $P(A \cap B)$ on pages 156 and 157. We have

$$P(A) \cdot P(B|A) = P(B) \cdot P(A|B)$$

and, hence,

$$P(B|A) = \frac{P(B) \cdot P(A|B)}{P(A)}$$

after dividing the expressions on both sides of the equation by $P(A)$.

EXAMPLE The loan officer of a bank knows that 5 percent of all loan applicants are bad risks, 92 percent of all loan applicants who are bad risks are also rated bad risks by a credit advisory service, and 2 percent of all loan applicants who are not bad risks are rated bad risks by the service. What is the probability that a loan applicant who is rated a bad risk by the service is actually a bad risk?

SOLUTION Letting A denote the event that the service rates a loan applicant a bad risk and B the event that a loan applicant is, in fact, a bad risk, we can translate the given percentages into probabilities and write $P(B) = 0.05$, $P(A|B) = 0.92$, and $P(A|B') = 0.02$.

Before we can calculate $P(B|A)$ by means of the formula above, we will first have to determine $P(A)$, and to this end let us look at the tree diagram of Figure 9.1. Here A is reached either along the branch

9.1

Tree diagram for calculating $P(A)$ in example.

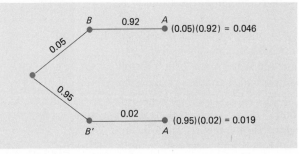

which passes through B or along the branch which passes through B', so the probabilities of reaching A are $(0.05)(0.92) = 0.046$ and $(1 - 0.05)(0.02) = 0.019$. Since the possibilities represented by the two branches are mutually exclusive, we get $P(A) = 0.046 + 0.019 = 0.065$, and substitution into the formula for $P(B|A)$ yields

$$P(B|A) = \frac{P(B) \cdot P(A|B)}{P(A)} = \frac{(0.05)(0.92)}{0.065} = 0.71$$

This is the probability that a loan applicant, rated a bad risk by the service, actually is a bad risk.

With reference to the tree diagram of Figure 9.1 we can say that $P(B|A)$ is the probability that event A is reached via the upper branch of the tree, and its value is given by the ratio of the probability associated with that branch to the sum of the probabilities associated with both branches of the tree. This argument can be generalized to the case where there are more than two branches leading to an event A. With reference to Figure 9.2 we can say that $P(B_i|A)$ is the

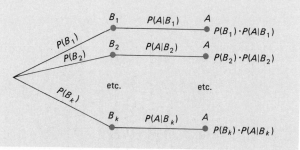

9.2

Tree diagram for
Bayes' theorem.

probability that event A is reached via the ith branch of the tree (for $i = 1$, $2, \ldots,$ or k), and it can be shown that the value of this probability is given by the ratio of the probability associated with the ith branch to the sum of the probabilities associated with all the branches leading to A. Symbolically, this result, called Bayes' theorem, is given by

Bayes' theorem

> If $B_1, B_2, \ldots,$ and B_k are mutually exclusive events one of which must occur, then
>
> $$P(B_i|A) = \frac{P(B_i) \cdot P(A|B_i)}{P(B_1) \cdot P(A|B_1) + P(B_2) \cdot P(A|B_2) + \cdots + P(B_k) \cdot P(A|B_k)}$$
>
> for $i = 1, 2, \ldots,$ or k.

The expression in the denominator here actually equals $P(A)$. $P(B_1) \cdot P(A|B_1)$ is the probability of reaching A via the first branch, $P(B_2) \cdot P(A|B_2)$ is the probability of reaching A via the second branch, . . . , $P(B_k) \cdot P(A|B_k)$ is the probability of reaching A via the kth branch, and the sum of all these probabilities equals $P(A)$. This rule, or procedure, for calculating $P(A)$ is often called the **Rule of Elimination**, or the **Rule of Total Probability**.

EXAMPLE   A management consultant is asked for her opinion as to why an executive's dissatisfied secretary quit her job. Unable to get any direct information about the secretary, she takes the following data from a large-scale corporate morale and motivation study: Among all dissatisfied secretaries, 20 percent are dissatisfied mainly because they dislike their work, 50 percent because they feel they are underpaid, and 30 percent because they dislike their boss. Furthermore, the corresponding probabilities that they will quit are 0.60, 0.40, and 0.90. Based on these figures, what are the probabilities that the secretary quit because of the work, because of the pay, or because of the boss?

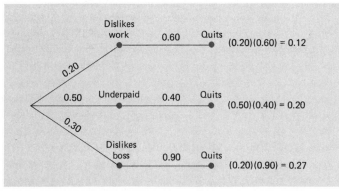

9.3

Tree diagram for example.

SOLUTION   Picturing this situation in Figure 9.3, we find that the probabilities associated with the three branches of the tree are $(0.20)(0.60) = 0.12$, $(0.50)(0.40) = 0.20$, and $(0.30)(0.90) = 0.27$, and that their sum is 0.59. Consequently, the probabilities that a dissatisfied secretary has quit because of the work, the pay, or the boss, are $\frac{0.12}{0.59} = 0.20$, $\frac{0.20}{0.59} = 0.34$, and $\frac{0.27}{0.59} = 0.46$. It follows that the secretary is most likely to have quit because of dislike for the boss.

To solve this problem by means of Bayes' formula, without reference to a tree diagram, we let A denote the event that a dissatisfied secretary quits, and B_1, B_2, and B_3 denote the events that she is dissatisfied with the work, the pay, or the boss. We can thus write $P(B_1) = 0.20$, $P(B_2) = 0.50$, $P(B_3) = 0.30$, $P(A|B_1) = 0.60$, $P(A|B_2) = 0.40$, and $P(A|B_3) = 0.90$, and the formula yields

$$P(B_1|A) = \frac{(0.20)(0.60)}{(0.20)(0.60) + (0.50)(0.40) + (0.30)(0.90)} = \frac{0.12}{0.59} = 0.20$$

The other two probabilities, $P(B_2|A) = 0.34$ and $P(B_3|A) = 0.46$, are obtained in the same way.

As the two examples of this section show, Bayes' formula is a relatively simple mathematical formula. The conditional probabilities calculated by means of the formula are sometimes called "probabilities of causes"—in the example directly above, for instance, what caused the secretary to quit? Dislike of the work, the pay, or the boss?

EXERCISES

9.1 The probability that a one-car accident is due to faulty brakes is 0.04, the probability that a one-car accident is correctly attributed to faulty brakes is 0.82, and the probability that a one-car accident is incorrectly attributed to faulty brakes is 0.03. What is the probability that a one-car accident attributed to faulty brakes was actually due to faulty brakes?

9.2 Use the information on the tree diagram of Figure 9.4 to determine the values of
(a) $P(B|A)$; (b) $P(B|A')$.

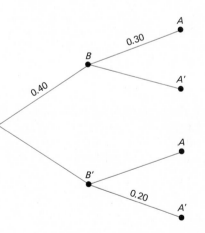

9.4

Tree diagram for
Exercise 9.2.

9.3 At an electronics plant, it is known from past experience that the probability is 0.86 that a new worker who has attended the company's training program will meet his production quota, and that the corresponding probability is 0.35

for a new worker who has not attended the company's training program. If 80 percent of all new workers attend the training program, what is the probability that a new worker will meet his production quota?

9.4 There are two parties in a large city mayoral election. A is nominated by one party and the probability that he will be elected is $\frac{3}{4}$ provided that B is not nominated by the other party. The probability that B will be nominated is $\frac{1}{3}$ and the probability that A will be elected if B is nominated is $\frac{2}{3}$.
(a) What is the probability that A will be elected?
(b) If A is elected, what is the probability that B was not nominated?

9.5 A shopper has bought two packs of clove gum, taking them at random either from bin K, which contains four fresh packs and two stale packs of clove gum, or from bin L, which contains two fresh packs and four stale packs of clove gum, but he is three times as likely to have taken them from bin L as from bin K.
(a) What is the probability that both packs the shopper got are stale?
(b) If both packs the shopper got are stale, what is the probability that he took them from bin L?
(c) If the shopper got at least one fresh pack, what is the probability that he took them from bin K?

9.6 In a cannery, production lines I, II, and III account for 50, 40, and 10 percent of the total output. If 0.5 percent of the cans from line I, 0.6 percent from line II, and 1.5 percent from line III have faulty seals, what are the probabilities that a can with a faulty seal, detected at final product inspection, was produced by
(a) line I;　　　　　　(b) line II;　　　　　　(c) line III?

9.7 It is known from experience that in a certain industry 55 percent of all labor–management disputes are over wages, 10 percent are over working conditions, and 35 percent are over fringe issues. Also, 40 percent of the disputes over wages are resolved without strikes, 70 percent of the disputes over working conditions are resolved without strikes, and 45 percent of the disputes over fringe issues are resolved without strikes. If a labor–management dispute in this industry is resolved without a strike, what are the odds that it was not over wages?

9.8 A wholesale drug company has four employees, $K, L, M,$ and N, who make mistakes in filling orders one time in 100, two times in 100, four times in 100, and six times in 100. Of all the orders filled, $K, L, M,$ and N fill 20, 40, 30, and 10 percent. If a mistake is found in a particular order, what are the probabilities that it was filled by $K, L, M,$ or N?

9.9 (From Miller, I., and Freund, J. E., *Probability and Statistics for Engineers*, 2nd ed. Englewood Cliffs, N.J.: Prentice-Hall, Inc., 1977.) An explosion in an LNG storage tank in the process of being repaired could have occurred as the result of static electricity, malfunctioning electrical equipment, an open flame in contact with the liner, or purposeful action (industrial sabotage). Interviews with engineers who were analyzing the risks involved led to estimates that such an explosion would occur with probability 0.25 as a result of static electricity, 0.20 as a result of malfunctioning electrical equipment, 0.40 as a result of an open flame, and 0.75 as a result of purposeful action. These interviews also yielded subjective estimates of the probabilities of the four causes of 0.30, 0.40, 0.15, and 0.15, respectively. Based on all this information, what is the most likely cause of the explosion?

9.2

Prior Probabilities and Posterior Probabilities

There are many situations in which we assign probabilities to events on the basis of whatever information or feelings we have about their likelihoods at the time. Later, however, we may get additional information which forces us to revise our earlier appraisals. Sometimes we learn or observe something which causes us to assign even higher probabilities to events we already considered very likely to happen; at other times we find that we must assign a probability of zero to an event which, at one time, we had felt was reasonably sure to happen. The mechanics of making logical revisions of earlier probability assignments is the subject matter of this section.

Suppose, for instance, that Adam and Brown are planning to form a corporation to build and operate a number of automatic car washes and that Adam feels that there is a fifty–fifty chance that any new unit will show a profit for the first year, but Brown feels that the odds on this are 2 to 1. Suppose, furthermore, that we may want to invest some money in this corporation and, hence, we would like to know who is right. If a business consultant feels that in view of Adam's experience he is three times as likely to be right as Brown, the consultant is, in fact, assigning the event B_1 that Adam is right the probability $P(B_1) = \frac{3}{4}$ and the event B_2 that Brown is right the probability $P(B_2) = \frac{1}{4}$. Since the entire operation is still in the planning stage, we have no direct information about the performance of the units, and we refer to $P(B_1) = \frac{3}{4}$ and $P(B_2) = \frac{1}{4}$ as the **prior probabilities** of events B_1 and B_2.

Now suppose that the corporation opens six new units and five of them show a profit for the first year and the sixth shows a loss. How does that affect our feelings about B_1 (that Adam is right) and B_2 (that Brown is right)? If we let A denote the event that five of the six new units show a first-year profit, the answer to this question is given by the probabilities $P(B_1 | A)$ and $P(B_2 | A)$. In order to find these we must first calculate the binomial probabilities $P(A | B_1)$ and $P(A | B_2)$, and then substitute the results together with the prior probabilities $P(B_1)$ and $P(B_2)$ into Bayes' formula. To find $P(A | B_1)$ we substitute $x = 5$, $n = 6$, and $p = \frac{1}{2}$ (corresponding to the fifty–fifty chance of a first-year profit) into the formula for the binomial distribution, and we get

$$P(A|B_1) = \binom{6}{5}\left(\frac{1}{2}\right)^5\left(1 - \frac{1}{2}\right)^{6-5} = \frac{3}{32}$$

Similarly, to find $P(A|B_2)$ we substitute $x = 5$, $n = 6$, and $p = \frac{2}{3}$ (corresponding to the 2 to 1 odds on a first-year profit) into the formula for the binomial distribution, and we get

$$P(A|B_2) = \binom{6}{5}\left(\frac{2}{3}\right)^5\left(1 - \frac{2}{3}\right)^{6-5} = \frac{64}{243}$$

Finally, substituting $P(B_1) = \frac{3}{4}$, $P(B_2) = \frac{1}{4}$, $P(A|B_1) = \frac{3}{32}$, and $P(A|B_2) = \frac{64}{243}$ into Bayes' formula, we find that

$$P(B_1|A) = \frac{\frac{3}{4}\cdot\frac{3}{32}}{\frac{3}{4}\cdot\frac{3}{32} + \frac{1}{4}\cdot\frac{64}{243}} = \frac{2,187}{4,235} = 0.52$$

It follows that $P(B_2|A) = 1 - 0.52 = 0.48$.

The probabilities $P(B_1|A)$ and $P(B_2|A)$ are called the **posterior probabilities** of events B_1 and B_2, and it is important to note how we have combined the direct evidence (five of the six new car washes showed a first-year profit) with the original subjective evaluation of the consultant. Note also how the weight of the direct evidence has given increased merit to Brown's claim. Whereas his claim was originally assigned a probability of $\frac{1}{4}$, it is now assigned a probability of 0.48, or almost $\frac{1}{2}$.

EXAMPLE Three engineers agree that the number of times a newly designed desktop computer will malfunction in a week of normal use is a random variable having a Poisson distribution, but they disagree on the expected (or mean) number of such malfunctions per week. The first engineer's calculations lead him to assert that the expected value is $\lambda = 1.4$, the second says it is $\lambda = 2.0$, and the third says it is $\lambda = 3.1$. If the three engineers are highly competent men whose judgments, in our considered opinion, are equally valid, and if in an experiment the computer actually malfunctions four times during one week of normal use, what posterior probabilities should we assign to the values of λ suggested by the three engineers?

SOLUTION If B_1 is the event that the first engineer is right, B_2 the event that the second engineer is right, and B_3 the event that the third engineer is right, the prior probabilities of these events are $P(B_1) = P(B_2) = P(B_3) = \frac{1}{3}$, since we feel that these judgments are equally valid. Then, if A is the event that the computer malfunctions four times in a week of normal use, substitution of $x = 4$ and the three suggested values of λ into the formula for the Poisson distribution yields

$$P(A|B_1) = \frac{(1.4)^4 \cdot e^{-1.4}}{4!} = 0.040$$

$$P(A|B_2) = \frac{(2.0)^4 \cdot e^{-2.0}}{4!} = 0.090$$

and

$$P(A|B_3) = \frac{(3.1)^4 \cdot e^{-3.1}}{4!} = 0.173$$

Substituting now in Bayes' formula, we find that for the first engineer the posterior probability is

$$P(B_1|A) = \frac{\frac{1}{3}(0.040)}{\frac{1}{3}(0.040) + \frac{1}{3}(0.090) + \frac{1}{3}(0.173)} = 0.13$$

Replacing the numerator by the second term in the denominator, we get $P(B_2|A) = 0.30$; then replacing it by the third term, we get $P(B_3|A) = 0.57$.

The effect of the additional information gained in the experiment is to increase the weight assigned originally to the third engineer's claim and to decrease the weights assigned to the claims of the others. This should not really come as a surprise. After the computer was observed to malfunction four times in one week, it seems entirely reasonable to give more weight to the claim of the engineer whose expectation was closest.

In the first example of this section we assumed that either Adam or Brown was right, and in the second example we assumed that one or another of the three engineers was right; no other possibilities were even considered to exist. This may not seem reasonable. Why, for instance, couldn't the probability of a first-year profit for one of the new car washes be $\frac{4}{5}$ instead of either $\frac{1}{2}$ or $\frac{2}{3}$? And why couldn't the computer malfunction on the average 3.5 times a week instead of either 1.4, 2.0, or 3.1 times? The answer is that assumptions of this sort are necessary in order to make the kind of analysis we have described. Of course, we used only two alternatives in the first example and three in the second to simplify the calculations; in actual practice, we may use 10 alternatives, 25 alternatives, or even more as the need may be.

9.3

Posterior Analysis

In Chapter 7 we introduced the Bayes decision rule, and we used it to maximize expected profits, minimize expected costs, or minimize expected opportunity losses, using only prior probabilities in each case. Now, if we combine prior information with direct sample evidence as in the preceding section, we can use the posterior probabilities instead of the prior probabilities, and we refer to this as **posterior analysis**.

Let us consider a very large company which routinely pays thousands of invoices submitted by its suppliers. They are collected in batches of 1,000 before payment is made, and since the company often receives erroneous invoices (the prices, the quantities, and the extensions have all been known to be in error on occasion) it must decide whether or not to have each invoice in a batch checked (or verified). Clearly, the proportion of erroneous invoices will vary from batch to batch, and for the sake of simplicity let us assume that for any given batch it can take on only the values $p = 0.001$, $p = 0.010$, or $p = 0.020$. Also, the company's statistical studies show that on the average an erroneous invoice overcharges the company by \$5, and it is known that it costs \$45 to eliminate all errors from a batch of 1,000 invoices by means of a computer.

All this information is summarized in the following payoff table, where the entries in the second column show that it costs the company \$45 to check and eliminate all errors from a batch of 1,000 invoices regardless of how many of them are in error. The entries in the first column are the expected costs (overcharges) which result from not checking a batch before payment is made. If $p = 0.001$, the expected number of erroneous invoices per batch is $(1,000)(0.001) = 1$, and hence, the expected cost (overcharge) is $1 \cdot 5 = \$5$. Similarly, for $p = 0.010$ the expected number of erroneous invoices per batch is $(1,000)(0.010) = 10$, and for $p = 0.020$ it is $(1,000)(0.020) = 20$, so the corresponding expected costs are $10 \cdot 5 = \$50$ and $20 \cdot 5 = \$100$.

		Cost or expected cost per batch	
		Without checking	With checking
Proportion of invoices containing errors	0.001	\$5	\$45
	0.010	\$50	\$45
	0.020	\$100	\$45

In this problem, as in some of those in Chapter 7, there are certain advantages to working with opportunity losses instead of payoffs; for one thing, the minimum expected opportunity loss, the minimum EOL, equals the expected value of perfect information, the EVPI, which we also call the cost of uncertainty. To convert the payoff table above into an opportunity-loss table, we subtract from both entries in each row the smaller of the two, and we get

| | | Opportunity losses | |
		Without checking	With checking
	0.001	0	$40
Proportion of invoices containing errors	0.010	$5	0
	0.020	$55	0

Now, if the value of p were known for each batch before it is decided whether or not it should be checked, there is no real problem and the right decision is obvious. When $p = 0.001$ the company should pay the invoices without having the batch checked, and when $p = 0.010$ or $p = 0.020$, the company should have each invoice checked before payment is made. In either case this minimizes the expected cost or the expected loss of opportunity.

In practice, of course, the value of p is not known in advance. This is precisely why there is a problem, and this is where the importance of probability in decision making becomes evident. If someone felt "pretty sure" that $p = 0.001$ for a given batch (that is, if he assigned $p = 0.001$ a very high subjective probability for this batch), he would not have the batch checked; conversely, if someone felt "pretty sure" that $p = 0.020$ for a given batch, he would have it checked. Since different "feelings" about the correct value of p can affect the decision on whether or not to have a batch checked, let us look more closely into the problem of assigning probabilities to the different values which p can take on.

One way to handle this problem is to assign objective, rather than subjective, prior probabilities—if they can be found—to the various events (in our example, the three possible values of p). Suppose that the company maintains a set of records extending back over a long period of time, and suppose that we find from these records that in 70 percent of all the many batches checked p was 0.001, in 20 percent of them p was 0.010, and in the remaining 10 percent p was 0.020. If we use 0.70, 0.20, and 0.10 as the prior probabilities of the three values of p, we find that the expected opportunity loss associated with the decision not to check the batch is

$$\$0(0.70) + \$5(0.20) + \$55(0.10) = \$6.50$$

and the expected opportunity loss associated with the decision to check the batch is

$$\$40(0.70) + \$0(0.20) + \$0(0.10) = \$28.00$$

Thus, the expected opportunity loss is minimized by deciding not to check the batch, and the expected value of perfect information is $6.50. This figure is the maximum amount the company should be willing to spend in collecting more information before making the final decision as to whether or not to have the entire batch checked.

If no further information having a bearing on this problem can be obtained, this procedure would seem to provide a reasonable solution. However, it may be observed that the weights assigned to the possible values of p (the prior probabilities, that is) are based on historical data which tell us what happened in the past but tell us nothing directly about the particular batch which is of concern. What we would like to do, if possible, is take a sample of the 1,000 invoices contained in the batch, observe the number of erroneous invoices, and revise the prior probabilities in the light of this direct evidence. So let us suppose that 25 invoices are randomly selected (at negligible cost) from the batch in question, and that only one of them contains an error. If we let A denote the event of getting one erroneous invoice in the sample of 25 invoices, B_1 denote the event that the actual proportion of erroneous invoices in the lot is 0.001, and B_2 and B_3 denote the events that the proportions are 0.010 and 0.020, we first want to determine the probabilities $P(A|B_1)$, $P(A|B_2)$, and $P(A|B_3)$, and then use Bayes' formula to find the posterior probabilities $P(B_1|A)$, $P(B_2|A)$, and $P(B_3|A)$. The sampling here is without replacement, as is usually the case, so the exact values of the first three probabilities are given by the formula for the hypergeometric distribution. However, since the sample of 25 invoices constitutes only a small portion of the entire batch, we can use the binomial distribution to approximate these probabilities, and we get

$$P(A|B_1) = \binom{25}{1}(0.001)^1(0.999)^{24} = 0.0244$$

$$P(A|B_2) = \binom{25}{1}(0.010)^1(0.990)^{24} = 0.1965$$

and

$$P(A|B_3) = \binom{25}{1}(0.020)^1(0.980)^{24} = 0.3080$$

Combining these probabilities with the prior probabilities $P(B_1) = 0.70$, $P(B_2) = 0.20$, and $P(B_3) = 0.10$, and substituting into Bayes' formula, we get

$$P(B_1|A) = \frac{(0.70)(0.0244)}{(0.70)(0.0244) + (0.20)(0.1965) + (0.10)(0.3080)}$$

$$= \frac{0.01708}{0.08718} = 0.196$$

$$P(B_2|A) = \frac{(0.20)(0.1965)}{0.08718} = 0.451$$

and

$$P(B_3|A) = \frac{(0.10)(0.3080)}{0.08718} = 0.353$$

Finally, using these posterior probabilities with the opportunity-loss table, we find that the expected opportunity loss associated with the decision not to check the batch is

$$\$0(0.196) + \$5(0.451) + \$55(0.353) = \$21.67$$

and the expected opportunity loss associated with the decision to check the batch is

$$\$40(0.196) + \$0(0.451) + \$0(0.353) = \$7.84$$

Thus, the posterior analysis shows that the expected opportunity loss is minimized by deciding to check the entire batch, and that the expected value of perfect information is $7.84.

It may seem surprising that the expected value of perfect information has increased from $6.50 without the sample data to $7.84 with the added information. This is due to the fact that the sample evidence does not support the prior considerations, where most of the weight (a probability of 0.70) was given to $p = 0.001$. The proportion of erroneous invoices in the sample was $\frac{1}{25} = 0.04$, which exceeds even $p = 0.020$, and the posterior analysis has, in fact, reversed the decision we arrived at by means of the prior analysis. If there had been no erroneous invoices among the 25 in the sample, this would have offered support for the prior considerations, the decision would not have been reversed, and the expected value of perfect information would have been reduced to $4.55 (see Exercise 9.15).

EXERCISES

9.10 In planning the operation of a new restaurant, one expert claims that only one out of 5 waitresses can be expected to stay with the establishment for more than a year, while a second expert claims that it would be correct to say 3 out of 10. In the past, the two experts have been about equally reliable, so that in the absence of direct information we would assign their judgments equal weight. What posterior probabilities would we assign to their claims if it were found that among 12 waitresses actually hired for the restaurant only one stayed for more than a year?

9.11 The landscaping plans for a new motel call for a row of palm trees along the driveway. The landscape designer tells the owner that if he plants *Washingtonia filifera*, 20 percent of the trees will fail to survive the first heavy frost, the manager of the nursery which supplies the trees tells the owner that 10 percent of the trees will fail to survive the first heavy frost, and the owner's wife tells him that 30 percent of the trees will fail to survive the first heavy frost.

(a) If the owner feels that in this matter the landscape designer is 10 times as reliable as his wife and the manager of the nursery is 9 times as reliable as his wife, what prior probabilities should he assign to these percentages?

(b) If 10 of these palm trees are planted and 3 fail to survive the first heavy frost, what posterior probabilities should the manager assign to the three percentages?

9.12 Discussing the sale of a large estate, one broker expresses the feeling that a newspaper ad should produce three serious inquiries about the estate, a second broker feels that it should produce five serious inquiries, and a third broker feels that it should produce six.

(a) If in the past the second broker has been twice as reliable as the first and the first has been three times as reliable as the third, what prior probabilities should we assign to their claims?

(b) How would these probabilities be affected if the ad actually produced two inquiries and it can be assumed that the number of inquiries is a random variable having the Poisson distribution with either $\lambda = 3, \lambda = 5$, or $\lambda = 6$ according to the three claims?

9.13 A coin dealer receives a shipment of five ancient gold coins from abroad, and, on the basis of past experience, he feels that the probabilities that 0, 1, 2, 3, 4, or all 5 of them are counterfeits are 0.76, 0.09, 0.02, 0.01, 0.02, and 0.10. Since modern methods of counterfeiting have become greatly refined, the cost of authentication has risen sharply, and the dealer decides to select one of the five coins at random and send it away for authentication. If it turns out that the coin is a forgery, what posterior probabilities should he assign to the possibilities that 0, 1, 2, 3, or all 4 of the remaining coins are counterfeits?

9.14 Light bulbs are usually packaged in pairs, and the manager of a supermarket knows from experience that the probabilities that such a package will contain 0, 1, or 2 defective bulbs are, respectively, 0.92, 0.07, 0.01. If he randomly selects a package of bulbs from his shelves and then randomly selects one of the two bulbs and finds that it is defective, what is the probability that the other bulb is also defective?

9.15 If there had been no errors in any of the 25 invoices in the example on page 248, the binomial probabilities $P(A|B_1)$, $P(A|B_2)$, and $P(A|B_3)$ would have been 0.975, 0.778, and 0.604, with A denoting the event of getting zero erroneous invoices.

(a) Calculate the corresponding posterior probabilities.

(b) Use the posterior probabilities obtained in part (a) and the opportunity-loss table on page 247 to determine the expected opportunity losses and verify that the Bayes decision rule leads to the decision not to check the entire batch of invoices.

(c) Verify that the expected value of perfect information is reduced to $4.55.

9.16 With reference to the example on page 244, suppose that someone wants to buy one of the desktop computers, but only if the expected number of breakdowns is less than 2.5 per week. Would he buy the computer if he based his decision on

(a) a prior analysis using the prior probabilities given on page 244;

(b) a posterior analysis using the posterior probabilities given on page 245?

9.17 A large electrical supply house is closing out a number of items, among them a lot of four generators of a certain kind at a price of $240 for the entire lot, "as is—all sales final." A man can resell all good generators at $120 each, but each defective generator represents a total loss. Based on his familiarity both with the generators and with the seller, the man assigns prior probabilities of 0.1, 0.5, 0.2, 0.1, and 0.1 to the events that there are 0, 1, 2, 3, or 4 defective generators in the lot.

(a) If no inspection is possible, should the man buy the lot if he wants to minimize his expected loss of opportunity?

(b) If the seller will let the man unpack and inspect one of the four generators and he finds that it is not defective, should he buy the lot if he wants to minimize his expected loss of opportunity?

(c) How did the inspection of part (b) affect the cost of uncertainty?

9.18 With reference to the preceding exercise, suppose that the man can resell the good generators for only $90 while everything else remains unchanged.

(a) If no inspection is possible, should the man buy the lot if he wants to minimize his expected loss of opportunity?

(b) If the seller will let the man unpack and inspect two of the generators and he finds that both are defective, should he buy the lot if he wants to minimize his expected loss of opportunity?

(c) How did the inspection of part (b) affect the cost of uncertainty?

9.19 An importer expects a shipment of ginger roots from Indonesia. If the shipment arrives safely he will make a profit of $1,200, but if it gets lost in transit he will lose $400. From past experience he knows that the odds are 4 to 1 against such a shipment getting lost.

(a) If someone offered the importer a $750 profit for the shipment (regardless of whether it will arrive), what should he do so as to minimize the EOL?

(b) Suppose that the importer knows from past experience that half the shipments that do not get lost arrive within a month. If the shipment of ginger roots has not arrived within a month and the offer of part (a) is repeated, what should the importer do so as to minimize the EOL?

9.20 A company has received a lot of 1,000 ball bearings which are used in mechanical devices which the company manufactures. For the sake of simplicity, we assume that the proportion of defective bearings in the lot is either $p = 0.01$, $p = 0.03$, or $p = 0.05$. The lot may be put directly into production without inspection, but each defective bearing entering production costs the company $5 (largely for labor to replace it with a good one at final inspection). Alternatively, the company can submit the lot to a rapid man–machine screening procedure at a cost of 7 cents per item. In this procedure, the lot is 100 percent inspected and all defective items found are replaced with good ones. It is considered that the efficiency of this screening is such that all screened lots entering production will be 99 percent free of defective items. Based on previous experience the company assigns probabilities of 0.6, 0.3, and 0.1 to the events that the lot contains 1 percent, 3 percent, and 5 percent defective bearings.

(a) In the absence of further information, what should the company do so as to minimize its EOL?

(b) Suppose that it is possible for the company to make a preliminary inspection of the lot prior to deciding how it should be handled. Specifically, suppose that the company inspects at a negligible cost a sample of 10 bearings from the lot and finds one defective bearing in the sample. Should the company put the lot directly in production or have it screened, so as to minimize its EOL? (Use Poisson approximation to the probabilities of finding one defective in the sample of ten bearings.)

(c) How did the information supplied by the sample affect the EVPI?

9.4

Check List of Key Terms

Bayes' theorem, 239
Posterior analysis, 246
Posterior probabilities, 244

Prior probabilities, 243
Rule of Elimination, 240
Rule of Total Probability, 240

9.5

Review Exercises

9.21 The records of a company selling encyclopedias door to door with a large sales force in a certain area show that, of the initial calls made in this area, 70 percent are made by salesmen and the rest by saleswomen. The records also show that, on the first call, women close the sale 3 percent of the time and men close the sale 2 percent of the time. If it is known that an encyclopedia was sold on the first call, what is the probability that the call was made by a woman?

9.22 A businessman wants to find out whether it would be profitable to establish a travel agency in a certain community. Experience in other locations suggest that if only 5 percent of the families in the community patronize the agency, he will lose $300 per week; if 10 percent of the families in the community patronize the agency, he will make a profit of $500 per week; and if 20 percent of the families in the community patronize the agency, he will make a profit of $1,200 per week. Also, he feels that the probabilities are 0.50, 0.40, and 0.10 that 5, 10, or 20 percent of the families in the community will patronize the agency.

(a) In view of these data, should the businessman establish the travel agency if he wants to minimize his expected loss of opportunity?

(b) Suppose that, in a random sample of 10 families in the given community, not one family says that it will patronize the agency. Find the posterior probabilities associated with the three percentages.

(c) Using also the sample information, should the businessman establish the agency if he wants to minimize his expected loss of opportunity?

(d) How did the information of part (b) affect the cost of uncertainty?

9.23 A resort hotel gets cars for its guests from three rental agencies, 20 percent from agency D, 20 percent from agency E, and 60 percent from agency F. If 10 percent of the cars from D, 4 percent from E, and 12 percent from F need tune-ups, what is the probability that a car needing a tune-up which is delivered to a guest of the resort came from rental agency F?

9.24 In planning the operation of a new school, one school board member claims that 4 out of 5 newly hired teachers will stay with the school for more than a year, while another school board member claims that it would be correct to say 7 out of 10. In the past, the two board members have been about equally reliable in their predictions, so that in the absence of direct information we would assign their judgments equal weight. What posterior probabilities would we assign to their claims if it were found that 11 of 12 newly hired teachers stayed with the school for more than a year?

9.25 On page 168 we gave an example where a manufacturer of office equipment must decide whether to expand his plant capacity now or wait at least another year, and where the payoffs corresponding to the events "economic conditions remain good" and "there is a recession" are as shown in the following table:

	Expand now	Delay expansion
Economic conditions remain good	$369,000	$180,000
There is a recession	−$90,000	$18,000

Also, on page 194 we showed that if he feels that the odds on a recession are 2 to 1, he will minimize the EOL if he delays expanding the capacity of his plant. Now suppose that the manufacturer consults an expert, who in the past has made correct predictions 80 percent of the time when economic conditions remained good, but only 50 percent of the time when there were recessions.

(a) If the expert tells him that economic conditions will remain good, how will this affect the probabilities which the manufacturer assigns to the two events?

(b) Based on the probabilities of part (a), what should the manufacturer do so as to minimize the expected loss of opportunity?

9.26 In a state where cars have to be tested for the emission of pollutants, 20 percent of all cars emit excessive amounts of pollutants. When tested, 98 percent of all cars that emit excessive amounts will fail the test, but 15 percent of the cars that do not emit excessive amounts will also fail. What is the probability that a car which fails the test actually emits excessive amounts of pollutants?

9.27 An advertising executive feels that 60 percent of all persons who have seen a certain television commercial will recall two days later what products were advertised, while a company official feels that it is only 30 percent. If, originally, we assign these two assessments equal weight, how would this assignment be affected if a random sample taken two days after the program was on the air shows that only five of 15 viewers remembered what products were advertised?

9.28 A doctor has taken a vaccine from either storage unit P (which contains 30 current and 10 outdated vaccines), from unit Q (which contains 20 current and 20 outdated vaccines), or from unit R (which contains 10 current and 30 outdated vaccines), but she is twice as likely to have taken it from unit P as from unit Q and twice as likely to have taken it from unit Q as from unit R.
(a) What is the probability that the vaccine selected is an outdated one?
(b) If the vaccine selected is outdated, what is the probability that it came from unit P?

9.29 The prior probabilities are 0.70, 0.20, and 0.10 that 0, 1, or 2 eggs in a carton of 12 eggs are cracked. If two of the eggs in a carton are selected at random and it is found that neither of them is cracked, how will this affect the probabilities that 0, 1, or 2 of the eggs in a carton are cracked?

9.30 The purchasing agent of a firm feels that the probability is 0.80 that any one of several shipments of steel recently received will meet specifications. The head of the firm's quality control department feels that this probability is 0.90, and the chief engineer feels that it is 0.60.
(a) If the managing director of the firm feels that in this matter the purchasing agent is 10 times as reliable as the chief engineer while the head of the quality control department is 14 times as reliable as the chief engineer, what prior probabilities would she assign to the three claims?
(b) If five of the shipments are inspected and only two meet specifications, how does this affect the probabilities obtained in part (a)?

9.31 A coin dealer offers to sell us a lot of 20 foreign silver coins, which we know contains either 10 percent, 20 percent, or 30 percent counterfeit coins. From past experience we consider the first of these three possibilities to be three times as likely as the second and the second to be three times as likely as the third. The coins are priced to us at $20 each, and we can resell all good coins for $25 each and all counterfeits (as curios) for $2 each.
(a) If no inspection is possible, what should we do in order to minimize our expected opportunity loss? Buy the lot? Not buy the lot? What is the expected value of perfect information?
(b) Suppose that the dealer makes us the following offer: Before deciding whether or not to buy the lot, we can inspect three of the coins drawn at random from the lot. The cost of this inspection is $5, which the dealer will return to us if we buy the lot but keep if we do not buy the lot. If we accept the dealer's offer and find two counterfeits among the three coins inspected, what should we do in order to minimize our expected opportunity loss? Buy the lot? Not buy the lot?

10

The Normal Distribution

ontinuous sample spaces arise whenever we deal with quantities that are measured on a continuous scale—for instance, when we measure the speed of a car, the net weight of a package of frozen food, the purity of a product, or the amount of tar in a cigarette. In cases like these there exist continuums of possibilities, and in practice what we are really interested in are probabilities associated with intervals or regions, not individual numbers or points, of a sample space. For instance, we might want to know the probability that at a given time a car is moving between 60 and 65 miles per hour (not at exactly 20π miles per hour), or that a package of frozen food weighs more than 7.95 ounces (not exactly $\sqrt{64.1} = 8.0062475605...$ ounces).

In this chapter we shall learn how to determine, and work with, probabilities relating to continuous sample spaces and continuous random variables. The concept of a continuous distribution will be introduced in Section 10.1, followed by that of a normal distribution in Section 10.2. Various applications of the normal distribution will be discussed in Sections 10.3 and 10.4.

255

Continuous Distributions

In histograms, the frequencies, percentages, or proportions associated with the various classes are given by the heights of the rectangles, or by their areas if the class intervals are all equal; this is true also for histograms showing the probabilities associated with the values of discrete random variables. In the continuous case, we also represent probabilities by means of areas, as is illustrated in Figure 10.1, but instead of areas of rectangles we use areas under continuous curves. The first diagram of Figure 10.1 represents the probability distribution of a random variable which takes on only the values 0, 1, 2, . . . , 9, and 10, and the probability that it will take on the value 3, for example, is given by the area of the white region. The second diagram refers to a continuous random variable which can take on any value on the interval from 0 to 10, and the probability that it will take on a value on the interval from 2.5 to 3.5 is given by the area of the white region under the curve. Similarly, the area of the dark region under the curve gives the probability that it will take on a value greater than 8.

Continuous curves such as the one shown in the right-hand diagram of Figure 10.1 are the graphs of functions called probability densities, or informally, continuous distributions. A probability density is characterized by the fact that

10.1

Histogram of probability distribution and graph of continuous distribution.

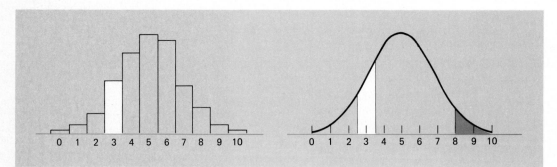

The area under the curve between any two values *a* and *b* (see Figure 10.2) gives the probability that a random variable having the continuous distribution will take on a value on the interval from *a* to *b*.

10.2

Continuous
distribution.

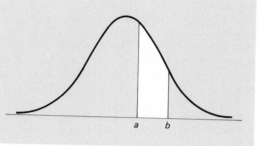

It follows from this that the total area under the curve (representing the certainty that a random variable must take on one of its values) is always equal to 1.

For instance, if we approximate a family income distribution with a smooth curve as in Figure 10.3, we can determine what proportion of the incomes falls into any given interval (or the probability that the income of a family, chosen at random, will fall into the interval) by looking at the corresponding area under the curve. By comparing the area of the white region of Figure 10.3 with the

10.3

Curve approximating
family income
distribution.

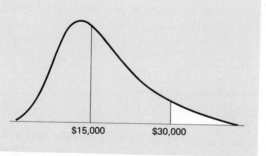

total area under the curve (representing 100 percent), we can judge by eye that roughly 10 to 12 percent of the families have incomes of $30,000 or more. Similarly, it can be seen that about 40 to 45 percent of the families have incomes of $15,000 or less.

Statistical descriptions of continuous distributions are as important as descriptions of probability distributions or distributions of observed data, but most of them, including the mean and the standard deviation, cannot be defined without the use of calculus. Nevertheless, we can always picture a continuous distribution as being approximated by a histogram of a probability distribution whose mean and standard deviation can be calculated (see Figure 10.4). Then,

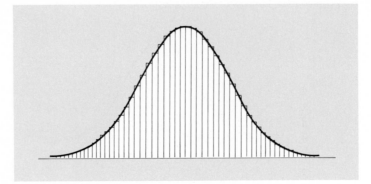

10.4

Continuous distribution approximated with histogram of probability distribution.

if we choose histograms with narrower and narrower classes, the means and the standard deviations of the corresponding probability distributions will approach the mean and the standard deviation of the continuous distribution. Actually, the mean and the standard deviation of a continuous distribution measure the same properties as the mean and the standard deviation of a probability distribution—the expected value of a random variable having the given distribution, and the expected value of the squared deviation from the mean. More intuitively, the mean μ of a continuous distribution is a measure of its "center" or "middle," and the standard deviation σ of a continuous distribution is a measure of its dispersion or spread.

10.2

The Normal Distribution

Among the many continuous distributions used in statistics, the **normal distribution** is by far the most important. Its study dates back to eighteenth-century investigations into the nature of experimental errors. It was observed that discrepancies between repeated measurements of the same physical quantity displayed a surprising degree of regularity; their patterns (distribution), it was found, could be closely approximated by a certain kind of continuous distribution curve, referred to as the "normal curve of errors" and attributed to the laws of chance.

The graph of a normal distribution is a bell-shaped curve that extends indefinitely in both directions. Although this may not be apparent from a small

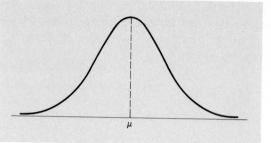

drawing such as Figure 10.5, the curve comes closer and closer to the horizontal axis without ever reaching it, no matter how far we go from the mean in either direction. Fortunately, it is seldom necessary to extend the tails of a normal distribution very far because the area under that part of the curve lying more than 4 or 5 standard deviations away from the mean is for most practical purposes negligible.

An important feature of a normal distribution is that its mathematical equation is such that we can determine the area under the curve between any two points on the horizontal scale if we know its mean and its standard deviation; in other words, there is one and only one normal distribution with a given mean μ and a given standard deviation σ.

Since the equation of the normal distribution depends on μ and σ, we get different curves and, hence, different areas for different values of μ and σ. For instance, Figure 10.6 shows the superimposed graphs of two normal distributions, one having $\mu = 10$ and $\sigma = 5$ and the other having $\mu = 20$ and $\sigma = 10$. The area under the curve, say, between 12 and 15, is obviously not the same for the two distributions.

In practice, we find areas under the graph of a normal distribution, or simply a normal curve, in special tables, such as Table I at the end of the book.

10.6

Two normal distributions.

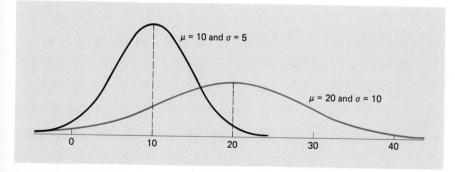

As it is physically impossible, and also unnecessary, to construct separate tables of normal-curve areas for all conceivable pairs of values of μ and σ, we tabulate these areas only for a normal distribution having $\mu = 0$ and $\sigma = 1$, the so-called **standard normal distribution**. Then, we obtain areas under any normal distribution by performing a simple change of scale (see Figure 10.7), in which we convert the units of measurement in the original, or x-scale, into **standard units**, **standard scores**, or **z-scores**, by means of the formula

Standard units

$$z = \frac{x - \mu}{\sigma}$$

In this new scale, the z-scale, z simply tells us how many standard deviations the corresponding x-value lies above or below the mean of its distribution.

The entries in Table I are the areas under the standard normal curve between the mean $z = 0$ and $z = 0.00, 0.01, 0.02, \ldots, 3.08$, and 3.09, and also $z = 4.00$, $z = 5.00$, and $z = 6.00$. In other words, the entries in Table I are areas under the standard normal distribution like the white area in Figure 10.8.

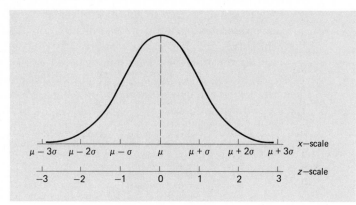

10.7

Change of scale.

10.8

Tabulated areas under the graph of the standard normal distribution.

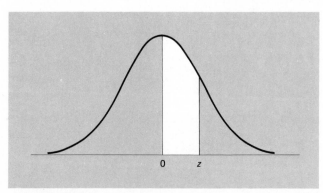

Table I has no entries corresponding to negative values of z, for these are not needed by virtue of the symmetry of any normal curve about its mean.

EXAMPLE Find the area under the standard normal curve between $z = -1.20$ and $z = 0$.

SOLUTION As can be seen from Figure 10.9, the area under the curve between $z = -1.20$ and $z = 0$ equals the area under the curve between $z = 0$ and $z = 1.20$. So, we look up the entry for $z = 1.20$ and get 0.3849.

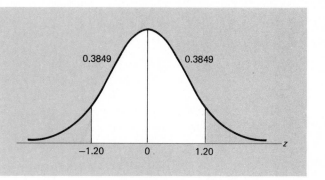

10.9

Area under normal curve.

Questions concerning areas under normal distributions arise in various ways, and the ability to find any desired area quickly can be a big help. Although the table gives only areas between the mean $z = 0$ and selected positive values of z, we often have to find areas to the left or to the right of given positive or negative values of z, or areas between two given values of z. This is easy, provided we remember exactly what areas are represented by the entries in Table I, and also that the standard normal distribution is symmetrical about $z = 0$, so that the area to the left of $z = 0$ and the area to the right of $z = 0$ are both equal to 0.5000.

EXAMPLE If a random variable has the standard normal distribution, what are the probabilities that it will take on a value
(a) less than 1.64;
(b) greater than -0.47;
(c) greater than 0.76;
(d) less than -1.35;
(e) between 0.95 and 1.36;
(f) between -0.45 and 0.65?

SOLUTION For each part see Figure 10.10. (a) The probability that the random variable will take on a value less than 1.64 (the area to the left of $z = 1.64$) is 0.5000 plus the entry in Table I corresponding to $z = 1.64$, or

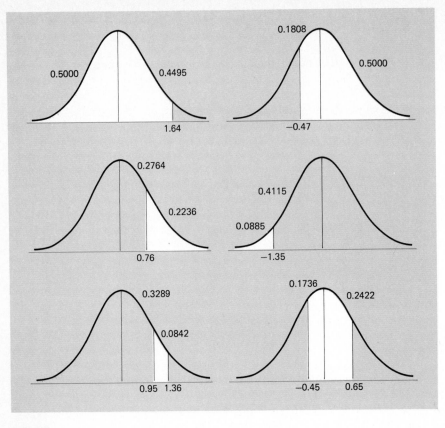

10.10

Areas under normal distributions.

$0.5000 + 0.4495 = 0.9495$; (b) the probability that it will take on a value greater than -0.47 (the area to the right of $z = -0.47$) is 0.5000 plus the entry in Table I corresponding to $z = 0.47$, or $0.5000 + 0.1808 = 0.6808$; (c) the probability that it will take on a value greater than 0.76 is 0.5000 minus the entry in Table I corresponding to $z = 0.76$, or $0.5000 - 0.2764 = 0.2236$; (d) the probability that it will take on a value less than -1.35 is 0.5000 minus the entry in Table I corresponding to $z = 1.35$, or $0.5000 - 0.4115 = 0.0885$; (e) the probability that it will take on a value between 0.95 and 1.36 is the difference between the entries in Table I corresponding to $z = 1.36$ and $z = 0.95$, or $0.4131 - 0.3289 = 0.0842$; (f) the probability that it will take on a value between -0.45 and 0.65 is the sum of the entries in Table I corresponding to $z = 0.45$ and $z = 0.65$, or $0.1736 + 0.2422 = 0.4158$.

For the two normal curves of Figure 10.6, find the area under the curve between 12 and 15 for the distribution with

(a) $\mu = 10$ and $\sigma = 5$;

(b) $\mu = 20$ and $\sigma = 10$.

SOLUTION (a) The values of z corresponding to $x = 12$ and $x = 15$ are

$$z = \frac{12 - 10}{5} = 0.40 \quad \text{and} \quad z = \frac{15 - 10}{5} = 1.00$$

the corresponding entries in Table I are 0.1554 and 0.3413, and the area under the curve between 12 and 15 (the area of the white region of the upper diagram of Figure 10.11) is $0.3413 - 0.1554 = 0.1859$; (b) the values of z corresponding to $x = 12$ and $x = 15$ are

$$z = \frac{12 - 20}{10} = -0.80 \quad \text{and} \quad z = \frac{15 - 20}{10} = -0.50$$

10.11

Areas under normal distributions.

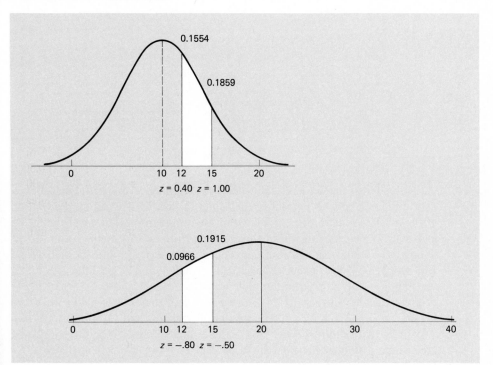

the corresponding entries in Table I are 0.2881 and 0.1915, and the area under the curve between 12 and 15 (the area of the white region of the lower diagram of Figure 10.11) is $0.2881 - 0.1915 = 0.0966$.

There are also problems in which we are given areas under normal curves and asked to find the corresponding values of z. The results of the example which follows will be used extensively in subsequent chapters.

EXAMPLE If z_α denotes the value of z for which the area under the standard normal curve to its right is equal to α (Greek lowercase alpha), find
(a) $z_{0.01}$;
(b) $z_{0.05}$.

SOLUTION (a) It can be seen from Figure 10.12 that $z_{0.01}$ corresponds to an entry of $0.5000 - 0.0100 = 0.4900$ in Table I; since the nearest entry is 0.4901 corresponding to $z = 2.33$, we have $z_{0.01} = 2.33$; (b) also from Figure 10.12, $z_{0.05}$ corresponds to an entry of $0.5000 - 0.0500 = 0.4500$ in Table I; since the two nearest entries are 0.4495 and 0.4505 corresponding to $z = 1.64$ and $z = 1.65$, we have $z_{0.05} = 1.645$.

10.12

Diagram for determination of z_α.

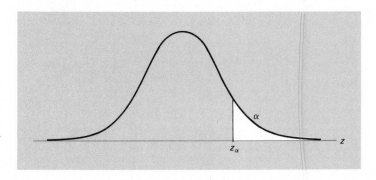

Table I also enables us to show that for reasonably symmetrical bell-shaped distributions, about 68 percent of the values fall within one standard deviation of the mean, about 95 percent fall within two standard deviations of the mean, and over 99 percent fall within three standard deviations of the mean. These figures are for normal distributions, and in parts (a), (b), and (c) of Exercise 10.6 on page 266 the reader will be asked to show that 0.6826 of the area under the standard normal distribution falls between $z = -1$ and $z = 1$, that 0.9544 of the area falls between $z = -2$ and $z = 2$, and that 0.9974 of the area falls between $z = -3$ and $z = 3$. The results of parts (d) and (e) of Exercise 10.6 also show that, although the "tails" extend indefinitely in both directions, the area under a standard normal curve beyond $z = 4$ or $z = 5$ is negligible.

10.1 Suppose that a continuous random variable takes on values on the interval from 2 to 10 and that the graph of its distribution, called a **uniform density**, is given by the horizontal line of Figure 10.13.

(a) What probability is represented by the white region of the diagram and what is its value?

(b) What is the probability that the random variable will take on a value less than 7? Is this probability the same as the probability that the random variable will take on a value less than or equal to 7?

(c) What is the probability that the random variable will take on a value between 2.7 and 8.8?

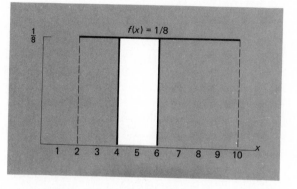

10.13

Uniform density.

10.2 Find the area under the standard normal curve which lies

(a) between $z = 0$ and $z = 0.86$;

(b) between $z = -1.63$ and $z = 0$;

(c) to the right of $z = 0.55$;

(d) to the right of $z = -0.38$;

(e) to the left of $z = 0.70$;

(f) to the left of $z = -0.15$.

10.3 Find the area under the standard normal curve which lies

(a) between $z = -0.55$ and $z = 0.55$;

(b) between $z = -0.93$ and $z = 2.21$;

(c) between $z = 0.56$ and $z = 0.96$;

(d) between $z = -0.95$ and $z = -0.66$.

10.4 Find the area under the standard normal curve which lies

(a) between $z = -0.45$ and $z = 0.45$;

(b) to the right of $z = -2.20$;

(c) to the left of $z = -1.35$;

(d) between $z = 2.25$ and $z = 2.65$;

(e) to the right of $z = 1.40$;

(f) between $z = -1.90$ and $z = -0.60$.

10.5 Find z if the normal-curve area

(a) between 0 and z is 0.4484;

(b) to the left of z is 0.9868;

(c) to the right of z is 0.8413;

(d) to the right of z is 0.3300;

(e) to the left of z is 0.3085;

(f) between $-z$ and z is 0.9700.

10.6 Find the normal-curve area between $-z$ and z if
    (a) $z = 1.00$;
    (b) $z = 2.00$;
    (c) $z = 3.00$;
    (d) $z = 4.00$;
    (e) $z = 5.00$.

10.7 Verify that
    (a) $z_{0.005} = 2.575$;
    (b) $z_{0.025} = 1.96$.

10.8 A random variable has a normal distribution with the mean $\mu = 80.0$ and the standard deviation $\sigma = 4.8$. What are the probabilities that this random variable will take on a value
    (a) less than 87.2;
    (b) greater than 76.4;
    (c) between 81.2 and 86.0;
    (d) between 71.6 and 88.4?

10.9 A normal distribution has the mean $\mu = 182.4$. If 20 percent of the area under the curve lies to the right of 199.2, find
    (a) the area to the right of 177.4;
    (b) the area between 192.4 and 202.4.

10.10 A random variable has a normal distribution with the standard deviation $\sigma = 10$. If the probability that the random variable will take on a value less than 82.5 is 0.8264, what is the probability that it will take on a value greater than 58.1?

10.11 Another continuous distribution, called the exponential distribution, has many important applications. If a random variable has an exponential distribution with mean μ, the probability that it will take on a value between 0 and any given positive value x is $1 - e^{-x/\mu}$ (see Figure 10.14). Here e is the constant which appears also in the formula for the Poisson distribution, and values of $e^{-x/\mu}$ can be obtained directly from Table X.
    (a) Find the probabilities that a random variable having an exponential distribution with $\mu = 10$ will take on a value between 0 and 4, a value greater than 6, and a value between 8 and 12.
    (b) The lifetime of a certain kind of battery is a random variable which has an exponential distribution with a mean of $\mu = 200$ hours. What is the probability that such a battery will last at most 100 hours? Between 400 and 600 hours?
    (c) In a certain brokerage house, the time a customer has to wait for confirmation of a transaction has an exponential distribution with $\mu = 30$ minutes. What is the probability that a customer will have to wait between 12 and 36 minutes for such a confirmation?

10.14

Exponential
distribution.

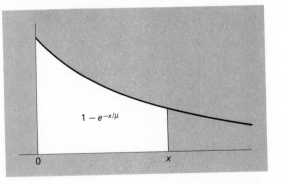

10.3

Applications of the Normal Distribution

Let us now consider some applied problems in which we shall assume that the distributions of the data, or the distributions of the random variables under consideration, can be approximated closely with normal curves.

EXAMPLE The lengths of the sardines received by a certain cannery have a mean of 4.62 inches and a standard deviation of 0.23 inch.
(a) What percentage of all these sardines are longer than 5.00 inches?
(b) What percentage of the sardines are between 4.35 and 4.85 inches long?

SOLUTION (a) The answer to this question is given by the area of the white region of Figure 10.15, that is, the area to the right of

$$z = \frac{5.00 - 4.62}{0.23} = 1.65$$

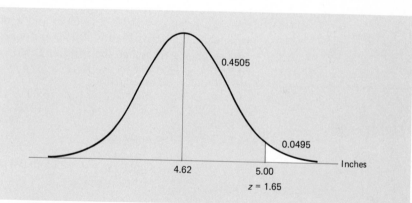

10.15

Normal distribution.

Since the entry in Table I corresponding to $z = 1.65$ is 0.4505, we find that $0.5000 - 0.4505 = 0.0495$; so 4.95 percent of the sardines are longer than 5.00 inches; (b) the answer is given by the area of the white region of Figure 10.16, the area under the curve between

$$z = \frac{4.35 - 4.62}{0.23} = -1.17 \quad \text{and} \quad z = \frac{4.85 - 4.62}{0.23} = 1.00$$

The corresponding entries in Table I are 0.3790 for $z = 1.17$ and 0.3413 for $z = 1.00$, and $0.3790 + 0.3413 = 0.7203$. Hence, a little

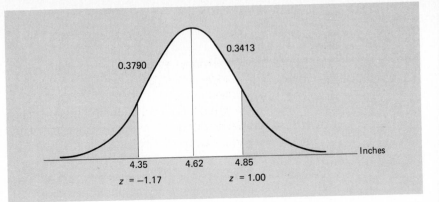

10.16

Normal distribution.

more than 72 percent of the sardines are between 4.35 and 4.85 inches long.

EXAMPLE The actual amount of instant coffee which a filling machine puts into "6-ounce" cans varies from can to can, and it may be looked upon as a random variable having a normal distribution with a standard deviation of 0.04 ounce. If only 2 percent of the cans are to contain less than 6 ounces of coffee, what must the mean fill of these cans be?

SOLUTION We are given $\sigma = 0.04$, $x = 6.00$, and a normal-curve area (that of the white region of Figure 10.17), and we are asked to find μ. Since the value of z for which the entry in Table I comes closest to $0.5000 - 0.0200 = 0.4800$ is 2.05, we have

$$-2.05 = \frac{6.00 - \mu}{0.04}$$

and, solving for μ, we get

$$6.00 - \mu = (-2.05)(0.04) = -0.082$$

10.17

Normal distribution.

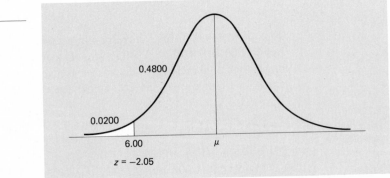

and then

$$\mu = 6.00 + 0.082 = 6.08 \text{ ounces}$$

Although, strictly speaking, the normal distribution applies to continuous random variables, it is often used to approximate distributions of discrete random variables, which can take on only a finite number of values or as many values as there are positive integers. This yields quite satisfactory results in many situations, provided that we make the continuity correction illustrated in the following example.

EXAMPLE   A baker knows that the daily demand for whole pecan pies is a random variable with a distribution which can be approximated closely by a normal distribution with the mean $\mu = 43.3$ and the standard deviation $\sigma = 4.6$. What is the probability that the demand for these pies will exceed 50 on any given day?

SOLUTION   The answer is given by the area of the white region of Figure 10.18; the area to the right of 50.5, not 50. The reason for this is that the number of pies the baker sells is a whole number. Hence, if we want to approximate this demand distribution with a normal curve, we must "spread" the values of this discrete random variable over a continuous scale, and we do this by representing each whole number k by the interval from $k - \frac{1}{2}$ to $k + \frac{1}{2}$. For instance, 20 is represented by the interval from 19.5 to 20.5, 25 is represented by the interval from 24.5 to 25.5, 50 is represented by the interval from 49.5 to 50.5, and the probability of a demand greater than 50 is given by the area under the curve to the right of 50.5. Accordingly, we get

$$z = \frac{50.5 - 43.3}{4.6} = 1.57$$

and it follows from Table I that the area of the white region of Figure

10.18

Normal distribution.

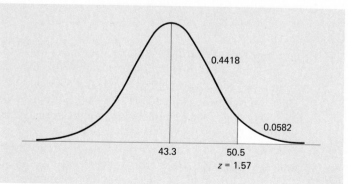

0.4418

0.0582

43.3   50.5
         z = 1.57

10.18—the probability of a demand for more than 50 pecan pies—is
0.5000 − 0.4418 = 0.0582.

10.4

The Normal Approximation to the Binomial Distribution

The normal distribution is sometimes introduced as a continuous distribution which provides a very close approximation to the binomial distribution when n, the number of trials, is very large and p, the probability of a success on an individual trial is close to 0.50. Figure 10.19 shows the histograms of binomial distributions having $p = 0.50$ and $n = 2, 5, 10,$ and 25, and it can be seen that with increasing n these distributions approach the symmetrical bell-shaped pattern of the normal distribution. In fact, a normal curve with the mean $\mu =$

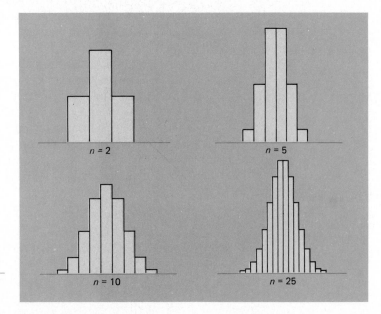

10.19

Binomial distributions
with $p = 0.50$.

np and the standard deviation $\sigma = \sqrt{np(1 - p)}$ can often be used to approximate a binomial distribution even when n is fairly small and p differs from 0.50 but is not too close to either 0 or 1. A good rule of thumb is to use this approximation only when np and $n(1 - p)$ are both greater than 5.

The following examples illustrate the normal approximation to the binomial distribution.

EXAMPLE Find the exact probability of getting 4 heads in 12 flips of a balanced coin, and also the normal approximation to this binomial probability.

To find the exact value, we substitute $n = 12$, $x = 4$, and $p = \frac{1}{2}$ into the binomial formula, getting

$$f(4) = \binom{12}{4}\left(\frac{1}{2}\right)^4\left(1 - \frac{1}{2}\right)^{12-4} = 495 \cdot \left(\frac{1}{2}\right)^{12}$$

$$= \frac{495}{4,096}$$

$$= 0.1208$$

To find the normal-curve approximation to this probability, we use the continuity correction and represent 4 heads by the interval from 3.5 to 4.5 (see Figure 10.20). Since $\mu = 12 \cdot \frac{1}{2} = 6$ and $\sigma = \sqrt{12 \cdot \frac{1}{2} \cdot \frac{1}{2}} =$

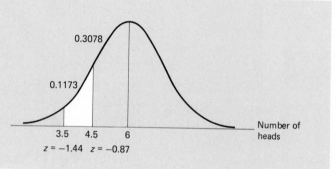

10.20

Normal-curve
approximation to
binomial distribution.

1.732, we have in standard units $z = \dfrac{3.5 - 6}{1.732} = -1.44$ for $x = 3.5$ and $z = \dfrac{4.5 - 6}{1.732} = -0.87$ for $x = 4.5$. The corresponding entries in Table I are 0.4251 and 0.3078, and the approximate probability is $0.4251 - 0.3078 = 0.1173$, only 0.0035 smaller than the exact probability.

The normal-curve approximation to the binomial distribution is particularly useful in problems where we would otherwise have to use the formula for the binomial distribution repeatedly to obtain the values of many different terms.

EXAMPLE What is the probability of getting at least 12 replies to questionnaires mailed to 100 persons, when the probability is 0.18 that any one of them will reply?

SOLUTION If we tried to solve this problem by using the formula for the binomial distribution, we would have to find the sum of the probabilities corresponding to 12, 13, 14, . . . , and 100 replies, or subtract from 1 the sum of the probabilities of 0, 1, 2, . . . , and 11 replies. This would obviously involve a tremendous amount of work, but using the normal-curve approximation, we need only find the white area of Figure 10.21, the area to the right of 11.5. We are again using the continuity correction according to which 12 is represented by the interval from 11.5 to 12.5, 13 is represented by the interval from 12.5 to 13.5, and so on.

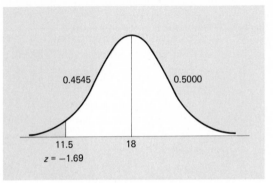

10.21

Normal-curve approximation to binomial distribution.

Since $\mu = 100(0.18) = 18$ and $\sigma = \sqrt{100(0.18)(0.82)} = 3.84$, we find that in standard units $x = 11.5$ becomes

$$z = \frac{11.5 - 18}{3.84} = -1.69$$

and that the probability is $0.4545 + 0.5000 = 0.9545$. This means that among many mailings of 100 questionnaires we can expect to get at least 12 replies about 95 percent of the time, provided the probability that any one person will reply is 0.18. The actual value of the probability, to four decimal places, is 0.9605, so the error of the approximation is only 0.0060.

EXERCISES

10.12 In an experiment to determine the amount of time required to assemble an "easy to assemble" toy, the assembly time was found to be a random variable having approximately a normal distribution with $\mu = 28.4$ minutes and $\sigma = 4.0$ minutes. What are the probabilities that this kind of toy can be assembled in
(a) less than 25.0 minutes;
(b) between 26.0 and 30.0 minutes?

10.13 The grapefruits grown in a large orchard have a mean weight of 19.3 ounces with a standard deviation of 2.2 ounces. Assuming that the distribution of the weight of these grapefruits has roughly the shape of a normal distribution, find
(a) what percentage of the grapefruits weigh less than 18.0 ounces;
(b) what percentage of the grapefruits weigh at least 20.0 ounces;

(c) what percentage of the grapefruits weigh between 18.5 and 20.5 ounces;

(d) the weight below which lies the lightest 15 percent of the grapefruits;

(e) the weight above which lies the heaviest 25 percent of the grapefruits.

10.14 In a photographic process, the developing time of prints may be looked upon as a random variable having a normal distribution with a mean of 15.28 seconds and a standard deviation of 0.12 second. Find the probabilities that the time required to develop a print is

(a) at least 15.50 seconds;

(b) at most 15.00 seconds;

(c) between 15.10 and 15.40 seconds;

(d) between 15.05 and 15.15 seconds.

10.15 A manufacturer must buy coil springs which will stand a load of at least 25 pounds. If supplier A can provide springs that stand an average load of 29.5 pounds with a standard deviation of 2.1 pounds and supplier B can provide springs that stand an average load of 28.3 pounds with a standard deviation of 1.6 pounds, and both load distributions are approximately normal, which of the two suppliers would provide the manufacturer with the smaller percentage of unsatisfactory springs?

10.16 With reference to the filling-machine example on page 268, show that if the variability of the machine is reduced so that $\sigma = 0.025$ ounce, this will lower the required average amount of coffee per can to $\mu = 6.05$ ounces, yet keep about 98 percent of the weights above 6 ounces.

10.17 An airline knows from experience that the number of suitcases it loses each week on a certain route is a random variable having approximately a normal distribution with the mean $\mu = 31.4$ and the standard deviation $\sigma = 4.5$. What are the probabilities that in any given week it will lose

(a) exactly 20 suitcases;

(b) at most 20 suitcases?

10.18 The number of days which guests stay at a large resort is a random variable having approximately a normal distribution with the mean $\mu = 8.5$ and the standard deviation $\sigma = 2.2$. Among 1,000 guests, how many can be expected to stay at the resort anywhere from 5 to 10 days?

10.19 A taxicab driver knows from experience that the number of fares he will pick up during an evening is a random variable having approximately a normal distribution with $\mu = 25.6$ and $\sigma = 4.2$. Find the probabilities that on one evening the driver will pick up

(a) more than 30 fares;

(b) less than 20 fares.

10.20 Use the normal-curve approximation to find the probability of getting 7 heads in 14 flips of a balanced coin, and compare the result with the value given in Table V.

10.21 If 20 percent of the loan applications received by a bank are refused, what is the probability that among 225 loan applications at least 50 will be refused?

10.22 What is the probability of getting fewer than 80 responses to 1,000 invitations sent out to promote a new land development, if the probability is 0.09 that any one person will respond?

10.23 A television network claims that its Monday night movie regularly has 36 percent of the total viewing audience. If this claim is correct, what is the probability that among 400 Monday night viewers more than 160 will be watching the network's movie?

10.24 A manufacturer knows that on the average 3 percent of the washing machines which he makes will require repairs within 60 days after they are sold. What is the probability that among 800 washing machines shipped by the manufacturer at least 20 will require repairs within 60 days after they are sold?

10.25 To avoid accusations of sexism, the authors of a statistics text flip a balanced coin to decide whether to use "he" or "she" whenever the occasion requires in exercises and examples. If they do this 80 times while revising one of their texts, what is the probability that they will use "she" at least 48 times?

10.26 To illustrate the Law of Large Numbers which we mentioned in connection with the frequency interpretation of probability and also on page 231, find the probabilities that the proportion of heads will be anywhere from 0.49 to 0.51 when a balanced coin is flipped
   (a) 100 times;
   (b) 10,000 times.

10.5

A Word of Caution

Although the normal distribution is the only continuous distribution which we have discussed in any detail in this chapter, we hope that this will not give the erroneous impression that the normal distribution is the only continuous distribution that matters in the study of statistics. The exponential distribution is an important one, too, and in later chapters we shall meet several other continuous distributions, among them the t distribution, the chi-square distribution, and the F distribution, which play important roles in problems of statistical inference.

It is true that the normal distribution plays a fundamental role in many statistical problems, but it is also true that its indiscriminate use can lead to very misleading results. There are various ways in which we can decide whether or not a distribution of observed data fits the overall pattern of a normal curve; one of these will be described in Exercise 14.54 on pages 384 and 385.

10.6

Check List of Key Terms

Continuity correction, 269
Continuous distribution, 256
Exponential distribution, 266
Normal approximation to binomial distribution, 270

Normal curve, 258
Normal distribution, 258
Probability density, 256
Standard normal distribution, 260
Standard units, 260

10.7

Review Exercises

10.27 The burning time of an experimental rocket is a random variable which has a normal distribution with $\mu = 4.26$ seconds and $\sigma = 0.04$ second. What is the probability that this kind of rocket will burn for
(a) less than 4.15 seconds;
(b) more than 4.30 seconds;
(c) 4.20 to 4.32 seconds?

10.28 Use the normal-curve approximation to find the probability of getting 9 heads in 15 flips of a balanced coin, and compare the result with the value given in Table V.

10.29 Find the area under the standard normal curve which lies
(a) between $z = 0$ and $z = 0.95$;
(b) to the left of $z = 2.50$;
(c) to the right of $z = -0.75$;
(d) to the right of $z = 1.44$;
(e) to the left of $z = -0.75$.

10.30 A random variable has a normal distribution with the mean $\mu = 112.4$ and the standard deviation $\sigma = 3.6$. What are the probabilities that this random variable will take on a value
(a) less than 117.8;
(b) greater than 109.7;
(c) between 116.9 and 120.5;
(d) between 106.1 and 114.2?

10.31 Find the values of
(a) $z_{0.02}$;
(b) $z_{0.10}$.

10.32 If 70 percent of all persons flying across the Atlantic Ocean feel the effect of the time difference for at least 24 hours, what is the probability that among 140 persons flying across the Atlantic, at least 100 will experience this feeling?

10.33 Suppose that a continuous random variable takes on values on the interval from 0 to 4 and that the graph of its distribution, called a **triangular density**, is given by the line of Figure 10.22. Find the probabilities that the random

10.22

Triangular density.

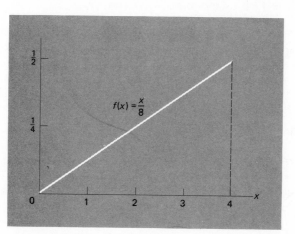

$$f(x) = \frac{x}{8}$$

variable will take on a value

(a) less than 1; (c) between 1.5 and 2.5.

(b) greater than 2;

10.34 The head of the complaint department of a department store knows from experience that the number of complaints he receives per day is a random variable with the mean $\mu = 33.4$ and the standard deviation $\sigma = 5.5$. Assuming that the distribution of the number of complaints has roughly the shape of a normal distribution, find the probabilities that he will receive in one day

(a) more than 40 complaints;

(b) at least 40 complaints;

(c) between 30 and 40 complaints, not inclusive.

10.35 A random variable has a normal distribution with the standard deviation $\sigma = 4.0$. If the probability is 0.9713 that the random variable will take on a value less than 87.6, what is the probability that it will take on a value between 75.0 and 78.0?

10.36 If 25 percent of all patients with high blood pressure have bad side effects from a certain kind of medicine, what is the probability that among 160 patients with high blood pressure who are treated with this medicine more than 45 have bad side effects?

10.37 Find the area under the standard normal curve which lies

(a) between $z = 0$ and $z = -1.25$;

(b) between $z = -0.81$ and $z = 0.81$;

(c) between $z = 0.47$ and $z = 0.88$;

(d) between $z = -1.90$ and $z = 0.95$;

(e) between $z = -2.22$ and $z = -1.82$.

10.38 If the speed of cars recorded at a certain checkpoint is a random variable having approximately a normal distribution with the mean $\mu = 48.5$ miles per hour and the standard deviation $\sigma = 4.7$ miles per hour, find the probabilities that the speed of a car passing the checkpoint will

(a) exceed the posted speed limit of 55 miles per hour;

(b) be under 40 miles per hour;

(c) be between 45 and 55 miles per hour.

10.39 Find z if the normal curve area

(a) between 0 and z is 0.2019; (c) to the right of z is 0.0336;

(b) to the right of z is 0.8810; (d) between $-z$ and z is 0.2662.

10.40 The average time required to perform job A is 85 minutes with a standard deviation of 16 minutes, and the average time required to perform job B is 110 minutes with a standard deviation of 11 minutes. Assuming normal distributions, what proportion of the time will job A take longer than the average job B, and what proportion of the time will job B take less time than the average job A?

The purpose of most statistical investigations is to make valid generalizations on the basis of samples about the populations from which the samples came. The whole problem of when and under what conditions samples permit such generalizations has no easy solution. For instance, if we want to estimate the average amount of money people spend on their vacations, we would hardly take as our sample the amounts spent by the deluxe-class passengers on a 92-day ocean cruise, nor would we attempt to estimate wholesale prices of all farm products on the basis of the prices of apricots alone. In both cases we would reject estimates based on these samples as ridiculous, but just which vacationers and which farm products we should include in our samples, and how many of them, is not intuitively clear.

In the theory which we shall study in most of the remainder of this book, it will be assumed that we are dealing with **random samples**. This attention to random samples is due to the fact that they permit valid, or logical, generalizations and hence are widely used in actual practice. As we shall see, however, random sampling is not always practical, feasible, or even desirable, and some other sampling procedures will be discussed briefly in Chapter 20.

In this chapter we begin with a formal definition of **random sampling** in Section 11.1. Then, in Section 11.2, we introduce the related concept of a **sampling distribution**, which tells us how quantities determined from samples may vary from sample to sample, and in Sections 11.3 and 11.4 we learn how such variations can be measured.

Sampling and Sampling Distributions

11

Random Sampling

Earlier in this book we distinguished between populations and samples, stating that a population consists of all conceivably or hypothetically possible instances (or observations) of a given phenomenon, while a sample is simply a part of a population. In preparation for the work which follows, let us now distinguish between two kinds of populations—**finite populations** and **infinite populations**. A finite population is one which consists of a finite number, or fixed number, of elements (items, objects, measurements, or observations). Examples of finite populations are the net weights of the 24,000 cans of rust remover in a production lot, the scores made by the 650-person entering class of a technical school on the engineer scale of the Strong Vocational Interest Blank, and the outside diameters of a lot of 10,000 precision ball bearings.

In contrast to finite populations, a population is said to be infinite if there is, at least hypothetically, no limit to the number of elements it can contain. The population which consists of the results obtained in all hypothetically possible rolls of a pair of dice is an infinite population, and so is the population which consists of all conceivably possible measurements which could be made of the length of a metal strip.

To introduce the idea of random sampling from a finite population, let us ask the following three questions: (1) "How many distinct samples of size n can be taken from a finite population of size N?" (2) "How is a random sample defined?" (3) "How can a random sample be taken in actual practice?"

To answer the first question, we refer to the rule for combinations according to which r objects can be selected from a set of n objects in $\binom{n}{r}$ ways. With a change of letters, we can say that the number of distinct samples of size n which can be drawn from a finite population of size N is $\binom{N}{n}$.

EXAMPLE How many different samples of size n can be drawn from a finite population of size N if
(a) $n = 2$ and $N = 10$;
(b) $n = 3$ and $N = 100$?

SOLUTION (a) $\binom{10}{2} = 45$; (b) $\binom{100}{3} = \dfrac{100 \cdot 99 \cdot 98}{3!} = 161{,}700$.

To answer the second question, we make use of the answer to the first one and define a **simple random sample** (or more briefly, a **random sample**) from a finite population as a sample which is chosen in such a way that each of the $\binom{N}{n}$ possible samples has the same probability, $1/\binom{N}{n}$, of being selected.

For instance, if a finite population consists of the $N = 5$ elements a, b, c, d, and e (which might be the repair costs on five ranges), there are $\binom{5}{3} = 10$ possible distinct samples of size $n = 3$; they consist of the elements abc, abd, abe, acd, ace, ade, bcd, bce, bde, and cde. If we choose one of these samples in such a way that each has the probability $\frac{1}{10}$ of being chosen, we call this sample a random sample.

With regard to the third question, how to take a random sample in actual practice, we could, in simple cases like the one above, write each of the $\binom{N}{n}$ possible samples on a slip of paper, put these slips in a hat, shuffle them thoroughly, and then draw one without looking. Such a procedure is obviously impractical, if not impossible, in more realistically complex problems of sampling; we mention it here only to make the point that the selection of a random sample must depend entirely on chance.

Fortunately, we can take a random sample without actually resorting to the tedious process of listing all possible samples. We can list instead the N individual elements of a finite population, then take a random sample by choosing the elements to be included in the sample one at a time without replacement, making sure that in each of the successive drawings each of the remaining elements of the population has the same chance of being selected. This leads to the same probability, $1/\binom{N}{n}$, for each possible sample. For instance, to take a random sample of 20 past-due accounts from a file of 257 such accounts, we could write each account number on a slip of paper, put the slips in a box and mix them thoroughly, then draw (without looking) 20 slips one after the other without replacement.

Even such a relatively simple procedure as this is often not necessary in practice, where the simplest way to take a random sample is to use a table of **random digits** (or **random numbers**). Published tables of random numbers consist of pages on which the decimal digits 0, 1, 2, . . . , and 9 are set down in much the same fashion as they might appear if they had been generated by a chance or gambling device giving each digit the same probability of $\frac{1}{10}$ of appearing at any given place in the table. Some early tables of random numbers were copied from pages of census data or from tables of 20-place logarithms, but they were found to be deficient in various ways. Nowadays, such tables are made with the use of electronic computers, but it would be possible to generate a table with a perfectly constructed spinner like that shown in Figure 11.1.

Table XIII is an excerpt from a published table of random numbers, and we shall illustrate its use by considering the problem of taking a random sample of 10 printing firms from the 562 firms listed in the Yellow Pages of a large city telephone directory. Numbering the firms on the alphabetical list 001, 002,

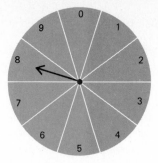

11.1

Spinner.

003, . . . , 561, and 562, we arbitrarily pick a starting place in the table and then move in any direction, reading out three-digit numbers. For instance, if we arbitrarily enter the table on page 582 and read out the digits in the 26th, 27th, and 28th columns starting with the 31st row and going down the page, we find that the sample consists of the 10 firms whose numbers are

$$187 \quad 155 \quad 388 \quad 320 \quad 281 \quad 88 \quad 520 \quad 275 \quad 480 \quad 273$$

In selecting these numbers, we ignored the tabled numbers greater than 562; also, had any number reoccurred, we would have ignored it, too.

When lists are available and items are, or can readily be, numbered, it is easy to take random samples from finite populations with the aid of random number tables. Unfortunately, however, it is often impossible to proceed in the way we have just described. For example, if we wanted to estimate from a sample the mean protein content of a carload of wheat, it would be impossible to number each of the millions of grains of wheat, choose random numbers, and then locate the corresponding grains. In this and in many similar situations, all one can do is proceed according to the dictionary definition of the word "random," namely, "haphazardly without definite aim or purpose." That is, we must not select or reject any element of a population because of its seeming typicalness or lack of it, nor must we favor or ignore any part of a population because of its accessibility or lack of it, and so on. Hopefully, such haphazard procedures will lead to samples which may be treated as though they were, in fact, random samples.

To this point we have been discussing only random samples from finite populations. The concept of a random sample from an infinite population is more difficult to define, but a few simple illustrations will help to explain the basic characteristics of such a sample. For instance, we consider 10 tosses of a coin as a sample from the hypothetically infinite population consisting of all possible tosses of the coin. Then, if the probability of getting heads is the same for each toss and the 10 tosses are independent, we say that the sample is random. Also, we would be sampling from an infinite population if we sample with replacement from a finite population, and our sample would be random if in each draw all elements of the population have the same probability of being selected, and successive draws are independent.

Generally speaking, we assert that the selection of each item in a random sample from an infinite population must be controlled by the same probabilities and that successive selections must be independent of one another. Unless these conditions are satisfied at least approximately, sets of observations drawn from infinite populations cannot legitimately be treated as random samples.

EXERCISES

11.1 How many different samples of size 2 can be selected from a finite population of
(a) size 5;   (b) size 12;   (c) size 25?

11.2 How many different samples of size 3 can be selected from a finite population of
(a) size 5;   (b) size 50;   (c) size 150?

11.3 List the 10 possible samples of size 2 which can be drawn from the finite population consisting of the elements $p, q, r, s,$ and t. If each of these samples is assigned the probability $\frac{1}{10}$, show that the probability is $\frac{2}{5}$ that any one specific element will be contained in the sample.

11.4 List all possible choices of four of the following six corporations: General Motors, IBM, Shell Oil, Coca-Cola, Polaroid, and American Airlines. If a person randomly selects four of these corporations, what is
(a) the probability of each possible sample;
(b) the probability that any particular one of these corporations (say, Shell Oil) will be included in the sample;
(c) the probability that any particular pair of these corporations (say, IBM and Polaroid) will be included in the sample?

11.5 Suppose that a newspaper reporter wants to interview 12 of the 600 persons attending a business forecasting conference. If these persons are listed in the program in alphabetic order, which ones (by number) would he select for an interview if he numbered them from 001 through 600 and then chose a random sample by using the first three columns of the table on page 579, beginning with the first row and going down the page?

11.6 The employees of a company have badges numbered serially from 1 through 643. Use the 16th, 17th, and 18th columns of the table on page 580, starting with the 11th row and going down the page, to select a random sample of six of the company's employees to serve on a committee.

11.7 A county assessor wants to review the assessments on a random sample of 20 of 6,327 one-family homes. If she numbers them $0001, 0002, \ldots, 6326,$ and 6327, which ones (by number) will she select, if she chooses them by using the 11th, 12th, 13th, and 14th columns of the table on page 581, going down the page starting with the seventh row?

11.8 Explain why each of the following samples may yield misleading information:
(a) In order to predict a municipal election, a public opinion poll telephones persons selected haphazardly from the city's telephone directory.
(b) To determine the proportion of improperly sealed cans of coffee, a quality-control inspector examines every 50th can coming off a production line.
(c) To determine the average annual income of its graduates 10 years after graduation, a college's alumni office sent questionnaires in 1981 to all

members of the class of 1971, and the estimate was based on the question-naires returned.

    (d) To ascertain facts about tooth-brushing habits, a sample of the residents of a community are asked how many times they brush their teeth each day.

    (e) To study executives' reaction to its copying machines, the Xerox corporation hires a research organization to ask executives the question: How do you like using Xerox copies?

    (f) A house-to-house survey is made to study consumer reaction to a new pudding mix, with no provisions for return visits in case no one is at home.

11.9 On page 279 we said that a random sample can be drawn from a finite population by choosing the elements to be included in the sample one at a time, making sure that in each of the successive drawings each of the remaining elements of the population has the same chance of being selected. To verify that this will give the correct probability to each sample, let us refer to the example on page 279, where we dealt with random samples of size 3 drawn from the finite population which consists of the elements a, b, c, d, and e. To find the probability of drawing any particular sample (say, b, c, and e), we can argue that the probability of getting one of these three elements on the first draw is $\frac{3}{5}$, the probability of then getting one of the remaining two elements on the second draw is $\frac{2}{4}$, and the probability of then getting the third element on the third draw is $\frac{1}{3}$. Multiplying these three probabilities, we find that the probability of getting the particular sample is $\frac{3}{5} \cdot \frac{2}{4} \cdot \frac{1}{3} = \frac{1}{10}$, and this agrees with the value obtained on page 279.

    (a) Use the same kind of argument to verify that for each possible random sample of size 3, drawn one at a time from a finite population of size 100, the probability is $1 / \binom{100}{3} = \frac{1}{161,700}$.

    (b) Use the same kind of argument to verify in general that for each possible random sample of size n, drawn one at a time from a finite population of size N, the probability is $1 / \binom{N}{n}$.

11.10 Making use of the fact that among the $\binom{N}{n}$ samples of size n which can be drawn from a finite population of size N there are $\binom{N-1}{n-1}$ which contain a specific element, show that the probability that any specific element of the population will be contained in a random sample of size n is $\frac{n}{N}$.

11.2

Sampling Distributions

Let us now introduce the concept of the **sampling distribution** of a statistic, probably the most basic concept of statistical inference. As we shall see, this concept is related to the idea of chance variation, or chance fluctuations, which we mentioned earlier to emphasize the need for measuring the variability of data. In this chapter we shall concentrate mainly on the sample mean and its sampling distribution, but later on we shall consider the sampling distributions of other statistics.

We can approach the study of sampling distributions in two ways. One, based on appropriate mathematical theory, leads to what is called a **theoretical sampling distribution**; the other, based on repeated samples from the same population, leads to what is called an **experimental sampling distribution**. The latter will prove to be very useful in our study because it provides experimental verification of some necessary theorems, which cannot be derived formally at the level of this book.

To introduce the idea of a theoretical sampling distribution, let us construct the one for the mean of random samples of size $n = 2$ from the finite population of size $N = 5$, whose elements are the numbers 1, 3, 5, 7, and 9. The mean of this population is

$$\mu = \frac{1 + 3 + 5 + 7 + 9}{5} = 5$$

its variance is

$$\sigma^2 = \frac{1}{5}[(1 - 5)^2 + (3 - 5)^2 + (5 - 5)^2 + (7 - 5)^2 + (9 - 5)^2]$$
$$= 8$$

and, hence, its standard deviation is $\sigma = \sqrt{8}$.

Now, there are $\binom{5}{2} = 10$ random samples of size $n = 2$ which can be drawn from this population; they are

1 and 3, 1 and 5, 1 and 7, 1 and 9, 3 and 5
3 and 7, 3 and 9, 5 and 7, 5 and 9, 7 and 9

and their means are 2, 3, 4, 5, 4, 5, 6, 6, 7, and 8. Since each sample has the probability $\frac{1}{10}$, we get the following theoretical sampling distribution of the mean for random samples of size $n = 2$ from the given population:

$\bar{x}$	Probability
2	1/10
3	1/10
4	2/10
5	2/10
6	2/10
7	1/10
8	1/10

A histogram of this distribution is shown in Figure 11.2.

An examination of this sampling distribution reveals some pertinent information relative to the problem of estimating the mean of the population on the

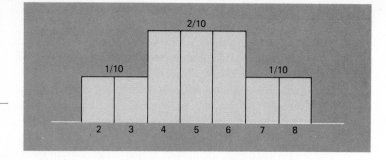

basis of a random sample of two items drawn from the population. For instance, we see that the probability is $\frac{6}{10}$ that a sample mean will not differ from the population mean by more than 1, and that the probability is $\frac{8}{10}$ that a sample mean will not differ from the population mean by more than 2; the first case corresponds to $\bar{x} = 4, 5,$ or 6, and the second to $\bar{x} = 3, 4, 5, 6,$ or 7.

We can get further useful information about this sampling distribution of the mean by calculating its mean $\mu_{\bar{x}}$ and its standard deviation $\sigma_{\bar{x}}$. (The subscript $\bar{x}$ is used here to distinguish these parameters from those of the original population.) Following the definitions of the mean and the variance of a probability distribution on pages 226 and 228, we get

$$\mu_{\bar{x}} = 2 \cdot \frac{1}{10} + 3 \cdot \frac{1}{10} + 4 \cdot \frac{2}{10} + 5 \cdot \frac{2}{10} + 6 \cdot \frac{2}{10} + 7 \cdot \frac{1}{10} + 8 \cdot \frac{1}{10}$$
$$= 5$$

and

$$\sigma_{\bar{x}}^2 = (2-5)^2\frac{1}{10} + (3-5)^2\frac{1}{10} + (4-5)^2\frac{2}{10} + (5-5)^2\frac{2}{10}$$
$$+ (6-5)^2\frac{2}{10} + (7-5)^2\frac{1}{10} + (8-5)^2\frac{1}{10}$$
$$= 3$$

so that $\sigma_{\bar{x}} = \sqrt{3}$. Thus, we find that the mean of the sampling distribution of the $\bar{x}$'s equals the mean of the population the samples came from, and although the exact relationship is by no means obvious, the standard deviation of the sampling distribution of $\bar{x}$ is smaller than that of the population.

The relationships demonstrated in this example are of fundamental importance, and we shall state them formally later. For now, let us merely take note of them and turn to the problem of constructing an experimental sampling distribution of the mean, hoping thereby to gain some further insight into the "behavior" of sample estimates $\bar{x}$ of a population mean μ. Specifically, let us suppose that, in connection with requests for additional service, a bus line operator wants to determine the average number of passengers carried on weekdays on a certain once-a-day route between two rural communities. Let us suppose, furthermore, that in a sample of five weekdays there were 19, 15, 11, 12, and 21 passengers. The mean of this sample is

$$\bar{x} = \frac{19 + 15 + 11 + 12 + 21}{5} = 15.6$$

and in the absence of any other information this figure may be used as an estimate of μ, the true average number of passengers carried on a weekday on this route. It stands to reason, however, that if the sample had consisted of the numbers of passengers carried on five other weekdays, the mean would most likely not have been 15.6. Indeed, if the operator took several samples, each consisting of the numbers of passengers carried on five weekdays, he might get such discrepant estimates of μ as 13.8, 18.6, 12.4, and 17.0.

We want to show now how the means of such samples vary purely as the result of chance, and to this end we did the following. First, we supposed that the daily number of passengers is a random variable having the Poisson distribution with $\lambda = 16$. Then we performed an experiment which consisted of drawing (in a way we shall describe in Section 11.5) 50 random samples from this kind of population. Each of these samples contains five observations—the numbers of passengers carried on five days. The results of the experiment are shown in the table below, where we see that in the first sample there were 16, 15, 14, 12, and 18 passengers carried on the five days, and so on.

Sample	Number of passengers	Sample	Number of passengers
1	16, 15, 14, 12, 18	26	10, 18, 19, 13, 20
2	20, 18, 16, 19, 14	27	14, 19, 16, 13, 21
3	21, 18, 17, 26, 25	28	18, 14, 23, 23, 14
4	16, 10, 9, 19, 15	29	17, 16, 11, 17, 11
5	17, 16, 20, 17, 7	30	16, 13, 10, 14, 20
6	20, 8, 17, 16, 13	31	13, 17, 22, 19, 18
7	22, 21, 16, 15, 13	32	15, 13, 16, 14, 21
8	11, 23, 12, 20, 14	33	16, 15, 8, 12, 23
9	17, 22, 21, 16, 20	34	16, 15, 11, 20, 13
10	18, 13, 15, 11, 12	35	17, 17, 16, 21, 14
11	15, 11, 14, 14, 18	36	18, 20, 14, 26, 18
12	22, 15, 13, 19, 11	37	13, 16, 17, 11, 6
13	20, 17, 11, 19, 15	38	17, 19, 15, 19, 16
14	15, 16, 16, 15, 17	39	11, 13, 18, 23, 18
15	15, 16, 17, 17, 16	40	12, 25, 21, 18, 8
16	13, 15, 15, 13, 18	41	21, 12, 14, 17, 16
17	12, 11, 19, 17, 16	42	19, 10, 15, 16, 18
18	15, 26, 19, 20, 15	43	20, 10, 15, 15, 19
19	15, 11, 21, 8, 17	44	16, 10, 26, 14, 20
20	12, 21, 10, 15, 16	45	18, 12, 13, 19, 9
21	10, 11, 9, 11, 11	46	15, 14, 21, 17, 11
22	12, 12, 24, 11, 5	47	12, 13, 13, 14, 12
23	16, 18, 14, 9, 11	48	18, 8, 21, 14, 15
24	17, 9, 18, 16, 9	49	20, 17, 16, 18, 19
25	11, 17, 19, 20, 17	50	19, 17, 16, 13, 15

Next we calculated the mean number of passengers per day for each of the 50 samples, getting

15.0	17.4	21.4	13.8	15.4	14.8	17.4	16.0	19.2	13.8
14.4	16.0	16.4	15.8	16.2	14.8	15.0	19.0	14.4	14.8
10.4	12.8	13.6	13.8	16.8	16.0	16.6	18.4	14.4	14.6
17.8	15.8	14.8	15.0	17.0	19.2	12.6	17.2	16.6	16.8
16.0	15.6	15.8	17.2	14.2	15.6	12.8	15.2	18.0	16.0

and finally we grouped these means into a table having the classes 9.5–10.5, 10.5–11.5, . . . , and 20.5–21.5. (There is no risk of ambiguity here, since division of integers by 5 always leaves a remainder of 0, 2, 4, 6, or 8 tenths, never 5 tenths.) The resulting distribution is shown as a histogram in Figure 11.3.

Now, inspection of this experimental sampling distribution of $\bar{x}$ tells us a great deal about the way in which the means of random samples drawn from the same population tend to scatter among themselves due to chance. For instance, we find that the smallest mean is 10.4 and the largest is 21.4; furthermore, 31 out of 50 (or 62 percent) of the means are between 14.5 and 17.5, and 48 out of 50 (or 96 percent) of the means are between 12.5 and 19.5. Since the mean of the population from which the samples came is $\mu = \lambda = 16$, we find that 62 percent of the sample means are "off" (differ from the population mean) by less than 1.5, and that 96 percent of the sample means are "off" by less than 3.5.

The two examples which we have given in this section are actually only learning devices designed to introduce the concept of a sampling distribution.

11.3

Experimental sampling distribution of the mean.

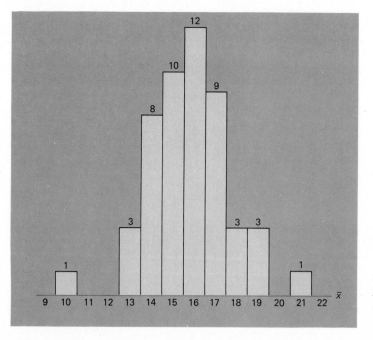

FIGURE 9.2

Experimental s tribution of the

In actual practice, we ordinarily take only one sample and, consequently, have only one estimate of the mean of a population. Indeed, there is really no need to construct a sampling distribution in order to estimate μ on the basis of one sample or to evaluate the goodness of such an estimate. What should be clearly understood at this point is that (1) this single mean is only one of many (maybe, infinitely many) possible means, (2) the distribution of the totality of these means is called a sampling distribution of the mean, and (3) the many means which form this distribution scatter more or less widely (and, as we shall see, predictably) about the population mean μ.

11.3

The Standard Error of the Mean

In most practical situations we can determine how close a sample mean might be to the mean of the population from which it came, by applying two theorems, one given below and the other on page 290, which express essential facts about sampling distributions of the mean. The first of these theorems expresses formally what we discovered in connection with the example on page 284: The mean of the sampling distribution of $\bar{x}$ equals the mean of the population sampled and the standard deviation of the sampling distribution is smaller than the standard deviation of the population. It may be phrased as follows: **For random samples of size n taken from a population having the mean μ and the standard deviation σ, the theoretical sampling distribution of $\bar{x}$ has the mean $\mu_{\bar{x}} = \mu$ and the standard deviation**

Standard error of the mean (finite population)

$$\sigma_{\bar{x}} = \frac{\sigma}{\sqrt{n}} \cdot \sqrt{\frac{N-n}{N-1}}$$

for finite populations of size N and

Standard error of the mean (infinite population)

$$\sigma_{\bar{x}} = \frac{\sigma}{\sqrt{n}}$$

for infinite populations.

It is customary to refer to $\sigma_{\bar{x}}$, the standard deviation of the sampling distribution of the mean, as the **standard error of the mean**. Its role in statistics is fundamental, since it measures the extent to which sample means can be expected to fluctuate, or vary, due to chance. Clearly, some knowledge of this variability is essential in determining how well $\bar{x}$ estimates the population mean μ. Intuition leads one (correctly) to feel that the smaller $\sigma_{\bar{x}}$ is (the less the $\bar{x}$'s are spread out) the better the estimate will be, and the larger $\sigma_{\bar{x}}$ is (the more the $\bar{x}$'s are spread out) the poorer the estimate will be. What determines the size of $\sigma_{\bar{x}}$, and hence the goodness of an estimate, can be seen from the formulas above. The formula for samples from finite populations shows among other things that (for fixed N) the standard error of the mean *increases* as the variability of the population increases, and *decreases* as the number of items in the sample increases. With respect to the latter, we note that substitution into the formula yields $\sigma_{\bar{x}} = \sigma$ for $n = 1$ and $\sigma_{\bar{x}} = 0$ for $n = N$; in other words, $\sigma_{\bar{x}}$ takes on values between 0 and σ and is 0 only when the sample includes the entire population.

EXAMPLE When we sample from an infinite population, what happens to the standard error of the mean (and, hence, to the size of the error we are exposed to when we use $\bar{x}$ as an estimate of μ) if the sample size is increased from $n = 50$ to $n = 200$?

SOLUTION The ratio of the two standard errors is

$$\frac{\dfrac{\sigma}{\sqrt{200}}}{\dfrac{\sigma}{\sqrt{50}}} = \frac{\sqrt{50}}{\sqrt{200}} = \frac{1}{2}$$

so that the standard error of the mean is divided by 2, or halved.

The factor $\sqrt{\dfrac{N-n}{N-1}}$ in the first formula for $\sigma_{\bar{x}}$ is called the **finite population correction factor,** for without it the two formulas for $\sigma_{\bar{x}}$ (for finite and infinite populations) are the same. It is usually ignored unless the sample constitutes at least 5 percent of the population, for otherwise it has little effect on the standard error $\sigma_{\bar{x}}$.

EXAMPLE Find the value of the finite population correction factor for $n = 100$ and $N = 10,000$.

SOLUTION Substituting $n = 100$ and $N = 10,000$, we get

$$\sqrt{\frac{N-n}{N-1}} = \sqrt{\frac{10,000-100}{10,000-1}} = 0.995$$

This is so close to 1 that the correction factor would ordinarily be ignored in practice.

To get a feeling for the two formulas for $\sigma_{\bar{x}}$, let us return to the two illustrations of the preceding section.

EXAMPLE   With reference to the illustration on page 284, verify that the formula for $\sigma_{\bar{x}}$ for a random sample from a finite population also yields $\sigma_{\bar{x}} = \sqrt{3}$.

SOLUTION   Substituting $n = 2$, $N = 5$, and $\sigma = \sqrt{8}$ into the first of the two formulas for $\sigma_{\bar{x}}$, we get

$$\sigma_{\bar{x}} = \frac{\sqrt{8}}{\sqrt{2}} \cdot \sqrt{\frac{5-2}{5-1}} = \sqrt{3}$$

EXAMPLE   With reference to the illustration on pages 285 and 286, use the standard deviation of the 50 sample means as an estimate of $\sigma_{\bar{x}}$, and compare it with the value given by the second formula for $\sigma_{\bar{x}}$ for random samples of size $n = 5$ from a population having the Poisson distribution with the mean $\lambda = 16$ and, hence, the standard deviation $\sigma = \sqrt{\lambda} = \sqrt{16} = 4$.

SOLUTION   The standard deviation of the 50 means is 1.95. Substitution into the formula for $\sigma_{\bar{x}}$ for random samples from infinite populations yields

$$\sigma_{\bar{x}} = \frac{4}{\sqrt{5}} = 1.79$$

and it should be observed that 1.95 (the experimental value) is quite close to 1.79 (the exact value). Thus, this small-scale sampling experiment provides good supporting evidence for the theoretical result. Observe also that the mean of the 50 sample means, which is 15.75, is quite close to the theoretical value $\mu = 16$.

11.4

The Central Limit Theorem

When we estimate the mean of a population, we usually attach a probability to a measure of the error of our estimate. Using Chebyshev's theorem, we can assert with a probability of at least $1 - 1/k^2$ that the mean of a random sample of size n will differ from the mean of the population from which it came by less than $k \cdot \sigma_{\bar{x}}$. In other words, when we use the mean of a random sample to estimate the mean of a population, we can assert with a probability of at least $1 - 1/k^2$ that our error will be less than $k \cdot \sigma_{\bar{x}}$.

EXAMPLE   Based on Chebyshev's theorem with, say, $k = 2$, what can we assert about the possible size of our error if we use the mean of a random sample of size $n = 64$ to estimate the mean of an infinite population with $\sigma = 20$?

Substituting $\sigma = 20$ and $n = 64$ into the second of the two formulas for the standard error of the mean, we get

$$\sigma_{\bar{x}} = \frac{20}{\sqrt{64}} = 2.5$$

and it follows that we can assert with a probability of at least $1 - \frac{1}{2^2} = 0.75$ that the error is less than $k \cdot \sigma_{\bar{x}} = 2(2.5) = 5$.

This shows that we can make probability statements about errors of estimates without having to go through the tedious (if not impossible) process of constructing the corresponding theoretical sampling distributions.

Chebyshev's theorem applies to any distribution, and it is always possible to use it as in the preceding example. However, there exists another basic theorem of statistics, the **central limit theorem,** which enables us in a great many instances to make much stronger probability statements then we can with Chebyshev's theorem. This is the second theorem referred to on page 287, and it may be stated as follows

Central limit theorem

> *If n (the sample size) is large, the theoretical sampling distribution of the mean can be approximated closely with a normal distribution.*

This theorem is of fundamental importance in statistics, since it justifies the use of normal-curve methods in a wide range of problems; it applies to infinite populations, and also to populations where n, though large, constitutes but a small portion of the population. It is difficult to say precisely how large n must be so that the central limit theorem applies, but unless the population has a very unusual shape, $n = 30$ is usually regarded as sufficiently large. The distribution of Figure 11.3 is fairly symmetrical and bell-shaped, even though the sample size is only $n = 5$. When the population we are sampling has, itself, roughly the shape of a normal curve, the sampling distribution of the mean can be approximated closely with a normal distribution regardless of the size of n.

The importance of the central limit theorem can be illustrated by reexamining the preceding example and also the illustration of Section 11.2.

EXAMPLE Based on the central limit theorem, what is the probability that the error will be less than 5 when we use the mean of a random sample of size $n = 64$ to estimate the mean of an infinite population with $\sigma = 20$?

SOLUTION The probability is given by the area of the white region under the curve in Figure 11.4; that is, by the normal-curve area between

$$z = \frac{-5}{20/\sqrt{64}} = -2 \quad \text{and} \quad z = \frac{5}{20/\sqrt{64}} = 2$$

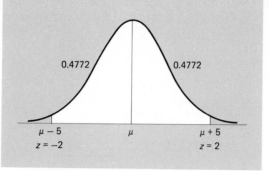

11.4

Sampling distribution
of the mean.

The entry in Table I corresponding to $z = 2.00$ is 0.4772, so this probability is $0.4772 + 0.4772 = 0.9544$. According to Chebyshev's theorem the probability is "at least 0.75," but we can actually make the much stronger statement that the probability is 0.9544 that the mean of a random sample of size $n = 64$ from the given population will differ from the mean of the population by less than 5.

EXAMPLE When we constructed the experimental sampling distribution in Section 11.2, we took 50 random samples of size $n = 5$ from a population with $\sigma = 4$ and found that 31 of their means (or 62 percent) differed from the mean of the population by less than 1.5. Based on the central limit theorem, what is the probability that the difference between the sample mean and the population mean will be less than 1.5, when we take a random sample of size $n = 5$ from an infinite population with $\sigma = 4$?

SOLUTION The probability is given by the area of the white region under the curve in Figure 11.5; that is, by the normal-curve area between

$$z = \frac{-1.5}{4/\sqrt{5}} = -0.84 \quad \text{and} \quad z = \frac{1.5}{4/\sqrt{5}} = 0.84$$

11.5

Sampling distribution
of the mean.

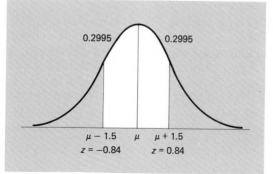

Since the entry in Table I corresponding to $z = 0.84$ is 0.2995, this probability is $0.2995 + 0.2995 = 0.5990$, or about 0.60. This is remarkably close to the 0.62 proportion which we got in the small sampling experiment of Section 11.2.

EXERCISES

11.11 Random samples of size 2 are taken from the finite population which consists of the numbers 6, 7, 8, 9, 10, and 11.
  (a) Show that the mean of this population is $\mu = 8.5$ and that its standard deviation is $\sigma = \sqrt{\frac{35}{12}}$.
  (b) List the 15 possible random samples of size 2 that can be taken from this finite population and calculate their means.
  (c) Use the results of part (b) to construct the sampling distribution of the mean for random samples of size 2 from the given finite population. Assign each possible sample a probability of $\frac{1}{15}$.
  (d) Calculate the mean and the variance of the probability distribution obtained in part (c) and verify the results with the use of the theorem on page 287.

11.12 Repeat parts (b), (c), and (d) of Exercise 11.11 for random samples of size 3 from the given population.

11.13 The finite population of Exercise 11.11 can be converted into an infinite population if we sample with replacement, that is, if we take a random sample of size 2 by first drawing one value and replacing it before drawing the second.
  (a) List the 36 possible samples of size 2 that can be drawn with replacement from the given population.
  (b) Determine the means of the 36 samples obtained in part (a) and, assigning each of the samples a probability of $\frac{1}{36}$, construct the sampling distribution of the mean for random samples of size 2 from this infinite population.
  (c) Calculate the mean and the standard deviation of the probability distribution obtained in part (b) and compare them with the corresponding values expected according to the theorem on page 287.

11.14 Convert the 50 samples on page 285 into 25 samples of size $n = 10$ by combining samples 1 and 26, samples 2 and 27, . . . , and samples 25 and 50. Calculate the mean of each of these samples of size 10 and determine their mean and their standard deviation. Compare this mean and this standard deviation with the corresponding values expected in accordance with the theorem on page 287.

11.15 Find the medians of the 50 samples on page 285, group them as we grouped the corresponding means, and construct a histogram like that of Figure 11.3. Comparing the standard deviation of this experimental sampling distribution of the median with that of the corresponding experimental sampling distribution of the mean (which was 1.95), what can we say about the relative reliability of the median and the mean in estimating the mean of the given population?

11.16 For random samples of size n from a population having the shape of a normal distribution, the standard error of the median (the standard deviation of the sampling distribution of the median) is given by $\sqrt{\frac{\pi}{2}} \cdot \frac{\sigma}{\sqrt{n}}$ or approximately

$1.25\dfrac{\sigma}{\sqrt{n}}$. How close is the value obtained for the standard deviation of the medians in Exercise 11.15 to the value expected according to this formula with $\sigma = 4$ and $n = 5$?

11.17 If μ is to be estimated by the median of a random sample of size n from a population having the shape of a normal curve, how large must n be so that this estimate is equally as reliable as another estimate of μ based on the mean of a random sample of size 64?

11.18 When we calculate the standard deviations of the 50 samples on page 285, we obtain the following experimental sampling distribution of s:

Sample standard deviation	Frequency
0.5–1.5	4
1.5–2.5	6
2.5–3.5	12
3.5–4.5	16
4.5–5.5	8
5.5–6.5	2
6.5–7.5	2

(Since none of the values is exactly 1.5, 2.5, ..., or 6.5, these overlapping class boundaries do not matter.)
(a) Calculate the mean and the standard deviation of this sampling distribution.
(b) For large samples, the formula $\sigma/\sqrt{2n}$ is sometimes used for the standard error of the standard deviation. Substituting $\sigma = 4$ and $n = 5$ into this formula, calculate the value of this standard error for the example and compare it with the value of the standard deviation obtained for the experimental sampling distribution in part (a). (Since $n = 5$ is not exactly a large sample, we should not be surprised if the results are not too close.)

11.19 When we sample from an infinite population, what happens to the standard error of the mean if the sample size is
(a) increased from 25 to 225;   (b) increased from 100 to 225;
(c) decreased from 750 to 30?

11.20 What is the value of the finite population correction factor when
(a) $n = 5$ and $N = 200$;   (c) $n = 100$ and $N = 5{,}000$?
(b) $n = 10$ and $N = 400$;

11.21 Show that if the mean of a random sample of size n is used to estimate the mean of an infinite population with the standard deviation σ, there is a fifty–fifty chance that the error (that is, the difference between $\bar{x}$ and μ) is less than $0.6745\dfrac{\sigma}{\sqrt{n}}$. This quantity is called the probable error of the mean and it is used now mainly in military applications.

11.22 What is the probable error of the mean (see Exercise 11.21), if a random sample of size $n = 40$ is taken from an infinite population with $\sigma = 18.5$?

11.23 When we constructed the experimental sampling distribution in Section 11.2, we took 50 random samples of size $n = 5$ from a population with $\sigma = 4$ and found that 48 of the sample means (or 96 percent) differed from the mean of the population by less than 3.5. Based on the central limit theorem, what is the probability that the difference between the sample mean and the population mean will be less than 3.5, when we take a random sample of size $n = 5$ from an infinite population with $\sigma = 4$? Compare this figure with the 0.96 proportion obtained in the sampling experiment.

11.24 The mean of a random sample of size $n = 225$ is used to estimate the mean of an infinite population having the standard deviation $\sigma = 3$. What can we assert about the probability that the error will be less than 0.25
  (a) using Chebyshev's theorem;
  (b) using the central limit theorem?

11.25 The mean of a random sample of size $n = 64$ is used to estimate the mean of a very large population, consisting of the lifetimes of certain light bulbs, which has a standard deviation of $\sigma = 60$ hours. If we use the central limit theorem, what can we say about the probability that our estimate will be "off" by
  (a) less than 6.0 hours;
  (b) less than 16.5 hours?

11.26 If the distribution of the weights of all men traveling by air between Dallas and El Paso has a mean of 163 pounds and a standard deviation of 18 pounds, what is the probability that the combined gross weight of 36 men traveling on a plane between these two cities is more than 6,012 pounds?

11.5

Technical Note
(Simulating Sampling Experiments)

A table of random numbers, like Table XIII, constructed in such a way that each of the digits 0, 1, 2, . . . , and 9 has probability $\frac{1}{10}$ of appearing at any given location in the table, would seem to have limited usefulness. After all, it is just a collection of independent observations of a particular random variable whose probability distribution is known. But, in fact, such a table constitutes a single basic supply of (random) observations which can be used to generate sequences of observations of any other random variable whose distribution is known. By selecting one-digit numbers from the table we get the integers from 0 to 9; by selecting two-digit numbers we get the integers from 00 to 99; by selecting three-digit numbers we get the integers from 000 to 999; and so on. Also, by putting a decimal point in front of the numbers we get the decimal fractions from .0 to .9; from .00 to .99, from .000 to .999; and so on. Now, these numbers, in one form or the other, can be used to sample populations having probability distributions like the binomial or Poisson distributions, populations having continuous distributions like the normal distribution, and also distributions of raw observed data having no known standard form.

For instance, we can play "heads or tails" without ever flipping a coin by letting the digits 0, 2, 4, 6, and 8 represent heads and the digits 1, 3, 5, 7, and 9 represent tails. Then using, for instance, the fourth column of the table on page 579, starting at the top and going down the page, we get 3, 9, 8, 1, 5, 1, 6, 3, 2, 4, . . . , and we interpret this as tail, tail, head, tail, tail, tail, head, tail, head, head, ...

Repeated flips of three coins can be simulated in the same way. If we use the first three columns of the table on page 580 starting at the top and going down the page, we read out the random numbers 486, 788, 194, 512, 558, 775, 776, 154, 140, 683, . . . , and we interpret them as 3, 2, 1, 1, 1, 0, 1, 1, 2, 2, . . . , heads. If we did not want to use three columns of random digits and count the number of even digits for this "experiment," we could make use of the fact that the probabilities of getting 0, 1, 2, or 3 heads when flipping three coins are $\frac{1}{8}$, $\frac{3}{8}$, $\frac{3}{8}$, and $\frac{1}{8}$, and use the coding

Number of heads	Random digits
0	0
1	1, 2, 3
2	4, 5, 6
3	7

The digits 8 and 9 are ignored whenever they occur, and we would interpret the random numbers 2, 7, 6, 5, 1, 1, 1, 7, 6, 0, . . . , in the 11th column on page 579 as representing 1, 3, 2, 2, 1, 1, 1, 3, 2, 0, . . . , heads in repeated tosses of three balanced coins. If we did not want to "waste" any digits, we could have performed this experiment also with three-digit random numbers and the coding shown in the following table:

Number of heads x	Probability of x heads	Probability of x or less heads	Random numbers
0	0.125	0.125	000–124
1	0.375	0.500	125–499
2	0.375	0.875	500–874
3	0.125	1.000	875–999

To facilitate the assignment of random numbers to the different values of x we have added the third column to the table. It should be apparent how the cumulative probabilities of x or less heads are used in arriving at the entries in the fourth column of the table. With this scheme, the random numbers 213, 109, 915, 657, and 359, for example, represent 1, 0, 3, 2, and 1 head in five flips of three coins.

Proceeding as in the last example, we took the 50 samples of Section 11.2 by simulating a random variable having the Poisson distribution with $\lambda = 16$. The following table shows the assignment of the random numbers to the values of the random variable:

Number of passengers	Probability	Random numbers
5	0.001	000
6	0.003	001–003
7	0.006	004–009
8	0.012	010–021
9	0.021	022–042
10	0.034	043–076
11	0.050	077–126
12	0.066	127–192
13	0.082	193–274
14	0.093	275–367
15	0.099	368–466
16	0.099	467–565
17	0.093	566–658
18	0.083	659–741
19	0.070	742–811
20	0.056	812–867
21	0.043	868–910
22	0.031	911–941
23	0.022	942–963
24	0.014	964–977
25	0.009	978–986
26	0.006	987–992
27	0.003	993–995
28	0.002	996–997
29	0.001	998
30	0.001	999

In recent years, simulation techniques, under the general heading of **Monte Carlo methods,** have found wide applications in business research. These methods are used to solve inventory problems, questions arising in connection with waiting lines, advertising, competition, the allocation of resources, scheduling of operations, and situations involving overall planning and organization. Very often, this eliminates the cost of building and operating expensive equipment or facilities, it avoids disrupting ongoing operations, and it enables us to perform experiments which would otherwise cost too much or require too much time.

EXERCISES

11.27 Using four random digits to represent the results obtained in tossing four balanced coins (letting 0, 2, 4, 6, and 8 represent heads and 1, 3, 5, 7, and 9 represent tails), simulate an experiment consisting of 100 tosses of four coins. Compare the observed numbers of times that 0, 1, 2, 3, and 4 heads occurred with the corresponding expected frequencies.

11.28 Repeat Exercise 11.27, letting 0, 1, 2, 3, and 4 heads be represented by the four-digit random numbers 0000–0624, 0625–3124, 3125–6874, 6875–9374, and 9375–9999.

11.29 Use one-digit random numbers (omitting 7, 8, 9, and 0) to simulate 240 rolls of a balanced die.

11.30 With reference to part (b) of Exercise 11.11 on page 292, label the 15 possible samples 01, 02, 03, ..., 14, and 15, and use random numbers to simulate an experiment in which 100 random samples of size 2 are taken from the given population. Compare the distribution of the means of these samples with the corresponding theoretical sampling distribution obtained in part (c) of Exercise 11.11 on page 292.

11.31 Use the scheme on page 296 to draw 25 random samples of size $n = 4$ from the given population (which has the mean $\mu = 16$ and the standard deviation $\sigma = 4$), calculate the mean and the standard deviation of their means, and compare the results with those expected in accordance with the theorem of Section 11.3.

11.32 Suppose that the probabilities are 0.45, 0.27, 0.16, 0.08, and 0.04 that it takes 1, 2, 3, 4, or 5 days in court to settle charges of embezzlement not exceeding $100,000.
 (a) Distribute the two-digit random numbers from 00 to 99 to the five values of this random variable, so that they can be used to simulate the time it takes to settle such charges.
 (b) Use the results of part (a) and Table XIII to simulate the time it takes to settle 25 such embezzlement charges.

11.33 Depending upon the availability of parts, a company can manufacture 3, 4, 5, or 6 units of a certain item per week with corresponding probabilities of 0.10, 0.40, 0.40, and 0.10. The probabilities of a weekly demand for 0, 1, 2, 3, ..., or 8 units are 0.05, 0.10, 0.30, 0.30, 0.10, 0.05, 0.05, 0.04, and 0.01. If a unit is sold during the week it is made, it will yield a profit of $200; this profit is reduced by $25 for each week that a unit has to be stored. Use random numbers to simulate the operations of this company for 52 weeks and estimate its expected weekly profit on this item.

11.6

A Word of Caution

A point worth repeating is that our examples of the theoretical sampling distribution based on all possible sample means and the experimental sampling distribution based on 50 sample means were meant to be teaching aids, designed to convey the idea of a sampling distribution. These examples do not reflect what we do in actual practice, where we ordinarily base an inference on one sample and not 50. In Chapter 12 and subsequent chapters we shall go further into the problem of translating theory concerning sampling distributions into methods of evaluating the goodness of an estimate or the merits or disadvantages of a statistical decision procedure.

Another fact worth noting concerns the $\sqrt{n}$ appearing in the denominator of the formula for the standard error of the mean. As we pointed out, as n becomes larger and larger and we gain more and more information, our generalizations should be subject to smaller errors and, in general, our results should be more reliable or more precise. However, the $\sqrt{n}$ in the formulas for $\sigma_{\bar{x}}$ illustrates the fact that gains in precision or reliability are not proportional to increases in the size of the sample. For instance, doubling the size of the sample does not double the reliability of $\bar{x}$ as an estimate of the mean of a population. As is apparent from the formula $\sigma_{\bar{x}} = \sigma/\sqrt{n}$ for samples from infinite populations, we must take four times as large a sample to cut the standard error in half, and nine times as large a sample to triple the reliability, that is, to divide the standard error by 3. This clearly illustrates the fact that it seldom pays to take excessively large samples. For instance, if we increase the sample size from 100 to 10,000 (probably at considerable expense), the size of the error we are exposed to is reduced only by a factor of 10. Similarly, if we increase the sample size from 50 to, say, 20,000, the chance fluctuations we are exposed to are reduced only by a factor of 20, and this may in no way be worth the cost of taking 19,950 additional observations.

11.7

Check List of Key Terms

Central limit theorem, 290
Experimental sampling distribution, 283
Finite population, 278
Finite population correction factor, 288
Infinite population, 278
Monte Carlo methods, 296
Probable error of the mean, 293
Random numbers, 279

Random sample, 279
Sampling distribution, 282
Simulation, 294
Standard error of the mean, 287
Standard error of the median, 292
Standard error of the standard deviation, 293
Theoretical sampling distribution, 283

11.8

Review Exercises

11.34 Random samples of size $n = 2$ are drawn from the finite population which consists of the numbers 2, 4, 6, and 8.
(a) Verify that the mean of this population is $\mu = 5$ and that its standard deviation is $\sigma = \sqrt{5}$.

(b) List the six possible samples of size $n = 2$ that can be drawn without replacement from this population, calculate their means and, assigning each of these values the probability $\frac{1}{6}$, construct the theoretical sampling distribution of the mean for random samples of size $n = 2$ from the population.

(c) Calculate the mean and the standard deviation of the sampling distribution obtained in part (b) and verify the results using the theorem of Section 11.3.

11.35 Suppose that in the preceding exercise sampling is with replacement, so that the population is (hypothetically) infinite; that is, there is no limit to the number of observations we could make.

(a) List the 16 possible samples of size $n = 2$ that can be drawn with replacement from the population consisting of the numbers 2, 4, 6, and 8 (counting 4 and 6, for example, and 6 and 4 as different samples), calculate their means and, assigning each of these values the probability $\frac{1}{16}$, construct the theoretical sampling distribution of the mean.

(b) Calculate the mean and the standard deviation of the sampling distribution obtained in part (a), and verify the results using the theorem of Section 11.3.

11.36 The probabilities are $0.12, 0.21, 0.36, 0.17, 0.10$, and 0.04 that a bank will receive 0, 1, 2, 3, 4, or 5 bad checks on any given day.

(a) Distribute the two-digit numbers from 00 through 99 to the six values of this random variable, so that the corresponding random numbers can be used to simulate the numbers of bad checks received by the bank for any period of days.

(b) Use the scheme of part (a) to simulate the numbers of bad checks received by the bank on twenty days.

11.37 What is the value of the finite population correction factor when
(a) $n = 30$ and $N = 120$;   (b) $n = 20$ and $N = 500$?

11.38 What is the probability of each possible sample, when a random sample of size $n = 4$ is drawn from a finite population of size $N = 120$?

11.39 When we sample from an infinite population, what happens to the standard error of the mean if the sample size is
(a) increased from 20 to 500;
(b) decreased from 490 to 40?

11.40 Use the scheme on page 296 to simulate the numbers of passengers carried on the bus route on twenty days. Also, calculate the mean and the standard deviation of these values and compare them with the corresponding population parameters, $\mu = 16$ and $\sigma = 4$.

11.41 How many different samples of size 4 can be selected from a finite population of
(a) size 18;   (b) size 30;   (c) size 100?

11.42 What is the probability of each possible sample if a random sample of size $n = 5$ is to be drawn from a finite population of size $N = 25$?

11.43 List all possible choices of two of the following grades of raw material: A+, A, A−, B+, B, and B −. Assigning each of these choices the same probability, show that the probability that any one specific grade will be included is $\frac{1}{3}$.

11.44 If measurements of the thickness of a synthetic fiber can be looked upon as a random sample from a population having a normal distribution with

$\sigma = 0.25$ mm, what is the probability that the mean of a random sample of size $n = 16$ will differ from the mean of the population by more than 0.10 mm?

11.45 When we sample from an infinite population, what happens to the standard error of the mean if the sample size is
(a) increased from 40 to 90;
(b) decreased from 500 to 80?

11.46 The mean of a random sample of size $n = 50$ is used to estimate the mean lifetime of certain light bulbs. Using the central limit theorem and assuming that $\sigma = 60$ hours for such data, with what probability can we assert that the error will be
(a) less than 10 hours;
(b) less than 20 hours?

11.47 A young entrepreneur can borrow power lawn mowers from his acquaintances at a cost of $1.00 per day. He, in turn, rents these power mowers to neighborhood home owners for $2.00 per day. Owing to the uncertain availability of lawn mowers to be borrowed from acquaintances, the probabilities that the entrepreneur will have 2, 3, 4, 5, or 6 lawn mowers available on any given day are 0.10, 0.25, 0.40, 0.15, and 0.10. Past experience leads him to believe that his customers' daily demand for rental lawn mowers can be approximated with a Poisson distribution with a mean of 4, so the probabilities of a demand of 0, 1, 2, ..., or 9 lawn mowers are 0.02, 0.07, 0.15, 0.20, 0.20, 0.16, 0.10, 0.06, 0.03, and 0.01.
(a) Simulate the entrepreneur's borrowing and rental of these lawn mowers on 100 days.
(b) Referring to the results of part (a) and making use of the fact that the enterpreneur incurs an (additional) loss of 10 cents in good will for each customer turned away because of insufficient inventory, estimate his average daily profit.

11.48 The mean of a random sample of size $n = 36$ is used to estimate the mean of a population having a normal distribution with the standard deviation $\sigma = 9$. With what probability can we assert that the error will be less than 4.5, if we use
(a) Chebyshev's theorem;
(b) the central limit theorem?

11.49 Suppose that a customs official wants to check 10 of 348 shipments listed on a ship's manifest. Which ones (by numbers) will she check, if she numbers the shipments from 001 through 348 and chooses a random sample by means of random numbers, using the last three columns of the table on page 582, starting with the first row and going down the page.

raditionally, statistical inference has been divided into problems of estimation, in which we estimate various unknown parameters (statistical descriptions) of populations, and tests of hypotheses, in which we either accept or reject specific assertions about populations or their parameters.

Problems of estimation arise everywhere—in business, in science, and in everyday life. In business, a large retailer may want to determine the average income of all families living within 2 miles of a proposed new branch store site, an industrial union official may want to know how much variation there is in the time it takes members to get to work, and a restaurant owner may want to know what proportion of his customers tip at least 15 percent. In science, a psychologist may want to determine the average time that it takes an adult to react to a given stimulus, an engineer may need to know how much variability to expect in the strength of a new alloy, and a biologist may want to determine what percentage of certain insects are born physically defective. Finally, in everyday life, we may want to know how long it takes on the average to iron a shirt, we may be interested in the variation we can expect in a child's performance in school, and we may want to find out what proportion of all one-car accidents is due to driver fatigue.

In each of these examples interest is centered on determining the "true" value of some quantity, and they are all problems of estimation. However, they would all have been hypothesis-testing problems instead, if the retailer had wanted to know whether the average income of families living within 2 miles of the site exceeds $14,400, if the union official had wanted to determine whether the variation in the time it takes members to get to work actually is $\sigma = 8.2$ minutes, if the restaurant owner had wanted to investigate the claim that fewer than three fourths of his customers tip at least 15 percent, and so on.

In each of these three sets of examples, the first example concerned an inference about a mean, the second an inference about a measure of variation, and the third an inference about a percentage or proportion. Since the statistical treatment of such inferences differs, we shall devote this chapter to inferences about means, and then take up inferences about measures of variation and inferences about proportions in Chapters 13 and 14. In particular, we shall devote Sections 12.1 through 12.3 to the estimation of means, Sections 12.4 through 12.7 to tests of hypotheses concerning means, and Sections 12.8 and 12.9 to tests of hypotheses concerning the means of two populations.

12

Decision Making: Inferences About Means

12.1

The Estimation of Means

Given as a single number, an estimate is a value intended to match some characteristic (parameter) of a population. When we say "intended to match" and not "which matches," we mean just that—it is possible for an estimate based on a sample to coincide with the population parameter it is intended to estimate, but this is the exception rather than the rule. This should be clear from our discussion of the sampling distribution of the mean in Chapter 11.

To illustrate some of the problems we face in the estimation of a mean, let us refer to a study in which it is desired to estimate the average daily emission, in tons, of sulfur oxides by a large industrial plant. Available for this purpose are the following data, considered to constitute a random sample, of the plant's emission of sulfur oxides on 40 days:

17	15	20	29	19	18	22	25	27	9
24	20	17	6	24	14	15	23	24	26
19	23	28	19	16	22	24	17	20	13
19	10	23	18	31	13	20	17	24	14

The mean of these figures is $\bar{x} = \frac{784}{40} = 19.6$ tons, and we could use it as an estimate of the plant's "true" average daily emission of sulfur oxides, and let it go at that.

An estimate of this type is called a **point estimate**, since it consists of a single number, or a single point on the real number scale. Although this is the most common way in which estimates are expressed, it leaves room for many questions. One might wonder, for instance, how large a sample the estimate is based on and how much variability there is in the daily emissions. Thus, we might supplement the estimate, $\bar{x} = 19.6$ tons, with the information that this is the mean of a random sample of size $n = 40$, whose standard deviation is 5.51 tons.

Reports often present sample means together with the values of n and s, but to be meaningful this requires that the "consumer" of the information have some knowledge of statistics. To make this supplementary information meaningful also to the layman, let us refer to the two theorems of Chapter 11 on the sampling distribution of the mean. According to the central limit theorem, the sampling distribution of the mean can, for large random samples, be approximated closely with a normal curve. Hence, we can assert with probability $1 - \alpha$ that a sample mean $\bar{x}$ will deviate from its population mean μ by less than

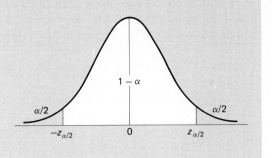

12.1

Normal distribution.

$z_{\alpha/2}$ standard errors of the mean. As before, $z_{\alpha/2}$ is the z value such that the area to its right under the standard normal curve is $\alpha/2$ (see Figure 12.1). And since the standard error of the mean is $\sigma_{\bar{x}} = \dfrac{\sigma}{\sqrt{n}}$ for random samples of size n from infinite or very large populations, we can make the following assertion: The probability is $1 - \alpha$ that $\bar{x}$ will differ from μ by less than $z_{\alpha/2} \cdot \dfrac{\sigma}{\sqrt{n}}$, and since $\bar{x} - \mu$ is the error we make when we use $\bar{x}$ as an estimate of μ, **the probability is $1 - \alpha$ that the size of this error will be less than**

Maximum error of estimate

$$E = z_{\alpha/2} \cdot \frac{\sigma}{\sqrt{n}}$$

The two most widely used values for $1 - \alpha$ are 0.95 and 0.99, and from Table I we find that $z_{\alpha/2} = z_{0.025} = 1.96$ and $z_{\alpha/2} = z_{0.005} = 2.575$.

The result we have obtained involves one complication. To be able to judge the size of the error we might make when we use $\bar{x}$ as an estimate of μ, we must know σ, the population standard deviation. Since this is not the case in most practical situations, we have no choice but to replace σ with an estimate, usually the sample standard deviation s. In general, this is considered to be reasonable provided the sample size is at least 30.

EXAMPLE

With reference to the air pollution example above, what can we assert with probability 0.95 about the maximum size of our error when we use $\bar{x} = 19.6$ tons as an estimate of the plant's true average daily emission of sulfur oxides?

SOLUTION

Substituting $n = 40$, $z_{0.025} = 1.96$, and $s = 5.51$ for σ into the formula for the maximum error, we find that we can assert with probability 0.95 that the error is less than

$$1.96 \cdot \frac{5.51}{\sqrt{40}} = 1.71 \text{ tons}$$

Of course, the error of this estimate is less than 1.71 tons or it is not, and we really do not know which, but the 0.95 probability implies that the method used to determine the maximum error (getting the sample data and using the formula for E) leads to correct results 95 percent of the time. To put it another way, "less than 1.71" may be right or wrong, but if we had to bet, 95 to 5 (or 19 to 1) would be fair odds that the error is, in fact, less than 1.71.

To emphasize the point that the 0.95 probability of the preceding example applies to the method used to find the maximum error and not directly to the parameter we are trying to estimate, many statisticians use the word confidence here instead of probability. In general, we make probability statements about future values of random variables (say, the error of an estimate) and confidence statements once the data have been obtained. Accordingly, we would say in our example that we are 95% confident that the error of the estimate is less than 1.71.

The formula for the maximum error we are exposed to with a specified probability, or confidence, can also be used to determine the sample size needed to attain a desired degree of precision in an estimate. Suppose that we want to use the mean of a random sample to estimate the mean of a population, and we want to assert with probability $1 - \alpha$ that the error of this estimate will be less than some prescribed quantity E. We write

$$E = z_{\alpha/2} \cdot \frac{\sigma}{\sqrt{n}}$$

and upon solving this equation for n we get

Sample size

$$n = \left[\frac{z_{\alpha/2} \cdot \sigma}{E} \right]^2$$

EXAMPLE   The personnel director of a manufacturing company wants to estimate the average mechanical aptitude (as measured by a certain test) of a large group of employees, and she wants this estimate to be in error by at most 2.0 with probability 0.99. If it is presumed from experience that $\sigma = 15.0$ for this test, how large a sample is required?

SOLUTION   Substituting $E = 2.0$, $\sigma = 15.0$, and $z_{0.005} = 2.575$ into the formula for n, we get

$$n = \left[\frac{(2.575)(15.0)}{2.0} \right]^2 = 373$$

rounded up to the nearest whole number. Thus, a random sample of size $n = 373$ is required for the estimate.

As can be seen from the formula, this method has the shortcoming that it cannot be used unless we know (at least approximately) the value of the population standard deviation.

Let us now introduce a different way of assessing the possible error in the estimate $\bar{x}$ of its population mean μ. In what follows, we shall make use of the fact that, for large random samples from infinite populations, the sampling distribution of the mean is approximately normal with the mean μ and the standard deviation $\sigma_{\bar{x}} = \dfrac{\sigma}{\sqrt{n}}$, so that

$$ z = \frac{\bar{x} - \mu}{\sigma/\sqrt{n}} $$

is a value of a random variable having approximately the standard normal distribution. Since the probability is $1 - \alpha$ that a random variable having the standard normal distribution will take on a value between $-z_{\alpha/2}$ and $z_{\alpha/2}$ (see Figure 12.1), or that

$$ -z_{\alpha/2} < z < z_{\alpha/2} $$

we can substitute the foregoing expression for z into this inequality and get

$$ -z_{\alpha/2} < \frac{\bar{x} - \mu}{\sigma/\sqrt{n}} < z_{\alpha/2} $$

If we now apply some simple algebra, we can rewrite this inequality as

Large-sample confidence interval for μ

$$ \bar{x} - z_{\alpha/2} \cdot \frac{\sigma}{\sqrt{n}} < \mu < \bar{x} + z_{\alpha/2} \cdot \frac{\sigma}{\sqrt{n}} $$

and we can assert with probability $1 - \alpha$ that it will be satisfied for any given sample. In other words, we can assert with $(1 - \alpha)100\%$ confidence that the interval from $\bar{x} - z_{\alpha/2} \cdot \dfrac{\sigma}{\sqrt{n}}$ to $\bar{x} + z_{\alpha/2} \cdot \dfrac{\sigma}{\sqrt{n}}$, determined on the basis of a large sample, contains the population mean we are trying to estimate. When σ is unknown and n is at least 30, we replace σ by the sample standard deviation s.

An interval like this is called a **confidence interval**, its endpoints are called **confidence limits**, and $1 - \alpha$ is called the **degree of confidence**. As before, the values most commonly used for the degree of confidence are 0.95 and 0.99, and the corresponding values of $z_{\alpha/2}$ are 1.96 and 2.575. In contrast to point estimates, estimates given in the form of confidence intervals are called **interval estimates**.

EXAMPLE   In the air pollution example on page 302, $n = 40$, $\bar{x} = 19.6$ tons, and $s = 5.51$ tons. Construct a 95% large-sample confidence interval for the plant's true average daily emission of sulfur oxides.

SOLUTION   Substituting into the confidence interval formula, we get

$$19.6 - 1.96 \cdot \frac{5.51}{\sqrt{40}} < \mu < 19.6 + 1.96 \cdot \frac{5.51}{\sqrt{40}}$$

$$17.89 < \mu < 21.31$$

for the true average value. Of course, the interval from 17.89 to 21.31 contains μ or it does not, but we are 95% confident that it does.

Had we calculated a 99% confidence interval in the preceding example, we would have obtained $17.36 < \mu < 21.84$, and it should be observed that this interval is wider than the 95% interval. This illustrates the important fact that "the surer we want to be, the less we have to be sure of." In other words, if we increase the degree of confidence, the confidence interval becomes wider and tells us less about the quantity we want to estimate.

12.2

The Estimation of Means (Small samples)

So far we have assumed not only that the sample size is large enough to treat the sampling distribution of the mean as if it were a normal distribution, but that (when necessary) σ can be replaced with s in the formula for the standard error of the mean. To develop a corresponding theory that applies also to small samples, we must now assume that the population we are sampling from has roughly the shape of a normal distribution. We can then base our methods on the statistic

$$t = \frac{\bar{x} - \mu}{s/\sqrt{n}}$$

whose sampling distribution is called the *t* **distribution**, (More specifically, it is called the **Student-*t* distribution**, as it was first investigated by W. S. Gosset, who published his writings under the pen name "Student.") As is shown in Figure 12.2, the shape of this distribution is very much like that of a normal distribution, and it is symmetrical with zero mean. The exact shape of the *t*

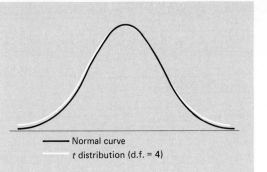

- Normal curve
- *t* distribution (d.f. = 4)

distribution depends on the quantity $n - 1$, the sample size less 1, called the
number of degrees of freedom.†

For the standard normal distribution, we defined $z_{\alpha/2}$ in such a way that
the area under the curve to its right equals $\alpha/2$ and, hence, the area under the
curve between $-z_{\alpha/2}$ and $z_{\alpha/2}$ equals $1 - \alpha$. As is shown in Figure 12.3, the
corresponding values for the *t* distribution are $-t_{\alpha/2}$ and $t_{\alpha/2}$. Since these values
depend on $n - 1$, the number of degrees of freedom, they must be looked up in a
special table, such as Table II at the end of this book; this table contains among
others the values of $t_{0.025}$ and $t_{0.005}$ for 1 through 29 degrees of freedom, and it
can be seen that $t_{0.025}$ and $t_{0.005}$ approach the corresponding values for the

†It is hard to explain at this time why one should want to assign a special name to $n - 1$.
However, we shall see later in this chapter that there are other applications of the *t* distribution, where the number of degrees of freedom is defined in a different way. The reason for the
term "degrees of freedom" lies in the fact that if we know $n - 1$ of the deviations from the
mean, then the *n*th is automatically determined (see argument on page 44.) Since the sample
standard deviation measures variation in terms of the squared deviations from the mean, we
can say that this estimate of σ is based on $n - 1$ independent quantities or that we have $n - 1$
degrees of freedom.

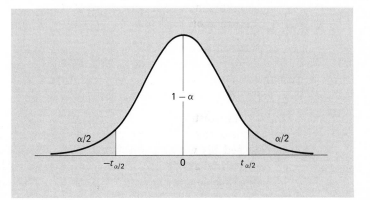

$1 - \alpha$

$\alpha/2$ $\alpha/2$

$-t_{\alpha/2}$ 0 $t_{\alpha/2}$

standard normal distribution as the number of degrees of freedom becomes large.

Since the t distribution, like the standard normal distribution, is symmetrical about its mean $\mu = 0$ (see Figure 12.3), we can now duplicate the argument on page 305 and thus arrive at the following $(1 - \alpha)100\%$ small-sample confidence interval for μ:

Small-sample
confidence
interval for μ

$$\bar{x} - t_{\alpha/2} \cdot \frac{s}{\sqrt{n}} < \mu < \bar{x} + t_{\alpha/2} \cdot \frac{s}{\sqrt{n}}$$

The only difference between this confidence interval formula and the large-sample formula (with s substituted for σ) is that $t_{\alpha/2}$ takes the place of $z_{\alpha/2}$.

EXAMPLE   To test the durability of a new paint for white center lines, a highway department has painted test strips across heavily traveled roads in eight different locations, and automatic counters showed that they deteriorated after having been crossed by (to the nearest hundred) 142,600, 167,800, 136,500, 108,300, 126,400, 133,700, 162,000, and 149,000 cars. Construct a 95% confidence interval for the average number of crossings this paint can withstand before it deteriorates.

SOLUTION   The mean and the standard deviation of these values are $\bar{x} = 140,800$ and $s = 19,200$ (to the nearest hundred), and since $t_{0.025}$ for $8 - 1 = 7$ degrees of freedom equals 2.365, substitution into the formula yields

$$140,800 - 2.365\frac{19,200}{\sqrt{8}} < \mu < 140,800 + 2.365\frac{19,200}{\sqrt{8}}$$

or

$$124,700 < \mu < 156,900$$

for the 95% confidence interval.

The method which we used earlier to measure the possible size of the error we make in using a sample mean to estimate the mean of a population can easily be adapted to small samples (provided the population we are sampling has roughly the shape of a normal distribution). All we have to do is substitute s for σ and $t_{\alpha/2}$ for $z_{\alpha/2}$ in the formula for the maximum error E on page 303.

EXAMPLE   In 12 test runs an experimental engine consumed on the average 12.9 gallons of gasoline per minute with a standard deviation of 1.6 gallons. What can we assert with 99% confidence about the maximum size of our error in the estimate $\bar{x} = 12.9$ gallons of the true average gasoline consumption of the engine?

SOLUTION   Substituting $s = 1.6$, $n = 12$, and $t_{0.005} = 3.106$ (the entry in Table II for 11 degrees of freedom) into this new formula for E, we get

$$E = t_{\alpha/2} \cdot \frac{s}{\sqrt{n}} = 3.106 \cdot \frac{1.6}{\sqrt{12}} = 1.43 \text{ gallons}$$

Thus, if we use the mean $\bar{x} = 12.9$ gallons per minute as an estimate of the true average gasoline consumption of the engine, we can be 99% confident that the error of this estimate is less than 1.43 gallons.

12.3

Bayesian Estimation

In recent years there has been mounting interest in methods of inference in which parameters (for example, the population mean μ or the population standard deviation σ) are looked upon as random variables having **prior distributions** which reflect how a person feels about the different values that a parameter can take on. Such prior considerations are then combined with direct sample evidence to obtain **posterior distributions** of the parameters, on which subsequent inferences are based. Since the method used to combine the prior considerations with the direct sample evidence is based on a generalization of Bayes' theorem of Section 9.1, we refer to such inferences as **Bayesian**.

In this section we shall present a Bayesian method of estimating the mean of a population. As we said, our prior feelings about the possible values of μ are expressed in the form of a prior distribution, and like any distribution, this kind of distribution has a mean and a standard deviation. We shall designate these values μ_0 and σ_0 and call them the **prior mean** and the **prior standard deviation.**

If we are sampling a population having the mean μ (which we want to estimate) and the standard deviation σ, if the sample is large enough to apply the central limit theorem (or if the population has a normal distribution), and if the prior distribution of μ has roughly the shape of a normal distribution, it can be shown that the posterior distribution of μ is also a normal distribution with the mean

Posterior mean

$$\mu_1 = \frac{\dfrac{n}{\sigma^2} \cdot \bar{x} + \dfrac{1}{\sigma_0^2} \cdot \mu_0}{\dfrac{n}{\sigma^2} + \dfrac{1}{\sigma_0^2}}$$

and the standard deviation σ_1 given by the formula

Posterior standard deviation

$$\frac{1}{\sigma_1^2} = \frac{n}{\sigma^2} + \frac{1}{\sigma_0^2}$$

Since μ_1, the posterior mean, may be used as an estimate of the mean of the population, let us examine some of its most important features. We note first that μ_1 is a weighted mean of $\bar{x}$ and μ_0, and that the weights are $\frac{n}{\sigma^2}$ and $\frac{1}{\sigma_0^2}$, the reciprocals of the variances of the distribution of $\bar{x}$ and the prior distribution of μ. We see also that when no direct information is available and $n = 0$, the weight assigned $\bar{x}$ is 0, the formula reduces to $\mu_1 = \mu_0$, and the estimate is based entirely on the subjective prior information. However, as more and more direct evidence becomes available (that is, as n becomes larger and larger), the weight shifts more and more toward the direct sample evidence, the sample mean $\bar{x}$. Finally, we see that when the subjective feelings about the possible values of μ are vague, that is, when σ_0 is relatively large, the estimate will be based to a greater extent on $\bar{x}$. On the other hand, when there is a great deal of variability in the population we are sampling from, that is, when σ is relatively large, the estimate will be based to a greater extent on μ_0.

EXAMPLE An investor who is planning to open a new bowling alley feels most strongly that he should net on the average $\mu_0 = \$2,600$ per month; also, the subjective prior distribution which he attaches to the various possible values of μ has roughly the shape of a normal distribution with the standard deviation $\sigma_0 = \$130$. If during nine months the operation of the bowling alley nets $2,810, $2,690, $2,350, $2,400, $2,320, $2,250, $2,430, $2,600, and $2,670, what is the posterior probability that the bowling alley will net on the average between $2,500 and $2,600 per month?

SOLUTION The mean and the standard deviation of the sample data are $\bar{x} = 2,502$ and $s = 195$. Using this sample standard deviation to estimate the unknown σ, and substituting $n = 9$, $\bar{x} = 2,502$, $s = 195$, $\mu_0 = 2,600$, and $\sigma_0 = 130$ into the formulas for the posterior mean and the posterior standard deviation, we get

$$\mu_1 = \frac{\frac{9}{195^2} \cdot 2,502 + \frac{1}{130^2} \cdot 2,600}{\frac{9}{195^2} + \frac{1}{130^2}} = \$2,522$$

and

$$\frac{1}{\sigma_1^2} = \frac{9}{195^2} + \frac{1}{130^2} \quad \text{and} \quad \sigma_1 = 58.1$$

Having found the mean and the standard deviation of the posterior distribution of μ pictured in Figure 12.4, we must now determine the area of the white region under the curve, the area under the standard normal distribution between $z = \dfrac{2,500 - 2,522}{58.1} = -0.38$ and $z = \dfrac{2,600 - 2,522}{58.1} = 1.34$. The entries corresponding to 0.38 and 1.34 in Table I are 0.1480 and 0.4099, so we get $0.1480 + 0.4099 = 0.5579$, or approximately 0.56, for the posterior probability that the bowling alley will net on the average between \$2,500 and \$2,600 per month.

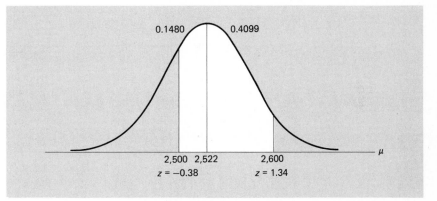

12.4

Posterior distribution of μ.

This very brief introduction to **Bayesian inference** should have served to bring out the following two points: (1) In Bayesian statistics the parameter about which an inference is to be made is looked upon as a random variable having a distribution of its own, and (2) this kind of inference permits the use of direct as well as collateral information. To clarify the last point, let us add that in the bowling-alley example the subjective prior distribution of the investor may have been based on a subjective evaluation of various factors (business conditions in general, for instance, and indirect information about other bowling alleys), and this evaluation was combined with the figures which were actually observed for the nine months.

EXERCISES

12.1 To estimate the average time required to complete a certain form, an efficiency expert timed a random sample of 40 persons in the performance of this task, getting a mean of 18.63 minutes and a standard deviation of −2.45 minutes. With 95 % confidence, what can he say about the possible size ‘of his error in estimating the true average time required to complete this form to be 18.63 minutes?

12.2 With reference to the preceding exercise, construct a 95 % confidence interval for the true average time required to complet e the form.

12.3 In a study of automobile collision insurance costs, a random sample of 80 body repair costs on a particular kind of damage had a mean of $432.56 and a standard deviation of $67.31.

    (a) Construct a 95% confidence interval for the true average cost of this kind of body repair.

    (b) What can we say with 99% confidence about the possible size of the error, if the sample mean $432.56 is used as an estimate of the true average cost of this kind of body repair?

12.4 The management of a restaurant found on the basis of a random sample of size $n = 50$ that it took its head chef on the average $\bar{x} = 14.7$ minutes to prepare a particular cooked-to-order entree; the standard deviation of the sample is $s = 4.8$ minutes. What can we assert with 98% confidence about the possible size of the error in the estimate $\bar{x} = 14.7$ minutes of the true average time it takes the chef to prepare this item?

12.5 With reference to the preceding exercise, construct a 90% confidence interval for the true average time that it takes the chef to prepare the item.

12.6 A power company takes a random sample from its very extensive files and finds that the amounts owed on 200 delinquent accounts have a mean of $21.44 and a standard deviation of $6.19. If $21.44 is used as an estimate of the true average amount owed on all the company's delinquent accounts, with what confidence can we assert that this estimate is off by at most $0.50?

12.7 If a sample constitutes an appreciable portion of a finite population (say, 5 percent or more), the various formulas given in the text should be modified by applying the finite population correction factor. For instance, the formula for E on page 303 becomes

$$E = z_{\alpha/2} \cdot \frac{\sigma}{\sqrt{n}} \sqrt{\frac{N - n}{N - 1}}$$

    (a) A sample of 81 scores on the Admission Test for Graduate Study in Business is randomly selected from a population of 400 such scores made by applicants for admission to a certain college in a given year. If the mean of the sample is $\bar{x} = 570$ and its standard deviation is $s = 99$, what can we assert with 99% confidence about the maximum size of our error if we estimate the average score for all 400 of these applicants to be 570?

    (b) A random sample of 36 drums of a wax-base floor cleaner, drawn from among 100 such drums whose weights have a standard deviation of 12 pounds, has a mean weight of 240 pounds. Construct a 95% confidence interval for the actual mean weight of all 100 drums.

12.8 Use the finite population correction factor $\sqrt{\frac{N - n}{N - 1}}$ to modify the confidence interval formula on page 305, and thus make it applicable to problems in which a sample constitutes an appreciable portion of a finite population.

    (a) In a reading achievement test, a random sample of 40 of the 300 fifth graders from a certain school district had a mean score of 83.4 with a standard deviation of 14.8. Construct a 95% confidence interval for the mean score which all the fifth graders from this school district would get if they took the test.

(b) A sample survey conducted in a certain town in 1980 showed that 200 families spent on the average $85.44 per week on food with a standard deviation of $9.12. Construct a 99% confidence interval for the actual average weekly food expenditures of the 1,000 families in this town.

12.9 In a study of television viewing habits, it is desired to estimate the average number of hours a person over 65 spends watching per week. Assuming that it is reasonable to use a standard deviation of 3.4 hours, how large a sample would be required if one wants to be able to assert with a probability of 0.95 that the sample mean will be "off" by at most a quarter hour?

12.10 Before bidding on a contract, a manufacturer wants to be "99 percent sure" that he is in error by less than 2 minutes in estimating the average time it takes to perform a certain task. If the standard deviation of the time it takes to perform the task can be assumed to equal 10 minutes, on how large a sample should he base his estimate?

12.11 It is desired to estimate the mean lifetime of a certain kind of electronic component. Given that $\sigma = 65$ days, how large a sample is needed to be able to assert with probability 0.95 that the estimate will be off by at most 20 days?

12.12 In a survey conducted in a retirement community, it was found that a random sample of 10 senior citizens visited a physician on the average 6.7 times per year with a standard deviation of 1.6. Construct a 95% confidence interval for the true average number of times a person in the population sampled visits a physician per year.

12.13 A major truck stop has kept extensive records on various transactions with its customers, and a random sample of 20 of these records shows average sales of 64.8 gallons of diesel fuel with a standard deviation of 2.9 gallons. Construct a 99% confidence interval for the mean of the population sampled.

12.14 In six test runs it took 12, 13, 17, 13, 15, and 14 minutes to assemble a certain mechanical device. If the mean of this sample is used to estimate the actual mean time it takes to assemble the device, what can we assert with 95% confidence about the maximum size of the error?

12.15 Five containers of a commercial solvent randomly selected from a large production lot weigh 24.5, 24.3, 25.0, 24.0, and 24.7 pounds.
(a) With 95% confidence, what can we assert about the maximum size of our error in estimating the population mean to be 24.5 pounds?
(b) Construct a 99% confidence interval for the mean weight of all the containers of the solvent from which this sample came.

12.16 A distributor of soft-drink vending machines feels that the prior distribution of the average number of drinks one of his machines will dispense per week has the mean 835 and the standard deviation 12.4. So far as any one of the machines is concerned, the number of drinks it dispenses varies from week to week, and this variation is measured by a standard deviation of 43.6. If he puts one of these vending machines into a new supermarket and it averages 917 drinks dispensed per week during the first 50 weeks, find a Bayesian estimate of the number of drinks this machine can be expected to dispense per week.

12.17 With reference to the preceding exercise, what is the posterior probability that the machine in the new supermarket will average between 890.0 and 910.0 drinks per week?

12.18 An actuary feels that the prior distribution of the average annual losses for a certain kind of liability coverage has the mean $\mu_0 = \$93.50$ and the standard deviation $\sigma_0 = \$4.30$. He also knows that for any one policy the losses vary from year to year with the standard deviation $\sigma = \$21.72$. If a policy like this averages losses of $175.36 per year over a period of five years, find a Bayesian estimate of its true average annual losses.

12.19 With reference to the preceding exercise, what is the posterior probability that the policy's true average annual losses are between $100.00 and $120.00?

12.4

Hypothesis Testing: Two Kinds of Errors

All the problems we have studied so far in this chapter were problems of estimation; now we shall consider problems in which we must decide whether or not a population parameter is equal to some prescribed value. We call such decision procedures **tests of hypotheses**. Instead of asking, for example, what the mean assessed value of all duplexes in a large city is, we may want to decide whether or not the mean assessed value equals some particular value, say, $28,950. At this point it may seem to make little difference how we state the problem, but it will soon become apparent that a number of considerations arise in connection with testing hypotheses that are not present in problems of estimation.

To illustrate the nature of the situation we face in testing a statistical hypothesis, suppose that a company manufactures a liquid kitchen cleaning wax which it sells in cans marked "300 grams net weight" (about 10.6 ounces). It is known from long experience that the variability of the process is stable and well established at $\sigma = 5$ grams. The cans are filled by machine, and the company makes every effort to control the mean net weight (or the mean "fill") at the 300-gram standard. However, small errors occur in the machine settings and parts wear, for instance, and the mean fill sometimes varies more or less widely from the desired 300 grams. Small departures—on the order of 1 gram or less—from standard are of no consequence, but increasingly larger departures in either direction are of increasingly more concern. Overfilling means giving away product, the value of which in time may amount to the profit on a large volume of sales. Underfilling results in a loss to consumers and also exposes the company to possible punitive action by standards enforcement agencies.

In trying to control the mean fill of the cans at 300 grams, the company has devised the following inspection procedure. Each hour during production runs, the company takes a random sample of 25 cans from the hour's production lot, calculates the sample mean weight $\bar{x}$, and decides on the basis of this value whether or not the process is "in control" (the mean fill is 300 grams, as it is supposed to be) or "out of control" (the mean fill is not 300 grams). To make this decision, the company must have some unambiguous criterion, or rule, to follow. Accordingly, the company has specified this criterion: Consider the process to be out of control if $\bar{x}$ is either less than 297 grams or greater than 303 grams, and consider it to be in control if $\bar{x}$ is between 297 and 303 grams (both values included). So far as action is concerned, when the process is judged to be out of control, it is shut down immediately and a plant engineer is sent in to find out what (if anything) is wrong with it and put it back in control; when the process is judged to be in control, it is allowed to continue in operation without interruption.

In the language of statistics, the company wishes to test, for each submitted lot, the hypothesis that the mean net weight of the wax (the mean fill) is 300 grams against the alternative that the mean fill is not 300 grams. And it will make this test on the basis of the following criterion: Reject the hypothesis (and accept the alternative) if $\bar{x} < 297$ grams or if $\bar{x} > 303$ grams; otherwise, accept the hypothesis.

We may call the hypothesis that the process is in control hypothesis H and write it as $H: \mu = 300$ grams, and the alternative that the process is out of control alternative A and write it as $A: \mu \neq 300$ grams. Clearly, the hypothesis is either true or false, and whenever it is tested the criterion will lead either to its acceptance or rejection. Unfortunately, though, on any given occasion the company may err in either one or the other of two ways (but not both). First, the company may decide that the process is out of control when, in fact, it is in control. This will happen if the mean (lot) can weight is actually 300 grams but the sample mean $\bar{x}$ is less than 297 grams or greater than 303 grams; the consequence of this error is that a process operating at the desired level is shut down while an engineer looks for nonexistent trouble. Second, the company may decide that the process is in control when, in fact, it is out of control. This will happen if the mean (lot) can weight is not 300 grams (is, say, only 290 grams) but the sample mean $\bar{x}$ is between 297 and 303 grams; the consequence of this error is that a process which is not operating at the desired standard is allowed to continue operating.

The situation described in the example is typical of testing a statistical hypothesis, and it may be summarized in the following table:

	Accept H	Reject H
H is true	Correct decision	Type I error
H is false	Type II error	Correct decision

If the hypothesis is true, the decision to accept it is the correct one; conversely, if the hypothesis is false, the decision to reject it is the correct one. On the other hand, if the hypothesis is true and it is rejected, an error has been committed;

the error which is committed when one rejects a true hypothesis is called a **Type I error** and the probability of committing it is designated by the Greek letter α (alpha). Conversely, if the hypothesis is false and it is accepted, an error has been committed; the error which is committed when one accepts a false hypothesis is called a **Type II error**, and the probability of committing it is designated by the Greek letter β (beta).

Let us now direct our attention to an investigation of the "goodness" of the liquid-cleaning-wax company's criterion. Specifically, let us calculate the probability α the company faces of rejecting H if it is true, and the probabilities β the company faces of accepting H if the mean fill is some one of various possible values other than 300 grams. If these probabilities are satisfactory from an operational standpoint, the criterion may be considered to be a "good" one in the sense that it provides the company with suitable protection against committing one or the other of the two errors.

Recalling the theory of Chapter 11, we observe that the probability α of rejecting the hypothesis if it is true is just the probability of getting a mean of less than 297 grams or more than 303 grams in a random sample of size 25 drawn from a population whose mean is $\mu = 300$ and whose standard deviation is $\sigma = 5$ grams. This probability is represented by the white area of Figure 12.5, and using the normal-curve approximation to the sampling distribution of the mean, it is easily found. According to the theorem on page 287, the standard deviation of this sampling distribution, the standard error of the mean $\sigma_{\bar{x}}$, is given by $\dfrac{\sigma}{\sqrt{n}}$. Hence, we calculate

$$z = \frac{297 - 300}{5/\sqrt{25}} = -3.00 \quad \text{and} \quad z = \frac{303 - 300}{5/\sqrt{25}} = 3.00$$

and it follows from Table I that the area in each tail of the sampling distribution shown in Figure 12.5 is $0.5000 - 0.4987 = 0.0013$. Therefore, the probability

12.5

Test criterion.

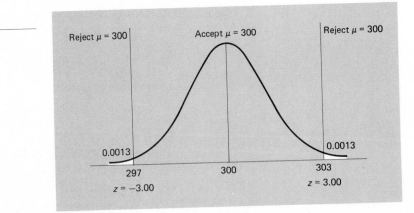

that $\bar{x}$ will either be less than 297 or greater than 303 is $0.0013 + 0.0013 = 0.0026$, and this is the probability α of erroneously rejecting the hypothesis that the process is in control. In other words, there are about 3 chances in 1,000 that the company will commit the error of shutting the process down and looking for trouble which does not exist.

Suppose now that, due to some equipment malfunction, the actual mean fill has shifted and is only 296 grams. The hypothesis is false and the process is out of control. In this case, the sample mean does not come from the distribution of Figure 12.5 but instead comes from the distribution of Figure 12.6. Thus, the probability of not detecting this shift and of accepting the false hypothesis that the process is in control is represented by the white area of Figure 12.6.

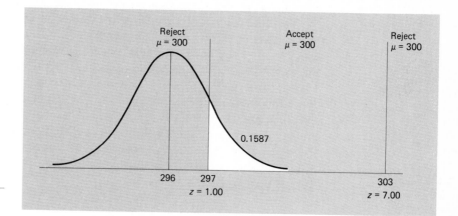

12.6

Test criterion.

It is the probability of getting a sample mean between 297 and 303 grams from a population whose mean is $\mu = 296$, and as before, the standard error of the mean is $\sigma_{\bar{x}} = \dfrac{5}{\sqrt{25}}$. Hence, we calculate

$$z = \frac{297 - 296}{5/\sqrt{25}} = 1.00 \quad \text{and} \quad z = \frac{303 - 296}{5/\sqrt{25}} = 7.00$$

and it follows from Table I that the white area of Figure 12.6 is $0.5000 - 0.3413 = 0.1587$ since the area to the right of $z = 7.00$ is negligible. Therefore, the probability β of accepting the false hypothesis that the process is in control when μ is actually 296 grams is 0.1587. In other words, there are about 16 chances in 100 that the company will commit the error of letting the process run, thinking the mean fill is 300 grams when it is really only 296 grams.

In computing the probability of accepting a false hypothesis, we supposed that the process mean had shifted from 300 to 296 grams. However, in this filling problem there are infinitely many other alternatives (infinitely many possible values other than 300 grams for the actual mean weight), and for each one of them there is a positive probability β that the company will accept the hypothesis when it is false. Thus, let us calculate the probability of accepting the hypothesis that the mean weight is 300 grams for several other alternatives, and examine these probabilities to see how well, under various circumstances, the criterion controls the risks facing the company. The procedure is precisely the same as in the case where we supposed that the mean had shifted to 296 grams. The probability of accepting the hypothesis that the mean weight is 300 grams when it is, in fact, say, 305 grams is just the probability that the mean of a sample of 25 weights drawn from a population with mean 305 grams and standard deviation 5 grams will lie between 297 and 303 grams. By drawing a figure like Figure 12.6, with $\mu = 305$, and proceeding as we did above, we find that there is a probability β of 0.0228 that the company will commit the error of deciding that the process is in control when the mean fill is actually 305 grams.

The third column of the table below shows, for 13 possible values of μ, the probabilities of accepting the hypothesis that $\mu = 300$ grams when μ is actually 294, 295, . . . , 305, and 306 grams. When the value of μ is 300, H is true and the probability of accepting it is the probability of not rejecting a true hypothesis: $1 - \alpha = 1 - 0.0026 = 0.9974$. Therefore, the probabilities shown of accepting H are the probabilities β of accepting a false hypothesis with the exception of the probability 0.9974, which is the probability of avoiding a Type I error. The probabilities of committing a Type II error, shown in the second column, are the same as the corresponding probabilities of accepting H except where $\mu = 300$ grams. In this case, H is true and there is no possibility of committing a Type II error.

Value of μ	Probability of Type II error	Probability of accepting H
294	0.0013	0.0013
295	0.0228	0.0228
296	0.1587	0.1587
297	0.5000	0.5000
298	0.8413	0.8413
299	0.9772	0.9772
300	—	0.9974
301	0.9772	0.9772
302	0.8413	0.8413
303	0.5000	0.5000
304	0.1587	0.1587
305	0.0228	0.0228
306	0.0013	0.0013

If we plot the probabilities of accepting H as in Figure 12.7 and fit a smooth curve, we get the **operating characteristic curve** of the test criterion, or simply the OC-curve. An operating characteristic curve provides a good overall picture of the merits of a test criterion. Examination of the curve of Figure 12.7 shows that the probability of accepting hypothesis H is greatest when it is true. For small departures from the 300-gram standard there is a high probability of accepting H; that is, when the process deviates only slightly from standard, it will most likely be allowed to continue to operate. But for larger and larger departures from standard in either direction, the probabilities of failing to detect them and accepting H in error become smaller and smaller. This is precisely what the company wants, and any test procedure which did not behave in this way would not be at all suited to its needs.

Of course, the OC-curve of Figure 12.7 applies only to the case where the hypothesis $\mu = 300$ grams is accepted if the mean of a random sample of size 25 falls between 297 and 303 grams and rejected otherwise, and σ is known to be 5 grams. If it wishes, the company can change the shape of the OC-curve (and hence, change the amount of protection it is getting against committing Type I and Type II errors) by changing the test criterion or the sample size. In fact, OC-curves can often be made to assume a particular desired shape by an appropriate choice of the sample size and/or the dividing lines of the test criterion.

A detailed study of OC-curves would go considerably beyond the scope of this text, and the purpose of our illustration was mainly to show how statistical methods can be used to measure and control the risks to which one is exposed in testing hypotheses. Of course, these methods are not limited to problems of quality control in industrial plants. This will become increasingly clear as we proceed.

12.7

Operating
characteristic curve.

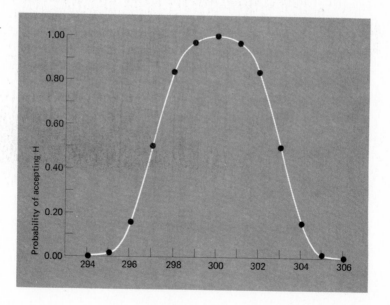

Hypothesis Testing:
Null Hypotheses and Significance Tests

In the example of the preceding section, we had less trouble with Type I errors than with Type II errors, because we formulated the hypothesis H as a simple hypothesis about the parameter μ; that is, we formulated the hypothesis H so that μ took on a single value and the probability of a Type I error could be calculated.† Had we formulated instead a composite hypothesis about μ, say, the composite hypothesis $\mu \neq 300$ grams, the composite hypothesis $\mu < 300$ grams, or the composite hypothesis $\mu > 300$ grams, where in each case μ can take on more than one possible value, we could not have calculated the probability of a Type I error without specifying how much μ differs from, is less than, or is greater than 300 grams.

To be able to calculate the probability of a Type I error (that is, to know what to expect when a hypothesis is true), it is customary to formulate hypotheses to be tested as simple hypotheses, and in many instances this requires that we hypothesize the opposite of what we hope to prove. For instance, if we want to show that a new copper-bearing steel has a higher yield strength than ordinary steel, we formulate the hypothesis that the two yield strengths are the same. Similarly, if we want to show that one method of teaching computer programming is more effective than another, we hypothesize that the two methods are equally effective; and if we want to show that the proportion of sales slips incorrectly written up in one department of a store is greater than that in another department, we formulate the hypothesis that the two proportions are identical. Since we hypothesize that there is no difference in the yield strength, no difference in the effectiveness of the two methods of teaching computer programming, and no difference between the two proportions, we call hypotheses such as these null hypotheses and denote them H_0. Nowadays, the term "null hypothesis" is used for any hypothesis set up primarily to see whether it can be rejected. Actually, the idea of setting up a null hypothesis is common even in nonstatistical thinking. It is precisely what we do in criminal proceedings, for example, where an accused is presumed to be innocent until his guilt is proved beyond a reasonable doubt. The presumption that the accused is not guilty is a null hypothesis.

Although a positive probability β of accepting a false hypothesis exists for all values of μ alternative to the test value, we can sometimes avoid a Type II

---

†Note that we are applying the term "simple hypothesis" to a hypothesis about a parameter; some statisticians use the term "simple hypothesis" only when the hypothesis completely specifies the population.

error altogether. To illustrate how this is done, suppose that a large title-search company knows from experience that the mean number of typing errors made on submitted copies of form A by typists preparing these forms is 2.3 per day with a standard deviation of 0.75. In trying to confirm its suspicion that one particular typist makes more errors on the average than the others, the company takes a random sample of nine forms prepared by this typist and tests the null hypothesis that there is no difference between the typist's performance and that of the others (that $\mu = 2.3$ applies to this typist also) by using the following criterion:

Reject the null hypothesis $\mu = 2.3$ (and accept the alternative $\mu > 2.3$) if the typist averages 2.7 or more errors per form; otherwise, reserve judgment pending further study.

With this criterion there is no possibility of committing a Type II error; when the hypothesis is tested it may be rejected outright (and the typist presumed to be less proficient than average), but otherwise, it will not actually be accepted.

The procedure we have just outlined is called a **significance test**. If the difference between what we expect under the hypothesis and what we observe in a sample is too large to be reasonably attributed to chance, we reject the null hypothesis. If the difference between what we expect and what we observe is so small that it might well be attributed to chance, we say that the result is **not statistically significant**. We then either accept the null hypothesis or reserve judgment depending on whether a definite decision one way or the other is required.

With reference to the criterion above, the company's suspicion that the typist is worse than average is confirmed if the average number of errors is 2.7 or more; in that case it is felt that the difference between the sample mean and $\mu = 2.3$ is too large to be attributed to chance (see also Exercise 12.28 on page 325). If the sample mean is less than 2.7, the company's suspicion is not confirmed. We do not say that the company's suspicion is wrong, or unjustified, when the sample mean is less than 2.7—we merely say that it is not confirmed. In this case, we may want to continue the investigation, perhaps, with a larger sample.

Returning to the filling example of the preceding section, we could convert the criterion on page 315 into that of a significance test by writing

Reject the hypothesis $\mu = 300$ grams (and consider the process to be out of control) if the mean of the 25 sample fills is less than 297 grams or greater 303 grams; otherwise, reserve judgment.

It should be obvious, though, that there is really no way to reserve judgment in this situation. Once the process is in operation, it must either be allowed to continue in operation or it must be shut down, and no matter how we phrase the criterion, so long as the filling process continues, there is no way possible to avoid the consequences of committing a Type II error.

Whether or not one can afford the luxury of reserving judgment in any given situation depends entirely on the nature of the situation. In general, whenever a decision must be reached one way or the other when a test is made, we are required either to reject or accept a hypothesis, and we have no way to avoid the risk of accepting a false one.

Since the general problem of testing hypotheses and constructing statistical decision criteria often seems confusing, it will help to proceed systematically as outlined in the following steps:

1. **We formulate a (null) hypothesis in such a way that the probability of a Type I error can be calculated.**†

2. **We formulate an alternative hypothesis so that the rejection of the null hypothesis is equivalent to the acceptance of the alternative hypothesis.**

In the filling example the null hypothesis is $\mu = 300$ grams, and the alternative is $\mu \neq 300$. We refer to this kind of alternative as a **two-sided alternative**. The company has specified this alternative because it wants protection against thinking the filling process is in control when it is either underfilling or overfilling the cans. In the typist example, the null hypothesis is $\mu = 2.3$ mistakes, and the alternative is $\mu > 2.3$. This is called a **one-sided alternative**, and the firm has chosen this alternative because it feels that the burden of proof is on it to show that the typist is poorer (makes more mistakes) than the average. (The firm could hardly argue that the typist is poorer than average if the average number of mistakes is less than 2.3 per form.) We can also write a one-sided alternative with the inequality sign pointing in the other direction. For instance, if we wanted to determine whether the average time required for messenger-pool employees to do a certain job is less than 40 minutes, we might take a random sample of, say, 20 of these times and test the null hypothesis $\mu = 40$ against the alternative $\mu < 40$.

As in the three examples of the preceding paragraph, alternative hypotheses usually specify that the population mean (or whatever other parameter may be of concern) is not equal to, greater than, or less than the value assumed under the null hypothesis. For any given problem, the choice of an appropriate alternative depends mostly on what we hope to be able to show, or better, perhaps, where we want to put the burden of proof.

EXAMPLE A shoe manufacturer is considering the purchase of a new automatic machine for stamping out uppers. If μ_1 is the average number of good uppers stamped out by the old machine per hour and μ_2 is the corresponding average for the new machine, the manufacturer wants to test

†See part (b) of Exercise 12.24 on page 325 for what may be done when this is not possible.

the null hypothesis $\mu_1 = \mu_2$ against a suitable alternative. What should the alternative be if

(a) the manufacturer does not want to buy the new machine unless it is definitely superior to the old one;
(b) the manufacturer wants to buy the new machine (which has some other nice features) unless tests prove it to be definitely inferior to the old one?

SOLUTION (a) The manufacturer should use the alternative hypothesis $\mu_1 < \mu_2$ and purchase the new machine only if the null hypothesis can be rejected; (b) the manufacturer should use the alternative hypothesis $\mu_1 > \mu_2$ and purchase the new machine unless the null hypothesis is rejected.

Having formulated a null hypothesis and a suitable alternative, we then proceed with the following step:

3. **We specify the probability of committing a Type I error; if possible, desired, or necessary, we may also make some specifications about the probabilities β of Type II errors for specific alternatives.**

The probability of committing a Type I error is usually referred to as the **level of significance** at which a test is performed. Usually, tests are performed at a level of significance of 0.05 or 0.01. Testing a hypothesis at a level of significance of, say, $\alpha = 0.05$ simply means that we are fixing the probability of rejecting the hypothesis if it is true at 0.05.

Once we have set up a hypothesis, the probability of committing a Type I error is absolutely under our control and can be made as small as we like. How small we actually make it in a particular case depends on the consequences (cost, inconvenience, embarrassment, etc.) of rejecting a true hypothesis. Ordinarily, the more serious the consequences which result from committing a Type I error, the smaller the risk we are willing to take of committing it. However, we are restrained in practice from setting very low probabilities α by the fact that, for a fixed sample size, the smaller we make the probability of rejecting a true hypothesis, the larger the probability β of accepting a false one becomes. For instance, if in the filling problem the probability of shutting down the process when it is in control were reduced from $\alpha = 0.0026$ to $\alpha = 0.0001$, the probability of not detecting that the process is out of control when the mean fill is actually either 296 or 304 grams would be increased from $\beta = 0.1587$ to $\beta = 0.4602$ (see Exercise 12.27 on page 325).

What we usually need in practice is some reasonable balance between the probabilities of committing the two kinds of errors, and (except in industrial quality control) long experience in practice seems to suggest that levels of significance of 0.05 or 0.01, or thereabouts, will often provide this balance.

4. **Using suitable statistical theory we construct a test criterion for testing the (null) hypothesis formulated in step 1 against the alternative hypothesis formulated in step 2 at the level of significance specified in step 3.**

In the filling example, we based the criterion on the normal-curve approximation to the sampling distribution of $\bar{x}$; in general, it depends on the statistic upon which we want to base the decision and on its sampling distribution. A good portion of the remainder of this book will be devoted to the construction of such criteria. As we shall see later, this usually involves choosing an appropriate statistic, specifying the sample size, and then determining the dividing lines, or **critical values**, of the criterion. In the previous examples, we used a **two-sided test** (or **two-tailed test**) with the two-sided alternative $\mu \neq 300$ in the filling example, rejecting the null hypothesis for either small or large values of $\bar{x}$; in the example of the number of mistakes made by a typist, we used a **one-sided test** (or **one-tailed test**) with the one-sided alternative $\mu > 2.3$, rejecting the null hypothesis only for large values of $\bar{x}$; and in the example of the time required to do a certain job, we used a one-sided test (or one-tailed test) with the one-sided alternative $\mu < 40$, rejecting the null hypothesis only for small values of $\bar{x}$. In general, a test is called two-sided (or two-tailed) if the null hypothesis is rejected when a value of the test statistic falls in either one or the other of the two tails of its sampling distribution, and one-sided (or one-tailed) if the null hypothesis is rejected when a value of the test statistic falls in just one specified tail of its sampling distribution.

5. **We specify whether the alternative to rejecting the hypothesis formulated in step 1 is to accept it or to reserve judgment.**

This, as we have said, depends on whether we must make a decision one way or the other on the basis of the test, or whether the circumstances of the problem are such that we can delay a decision pending further study. Sometimes we may accept a null hypothesis with the hope that we are not exposing ourselves to excessively high risks of committing serious Type II errors. Of course, if it is necessary and we have enough information, we can calculate the probabilities needed to get an overall picture from the OC-curve of the test criterion.

Before we discuss various special tests about means in the next few sections, let us point out that the concepts we have introduced here apply equally well to hypotheses concerning proportions, standard deviations, the randomness of samples, relationships among several variables, trends of time series, and so on.

12.20 Suppose that on the basis of a sample we want to test the hypothesis that the average noise level of a certain kind of vacuum cleaner is 72 decibels. Explain under what conditions we would commit a Type I error and under what conditions we would commit a Type II error.

12.21 Suppose that we want to test the hypothesis that an antipollution device for cars is effective. Explain under what conditions we would commit a Type I error and under what conditions we would commit a Type II error.

12.22 Whether an error is a Type I error or a Type II error depends on how we formulate the hypothesis we want to test. To illustrate this, rephrase the hypothesis of the preceding exercise so that the Type I error becomes a Type II error, and vice versa.

12.23 With reference to the filling example, verify the values of the probabilities of Type II errors given in the middle column of the table on page 318.

12.24 Suppose that for a given population with $\sigma = 8.4$ inches we want to test the null hypothesis $\mu = 75.0$ inches against the alternative hypothesis $\mu < 75.0$ inches on the basis of a random sample of size $n = 100$.
   (a) If the null hypothesis is rejected when $\bar{x} < 73.0$ inches and otherwise it is accepted, find the probability of a Type I error.
   (b) If the hypothesis we want to test had been $\mu \geq 75.0$ inches and the criterion is the same as in part (a), what could we have said about the probability of a Type I error?

12.25 Suppose that in the filling example the criterion is changed so that the hypothesis $\mu = 300$ grams is accepted if the sample mean falls between 298 and 302 grams; otherwise, the hypothesis is rejected.
   (a) Show that this will increase the probability of a Type I error from 0.0026 to 0.0456.
   (b) Show that this will decrease the probability of a Type II error when $\mu = 296$ from 0.1587 to 0.0228.

12.26 Suppose that in the filling example σ had been 6 grams instead of 5 grams but that everything else remained the same.
   (a) Show that this will increase the probability of a Type I error from 0.0026 to 0.0124.
   (b) Show that this will increase the probability of a Type II error when $\mu = 296$ from 0.1587 to 0.2033.

12.27 With reference to the filling example, use $z_{0.00005} = 3.9$ and verify that the probability of not detecting that the process is out of control is $\beta = 0.4602$ when $\mu = 296$ and $\alpha = 0.0001$.

12.28 With reference to the typist example on page 321, verify that the probability of committing a Type I error is 0.055.

12.29 An automatic machine in a food-processing plant is supposed to set the lids on pint jars of mayonnaise so that the average "twist" required for a person to loosen the lids ("break the sets") is 30 inch-pounds. It is known from long experience that the variability of the sets is stable and given by $\sigma = 2.0$ inch-pounds. The processor does not want the lids set too loosely since this may cause discoloration and spoilage of the mayonnaise, and he does not want them set too tightly since people resent having to struggle with stubborn lids. Consequently, the processor sets up a hypothesis that the lids are set at 30 inch-pounds on the average (the process is in control) and an alternative that the lids are not set at 30 inch-pounds (the process is out of control). The hypothesis is tested periodically by taking from production lots of sealed jars random samples of 36 jars, determining the mean twist $\bar{x}$ required to break the sets, accepting the hypothesis if $\bar{x}$ is between 29.2 inch-pounds and 30.8 inch-pounds and rejecting it if $\bar{x}$ is either less than 29.2 inch-pounds or greater than 30.8 inch-pounds.

(a) Find the probability of a Type I error.

(b) Find the probabilities of Type II errors for alternative mean sets of 28.5, 29.0, 29.5, 30.5, 31.0, and 31.5 inch-pounds.

(c) Plot the OC-curve.

12.30 A city police department is considering replacing the tires on its cars with radial tires. If μ_1 is the average number of miles the old tires last and μ_2 is the average number of miles the new tires will last, the null hypothesis to be tested is $\mu_1 = \mu_2$.

(a) What alternative hypothesis should the department use if it does not want to buy the radial tires unless they are definitely proved to give better mileage? In other words, the burden of proof is put on the radial tires and the old tires are to be kept unless the null hypothesis can be rejected.

(b) What alternative hypothesis should the department use if it is anxious to get the new tires (which have some other good features) unless they actually give poorer mileage than the old tires? Note that now the burden of proof is on the old tires, which will be kept only if the null hypothesis can be rejected.

(c) What alternative hypothesis would the department have to use so that the rejection of the null hypothesis can lead either to keeping the old tires or to buying the new ones?

12.31 Suppose that a large supermarket has one checkout clerk whom it suspects of making more mistakes than the average of all its clerks.

(a) If the market decides that it will let the clerk go, provided this suspicion is confirmed on the basis of observations made on the clerk's performance, what hypothesis and alternative should the market set up?

(b) If the market decides to let the clerk go unless he can prove himself significantly better than the average of all clerks, what hypothesis and alternative should the market set up?

12.32 With reference to the preceding exercise, suppose it is known that the average number of register mistakes per day per clerk is 18 and the standard deviation is 5, and the market decides to fire the clerk only if in a random sample of 40 days of work he averages more than 20 mistakes.

(a) What is the probability of firing the clerk when his work is, in fact, of average quality?

(b) What is the probability of keeping the clerk on if he averages 21 mistakes per day?

12.33 In hypothesis testing, a function whose values are the probabilities of rejecting a given hypothesis (and accepting the alternative) for various values of the parameter under consideration is called a **power function**. Thus, in the filling example, for all values of μ other than the hypothesized 300 grams, the power function gives the probabilities $1 - \beta$ of not committing a Type II error; for $\mu = 300$, however, it gives the probability α of committing a Type I error. Obviously, the values of the power function are 1 minus the corresponding values of the OC-curve. Plot the graph of the power function for the filling example and describe this curve.

12.6

Tests Concerning Means

Having used tests concerning means to illustrate the basic principles of hypothesis testing, let us now consider how to proceed in actual practice. Suppose, for instance, that we want to determine, on the basis of the mean $\bar{x}$ of a random sample of size 100, whether or not the average weekly food expenditure of families of three within a certain income range is $85.00. From information gathered in other pertinent studies, we assume that the variability of such expenditures is given by a standard deviation of $\sigma = \$12.20$.

Beginning with steps 1 and 2 on page 322, we formulate the null hypothesis to be tested and the two-sided alternative as

Null hypothesis: $\quad \mu = \$85.00$
Alternative hypothesis: $\mu \neq \$85.00$

That is, we will consider as evidence against the hypothesis $\mu = \$85.00$ values of $\bar{x}$ which are either significantly less than or significantly greater than $85.00. So far as step 3 is concerned, suppose that we fix the probability of rejecting the null hypothesis if it is true at $\alpha = 0.05$.

Next, in step 4, we shall depart slightly from the procedure used in the examples given earlier in Section 12.5. In these examples we stated the test criterion in terms of values of $\bar{x}$; now we shall base it on the statistic

Statistic for test concerning mean

$$z = \frac{\bar{x} - \mu_0}{\sigma/\sqrt{n}}$$

in which μ_0 is, by hypothesis, the mean of the population from which the sample comes. The reason for working with standard units, or z-values, is that it enables us to formulate criteria which are applicable to a great variety of problems, not just one.

If we approximate the sampling distribution of the mean, as before, with a normal distribution, we can now use the test criteria shown in Figure 12.8; depending on the choice of the alternative hypothesis, the dividing line (or lines) of the criterion is $-z_\alpha$ or z_α for the one-sided alternatives, and $-z_{\alpha/2}$ and $z_{\alpha/2}$ for the two-sided alternative. Once again, z_α and $z_{\alpha/2}$ are z-values such that the area to their right under the standard normal distribution is α and $\alpha/2$. Sym-

12.8

Test criteria.

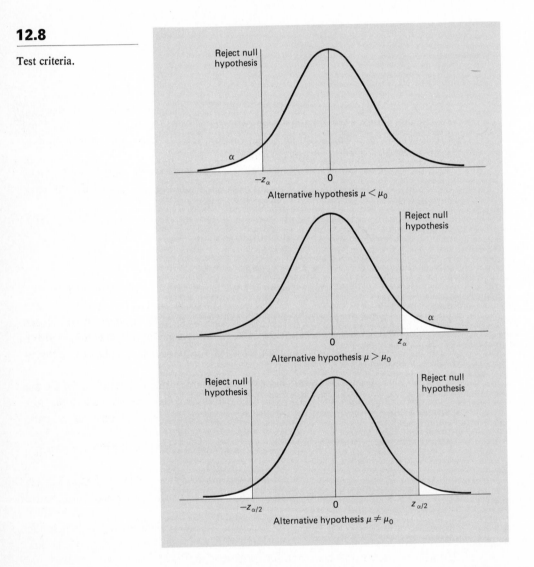

bolically, we can formulate these criteria as follows:

Alternative hypothesis	Reject the null hypothesis if	Accept the null hypothesis or reserve judgment if
$\mu < \mu_0$	$z < -z_\alpha$	$z \geq -z_\alpha$
$\mu > \mu_0$	$z > z_\alpha$	$z \leq z_\alpha$
$\mu \neq \mu_0$	$z < -z_{\alpha/2}$ or $z > z_{\alpha/2}$	$-z_{\alpha/2} \leq z \leq z_{\alpha/2}$

If $\alpha = 0.05$, the dividing lines, or **critical values**, of the criteria are -1.645 or 1.645 for the one-sided alternatives, and -1.96 and 1.96 for the two-sided alternative; if $\alpha = 0.01$, the dividing lines of the criteria are -2.33 or 2.33 for the one-sided alternatives, and -2.575 and 2.575 for the two-sided alternative. All these values come directly from Table I.

Returning now to the food-expenditures example, suppose that the mean of the sample is $87.80. Substituting this value together with $\mu_0 = \$85.00$, $\sigma = \$12.20$, and $n = 100$ into the formula for z, we get

$$z = \frac{87.80 - 85.00}{12.20/\sqrt{100}} = 2.30$$

Since this exceeds 1.96, we reject the null hypothesis and conclude that the average weekly food expenditure of the families in question does not equal $85.00. In other words, the difference of $2.80 between the observed mean and the hypothetical value of μ is too large to be accounted for by chance. It appears from the data that the true average weekly food expenditure is actually greater than $85.00.

If we had used the 0.01 level of significance instead of the 0.05 level, we could not have rejected the null hypothesis $\mu = \$85.00$ since $z = 2.30$ does not exceed 2.575. This illustrates the important point that the level of significance should always be specified before a significance test is actually performed. This will spare us the temptation of later choosing a level of significance which happens to suit our particular objectives.

In problems like this, some research workers accompany the calculated values of z with corresponding **tail probabilities**, or *p*-values, namely, the probabilities of getting a difference between $\bar{x}$ and μ_0 greater than or equal to that actually observed. For instance, in the food-expenditures example, the tail probability is given by the total area under the standard normal curve to the left of -2.30 and to the right of 2.30, and it equals $2(0.5000 - 0.4893)$ or

0.0214. This value falls between 0.01 and 0.05, which agrees with our earlier results, but observe that giving a tail probability does not relieve us of the responsibility of specifying the level of significance before the test is actually performed.

The test we have described in this section is essentially a **large-sample test**; it is exact only when the population we are sampling has a normal distribution. Also, since σ is unknown in many practical applications, we often have no choice but to make the further approximation of substituting for it the sample standard deviation s.

EXAMPLE A large retailer wants to determine whether the mean income of families living within two miles of a proposed building site exceeds $14,400. What can he conclude at the 0.05 level of significance, if the mean income of a random sample of $n = 60$ families living within two miles of the proposed site is $\bar{x} = \$14,524$ and the standard deviation is $s = \$763$?

SOLUTION The null hypothesis to be tested and the alternative hypothesis are

> Null hypothesis: $\mu = \$14,400$
> Alternative hypothesis: $\mu > \$14,400$

and we shall have to see whether the value we get for z by means of the formula on page 327 (with σ replaced by s) exceeds 1.645. Substituting $\bar{x} = 14,524$, $s = 763$, $n = 60$, and $\mu_0 = 14,400$ into the formula, we get

$$z = \frac{14,524 - 14,400}{763/\sqrt{60}} = 1.26$$

and since this value does not exceed 1.645, the null hypothesis cannot be rejected. In other words, the difference between $14,524 and $14,400 is not large enough to provide evidence at the 0.05 level of significance that average family income in the area exceeds $14,400. In this instance, the retailer may well decide to reserve judgment pending further study rather than to accept the null hypothesis and thus risk the possibility of committing a Type II error.

EXAMPLE A trucking firm suspects that the average lifetime of 25,000 miles claimed for certain tires is too high. To test the claim, the firm puts a random sample of 40 of these tires on its trucks and later finds that their mean lifetime is 24,421 miles and the standard deviation is 1,349 miles. What can it conclude at the 0.01 level of significance, if it tests the null hypothesis $\mu = 25,000$ miles against the alternative hypothesis $\mu < 25,000$ miles?

SOLUTION Substituting the given values of $\bar{x}$, μ_0, n, and s (for σ) into the formula for z, we get

$$z = \frac{24{,}421 - 25{,}000}{1{,}349/\sqrt{40}} = -2.71$$

and since this is less than -2.33, we find that the null hypothesis must be rejected. In other words, the trucking firm can conclude that the average lifetime of the tires is not as high as claimed. In order to find the p-value corresponding to $z = -2.71$, we must find the area under the standard normal curve to the left of -2.71. Since the entry in Table I corresponding to $z = 2.71$ is 0.4966, it follows that the p-value is $0.5000 - 0.4966 = 0.0034$.

12.7

Tests Concerning Means (Small Samples)

When we do not know the value of the population standard deviation and the sample is small, $n < 30$, we shall assume again that the population we are sampling from has roughly the shape of a normal distribution, and base our decision on the statistic

Statistic for small-sample test concerning mean

$$t = \frac{\bar{x} - \mu_0}{s/\sqrt{n}}$$

whose sampling distribution is the t distribution with $n - 1$ degrees of freedom. The criteria for small-sample tests concerning means based on the t statistic are those of Figure 12.8 and the table on page 329 with z replaced by t and z_α and $z_{\alpha/2}$ replaced by t_α and $t_{\alpha/2}$. Here, again, t_α and $t_{\alpha/2}$ are values for which the area to their right under the t distribution is equal to α and $\alpha/2$. All the dividing lines of these tests may be read from Table II, with the number of degrees of freedom equal to $n - 1$.

EXAMPLE Suppose that we want to decide, on the basis of a random sample of five specimens, whether the fat content of a certain kind of ice cream is less than 12.0 percent. Can we reject the null hypothesis $\mu = 12.0$ percent against the alternative hypothesis $\mu < 12.0$ percent at the level of significance $\alpha = 0.01$, if the sample has the mean $\bar{x} = 11.3$ percent and the standard deviation $s = 0.38$ percent?

SOLUTION Substituting the given values of $\bar{x}$, μ_0, s, and n into the formula for t, we get

$$t = \frac{11.3 - 12.0}{0.38/\sqrt{5}} = -4.12$$

and since this is less than -3.747, the value of $-t_{0.01}$ for $5 - 1 = 4$ degrees of freedom, we find that the null hypothesis must be rejected. In other words, we conclude that the average fat content of the ice cream is, indeed, less than 12.0 percent. The exact tail probability, or p-value, cannot be determined from Table II, but it is 0.0073.

12.8

Differences Between Means

There are many statistical problems in which we must decide whether an observed difference between two sample means can be attributed to chance. We may want to decide, for instance, whether there is really a difference in the average gasoline consumption of two kinds of compact cars, if sample data show that one kind averaged 24.8 miles per gallon while, under the same conditions, the other kind averaged 25.6 miles per gallon. Similarly, we may want to decide on the basis of samples whether there actually is a difference in the size of delinquent charge accounts in two branches of a department store, whether men can perform a given task faster than women, whether one kind of television tube lasts longer than another, and so on.

The method we shall employ to test whether an observed difference between two sample means can be attributed to chance is based on the following theory: If $\bar{x}_1$ and $\bar{x}_2$ are the means of two large independent random samples of size n_1 and n_2, the sampling distribution of the statistic $\bar{x}_1 - \bar{x}_2$ can be approximated closely with a normal curve having the mean $\mu_1 - \mu_2$ and the standard deviation

$$\sqrt{\frac{\sigma_1^2}{n_1} + \frac{\sigma_2^2}{n_2}}$$

where μ_1, μ_2, σ_1, and σ_2 are the means and the standard deviations of the two populations from which the two samples came. It is customary to refer to the standard deviation of this sampling distribution as the **standard error of the difference between two means**.

By "independent" samples we mean that the selection of one sample is in no way affected by the selection of the other. Thus, the theory does not apply to "before and after" kinds of comparisons, nor does it apply, say, to the comparison of the IQs of husbands and wives (but see Exercise 12.55 on page 339).

In most practical situations σ_1 and σ_2 are unknown, but if we limit ourselves to large samples (neither n_1 nor n_2 should be less than 30), we can use the sample standard deviations s_1 and s_2 as estimates of σ_1 and σ_2, and base the test of the null hypothesis $\mu_1 - \mu_2 = 0$ on the statistic

Statistic for large-sample test concerning difference between two means

$$z = \frac{\bar{x}_1 - \bar{x}_2}{\sqrt{\dfrac{s_1^2}{n_1} + \dfrac{s_2^2}{n_2}}}$$

which has approximately the standard normal distribution. No matter how it looks, this is actually a z-value; we calculate it by subtracting from $\bar{x}_1 - \bar{x}_2$ the mean of its sampling distribution, which under the null hypothesis is $\mu_1 - \mu_2 = 0$, and then dividing this difference by the (estimated) standard error of the difference between two means.

Depending on whether the alternative hypothesis is $\mu_1 - \mu_2 < 0$, $\mu_1 - \mu_2 > 0$, or $\mu_1 - \mu_2 \neq 0$, the criteria we base the actual tests on are again those shown in Figure 12.8 and also in the table on page 329, with $\mu_1 - \mu_2$ substituted for μ and 0 for μ_0.

EXAMPLE In a department store's study designed to test whether the mean balance outstanding on 30-day charge accounts is the same in its two suburban branch stores, random samples yielded the following results:

$$n_1 = 80 \qquad \bar{x}_1 = \$64.20 \qquad s_1 = \$16.00$$
$$n_2 = 100 \qquad \bar{x}_2 = \$71.41 \qquad s_2 = \$22.13$$

where the subscripts denote branch store 1 and branch store 2. Use the 0.05 level of significance to test the null hypothesis $\mu_1 - \mu_2 = 0$ against the alternative hypothesis $\mu_1 - \mu_2 \neq 0$, where μ_1 and μ_2 are the actual mean balances outstanding on all 30-day charge accounts in branch stores 1 and 2.

SOLUTION Substituting the given values of $n_1, n_2, \bar{x}_1, \bar{x}_2, s_1$, and s_2 into the preceding formula for z, we get

$$z = \frac{64.20 - 71.41}{\sqrt{\dfrac{(16.00)^2}{80} + \dfrac{(22.13)^2}{100}}} = -2.53$$

and since this value is less than -1.96, it follows that the observed difference of \$7.21 between the average outstanding balances in the two branch stores is significant. That is, the difference is too large to be accounted for by chance, and it can be concluded that there is a real difference between the two population means. The p-value for $z = -2.53$ is 0.0057.

12.9

Differences Between Means (Small Samples)

The significance test for the difference between two means described above applies only to large independent random samples. However, a small-sample criterion for tests concerning the difference between two means is provided by the t distribution. To use this criterion we must assume that the two populations we are sampling have roughly the shape of normal distributions with equal variances. Specifically, we test the null hypothesis $\mu_1 - \mu_2 = 0$ against an appropriate one-sided or two-sided alternative with the statistic

Statistic for small-sample test concerning difference between two means

$$t = \cfrac{\bar{x}_1 - \bar{x}_2}{\sqrt{\cfrac{\sum (x_1 - \bar{x}_1)^2 + \sum (x_2 - \bar{x}_2)^2}{n_1 + n_2 - 2} \cdot \left(\cfrac{1}{n_1} + \cfrac{1}{n_2}\right)}}$$

where $\sum (x_1 - \bar{x}_1)^2$ is the sum of the squared deviations from the mean of the first sample and $\sum (x_2 - \bar{x}_2)^2$ is the sum of the squared deviations from the mean of the second sample. Since, by definition, $\sum (x_1 - \bar{x}_1)^2 = (n_1 - 1) \cdot s_1^2$ and $\sum (x_2 - \bar{x}_2)^2 = (n_2 - 1) \cdot s_2^2$, the above formula can be simplified somewhat when the two sample variances have already been calculated from the data.

Under these assumptions, it can be shown that the sampling distribution of this t statistic is the Student-t distribution with $n_1 + n_2 - 2$ degrees of freedom. Depending on whether the alternative hypothesis is $\mu_1 - \mu_2 < 0$, $\mu_1 - \mu_2 > 0$, or $\mu_1 - \mu_1 \neq 0$, the criteria on which we base the small-sample tests for the significance of the difference between two means are again those shown in Figure 12.8 and also in the table on page 329, with t substituted throughout for z, $\mu_1 - \mu_2$ substituted for μ, and 0 for μ_0.

EXAMPLE The following are measurements of the heat-producing capacity (in millions of calories per ton) of random samples of five specimens each of coal from two mines:

Mine 1: 8,380, 8,210, 8,360, 7,840, 7,910

Mine 2: 7,540, 7,720, 7,750, 8,100, 7,690

Use the 0.05 level of significance to test whether the difference between the means of these two samples is significant.

The sample means are $\bar{x}_1 = 8{,}140$ and $\bar{x}_2 = 7{,}760$, and to calculate t in accordance with the formula given above, we first determine

$$\sum (x_1 - \bar{x}_1)^2 = (8{,}380 - 8{,}140)^2 + \cdots + (7{,}910 - 8{,}140)^2$$
$$= 253{,}800$$

and

$$\sum (x_2 - \bar{x}_2)^2 = (7{,}540 - 7{,}760)^2 + \cdots + (7{,}690 - 7{,}760)^2$$
$$= 170{,}600$$

Now, substituting these sums together with $n_1 = 5$, $n_2 = 5$, $\bar{x}_1 = 8{,}140$, and $\bar{x}_2 = 7{,}760$ into the formula for t, we get

$$t = \frac{8{,}140 - 7{,}760}{\sqrt{\dfrac{253{,}800 + 170{,}600}{5 + 5 - 2} \cdot \left(\dfrac{1}{5} + \dfrac{1}{5}\right)}} = 2.61$$

Since this exceeds 2.306, the value of $t_{0.025}$ for $5 + 5 - 2 = 8$ degrees of freedom, it follows that the null hypothesis $\mu_1 - \mu_2 = 0$ must be rejected in favor of the alternative hypothesis $\mu_1 - \mu_2 \neq 0$. In other words, we conclude that the average heat-producing capacity of the coal from the two mines is not the same.

EXERCISES

12.34 The security department of a factory wants to know whether the true average time required by the night watchman to walk his round is 30 minutes. If, in a random sample of 45 rounds, the night watchman averaged 29.2 minutes with a standard deviation of 1.6 minutes, determine at the level of significance 0.05 whether this is sufficient evidence to reject the null hypothesis $\mu = 30$ minutes.

12.35 In trying to rent a location for a restaurant in a shopping mall, a rental agent assures a restaurant owner that on an average business day at least 3,600 persons pass through the mall. Being a cautious businessman, the restaurant owner hires a firm to conduct a survey, and this survey shows that on the average 3,453 persons, with a standard deviation of 428 persons, passed through the mall on 32 randomly selected days. Use the 0.05 level of significance to test the null hypothesis $\mu = 3{,}600$ against the alternative hypothesis $\mu < 3{,}600$.

12.36 Data gathered from many male users of a certain type of electric razor suggest that the time it takes men to shave with this razor is a random variable having a mean of 4.9 minutes and a standard deviation of 0.8 minute. A group of 50 users of the razor is randomly selected and shown a film on how to shave comfortably and effectively with an electric razor. Following the film each member of this group is asked to shave, and it is found that they average 4.6 minutes. Does this constitute evidence (at the level of significance $\alpha = 0.01$)

that viewing the film reduces the mean time required to shave with this type of razor?

12.37 During the investigation of an alleged unfair trade practice, the Federal Trade Commission takes a random sample of 49 "9-ounce" candy bars from a large shipment. The mean of the sample weights is 8.94 ounces and the standard deviation is 0.12 ounce.

(a) Show that, at a level of significance of 0.01, the commission has grounds upon which to proceed against the manufacturer on the unfair practice of short-weight selling.

(b) If the level of significance were set at, say, 0.0001, the commission would have no grounds for such a proceeding (the critical value of z is -3.72). What would be the advantages or disadvantages to anyone in setting the level of significance at such an extremely low figure?

12.38 According to specifications, the mean time required to inflate a rubber life raft is to be 8.5 seconds. What can one conclude at the level of significance $\alpha = 0.05$ about a particular shipment of life rafts, if a random sample of 64 of the rafts has a mean inflation time of 8.6 seconds with a standard deviation of 0.7 second?

12.39 The police chief of a large city claims that the mean age of bicycle thieves is 9.5 years. A research worker who feels that this figure is too low takes a random sample of 80 cases of bicycle theft from police files, and finds the mean age of the thieves to be 10.8 years with a standard deviation of 4.3 years. What can the research worker conclude from these figures at the level of significance $\alpha = 0.01$?

12.40 A production process is designed to fill No. $1\frac{1}{4}$ cans with 14.5 ounces net weight of sliced pineapple. The mean weight varies from time to time, but the standard deviation is considered to be stable and well established at 0.64 ounce. In order to test incoming lots for weight, a large institutional buyer takes a random sample of 30 cans from each lot and determines the mean net weight.

(a) For what values of $\bar{x}$ should the buyer reject the null hypothesis $\mu = 14.5$ ounces, if he uses a two-sided alternative and a level of significance of 0.01?

(b) Using the criterion established in part (a), what is the probability that the buyer will fail to detect a lot whose mean net weight is only 14.2 ounces?

12.41 Repeat the preceding exercise using the one-sided alternative hypothesis $\mu < 14.5$ ounces, which seems more appropriate for a buyer who feels he needs protection only against accepting lots which are underfilled. ("Buyers" often feel they need protection against getting either less than or more than a guaranteed standard. Can you think of an example of this?)

12.42 If we wish to test the null hypothesis $\mu = \mu_0$ in such a way that the probability of a Type I error is α, and the probability of a Type II error is β for the specific alternative $\mu = \mu_A$, we must take a random sample of size n, where

$$ n = \frac{\sigma^2(z_\alpha + z_\beta)^2}{(\mu_A - \mu_0)^2} $$

if the alternative is one-sided, and

$$n = \frac{\sigma^2(z_{\alpha/2} + z_\beta)^2}{(\mu_A - \mu_0)^2}$$

if the alternative is two-sided.

Suppose, for instance, that for a population with $\sigma = 5$ we want to test the null hypothesis that its mean is 200 pounds against the alternative that its mean is less than 200 pounds. The probability of a Type I error is to be 0.05, and the probability of a Type II error when the mean of the population is 198 pounds is to be 0.20. Substituting into the first of the formulas for n above, we get

$$n = \frac{(5^2)(1.645 + 0.84)^2}{(198 - 200)^2} = 38.6$$

so that we must take a sample of size 39, rounding up.

(a) Suppose that we want to test the hypothesis $\mu = \$400$ against the alternative $\mu \neq \$400$ for a population whose standard deviation is $12. If this hypothesis is true, we want to be 95 percent sure of accepting it, and if the true mean differs from $400 by $5 in either direction, we want to be 90 percent sure of rejecting the hypothesis. What is the required sample size? For what values of $\bar{x}$ should the hypothesis be rejected?

(b) Suppose that we want to test the null hypothesis $\mu = 128$ feet against the alternative hypothesis $\mu > 128$ feet for a population whose standard deviation is 6 feet. How large a sample will be required if the probability of a Type I error is to be 0.05 and the probability of a Type II error is to be 0.01 when $\mu = 130$ feet? For what values of $\bar{x}$ should the hypothesis be rejected?

12.43 In 10 test runs, a truck operated for 10, 12, 12, 9, 11, 14, 12, 10, 9, and 11 miles with one gallon of a certain gasoline. Is this evidence at the 0.01 level of significance that the truck is not operating at an average of 12.5 miles per gallon with this gasoline?

12.44 A random sample from a company's very extensive files shows that eight orders for a certain piece of machinery were filled in 12, 10, 17, 14, 13, 18, 11, and 9 days. Use the 0.01 level of significance to test the claim that on the average such orders are filled in 9.5 days. Choose the alternative hypothesis in such a way that rejection of the null hypothesis $\mu = 9.5$ days implies that it takes longer than that.

12.45 A manufacturer guarantees a certain ball bearing to have a mean outside diameter of 0.8525 inch with standard deviation of 0.0003 inch. If a random sample of nine bearings from a large lot of these bearings has a mean outside diameter of 0.8529 inch, at the 1 percent level of significance does this lot meet the manufacturer's guarantee on the mean outside diameter?

12.46 Advertisements claim that the average nicotine content of a certain kind of cigarette is 0.30 milligram. Suspecting that this figure is too low, a consumer protection service takes a random sample of 15 of these cigarettes from different production lots and finds that their nicotine content has a mean of 0.33 milligram with a standard deviation of 0.018 milligram. Use the 0.05 level of significance to test the null hypothesis $\mu = 0.30$ against the alternative hypothesis $\mu > 0.30$.

12.47 A soft-drink vending machine is set to dispense 6 ounces per cup. If the machine is tested eight times, yielding a mean cup fill of 5.8 ounces with a standard deviation of 0.16 ounce, is this evidence at the level of significance $\alpha = 0.05$ that the machine is underfilling cups?

12.48 A sample study was made of the number of lunches that executives claim as business expenses in a given month. If 40 executives in the insurance industry averaged 9.8 such lunches with a standard deviation of 1.8 and 50 bank executives averaged 8.4 such lunches with a standard deviation of 2.2, test at the level of significance $\alpha = 0.01$ whether the difference between these two sample means is significant.

12.49 A company wants to compare the lifetimes of two stones used in an abrasive process (called "superfinishing"), which produces rapidly a fine microfinish on machined surfaces. In laboratory tests, hot-rolled-steel driveshafts of the same degree of surface roughness are processed for 2 minutes each under specified conditions using both stones. If the average lifetime of 10 stones of the first kind was 60 pieces with a standard deviation of 5 pieces, while the average lifetime of 10 stones of the second kind was 64 pieces with a standard deviation of 3 pieces, is the difference between these two means significant at a level of significance of 0.05?

12.50 Suppose that we want to investigate whether men and women earn comparable wages in a certain industry. If sample data show that 60 men earn on the average $262.50 per week with a standard deviation of $15.60, while 60 women earn on the average $246.10 per week with a standard deviation of $18.20, test the null hypothesis $\mu_1 - \mu_2 = 0$ against the alternative hypothesis $\mu_1 - \mu_2 > 0$ at the level of significance $\alpha = 0.01$.

12.51 The following are the numbers of sales which a random sample of nine salesmen of industrial chemicals in Detroit and a random sample of six salesmen of industrial chemicals in Chicago made over a fixed period of time:

Detroit: 54, 63, 39, 66, 58, 41, 64, 49, 43
Chicago: 45, 26, 57, 47, 65, 36

Use the level of significance $\alpha = 0.01$ to test whether the difference between the means of these two samples is significant.

12.52 As part of an industrial training program, some trainees are taught by method A, which is straight teaching-machine instruction, and some are taught by method B, which also involves the personal attention of an instructor. Random samples of size 10 are taken from large groups of trainees taught by each of these two methods, and the following are their scores on an appropriate achievement test:

Method A: 74, 78, 68, 72, 76, 69, 71, 74, 77, 71
Method B: 75, 80, 87, 81, 72, 73, 80, 76, 68, 78

Use the level of significance $\alpha = 0.05$ to test the claim that method B is superior to method A.

12.53 In the comparison of two kinds of paint, a consumer-testing service found that four 1-gallon cans of brand A covered on the average 514 square feet with a standard deviation of 32 square feet, while four 1-gallon cans of brand B covered on the average 487 square feet with a standard deviation of 27 square feet. Use the level of significance $\alpha = 0.05$ to test whether the difference between the two sample means is significant.

12.54 In some problems we are interested in testing whether the difference between the means of two populations is equal to, less than, or greater than a given constant. So, we test the null hypothesis $\mu_1 - \mu_2 = \delta$ (delta), where δ is the given constant, against an appropriate alternative hypothesis. To perform this kind of test, we substitute $\bar{x}_1 - \bar{x}_2 - \delta$ for $\bar{x}_1 - \bar{x}_2$ in the numerator of the z-statistic on page 333, or the t-statistic on page 334, and otherwise proceed in the same way as before.

(a) Sample surveys conducted in a large county in 1950 and again in 1970 showed that in 1950 the average height of 400 ten-year-old boys was 53.2 inches with a standard deviation of 2.4 inches, while in 1970 the average height of 500 ten-year-old boys was 54.5 inches with a standard deviation of 2.5 inches. Use the level of significance $\alpha = 0.05$ to test whether the true average increase in height is at most 0.5 inch.

(b) To test the claim that the resistance of electric wire can be reduced by more than 0.050 ohm by alloying, 25 values obtained for alloyed wire yielded $\bar{x}_1 = 0.083$ ohm and $s_1 = 0.003$ ohm, and 25 values obtained for standard wire yielded $\bar{x}_2 = 0.136$ ohm and $s_2 = 0.002$ ohm. Use the level of significance $\alpha = 0.05$ to determine whether the claim has been substantiated.

12.55 If we want to study the effectiveness of a new diet on the basis of weights "before and after," or if we want to study whatever differences there may be between the IQ's of husbands and wives, the methods introduced in this chapter cannot be used. The samples are not independent; in fact, in each case the data are *paired*. To handle data of this kind, we work with the (signed) differences of the paired data and test whether these differences may be looked upon as a sample from a population for which $\mu = 0$. If the sample is small, we use the t test; otherwise, we use a large-sample test. Apply this technique to determine the effectiveness of an industrial safety program on the basis of the following data (collected over a period of one year) on the average weekly loss of man-hours due to accidents in 12 plants "before and after" the program was put into operation:

37 and 28	72 and 59	26 and 24	125 and 120
45 and 46	54 and 43	13 and 15	79 and 75
12 and 18	34 and 29	39 and 35	26 and 24

Use the level of significance $\alpha = 0.05$ to decide whether the safety program is effective.

12.56 In a study of the effectiveness of physical exercise in weight reduction, a group of 16 persons engaged in a prescribed program of physical exercise for one month showed the following results:

Weight before (pounds)	Weight after (pounds)	Weight before (pounds)	Weight after (pounds)
209	196	170	164
178	171	153	152
169	170	183	179
212	207	165	162
180	177	201	199
192	190	179	173
158	159	243	231
180	180	144	140

Use the 0.01 level of significance to test the null hypothesis that the prescribed program of exercise is not effective in reducing weight.

12.57 In industrial quality control it is often necessary to test the same hypothesis over and over again at regular intervals of time. Suppose, for example, that a process for making steel pipe is in control if the diameter of the pipe has a mean of $\mu = 3.0$ inches; also, it is known from past experience that the standard deviation of the diameter of the pipe produced by this process is $\sigma = 0.02$ inch. In order to test whether the process is in control so far as μ is concerned, random samples of size n are taken, say, every hour, and it is decided in each case on the basis of $\bar{x}$ whether to accept or reject the null hypothesis $\mu = 3.0$ inches. To simplify this task, quality-control engineers use **control charts** such as that of Figure 12.9, where the vertical scale is the scale of measurement for $\bar{x}$, the **central line** is at $\mu = \mu_0$, and the **upper and lower control limits** are at $\mu_0 + 3 \cdot \dfrac{\sigma}{\sqrt{n}}$ and $\mu_0 - 3 \cdot \dfrac{\sigma}{\sqrt{n}}$. Each sample mean is plotted on this chart and the process is considered to be in control so long as the $\bar{x}$'s fall between the control limits. If the data constitute random samples from a population having roughly the shape of a normal distribution, the probability of a Type I error is less than 0.003.

12.9

Control chart.

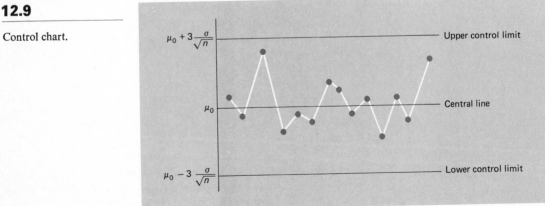

(a) Use the standards $\mu = 3.0$ inches and $\sigma = 0.02$ inch to construct a control chart for the mean of random samples of size $n = 5$.

(b) Plot on the chart obtained in part (a) the following data, constituting the means of random samples of size $n = 5$ taken at hourly intervals during the operation of the process: 3.010, 2.925, 3.021, 3.005, 3.024, 3.020, 2.988, 2.965, 3.029, 2.944, 3.031, 3.002, 2.990, 3.031, 2.975, 2.978, 3.022, 2.991, 2.996, and 2.982. Was the process ever out of control?

12.58 In the production of certain piston rings, a process is said to be in control if their weights have a mean of $\mu = 23.250$ grams and a standard deviation of $\sigma = 0.002$ gram.

(a) Draw a control chart for the means of random samples of size $n = 4$.

(b) Plot on the chart drawn in part (a) the following means of 25 such random samples taken at regular intervals of time:

23.2514	23.2522	23.2495	23.2506	23.2485	23.2502	23.2516	23.2485
23.2492	23.2528	23.2516	23.2500	23.2501	23.2495	23.2503	23.2512
23.2488	23.2475	23.2511	23.2492	23.2502	23.2514	23.2476	23.2503
23.2515							

Was the process ever out of control?

12.10

A Word of Caution

We should note that in statistics the term "significant" is used in a technical sense. If we say that something is "statistically significant," we do not mean to imply that it is necessarily of any practical significance or importance. For instance, a battery manufacturer may find from sample data that a new, expensive, additive produces a statistically significant increase in the average useful lifetime of his batteries. However, this average increase in the lifetimes of the batteries may be so small that there is no economic justification for using the costly new additive.

12.11

Check List of Key Terms

Alternative hypothesis, 322
Bayesian inference, 309, 311
Composite hypothesis, 320
Confidence, 304
Confidence interval, 305
Confidence limits, 305
Control chart, 340

Critical values, 324
Degree of confidence, 305
Degrees of freedom, 307
Estimation, 302
Interval estimate, 305
Level of significance, 323, 329
Null hypothesis, 320

12.12

Review Exercises

12.59 A college professor is making up a final examination in economics which is to be given to a large group of students. His feelings about the average grade they should get is expressed subjectively by a distribution which has the mean $\mu_0 = 64$ and the standard deviation $\sigma_0 = 1.4$. If, subsequently, the examination is tried on a random sample of 50 students whose grades have a mean of 73.9 and a standard deviation of 7.6, find
(a) the posterior mean and the posterior standard deviation;
(b) the posterior probability that μ lies between 70.0 and 72.0.

12.60 To estimate the time required for certain repairs, a random sample of 50 automobile mechanics is timed in the performance of this task. The mean is 24.85 minutes and the standard deviation is 3.22 minutes. What can we say with 95% confidence about the maximum size of the error, if we use $\bar{x} = 24.85$ minutes as an estimate of the true average time it takes a mechanic to perform this task?

12.61 In a study of new sources of food, it is reported that 1 pound of a certain kind of fish will yield on the average 2.41 ounces of FPC (fish-protein concentrate), which is used to enrich flour and other food products. Is this figure supported by a study in which a random sample of 32 such fish yielded on the average 2.38 ounces of FPC per pound with a standard deviation of 0.07 ounce? Use the level of significance $\alpha = 0.01$ to test the null hypothesis $\mu = 2.41$ ounces.

12.62 Suppose that we want to test the hypothesis that stock X is a better investment than stock Y. Explain under what conditions we would be committing a Type I error and under what conditions we would be committing a Type II error.

12.63 For a very large population of lengths of metal strips whose standard deviation is assumed to be 0.10 inch, a buyer wants to test (on the basis of a random sample of strips drawn from the lot) the hypothesis that the true mean length of the strips is 4 inches against the hypothesis that it is less than 4 inches. Consequences of rejecting the hypothesis if it is true and of accepting it if the mean length is actually 3.95 inches are considered to be equally serious, and their risks are both set at 0.02. For what values of $\bar{x}$ should the buyer reject the hypothesis?

12.64 Five measurements of the tar content of a certain kind of cigarette yielded 14.5, 14.2, 14.4, 14.3, and 14.6 mg/cig (milligrams per cigarette). Show that the difference between the mean of this sample, $\bar{x} = 14.4$, and the average tar content claimed by the cigarette manufacturer, $\mu = 14.0$, is significant at $\alpha = 0.05$.

12.65 Suppose that in the preceding exercise the first measurement is recorded incorrectly as 16.0 instead of 14.5. Show that now the difference between the mean of the sample, $\bar{x} = 14.7$, and the average tar content claimed by the cigarette manufacturer, $\mu = 14.0$, is not significant at $\alpha = 0.05$. Explain the apparent paradox that even though the difference between $\bar{x}$ and μ has increased, it is no longer significant.

12.66 In establishing the authenticity of an ancient coin, its weight is often of critical importance. If four experts weighed a Phoenician tetradrachm and obtained $\bar{x} = 14.30$ grams and $s = 0.037$ gram, construct a 99% confidence interval for the true weight of the coin.

12.67 Fifteen randomly selected mature citrus trees of one variety have a mean height of 14.8 feet with a standard deviation of 1.3 feet, while twelve randomly selected citrus trees of another variety have a mean height of 13.6 feet with a standard deviation of 1.5 feet. Test at the 0.01 level of significance whether the difference between the two sample means is significant.

12.68 A random sample of 50 cans of pear halves has a mean weight of 29.0 ounces and a standard deviation of 0.8 ounce. If this mean of 29.0 ounces is used as an estimate of the true mean weight of all the cans of pear halves from which this sample came, with what probability can we assert that this estimate is "off" by at most 0.2 ounce?

12.69 In an air pollution study, an experiment station obtained a mean of 2.28 micrograms of suspended benzene-soluble organic matter per cubic meter with a standard deviation of 0.54 from a random sample of eight different specimens. If the mean of this sample is used to estimate the corresponding true mean, what can the station assert with 99% confidence about the possible size of the error?

12.70 A study conducted by an airline at a certain airport showed that 120 of its passengers had to wait on the average 11.45 minutes with a standard deviation of 1.94 minutes to get their luggage. Construct a 95% confidence interval for the true average time it takes one of the airline's passengers to get his or her luggage when disembarking at this airport.

12.71 The average drying time of a manufacturer's paint is 20 minutes. Investigating the effectiveness of a modification in the chemical composition of his paint, the manufacturer wants to test the null hypothesis $\mu = 20$ minutes against a suitable alternative, where μ is the average drying time of the new paint.
  (a) What alternative hypothesis should the manufacturer use if he does not want to make the modification in the chemical composition of the paint unless it is definitely superior with respect to drying time?
  (b) What alternative hypothesis should the manufacturer use if the new process is actually cheaper and he wants to make the modification unless it increases the drying time of the paint?

12.72 To compare freshmen's knowledge of economics in two universities, samples of 50 freshmen from each of the two universities were given a special test. If those from the first university had an average score of 72.4 with a standard deviation of 5.0, while those from the second university had an average score of 67.8 with a standard deviation of 4.6, test at the 0.05 level of significance whether the difference between the two sample means is significant.

12.73 It is desired to estimate the average number of days of continuous use until a new refrigerator of a certain kind will first require repairs. If it can be assumed that $\sigma = 242$ days, how large a sample is needed so that it can be asserted with probability 0.95 that the sample mean will be off by less than 30 days?

12.74 With reference to the illustration of Section 12.9, suppose that we had wanted to test whether the heat-producing capacity of the coal from the first mine exceeds that of the coal from the second mine by more than 50 million calories per ton. Use the data on page 334 to test the null hypothesis $\mu_1 - \mu_2 = 50$ against the alternative hypothesis $\mu_1 - \mu_2 > 50$ at the level of significance $\alpha = 0.05$.

12.75 It is desired to test the null hypothesis $\mu = 100$ pounds against the alternative hypothesis $\mu > 100$ pounds on the basis of a random sample of size $n = 64$ from a population with $\sigma = 12$.
(a) If $\alpha = 0.05$, for what values of $\bar{x}$ will the null hypothesis be rejected?
(b) Calculate β for $\mu = 102, 104$, and 106 pounds, and draw a rough sketch of the OC-curve.

12.76 A random sample of six daily scrap records (where scrap is expressed as a percentage of material requisitioned) shows 3.5, 4.1, 3.9, 6.1, 5.5, and 4.5 percent scrap. If the mean of this sample is used to estimate the mean percentage of the corresponding population, what can we assert with 90% confidence about the possible size of our error?

12.77 In an experiment with a new tranquilizer, the pulse rate of 18 patients was measured before they were given the tranquilizer and again 5 minutes later, and their pulse rate was found to be reduced on the average by 7.8 beats with a standard deviation of 1.9. Using the level of significance $\alpha = 0.05$, what can we conclude about the claim that this tranquilizer will reduce the pulse rate on the average by at least 9 beats in 5 minutes?

12.78 A law student, who wants to check the claim that convicted embezzlers spend on the average 12.8 months in jail, takes a sample of 40 such cases from very extensive court files. Using her results, $\bar{x} = 11.1$ months and $s = 3.4$ months, test the null hypothesis $\mu = 12.8$ months against the alternative hypothesis $\mu \neq 12.8$ months at the 0.01 level of significance.

In Chapter 12 we learned to construct confidence intervals for means and to perform tests of hypotheses concerning the means of one and of two populations. As we shall see in this and some of the following chapters, very similar methods can be used for inferences about other population parameters. By studying the sampling distributions of appropriate statistics, statisticians have developed methods of inference about population proportions, standard deviations, medians, quartiles, and the like. In principle, the ideas are always the same, but some of the sampling distributions are mathematically quite involved. Fortunately, this difficulty is resolved by the important result that for large samples, many of these sampling distributions can be approximated with normal curves.

In this chapter we shall concentrate on population standard deviations, and population variances, which are not only important in their own right, but which must sometimes be estimated before we can make inferences about other parameters. This is the case, for example, when we make inferences about population means and must know or estimate the value of σ.

In Section 13.1 we shall be concerned with the estimation of σ, and in Sections 13.2 and 13.3 we shall study tests concerning the standard deviation of one population and the standard deviations of two populations.

Decision Making: Inferences About Standard Deviations

13

13.1

The Estimation of σ

Although there are other methods of estimating the standard deviation of a population (see, for example, Exercise 13.7 on page 349), the sample standard deviation is the most widely used estimator of this parameter. Limiting our discussion to problems in which we use s to make inferences about σ (or s^2 to make inferences about σ^2), let us begin by constructing a confidence interval for σ based on the standard deviation of a random sample of size n. The theory on which such an interval is based requires that the population sampled has roughly the shape of a normal distribution, in which case the statistic

Chi-square statistic

$$\chi^2 = \frac{(n-1)s^2}{\sigma^2}$$

called "chi-square," has as its sampling distribution an important continuous distribution called the **chi-square distribution**. The mean of this distribution is $n-1$ and, as with the t distribution, we call this quantity the number of degrees of freedom, or simply the **degrees of freedom**. An example of a chi-square distribution is shown in Figure 13.1; unlike the normal and t distributions, its domain is restricted to the nonnegative real numbers.

As we did with z_α and t_α, we now define χ^2_α as the value for which the area to its right under the chi-square distribution is equal to α. Thus, $\chi^2_{\alpha/2}$ is such

13.1

Chi-square distribution.

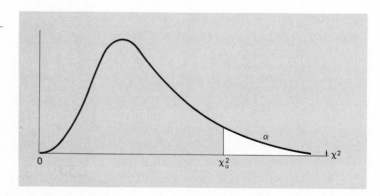

that the area to its right under the curve is $\alpha/2$, while $\chi^2_{1-\alpha/2}$ is such that the area to its left under the curve is $\alpha/2$ (see also Figure 13.2). We make this distinction because the chi-square distribution is not symmetrical. Among others, values of $\chi^2_{0.995}$, $\chi^2_{0.975}$, $\chi^2_{0.025}$, and $\chi^2_{0.005}$ are given in Table III at the end of the book for 1, 2, 3, . . . , and 30 degrees of freedom.

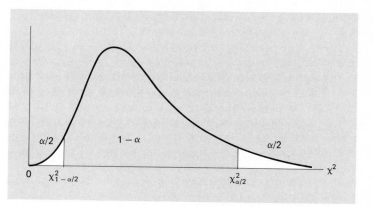

13.2

Chi-square
distribution.

Referring to Figure 13.2, we find that we can assert with probability $1 - \alpha$ that a random variable having the chi-square distribution will take on a value between $\chi^2_{1-\alpha/2}$ and $\chi^2_{\alpha/2}$. Applying this result to the χ^2 statistic given above, we can assert with probability $1 - \alpha$ that

$$\chi^2_{1-\alpha/2} < \frac{(n-1)s^2}{\sigma^2} < \chi^2_{\alpha/2}$$

This double inequality can be rewritten as

*Confidence
interval for σ^2*

$$\frac{(n-1)s^2}{\chi^2_{\alpha/2}} < \sigma^2 < \frac{(n-1)s^2}{\chi^2_{1-\alpha/2}}$$

which is a $(1-\alpha)100\%$ confidence interval for σ^2, the population variance. Also, if we take the square root of each of the three terms of this double inequality, we get a $(1-\alpha)100\%$ confidence interval for σ, the population standard deviation.

EXAMPLE A random sample of $n = 5$ specimens of a certain kind of ice cream has a mean fat content of $\bar{x} = 11.3$ percent and a standard deviation of $s = 0.38$ percent. Construct a 95% confidence interval for σ, the standard deviation of the population sampled.

Since we have $5 - 1 = 4$ degrees of freedom, we find from Table III that $\chi^2_{0.975} = 0.484$ and $\chi^2_{0.025} = 11.143$. Substituting these values together with $n = 5$ and $s = 0.38$ into the confidence-interval formula for σ, we get

$$\sqrt{\frac{4(0.38)^2}{11.143}} < \sigma < \sqrt{\frac{4(0.38)^2}{0.484}}$$

and

$$0.23\% < \sigma < 1.09\%$$

The kind of confidence interval we have just described is often referred to as a **small-sample confidence interval**, since it is used mainly when n is small (and, of course, only when we can assume that the population from which we are sampling has roughly the shape of a normal distribution). Otherwise, we make use of the fact that for large samples of 30 or more the sampling distribution of s can be approximated with a normal distribution having the mean σ and the standard deviation $\dfrac{\sigma}{\sqrt{2n}}$ (see Exercise 11.18 on page 293). So, we can assert with probability $1 - \alpha$ that

$$-z_{\alpha/2} < \frac{s - \sigma}{\dfrac{\sigma}{\sqrt{2n}}} < z_{\alpha/2}$$

and simple algebra leads to the following $(1 - \alpha)100\%$ large-sample confidence interval for the population standard deviation σ:

Large-sample confidence interval for σ

$$\frac{s}{1 + \dfrac{z_{\alpha/2}}{\sqrt{2n}}} < \sigma < \frac{s}{1 - \dfrac{z_{\alpha/2}}{\sqrt{2n}}}$$

EXAMPLE With reference to the example on page 302, where we calculated $s = 5.51$ tons for a large industrial plant's emission of sulfur oxides on $n = 40$ days, construct a 95% confidence interval for the standard deviation of the population sampled.

SOLUTION Substituting $n = 40$, $s = 5.51$, and $z_{\alpha/2} = 1.96$ into the above confidence-interval formula, we get

$$\frac{5.51}{1 + \dfrac{1.96}{\sqrt{80}}} < \sigma < \frac{5.51}{1 - \dfrac{1.96}{\sqrt{80}}}$$

and

$$4.52 < \sigma < 7.06$$

This means that we are 95% confident that the interval from 4.52 tons to 7.06 tons contains σ, the true standard deviation of the plant's daily emission of sulfur oxides.

EXERCISES

13.1 With reference to Exercise 12.12 on page 313, construct a 95% confidence interval for σ, the true standard deviation of the number of times that senior citizens in the given retirement community visit a physician per year.

13.2 With reference to Exercise 12.13 on page 313, construct a 99% confidence interval for the true standard deviation of the amounts of diesel fuel sold at the truck stop.

13.3 With reference to Exercise 12.14 on page 313, construct a 95% confidence interval for the true variance of the time it takes to assemble the mechanical device.

13.4 With reference to Exercise 12.1 on page 311, construct a 95% confidence interval for σ, the true standard deviation of the amount of time required to complete the form.

13.5 With reference to Exercise 12.3 on page 312, construct a 99% confidence interval for the true standard deviation of the body repair costs.

13.6 With reference to Exercise 12.4 on page 312, construct a 99% confidence interval for the true variance of the time it takes the chef to prepare the entree.

13.7 When we deal with very small samples, good estimates of the population standard deviation can often be obtained on the basis of the sample range (the largest sample value minus the smallest). Such quick estimates of σ are given by the sample range divided by the divisor d, which depends on the size of the sample; for samples from populations having roughly the shape of a normal distribution, its values are shown in the following table:

n	2	3	4	5	6	7	8	9	10	11	12
d	1.13	1.69	2.06	2.33	2.53	2.70	2.85	2.97	3.08	3.17	3.26

For instance, in the illustration on page 308, which deals with the durability of a paint for highway center lines, we have $n = 8$ and a sample range of $167,800 - 108,300 = 59,500$ crossings. Since $d = 2.85$ for $n = 8$, we find that we can estimate σ, the true standard deviation of the population sampled, as

$$\frac{59,500}{2.85} = 20,878 \text{ crossings}$$

This is somewhat higher than the sample standard deviation $s = 19,200$ crossings, but not knowing the true value of σ, we cannot say which of the two estimates is actually closer.

(a) With reference to Exercise 12.14 on page 313, use this method to estimate the true standard deviation of the amount of time it takes to assemble the mechanical device, and compare the result with the sample standard deviation s.

(b) With reference to Exercise 12.15 on page 313, use this method to estimate the true standard deviation of the weights of the containers of solvent, and compare the result with the sample standard deviation s.

13.2

Tests Concerning σ and σ^2

In this section we shall consider the problem of testing the null hypothesis that a population standard deviation equals a specified constant σ_0, or that a population variance equals σ_0^2. This kind of test is required whenever we want to test the uniformity of a product, process, or operation. For instance, we may want to test whether a certain kind of glass is sufficiently homogeneous for making delicate optical equipment, whether the variation in the outside diameter of mass produced copper tubing is within permissible limits, whether a lack of uniformity in certain workers' performance may call for stricter supervision, and so on.

The test of the null hypothesis $\sigma = \sigma_0$, the hypothesis that a population standard deviation equals a specified constant, is based on the same assumptions, the same statistic, and the same sampling theory as the small-sample confidence interval for σ. Assuming that our random sample comes from a population having roughly the shape of a normal distribution, we base our decision on the statistic

Statistic for test concerning standard deviation

$$\chi^2 = \frac{(n-1)s^2}{\sigma_0^2}$$

where n and s^2 are the sample size and the sample variance, and σ_0 is the value of the population standard deviation assumed under the null hypothesis. The sampling distribution of this statistic is the chi-square distribution with $n-1$ degrees of freedom; hence, the criteria for testing the null hypothesis $\sigma = \sigma_0$ against the alternative hypothesis $\sigma < \sigma_0, \sigma > \sigma_0$, or $\sigma \neq \sigma_0$ are as shown in Figure 13.3. For the one-sided alternative $\sigma < \sigma_0$, we reject the null hypothesis for values of χ^2 falling into the left-hand tail of its sampling distribution; for the one-sided alternative $\sigma > \sigma_0$, we reject the null hypothesis for values of χ^2 falling into the right-hand tail of its sampling distribution; and for the two-sided alternative $\sigma \neq \sigma_0$, we reject the null hypothesis for values of χ^2 falling

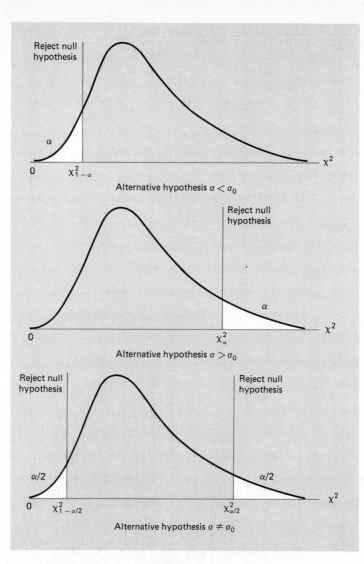

Alternative hypothesis $\sigma < \sigma_0$

Alternative hypothesis $\sigma > \sigma_0$

Alternative hypothesis $\sigma \neq \sigma_0$

13.3
_____

Test criteria.

into either tail of its sampling distribution. The quantities χ_α^2, $\chi_{1-\alpha}^2$, $\chi_{\alpha/2}^2$, and $\chi_{1-\alpha/2}^2$ may be read from Table III.

EXAMPLE   Specifications for mass-produced bearings of a certain type require, among other things, that the standard deviation of their outside diameters should not exceed 0.0050 cm. Use the level of significance 0.01 to test the null hypothesis $\sigma = 0.0050$ against the alternative hypothesis $\sigma > 0.0050$ on the basis of a random sample of size $n = 12$ for which $s = 0.0077$ cm.

Substituting $n = 12$, $s = 0.0077$, and $\sigma_0 = 0.0050$ into the formula for χ^2, we get

$$\chi^2 = \frac{11(0.0077)^2}{(0.0050)^2} = 26.09$$

and since this exceeds 24.725, the value of $\chi^2_{0.01}$ for $12 - 1 = 11$ degrees of freedom, we find that the null hypothesis must be rejected. In other words, it appears that the bearings do not meet specifications with regard to the variability in their diameters.

When n is large, $n \geq 30$, we can base tests of the null hypothesis $\sigma = \sigma_0$ on the same theory we used in constructing large-sample confidence intervals in the preceding section. That is, we use the statistic

Statistic for large-sample test concerning standard deviation

$$z = \frac{s - \sigma_0}{\sigma_0 / \sqrt{2n}}$$

whose sampling distribution is approximately the standard normal distribution, and the criteria of Figure 12.8.

EXAMPLE The variability of a store's sales in a random sample of 50 days is measured by the standard deviation $s = \$2,250$. Use the 0.01 level of significance to test the null hypothesis $\sigma = \$3,000$ against the alternative hypothesis $\sigma < \$3,000$ for the population sampled.

SOLUTION Substituting $n = 50$, $s = 2,250$, and $\sigma_0 = 3,000$ into the formula for z, we get

$$z = \frac{2,250 - 3,000}{3,000 / \sqrt{100}} = -2.50$$

and since this is less than $-z_{0.01} = -2.33$, the null hypothesis must be rejected. In other words, we conclude that sales are less variable than assumed under the null hypothesis.

13.3

Tests Concerning Two Standard Deviations

In this section we shall discuss a test concerning the equality of the standard deviations, or variances, of two populations. This test is often used in connection with the small-sample test of the difference between two means, which requires that the variances of the two populations be equal. For instance, in the example on page 334 dealing with the heat-producing capacity of the coal from two mines, the two samples had variances of $\frac{253,800}{4} = 63,450$ and $\frac{170,600}{4} = 42,650$, and despite what may seem to be a large difference, we assumed that the population variances were, indeed, equal. We could not discuss the rationale of this assumption at that time, but we actually tested—and were unable to reject—the null hypothesis that the populations had equal variances before we performed the t test for the significance of the difference between the two sample means.

Given independent random samples of size n_1 and n_2 from two populations, we usually base tests of the equality of the two population standard deviations (or variances) on the ratios s_1^2/s_2^2 or s_2^2/s_1^2, where s_1 and s_2 are the two sample standard deviations. Assuming that the populations from which the samples came have roughly the shape of normal distributions, it can be shown that the sampling distribution of such a ratio, appropriately called a **variance ratio**, is a continuous distribution called the *F* **distribution**. This distribution depends on the two parameters $n_1 - 1$ and $n_2 - 1$, the number of degrees of freedom in the sample estimates, s_1^2 and s_2^2, of the unknown population variances. One difficulty with this distribution is that most tables give only values of $F_{0.05}$ (defined in the same way as $z_{0.05}$, $t_{0.05}$, and $\chi^2_{0.05}$) and $F_{0.01}$, so we can work only with the right-hand tail of the distribution. For this reason we base our decision on the equality of two population standard deviations σ_1 and σ_2 (or variances σ_1^2 and σ_2^2) on the statistic

Statistic for test concerning the equality of two standard deviations

$$F = \frac{s_1^2}{s_2^2} \ \text{ or } \ \frac{s_2^2}{s_1^2} \ \text{ whichever is larger}$$

With this statistic we reject the null hypothesis $\sigma_1 = \sigma_2$ and accept the alternative $\sigma_1 \neq \sigma_2$ when the observed value of F exceeds $F_{\alpha/2}$, where α is the level of significance (see also Figure 13.4). By using a right-hand tail area of $\alpha/2$ instead

13.4

F distribution.

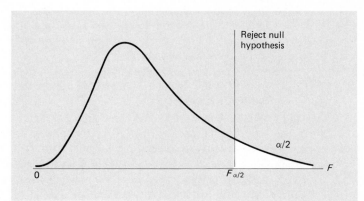

of α, we compensate for the fact that we always use the larger of the two variance ratios. The necessary values of $F_{\alpha/2}$ for $\alpha = 0.02$ or 0.10, $F_{0.01}$ and $F_{0.05}$, are given in Table IV at the end of the book, where the number of degrees of freedom for the numerator is $n_1 - 1$ or $n_2 - 1$, depending on whether we are using the ratio s_1^2/s_2^2 or the ratio s_2^2/s_1^2; correspondingly, the number of degrees of freedom for the denominator is $n_2 - 1$ or $n_1 - 1$.

EXAMPLE In the example on the coal from two mines on page 334, we used a t test to test the significance of the difference between two sample means. Test at the 0.02 level of significance whether there is any real evidence that the standard deviations of the two populations are not equal.

SOLUTION The sample sizes were $n_1 = 5$ and $n_2 = 5$ and the sample variances were $s_1^2 = \dfrac{253{,}800}{4} = 63{,}450$ and $s_2^2 = \dfrac{170{,}600}{4} = 42{,}650$, so that

$$F = \frac{s_1^2}{s_2^2} = \frac{63{,}450}{42{,}650} = 1.49$$

Since this falls short of 16.0, the value of $F_{0.01}$ for $5 - 1 = 4$ and $5 - 1 = 4$ degrees of freedom, we find that the null hypothesis $\sigma_1 = \sigma_2$ cannot be rejected at the 0.02 level of significance.

EXERCISES

13.8 An investigation of certain air-conditioning equipment showed that 12 failures of a given kind took on the average 43.2 minutes to repair with a standard deviation of 9.5 minutes. Use the 0.05 level of significance to test the claim that $\sigma = 8.0$ minutes for the population sampled.

13.9 Past data indicate that the standard deviation of measurements made on sheet metal stampings by experienced inspectors is 0.40 inch. If a new inspector measures 25 stampings with a standard deviation of 0.54 inch, test at the 0.01 level of significance whether the inspector is making satisfactory measurements; that is, test the null hypothesis $\sigma = 0.40$ against the alternative hypothesis $\sigma > 0.40$.

13.10 In a random sample, the time which 20 women took to complete the written test for their driver's license had a variance of 6.2 minutes^2. Test the null hypothesis $\sigma^2 = 8$ minutes^2 against the alternative hypothesis $\sigma^2 \neq 8$ minutes^2 at the level of significance $\alpha = 0.05$.

13.11 With reference to Exercise 12.46 on page 337, use the 0.05 level of significance to test the null hypothesis $\sigma = 0.010$ against the alternative hypothesis that this figure is too low.

13.12 With reference to Exercise 12.34 on page 335, use the large-sample test at the level of significance $\alpha = 0.01$ to test the null hypothesis $\sigma = 2.2$ minutes against the alternative hypothesis $\sigma < 2.2$ minutes.

13.13 With reference to Exercise 12.38 on page 336, use the large-sample test at the 0.05 level of significance to test the null hypothesis $\sigma = 0.5$ second against the alternative hypothesis $\sigma \neq 0.5$ second.

13.14 The specifications for the mass production of certain springs require, among other things, that the standard deviation of their compressed lengths should not exceed 0.040 cm. Use the level of significance $\alpha = 0.05$ to test the null hypothesis $\sigma = 0.040$ against the alternative hypothesis $\sigma > 0.040$ on the basis of a random sample of size $n = 35$ for which $s = 0.051$.

13.15 Two different techniques of lighting a store's window displays are compared by measuring the intensity of light at selected locations in areas lighted by the two methods. If a random sample of 12 measurements of the intensity of light provided by the first technique has a standard deviation of 2.6 foot-candles and a random sample of 16 measurements of the intensity of light provided by the second technique has a standard deviation of 4.4 foot-candles, test the null hypothesis $\sigma_1 = \sigma_2$ against the alternative hypothesis $\sigma_1 \neq \sigma_2$ at the 0.10 level of significance.

13.16 With reference to Exercise 12.51 on page 338, test at the 0.02 level of significance whether it is reasonable to assume that the two population standard deviations are equal.

13.17 With reference to Exercise 12.53 on page 339, test at the 0.10 level of significance whether it is reasonable to assume that the two population standard deviations are equal.

13.4

Check List of Key Terms

Chi-square distribution, 346
Chi-square statistic, 346
Degrees of freedom, 346

F distribution, 353
Variance ratio, 353

13.5

Review Exercises

13.18 In a random sample of the scores on 10 rounds of golf played on his home course, a golf professional averaged 71.8 with a standard deviation of 1.46. Test the null hypothesis that the consistency of his game on his home course is actually measured by $\sigma = 1.20$, using the 0.01 level of significance and the alternative hypothesis that his game is actually less consistent.

13.19 Given that a random sample of 15 one-gallon cans of a certain exterior paint covered on the average 463.2 square feet with a standard deviation of 17.9 square feet, construct a 95% confidence interval for σ.

13.20 In a random sample of 12 supermarkets in Atlanta the prices of a pound of lean ground beef have a standard deviation of $0.14, and in a random sample of 10 supermarkets in New Orleans the prices of a pound of lean ground beef have a standard deviation of $0.18. Use the 0.10 level of significance to test the null hypothesis that the prices of a pound of lean ground beef are equally variable in the two cities.

13.21 The following are the attendance figures in a random sample of four of a minor league baseball team's home games: 4,512, 5,633, 4,280, and 5,139.

    (a) Calculate s for these data as an estimate of the standard deviation of the population sampled.

    (b) Use the range of the data and the table of Exercise 13.7 to obtain another estimate of this population standard deviation, and compare it with the one obtained in part (a).

13.22 In a random sample of 200 tax returns, the amount due after an audit averaged $275.50 with a standard deviation of $31.62. Construct a 95% confidence interval for the true standard deviation of such amounts.

13.23 With reference to Exercise 12.67 on page 343, test at the 0.10 level of significance whether it is reasonable to assume that the two populations sampled have equal standard deviations.

13.24 In a random sample, 40 accountants had an average annual income of $32,460 with a standard deviation of $4,190. Use the 0.05 level of significance to test the claim that for the population sampled σ is at least $6,000.

14

Decision Making: Inferences About Proportions

n principle, the work of this chapter will be very similar to that of Chapters 12 and 13. In problems of estimation we shall again construct confidence intervals and determine, or control, the possible size of our error. In tests of hypotheses we shall again formulate null hypotheses and their alternatives, decide between one-tailed tests and two-tailed tests, choose levels of significance, and so forth. The main difference is that we will be concerned with other parameters. Instead of population means, variances, and standard deviations, we will deal with population proportions, percentages, or probabilities.

Section 14.1 deals with the estimation of proportions; Section 14.2 deals with tests concerning proportions; Sections 14.3 and 14.4 deal with tests concerning two or more proportions; in Section 14.5 we shall learn how to analyze data tallied into two-way classifications; and in Section 14.6 we shall learn how to judge whether differences between the frequencies of an observed distribution and corresponding expectations can be attributed to chance.

The Estimation of Proportions

The information that is usually available for the estimation of a true proportion (percentage, or probability) is a **sample proportion** $\frac{x}{n}$, where x is the number of times that an event has occurred in n trials. For instance, if in a random sample of 800 purchases at a department store 424 were charged to credit cards, then $\frac{x}{n} = \frac{424}{800} = 0.53$, and we can use this figure as a point estimate of the true proportion of purchases made at this store that are charged to credit cards. Similarly, a large finance company might estimate the proportion of its debtors who are at least one installment behind as 0.06, if a random sample of 400 accounts included 24 which were at least one installment behind.

Throughout this section it will be assumed that the situations satisfy (at least approximately) the conditions underlying the binomial distribution; that is, our information will consist of the number of successes observed in a given number of independent trials, and it will be assumed that for each trial the probability of a success—the parameter we want to estimate—has the constant value p. Thus, the sampling distribution of the counts our methods will be based on is the binomial distribution with the mean $\mu = np$ and the standard deviation $\sigma = \sqrt{np(1-p)}$, and we know that this distribution can be approximated with a normal distribution when n is large.† It follows that, for large values of n, the statistic

$$z = \frac{x - np}{\sqrt{np(1-p)}}$$

has approximately the standard normal distribution. If we substitute this expression for z into the double inequality $-z_{\alpha/2} < z < z_{\alpha/2}$ (as on page 305) and use some simple algebra, we arrive at

$$\frac{x}{n} - z_{\alpha/2}\sqrt{\frac{p(1-p)}{n}} < p < \frac{x}{n} + z_{\alpha/2}\sqrt{\frac{p(1-p)}{n}}$$

---

†On page 270 we said that the normal approximation to the binomial distribution may be used when np and $n(1-p)$ are both greater than 5. So, when $n = 50$, for example, the normal-curve methods discussed here and later in this chapter may be used if p lies between 0.10 and 0.90; when $n = 100$ they may be used if p lies between 0.05 and 0.95; and when $n = 200$ they may be used if p lies between 0.025 and 0.975. This illustrates what we mean here by "n being large."

This may look like a confidence-interval formula for p and, indeed, the inequalities will be satisfied with probability $1 - \alpha$, but it cannot be used in this form because the unknown parameter p itself appears in $\sqrt{\dfrac{p(1-p)}{n}}$ to the left of the first inequality sign and to the right of the other. The quantity $\sqrt{\dfrac{p(1-p)}{n}}$ is called the **standard error of a proportion**, as it is, in fact, the standard deviation of the sampling distribution of a sample proportion (see Exercise 14.14 on page 363). To get around this difficulty, we substitute the sample proportion $\dfrac{x}{n}$ for p in $\sqrt{\dfrac{p(1-p)}{n}}$, and we thus arrive at the following $(1-\alpha)100\%$ large-sample confidence interval for p:

Large-sample confidence interval for p

$$\frac{x}{n} - z_{\alpha/2}\sqrt{\frac{\frac{x}{n}\left(1 - \frac{x}{n}\right)}{n}} < p < \frac{x}{n} + z_{\alpha/2}\sqrt{\frac{\frac{x}{n}\left(1 - \frac{x}{n}\right)}{n}}$$

EXAMPLE In a random sample of 400 cars stopped at a roadblock, 152 of the drivers were wearing their seat belts. Construct a 95% confidence interval for the corresponding true proportion in the population sampled.

SOLUTION Substituting $n = 400$, $\dfrac{x}{n} = \dfrac{152}{400} = 0.38$, and $z_{0.025} = 1.96$ into the large-sample confidence interval formula, we get

$$0.38 - 1.96\sqrt{\frac{(0.38)(0.62)}{400}} < p < 0.38 + 1.96\sqrt{\frac{(0.38)(0.62)}{400}}$$

$$0.33 < p < 0.43$$

Clearly, this interval either contains the true proportion p or it does not, and we really don't know which. However, the 95% confidence implies that the interval was constructed by a method which leads to correct results 95 percent of the time.

The theory presented here can also be used to judge the possible size of the error we may make when we use a sample proportion as a point estimate of a population proportion p. In this case, we can assert with $(1 - \alpha)100\%$ confidence that the size of our error is less than

Maximum error of estimate

$$E = z_{\alpha/2}\sqrt{\frac{p(1-p)}{n}} \quad \text{or approximately} \quad E = z_{\alpha/2}\sqrt{\frac{\frac{x}{n}\left(1 - \frac{x}{n}\right)}{n}}$$

The first of these two formulas cannot be used in practice since p is the quantity we are trying to estimate, but the second formula can be used provided that n is large enough to justify the normal-curve approximation to the binomial distribution.

EXAMPLE   In a random sample of 200 vacationers interviewed at a resort, 142 said that they chose the resort mainly because of its climate. With 99% confidence, what can we say about the maximum size of our error, if we use $\dfrac{x}{n} = \dfrac{142}{200} = 0.71$ as an estimate of the true proportion of vacationers who choose the resort mainly because of its climate?

SOLUTION   Substituting $n = 200$, $\dfrac{x}{n} = 0.71$, and $z_{0.005} = 2.575$ into the formula for E, we get

$$E = 2.575 \sqrt{\frac{(0.71)(0.29)}{200}} = 0.08$$

rounded to two decimals.

As in the estimation of means, we can use the expression for the maximum error to determine how large a sample is needed to attain a desired degree of precision. If we want to assert with probability $1 - \alpha$ that a sample proportion will differ from the true proportion p by less than some quantity E, we can solve the equation

$$E = z_{\alpha/2} \sqrt{\frac{p(1 - p)}{n}}$$

for n and get

Sample size

$$n = p(1 - p) \left[\frac{z_{\alpha/2}}{E} \right]^2$$

Since this formula involves p, it cannot be used unless we have some information about the possible values that p might assume. Without such information, we make use of the fact that $p(1 - p)$ equals $\frac{1}{4}$ when $p = \frac{1}{2}$ and is smaller than $\frac{1}{4}$ for all other values of p. Hence, if we use the formula

Sample size

$$n = \frac{1}{4} \left[\frac{z_{\alpha/2}}{E} \right]^2$$

our sample size may be larger than necessary, but we can account for this by asserting with a probability of *at least* $1 - \alpha$ that the error in our estimate will be less than E. In case we do have some information about the possible range of values p might assume in a given problem, we can take this into account in determining n. We substitute for p in the first of the two sample size formulas above whichever value within that range is closest to $\frac{1}{2}$.

EXAMPLE   Suppose that we want to estimate what proportion of gift items purchased at a department store are returned for a refund, and that we want to be "at least 95% sure" that the error of our estimate will be less than 0.05. How large a sample will we need if
(a) we have no idea what the true proportion might be;
(b) we know that the true proportion is between 0.01 and 0.20?

SOLUTION   (a) Substituting $E = 0.05$ and $z_{0.025} = 1.96$ into the second of the two formulas for n, we get

$$n = \frac{1}{4}\left(\frac{1.96}{0.05}\right)^2 = 385$$

rounded up to the nearest integer; (b) substituting these same values together with $p = 0.20$ into the first formula, we get

$$n = (0.20)(0.80)\left(\frac{1.96}{0.05}\right)^2 = 246$$

rounded up to the nearest integer.

The methods which we have discussed in this section are all large-sample techniques. For small samples, confidence intervals for proportions can be found in special tables, such as the ones in earlier editions of this text.

EXERCISES

14.1 In a (random) sample survey, 96 of 200 persons interviewed in a large city said that they oppose the construction of any new freeways. Construct a 95% confidence interval for the corresponding population proportion.

14.2 In a (random) sample survey, 300 persons with annual incomes of at least $15,000 were asked "Where would you be most likely to find out all you want to know about some news in which you are very much interested?" If 168 replied "television," construct a 99% confidence interval for the true proportion of persons in this income group who would respond in this way to the question.

14.3 In the sample survey of Exercise 14.2, another 100 persons in the same income group were asked where they are most likely to find advertising that can be trusted. If 43 replied "in newspapers," construct a 95% confidence interval for the corresponding population proportion.

14.4 In a study conducted by a trading-stamp company, it was found that among 800 randomly selected male shoppers who were deliberately not offered trading stamps, 664 asked for them. Construct a 99% confidence interval for the actual percentage of male shoppers in the population sampled who would ask for trading stamps.

14.5 In a random sample of 400 clerical employees in a large multinational corporation, 284 said they would like for the company to adopt a flexible scheduling policy, which would permit employees to decide when and how much they would work. Construct a 95% confidence interval for the corresponding true proportion for the population sampled.

14.6 In a random sample of 450 women, 369 of them said that, as a result of recent shortages and rising prices, their families were "making the home more the object and the center of their activities." What can we assert with 99% confidence about the maximum size of our error, if we use the sample proportion, $\frac{369}{450} = 0.82$, as an estimate of the actual population proportion of women who feel this way?

14.7 In a random sample of 160 persons who represented themselves in federal court bankruptcy proceedings but received help from cut-rate "do-it-yourself" services, it was found that 32 persons either lost money or assets that would otherwise not have been lost if they had been given competent legal advice. Construct a 95% confidence interval for the proportion of persons who suffer such losses in the population sampled.

14.8 In a study of consumer buying habits, 88 women in a random sample of 400 women, drawn from a large population, reported that they "automatically" buy (when available) the economy size of whatever they need. What can we assert with 98% confidence about the maximum size of our error, if we use the sample proportion, $\frac{88}{400} = 0.22$, as an estimate of the corresponding proportion in the population sampled?

14.9 In a random sample of 240 family doctors, 156 felt that there should be a ban on the sales of cigarettes in hospitals and nursing homes. If $\frac{156}{240} = 0.65$ is used as an estimate of the corresponding true proportion in the population sampled, what can one assert with 80% confidence about the maximum size of the error?

14.10 A life insurance company, considering a "nonsmoker plan" (with reduced rates for persons who do not smoke), wants to estimate what proportion of its many policyholders would qualify for the plan. How large a sample will it have to take from its files to be at least 95% confident that the sample proportion and the true proportion will differ by less than 0.02?

14.11 A large finance company wants to estimate from a sample of its thousands of accounts what percentage of all its customers plan to buy either an automobile, furniture, or major household appliance on credit during the coming year. How large a sample will be needed so that the company can be at least 99% confident that the sample percentage will be in error by less than 4 percent?

14.12 With reference to the preceding exercise, suppose the finance company has reason to believe that the actual percentage is somewhere between 60 and 70 percent. How large a sample will the company need to be at least 99% confident that the sample percentage will be in error by less than 4 percent?

14.13 In a study of advertising campaigns, a national manufacturer wants to determine what proportion of shirts purchased for wear by men are actually purchased by women. How large a sample will the manufacturer need to be at least 98% confident that the sample proportion will be within 0.03 of the population proportion?

14.14 Since the proportion of successes is simply the number of successes divided by n, the mean and the standard deviation of the sampling distribution of the proportion of successes may be obtained by dividing the mean and the standard deviation of the sampling distribution of the number of successes by n. Use this argument to verify the standard error formula given on page 359.

14.15 If a sample constitutes at least 5 percent of a population, and the sample itself is large, we can use the finite population correction factor to reduce the width of confidence intervals for p. If we make this correction, the large-sample confidence limits for p become

$$\frac{x}{n} \pm z_{\alpha/2} \sqrt{\frac{\frac{x}{n}\left(1 - \frac{x}{n}\right)}{n}} \cdot \sqrt{\frac{N - n}{N - 1}}$$

where N is, as before, the size of the population sampled.

(a) Among the 400 families living in a large apartment complex a random sample of 200 is interviewed, and it is found that 68 of these have children of college age. Construct a 95% confidence interval for the actual proportion of all families living in the complex who have children of college age.

(b) A hosiery manufacturer feels that unless its employees agree to a 10 percent wage reduction it cannot stay in business. If in a random sample of 80 of the 400 employees only 24 felt "kindly disposed" to the reduction, find a 99% confidence interval for the corresponding proportion for all the employees.

14.2

Tests Concerning Proportions

In this section we shall be concerned with tests of hypotheses which enable us to decide, on the basis of sample data, whether the true value of a proportion (percentage, or probability) equals, is greater than, or is less than a given constant. They will make it possible, for example, to determine whether the true proportion of shoppers who can identify a highly advertised trade mark is 0.40, whether it is true that 10 percent of the shirts "cleaned" by a certain laundry are rejected at final inspection because of inferior work, or whether the true probability is 0.70 that a person plans to buy his next car from the same dealer who sold him his last car.

Questions of this kind are usually decided on the basis of either the observed number or the proportion of successes in what are assumed to be n independent trials, each of which has the same probability p of success. In other words, we shall assume that we can use the binomial distribution and that we are, in fact, testing hypotheses about its parameter p.

When n is small, tests concerning true proportions can be based directly on tables of binomial probabilities such as Table V at the end of the book. Suppose, for instance, that we want to investigate the claim that at least 60 percent of the juniors and seniors attending a large university prefer to live off campus. Specifically, we want to test the null hypothesis $p = 0.60$ against the alternative hypothesis $p < 0.60$ on the basis of a random sample of size $n = 14$ and at the 0.05 level of significance.

In all the test criteria of Chapters 12 and 13, where we studied only continuous sampling distributions, we drew the dividing lines so that the probabilities associated with the "tails" (regions of rejection) were exactly α or $\alpha/2$. Since this cannot always be done when we deal with binomial distributions, we modify the criteria as follows. We draw the dividing lines so that the probability of getting a value in the "tail" is as close as possible to the level of significance α (or to $\alpha/2$ in a two-tailed test) without exceeding it. Thus, for our example we observe (from Table V) that for $p = 0.60$ and $n = 14$ the probability of getting at most 4 successes is

$$0.001 + 0.003 + 0.014 = 0.018$$

and the probability of getting at most 5 successes is

$$0.001 + 0.003 + 0.014 + 0.041 = 0.059$$

Since the first of these probabilities is less than 0.05 and the second probability exceeds it, we find that we must reject the null hypothesis $p = 0.60$ (and accept the alternative hypothesis $p < 0.60$) when in a random sample of $n = 14$ juniors or seniors attending the given university there are at most 4 who prefer to live off campus (see also Figure 14.1).

14.1

Binomial distribution with $p = 0.60$ and $n = 14$.

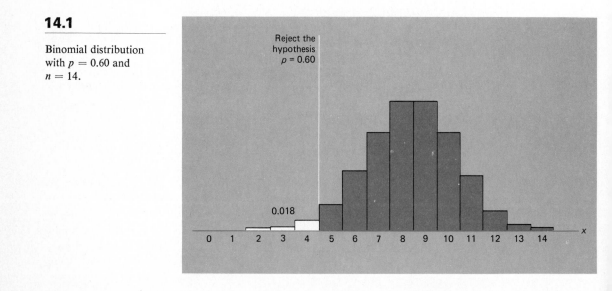

The example which follows illustrates the use of a two-tailed criterion in a test concerning a proportion.

EXAMPLE It has been claimed that 40 percent of all shoppers can identify a highly advertised trade mark. If only two shoppers can identify the trade mark in a random sample of $n = 12$, test whether the null hypothesis $p = 0.40$ can be rejected against the alternative hypothesis $p \neq 0.40$ at the 0.05 level of significance.

SOLUTION Table V shows that the probability of getting 0, 1, or 2 successes is

$$0.002 + 0.017 + 0.064 = 0.083$$

and since this value exceeds $\alpha/2 = 0.025$, we find that the null hypothesis cannot be rejected. As the reader will be asked to verify in Exercise 14.16 on page 367, the whole test criterion for this example is as shown in Figure 14.2.

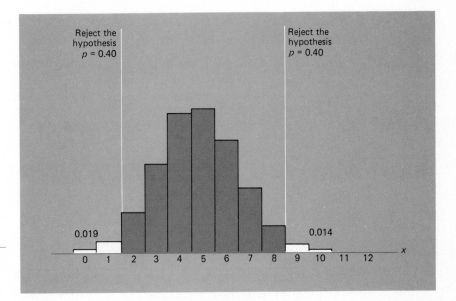

14.2

Binomial distribution with $p = 0.40$ and $n = 12$.

When n is large, tests concerning true proportions (percentages, or probabilities) are usually based on the normal-curve approximation to the binomial distribution. Using again the z statistic which led to the large-sample confidence interval for p, we base tests of the null hypothesis $p = p_0$ on the values of

Statistic for large-sample test concerning proportion

$$z = \frac{x - np_0}{\sqrt{np_0(1 - p_0)}}$$

which has approximately the standard normal distribution.† The actual test criteria are again those of Figure 12.8 on page 328: for the one-sided alternative $p < p_0$ we reject the null hypothesis when $z < -z_\alpha$; for the one-sided alternative $p > p_0$ we reject the null hypothesis when $z > z_\alpha$; and for the two-sided alternative $p \neq p_0$ we reject the null hypothesis when $z < -z_{\alpha/2}$ or $z > z_{\alpha/2}$. As before, α is the level of significance.

<hr>

EXAMPLE

It has been claimed that at least 30 percent of all families moving away from California move to Arizona. If a random sample of the records of several large van lines shows that the belongings of 153 of 600 families moving away from California were shipped to Arizona, test the null hypothesis $p = 0.30$ against the alternative hypothesis $p < 0.30$ at the 0.01 level of significance.

SOLUTION

Substituting $x = 153$, $n = 600$, and $p_0 = 0.30$ into the foregoing formula for z, we get

$$z = \frac{153 - 600(0.30)}{\sqrt{600(0.30)(0.70)}} = -2.41$$

Since this is less than $-z_{0.01} = -2.33$, we find that the null hypothesis must be rejected. In other words, the evidence contradicts the claim and we conclude that less than 30 percent of all families moving away from California move to Arizona.

<hr>

EXAMPLE

A large national brokerage firm claims that 80 percent of its customers who sell stock to establish tax losses immediately reinvest the proceeds in other stocks. If a random sample of 320 sales (chosen from a great number of sales known to have been made to establish losses) includes 245 in which the proceeds were immediately reinvested in other stocks, test the null hypothesis $p = 0.80$ against the alternative hypothesis $p \neq 0.80$ at the level of significance $\alpha = 0.05$.

SOLUTION

Substituting $x = 245$, $n = 320$, and $p_0 = 0.80$ into the formula for z, we get

$$z = \frac{245 - 320(0.80)}{\sqrt{320(0.80)(0.20)}} = -1.54$$

Since this is not less than $-z_{0.025} = -1.96$, we find that the null hypothesis cannot be rejected.

<hr>

†Many statisticians make a continuity correction here by substituting $x - \frac{1}{2}$ or $x + \frac{1}{2}$ for x in the formula for z, whichever makes z smaller. However, when n is large the effect of this correction is usually negligible.

EXERCISES

14.16 With reference to the example on page 365, verify that the null hypothesis must be rejected for $x = 0, 1, 9, 10, 11,$ or 12, where x is the number of shoppers who can identify the trade mark.

14.17 Suppose that we want to test the null hypothesis $p = 0.30$ against the alternative hypothesis $p > 0.30$ at the 0.01 level of significance. How many successes must be observed in 15 trials for the null hypothesis to be rejected?

14.18 Suppose that we want to decide on the basis of 15 flips of a coin whether it may be regarded as fair. How many heads would we have to get to be able to reject the null hypothesis that the probability of heads is 0.50 at the level of significance $\alpha = 0.05$?

14.19 A doctor claims that at most 5 percent of all persons exposed to a certain amount of radiation will suffer any ill effects. If 13 persons are exposed to this much radiation, how many of them must suffer ill effects from the exposure before the null hypothesis $p = 0.05$ can be rejected at the level of significance $\alpha = 0.05$?

14.20 A television critic claims that 70 percent of all viewers find the noise level of a certain commercial objectionable. If 4 of 10 persons shown this commercial object to the noise level, can we reject the null hypothesis $p = 0.70$ against the alternative hypothesis $p \neq 0.70$ at the 0.05 level of significance?

14.21 The manufacturer of a spot remover claims that his product removes at least 80 percent of all spots. In a random sample of 12 spots, at least how many will this product have to remove so that the null hypothesis $p = 0.80$ cannot be rejected at the level of significance $\alpha = 0.01$?

14.22 A fund-raising organization claims that it gets an 8 percent response to its mail solicitations for contributions to charitable organizations. Test this claim, at the 0.05 level of significance, against the alternative that this figure is too high, if only 137 responses are received to 2,000 letters sent out to raise funds for a certain charity.

14.23 In a random sample of 300 industrial accidents, it was found that 183 were due at least in part to unsafe working conditions. Use the level of significance $\alpha = 0.05$ to decide whether this supports the claim that 65 percent of such accidents are due at least in part to unsafe working conditions.

14.24 A food processor wants to know whether the probability is really 0.65 that a customer will prefer a new kind of packaging to the old kind. What can he conclude at the level of significance $\alpha = 0.05$ if only 44 of 80 randomly selected customers prefer the new kind of packaging and
(a) the alternative hypothesis is $p \neq 0.65$;
(b) the alternative hypothesis is $p < 0.65$?

14.25 In an energy-use study, a random sample of 150 households drawn from an area of northern New England showed 102 households heating at least partly with wood. Test, at the 0.01 level of significance, the null hypothesis that the corresponding population proportion is $p = 0.55$ against the alternative hypothesis $p > 0.55$.

14.26 In a random sample of 800 retired persons, 627 stated that they preferred living in an apartment to living in a one-family home. Test the null hypothesis that the true proportion of retired persons in the population sampled who feel this way is 0.75 against the alternative that this figure is incorrect, using

(a) the level of significance $\alpha = 0.05$.

(b) the level of significance $\alpha = 0.01$.

14.27 In the construction of tables of random numbers (see discussion on page 279), there are various ways of detecting possible departures from randomness. For instance, there should be about as many even digits (0, 2, 4, 6, or 8) as there are odd digits (1, 3, 5, 7, or 9). Count the number of even digits among the 350 digits constituting the first 10 rows of the table on page 582, and test at the level of significance $\alpha = 0.05$ whether, on the basis of this criterion, we should be concerned about the possibility that these random numbers are, in fact, not random?

14.28 In order to control the proportion of defectives or other characteristics (attributes) of mass-produced items, quality-control engineers take random samples of size n at regular intervals of time and plot the sample proportions on a control chart such as that of Figure 14.3. If the production process is considered

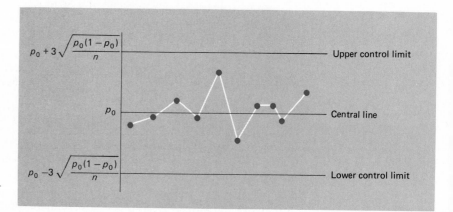

14.3

Control chart.

to be in control when the true proportion of defectives is p_0, the **central line** of the control chart for the proportion of defectives is at p_0, and the **3-sigma upper and lower control limits** are at

$$p_0 + 3\sqrt{\frac{p_0(1 - p_0)}{n}} \quad \text{and} \quad p_0 - 3\sqrt{\frac{p_0(1 - p_0)}{n}}$$

Now, a process is assumed to be in control as long as the sample proportions, plotted on the control chart, remain between the two control limits.

(a) Construct a control chart for the proportion of defectives obtained in repeated random samples of size 100 from a process which is considered to be in control when $p = 0.20$.

(b) Given that 25 consecutive samples of size 100 contained 21, 16, 28, 24, 19, 22, 20, 12, 17, 22, 13, 23, 19, 20, 21, 17, 23, 25, 14, 18, 22, 17, 25, 19, and 24 defectives, plot the sample proportions on the control chart constructed in part (a) and comment on the performance of the process.

14.29 Repeated random samples of size 150 are taken from a production process which is considered to be in control when $p = 0.10$.

    (a) Construct a control chart (see Exercise 14.28) for the proportion of defectives observed in these samples.

    (b) Given that 20 consecutive samples of size 150 contained 19, 9, 21, 13, 18, 22, 25, 14, 8, 20, 27, 29, 24, 18, 20, 12, 15, 16, 9, and 11 defectives, plot the sample proportions on the control chart constructed in part (a) and comment on the performance of the process.

14.3

Differences Between Proportions

There are many problems in which we must decide whether the observed difference between two sample proportions can be attributed to chance, or whether it is indicative of the fact that the two corresponding population proportions are unequal. For instance, we may want to decide on the basis of sample data whether one kind of mail-order solicitation will actually yield a greater response than another, or we may want to test on the basis of samples whether two manufacturers of electronic equipment ship equal proportions of defectives.

The method we shall use to test whether an observed difference between two sample proportions can be attributed to chance, or whether it is statistically significant, is based on the following theory: If x_1 and x_2 are the numbers of successes obtained in n_1 trials of one kind and n_2 of another, the trials are all independent, and the corresponding probabilities of a success are p_1 and p_2, then the sampling distribution of $\dfrac{x_1}{n_1} - \dfrac{x_2}{n_2}$ has the mean $p_1 - p_2$ and the standard deviation, called the **standard error of the difference between two proportions**,

$$\sqrt{\frac{p_1(1 - p_1)}{n_1} + \frac{p_2(1 - p_2)}{n_2}}$$

When we test the null hypothesis $p_1 = p_2 \ (= p)$ against an appropriate alternative hypothesis, the mean of the sampling distribution of the difference between two proportions is $p_1 - p_2 = 0$, and the standard error can be written

$$\sqrt{p(1 - p)\left(\frac{1}{n_1} + \frac{1}{n_2}\right)}$$

where p is usually estimated by **pooling** the data; that is, by substituting for it the combined sample proportion $\dfrac{x_1 + x_2}{n_1 + n_2}$. Then, since for large samples the sampling distribution of the difference between two proportions can be approximated closely with a normal distribution, we base the test on the statistic

Statistic for test concerning difference between two proportions

$$z = \frac{\dfrac{x_1}{n_1} - \dfrac{x_2}{n_2}}{\sqrt{p(1-p)\left(\dfrac{1}{n_1} + \dfrac{1}{n_2}\right)}} \quad \text{with} \quad p = \frac{x_1 + x_2}{n_1 + n_2}$$

which has approximately the standard normal distribution. The test criteria are again those of Figure 12.8 on page 328 with $p_1 - p_2$ substituted for μ and 0 substituted for μ_0. For the one-sided alternative $p_1 < p_2$ we reject the null hypothesis if $z < -z_\alpha$, for the one-sided alternative $p_1 > p_2$ we reject the null hypothesis if $z > z_\alpha$, and for the two-sided alternative $p_1 \neq p_2$ we reject the null hypothesis if $z < -z_{\alpha/2}$ or $z > z_{\alpha/2}$.

EXAMPLE One production process yielded 28 defective pieces in a random sample of size 400 while another yielded 15 defective pieces in a random sample of size 300. Test the null hypothesis $p_1 = p_2$ (that the two processes yield equal proportions of defectives) against the alternative hypothesis $p_1 \neq p_2$ at the 0.05 level of significance.

SOLUTION Substituting $x_1 = 28$, $n_1 = 400$, $x_2 = 15$, $n_2 = 300$, and $\dfrac{28 + 15}{400 + 300} = 0.061$ for p into the formula for z above, we get

$$z = \frac{\dfrac{28}{400} - \dfrac{15}{300}}{\sqrt{(0.061)(0.939)\left(\dfrac{1}{400} + \dfrac{1}{300}\right)}} = 1.10$$

Since this value falls between $-z_{0.025} = -1.96$ and $z_{0.025} = 1.96$, we find that the null hypothesis cannot be rejected. In other words, we cannot conclude that there is a real difference between the true proportions of defectives.

14.4

Differences Among k Proportions

There are also many problems in which we must decide whether observed differences among more than two sample proportions can be attributed to chance, or whether they are indicative of the fact that the corresponding population proportions are not all equal. For instance, if 26 of 200 brand A tires, 21 of 200 brand B tires, 17 of 200 brand C tires, and 34 of 200 brand D tires failed to last 30,000 miles, we may want to decide whether the differences among

$\frac{26}{200} = 0.13$, $\frac{21}{200} = 0.105$, $\frac{17}{200} = 0.085$, and $\frac{34}{200} = 0.17$ are significant, or whether they may be due to chance.

To illustrate the method we use to analyze this kind of data, suppose that a survey in which independent random samples of 100 men, 130 women, and 90 children were asked whether or not they like the flavor of a new toothpaste, yielded the results shown in the following table:

	Men	Women	Children
Like the flavor	60	67	49
Do not like the flavor	40	63	41
Total	100	130	90

The proportions of persons who like the flavor are $\frac{60}{100} = 0.60$, $\frac{67}{130} = 0.52$, and $\frac{49}{90} = 0.54$ for the three groups, and we want to decide at the 0.05 level of significance whether the differences among them can be attributed to chance.

If we let $p_1, p_2,$ and p_3 denote the true proportions of men, women, and children who like the flavor in the populations sampled, the null hypothesis we want to test is $p_1 = p_2 = p_3$ and the alternative hypothesis is that $p_1, p_2,$ and p_3 are not all equal. If the null hypothesis is true, the three samples come from populations having a common proportion p, and we can combine the three samples and look on them as one sample from one population. Also, we can pool the data, as in the preceding section, and estimate the common proportion of persons who like the flavor to be

$$\frac{60 + 67 + 49}{100 + 130 + 90} = 0.55$$

With this estimate we would expect $100(0.55) = 55$ of the men, $130(0.55) = 71.5$ of the women, and $90(0.55) = 49.5$ of the children to like the flavor of the toothpaste. Subtracting these figures from the totals of their samples, we find that $100 - 55 = 45$ of the men, $130 - 71.5 = 58.5$ of the women, and $90 - 49.5 = 40.5$ of the children would be expected not to like the flavor. These results are summarized in the following table, where the **expected frequencies** are shown in parentheses below the **observed frequencies**:

	Men	Women	Children
Like the flavor	60 (55)	67 (71.5)	49 (49.5)
Do not like the flavor	40 (45)	63 (58.5)	41 (40.5)

To test the null hypothesis that the p's are all equal in problems like this, we compare the frequencies which were actually observed with the frequencies we would expect if the null hypothesis were true. It stands to reason that the null hypothesis should be accepted if the discrepancies between the observed and the expected frequencies are small. On the other hand, if the discrepancies between the two sets of frequencies are large, the observed frequencies depart substantially from what we would expect to observe, and this suggests that the null hypothesis must be false.

Using the letter o for the observed frequencies and the letter e for the expected frequencies, we base their comparison on the following χ^2 (chi-square) statistic:

Statistic for test concerning differences among proportions

$$\chi^2 = \sum \frac{(o - e)^2}{e}$$

In words, χ^2 is the sum of the quantities obtained by dividing $(o - e)^2$ by e separately for each **cell** of the table, and for our example we get

$$\chi^2 = \frac{(60 - 55)^2}{55} + \frac{(67 - 71.5)^2}{71.5} + \frac{(49 - 49.5)^2}{49.5} + \frac{(40 - 45)^2}{45}$$

$$+ \frac{(63 - 58.5)^2}{58.5} + \frac{(41 - 40.5)^2}{40.5}$$

$$= 1.65$$

It remains to be seen whether this value is large enough to reject the null hypothesis $p_1 = p_2 = p_3$.

If the null hypothesis that the p's are all equal is true, the sampling distribution of the χ^2 statistic is approximately the chi-square distribution. Since the null hypothesis will be rejected only when the value obtained for χ^2 is too large to be accounted for by chance, we base our decision on the criterion shown in Figure 14.4, where χ_α^2 is such that the area under the chi-square distribution to its right equals α. The parameter of the chi-square distribution, the **number of degrees of freedom**, equals $k - 1$ when we compare k sample proportions. Intuitively, we can justify this formula with the argument that once we have calculated $k - 1$ of the expected frequencies in either row of the table, all of the other expected frequencies can be obtained by subtraction from the totals of the rows and columns (see Exercise 14.41 and also the discussion on page 378).

Returning to our illustration, we find that $\chi^2 = 1.65$ does not exceed 5.991, the value of $\chi_{0.05}^2$ for $3 - 1 = 2$ degrees of freedom. Consequently, we either reserve judgment or we accept the hypothesis that equal proportions of men,

14.4

Test criterion.

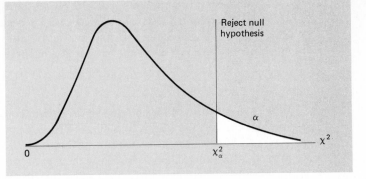

women, and children like the flavor of the new toothpaste. If there are, in fact, differences among the true proportions for the three groups, we have no evidence of it at the 0.05 level of significance.

In general, if we want to compare k sample proportions, we first combine the data and get the following estimate of p:

Estimate of common population proportion

$$\frac{x_1 + x_2 + \cdots + x_k}{n_1 + n_2 + \cdots + n_k}$$

where the n's are the sample sizes, and the x's the numbers of successes, in the k samples. We then multiply the n's by this estimate of p to get the expected frequencies for the first row of the table; after that we subtract these values from the totals of the corresponding samples to get the expected frequencies for the second row of the table. We can also get the expected frequency for any one of the cells by multiplying the total of the column to which it belongs by the total of the row to which it belongs, and dividing by the **grand total**, $n_1 + n_2 + \cdots + n_k$, for the entire table (see also page 377). Next, we calculate χ^2 as just defined, with $\frac{(o - e)^2}{e}$ determined separately for each of the $2k$ cells of the table, and reject the null hypothesis $p_1 = p_2 = \cdots = p_k$ if this value of χ^2 exceeds χ^2_α for $k - 1$ degrees of freedom.

When we calculate the expected frequencies, we usually round them to the nearest integer or to one decimal. The entries in Table III are given to three decimals, but there is seldom any need to carry more than two decimals in calculating the value of the χ^2 statistic, itself. Also, the test we have been discussing is an approximate test which should not be used when one (or more) of the expected frequencies is less than 5. If this is the case, we can sometimes combine some of the samples in such a way that none of the e's is less than 5.

It is of interest to note that for $k = 2$ the χ^2 statistic of this section actually equals the square of the z statistic of Section 14.3 (see Exercises 14.34 and 14.36 below). Thus, for $k = 2$ the two tests are equivalent so long as the alternative hypothesis is $p_1 \neq p_2$; when the alternative hypothesis is $p_1 < p_2$ or $p_1 > p_2$ the method of this section cannot be used.

EXERCISES

14.30 If one method of seeding clouds was successful in 54 of 150 attempts, while another method was successful in 31 of 100 attempts, can we conclude at the 0.05 level of significance that the first method is better than the second?

14.31 One mail solicitation for a charity brought 412 responses to 5,000 letters and another, more expensive, mail solicitation brought 311 responses to 3,000 letters. Use the level of significance $\alpha = 0.01$ to test the null hypothesis that the two solicitations are equally effective against the alternative that the more expensive one is more effective.

14.32 In a grape-planting investigation a random sample of 100 cuttings of root stock Saltcreek (which root with difficulty) was totally immersed in water for 1 day prior to planting, and another random sample of 100 cuttings was completely soaked for 5 days prior to planting. At the end of a given time after planting, it was found that 68 of the cuttings soaked 5 days and 62 of the cuttings soaked 1 day had rooted successfully. Is this evidence, at the 0.01 level of significance, that soaking the stock for 5 days increases the proportion of successful rootings?

14.33 The manager of a motel, in trying to decide which of two supposedly equally good cigarette-vending machines to install, tests each machine 500 times, and he finds that machine I fails to work (neither delivers the cigarettes nor returns the money) 26 times and machine II fails to work 12 times. Based on the method of Section 14.3, can he conclude at the 0.05 level of significance that the two machines are not equally good?

14.34 Rework the preceding exercise using the method of Section 14.4, and verify that the value of the χ^2 statistic equals the square of the value of the z statistic.

14.35 In random samples of visitors to a famous tourist attraction, 84 of 250 men and 154 of 250 women bought souvenirs. Use the method of Section 14.3 and the level of significance $\alpha = 0.01$ to test the null hypothesis $p_1 = p_2$ against the alternative hypothesis $p_1 \neq p_2$.

14.36 Rework the preceding exercise using the method of Section 14.4, and verify that the value obtained for the χ^2 statistic equals the square of the value obtained for the z statistic.

14.37 In four random samples, 26 of 200 brand A tires, 21 of 200 brand B tires, 17 of 200 brand C tires, and 34 of 200 brand D tires failed to last 30,000 miles. Using the 0.05 level of significance, test the null hypothesis that there is no difference in the true proportions of these four brands of tires which fail to last 30,000 miles.

14.38 The following table shows the results of a survey in which random samples of the members of three large labor unions were asked whether they are for or against a certain piece of legislation:

	Union X	Union Y	Union Z
For the legislation	88	66	104
Against the legislation	32	34	76

Use the level of significance $\alpha = 0.01$ to test the null hypothesis that there are no differences among the population proportions.

14.39 Among 300 business executives interviewed in each of three major cities, there were 86, 119, and 92 who predicted that the nation's business failure rate would rise in the coming year. At the 0.01 level of significance, are the differences among the corresponding sample proportions significant?

14.40 To determine the attitude of its field personnel toward whether the company should follow its present policy of maintaining profit levels on unit sales or adopt a proposed new high-volume, low-price policy, a national manufacturer took random samples of its salesmen in four broad geographic areas with the following results:

	North	South	East	West
Maintain old policy	41	22	54	53
Adopt new policy	49	38	146	77

Use the level of significance $\alpha = 0.05$ to test the null hypothesis that the true proportions of field men who favor the present policy are the same in the four geographic areas.

14.41 Make use of the fact that the expected numbers of successes for the k samples are obtained by multiplying $\dfrac{x_1 + x_2 + \cdots + x_k}{n_1 + n_2 + \cdots + n_k}$, respectively, by $n_1, n_2, \ldots,$ and n_k, to show that the sum of the expected numbers of successes for the k samples equals the sum of the observed numbers of successes.

14.5

Contingency Tables

The χ^2 statistic plays an important role in many other problems where information is obtained by counting or enumerating, rather than measuring. The method we shall describe here for analyzing such **count data** is an extension of the method of the preceding section, and it applies to two distinct kinds of problems which differ conceptually but are analyzed in the same way.

In the first kind of problem we deal with trials permitting more than two possible outcomes. For instance, in the illustration of the preceding section each person might have been asked whether he or she likes the flavor of the new toothpaste, dislikes it, or is indifferent to it, and this might have resulted in the following table:

	Men	Women	Children
Like the flavor	52	56	45
Indifferent	15	23	11
Do not like the flavor	33	51	34
Total	100	130	90

We refer to this kind of table as a 3×3 table, because it contains 3 rows and 3 columns; more generally, when there are c samples and each trial permits r alternatives, we refer to the resulting table as an $r \times c$ **table** (where $r \times c$ is read "r by c"). Here, as in the example of the preceding section, the column totals (the sizes of the different samples) are fixed. On the other hand, the row totals ($52 + 56 + 45 = 153$, $15 + 23 + 11 = 49$, and $33 + 51 + 34 = 118$) depend on the responses of the persons interviewed, and hence, on chance.

In the second kind of problem, the column totals as well as the row totals are left to chance; in other words, they are contingent on circumstances beyond our control. Suppose, for instance, that we want to investigate whether there really is a relationship between a man's performance in a company training program and his ultimate success in the job. Suppose, furthermore, that a random sample of 400 cases taken from the company's very extensive files of employees who completed the program, yielded the results shown in the following table:

	Performance in training program			
	Below average	Average	Above average	Total
Poor	23	60	29	112
Success in job (Employer's rating) — Average	28	79	60	167
Very good	9	49	63	121
Total	60	188	152	400

This is also a 3×3 table, and it is mainly in connection with problems like this that $r \times c$ tables are referred to as **contingency tables**.

Before we demonstrate how $r \times c$ tables are analyzed, let us examine the null hypotheses we want to test. In the problem dealing with the flavor of the toothpaste, we want to test the null hypothesis that the probabilities of getting a favorable reaction, indifference, or an unfavorable reaction are the same for each group. In other words, we want to test the null hypothesis that a person's reaction to the flavor of the toothpaste is independent of his or her being a man, woman, or child. In the problem directly above we are also concerned with a null hypothesis of independence; specifically, the null hypothesis that employees' success in their jobs is independent of their performance in the training program.

To illustrate the analysis of an $r \times c$ table, let us refer to the second example and begin by calculating the **expected cell frequencies**. If the null hypothesis of independence is true, the probability of randomly selecting an employee whose performance in the program was below average and who is doing poorly in his job is given by the product of the probability that his performance in the program was below average and the probability that he is doing poorly in his job. Using the totals of the first column and the first row to estimate these two probabilities, we get $\frac{23 + 28 + 9}{400} = \frac{60}{400}$ for the probability that an employee (randomly selected from among the employees who completed the training program) performed below average in the program, and $\frac{23 + 60 + 29}{400} = \frac{112}{400}$ for the probability that he is doing poorly in his job. Hence, we estimate the probability of choosing an employee whose performance was below average in the program and who is doing poorly in his job to be $\frac{60}{400} \cdot \frac{112}{400}$, and in a sample of size 400 we would expect to find

$$400 \cdot \frac{60}{400} \cdot \frac{112}{400} = \frac{60 \cdot 112}{400} = 16.8$$

employees who fit this description.

In this last result, $\frac{60 \cdot 112}{400}$ is just the product of the total of the first column and the total of the first row divided by the grand total for the entire table. In general

> The expected frequency for any cell of a contingency table may be obtained by multiplying the total of the column to which it belongs by the total of the row to which it belongs and then dividing by the grand total for the whole table.

With this rule we get an expected frequency of $\frac{188 \cdot 112}{400} = 52.6$ for the second cell of the first row, and $\frac{60 \cdot 167}{400} = 25.0$ and $\frac{188 \cdot 167}{400} = 78.5$ for the first two cells of the second row.

It is not necessary to calculate all the expected cell frequencies in this way, since it can be shown that the sum of the expected frequencies for any row or column must equal the sum of the corresponding observed frequencies. Therefore, we can get some of the expected cell frequencies by subtraction from row and column totals. For instance, for our illustration we get by subtraction

$$112 - 16.8 - 52.6 = 42.6$$

for the third cell of the first row,

$$167 - 25.0 - 78.5 = 63.5$$

for the third cell of the second row, and

$$60 - 16.8 - 25.0 = 18.2$$
$$188 - 52.6 - 78.5 = 56.9$$

and

$$152 - 42.6 - 63.5 = 45.9$$

for the three cells of the third row. These results are summarized in the following table, where the expected frequencies are shown in parentheses below the corresponding observed frequencies:

| | | Performance in training program | | |
		Below average	Average	Above average
	Poor	23 (16.8)	60 (52.6)	29 (42.6)
Success in job (Employer's rating)	Average	28 (25.0)	79 (78.5)	60 (63.5)
	Very good	9 (18.2)	49 (56.9)	63 (45.9)

From here on the work is like that of the preceding section; we calculate the χ^2 statistic according to the formula

Statistic for test of independence in contingency table

$$\chi^2 = \sum \frac{(o - e)^2}{e}$$

with $\frac{(o - e)^2}{e}$ calculated separately for each cell of the table. Then we reject the null hypothesis of independence at the level of significance α if the value obtained for χ^2 exceeds the value of χ^2_α for $(r - 1)(c - 1)$ degrees of freedom, where r is the number of rows and c the number of columns. In our example the number of degrees of freedom is $(3 - 1)(3 - 1) = 4$, and it should be observed that after we had calculated four of the expected cell frequencies, we got all the others by subtraction from the totals of appropriate rows and columns.

Returning to our numerical illustration, we find from the table directly above that

$$\chi^2 = \frac{(23 - 16.8)^2}{16.8} + \frac{(60 - 52.6)^2}{52.6} + \frac{(29 - 42.6)^2}{42.6}$$
$$+ \frac{(28 - 25.0)^2}{25.0} + \frac{(79 - 78.5)^2}{78.5} + \frac{(60 - 63.5)^2}{63.5}$$
$$+ \frac{(9 - 18.2)^2}{18.2} + \frac{(49 - 56.9)^2}{56.9} + \frac{(63 - 45.9)^2}{45.9}$$
$$= 20.34$$

Since this exceeds 13.277, the value of $\chi^2_{0.01}$ for 4 degrees of freedom, we reject the null hypothesis. In other words, we have shown that at the 0.01 level of significance there is a dependence (or relationship) between an employee's performance in the training program and his success in the job.

The method we have used here to analyze the contingency table applies also when the column totals are fixed sample sizes (as in the toothpaste problem) and do not depend on chance. The rule by which we multiply the row total by the column total and then divide by the grand total must be justified in a different way, but this is of no consequence—the expected cell frequencies are determined in exactly the same way (see Exercise 14.48 on page 383).

Since the χ^2 statistic we are using here has only approximately a chi-square distribution, it should not be used in cases where any of the expected cell frequencies are less than 5. When there are expected frequencies smaller than 5, it is often possible to combine some of the cells, subtract 1 degree of freedom for each cell eliminated, and then perform the test just as it has been described.

14.6

Goodness of Fit

In this section we shall treat a further application of the χ^2 criterion, in which we compare an observed frequency distribution with a distribution we might expect according to theory or assumptions. We refer to such a comparison as a test of **goodness of fit**.

To illustrate, let us consider Table XIII, the table of random digits, which is supposed to have been constructed in such a way that each digit is a value of a random variable which takes on the values 0, 1, 2, 3, 4, 5, 6, 7, 8, and 9 with equal probabilities of 0.10. To determine whether it is reasonable to conclude that this is, indeed, the case, we might count how many times each digit appears in the table, or part of the table; specifically, we shall take the 250 digits in the first five columns on page 580. This yields the values shown in the "Observed frequency" column of the following table:

Digit	Probability	Observed frequency o	Expected frequency e
0	0.10	23	25
1	0.10	25	25
2	0.10	20	25
3	0.10	23	25
4	0.10	23	25
5	0.10	22	25
6	0.10	29	25
7	0.10	25	25
8	0.10	33	25
9	0.10	27	25

We got the expected frequencies in the right-hand column by multiplying each of the probabilities of 0.10 by 250, the total number of digits counted.

To test whether the discrepancies between the observed and expected frequencies can be attributed to chance, we use the same chi-square statistic as in the two preceding sections,

Statistic for test of goodness of fit

$$\chi^2 = \sum \frac{(o - e)^2}{e}$$

calculating $\frac{(o - e)^2}{e}$ separately for each class of the distribution. Then, if the value we get for χ^2 exceeds χ^2_α, we reject the null hypothesis on which the expected frequencies are based at the level of significance α; the number of degrees of freedom is $k - m$, where k is the number of terms $\frac{(o - e)^2}{e}$ added in the formula for χ^2, and m is the number of quantities we must obtain from the observed data to calculate the expected frequencies.

For the illustration dealing with the random digits, we get

$$\chi^2 = \frac{(23 - 25)^2}{25} + \frac{(25 - 25)^2}{25} + \frac{(20 - 25)^2}{25} + \frac{(23 - 25)^2}{25}$$
$$+ \frac{(23 - 25)^2}{25} + \frac{(22 - 25)^2}{25} + \frac{(29 - 25)^2}{25} + \frac{(25 - 25)^2}{25}$$
$$+ \frac{(33 - 25)^2}{25} + \frac{(27 - 25)^2}{25}$$

$$= 5.20$$

and since this does not exceed 16.919, the value of $\chi^2_{0.05}$ for $10 - 1 = 9$ degrees of freedom, we find that the null hypothesis cannot be rejected. In other words, the table of random numbers has "passed the test." The number of degrees of freedom is 9, because there are 10 terms in the formula for χ^2, and the only quantity needed from the observed data to calculate the expected frequencies is the total frequency of 250.

The method we have illustrated in this section is used quite generally to test how well distributions we expect (on the basis of theory or assumptions) fit, or describe, observed data. In some of the exercises which follow, we shall test whether it is reasonable to treat an observed distribution as if it had (at least approximately) the shape of a normal distribution, and we shall also test whether given sets of data fit the pattern of binomial and Poisson distributions. As in the tests of the preceding sections, the sampling distribution of the χ^2 statistic is only approximately a chi-square distribution when it is used for tests of goodness of fit. So, if any of the expected frequencies is less than 5, we must again combine some of the data; in this case, we combine adjacent classes of the distribution.

EXERCISES

14.42 Analyze the 3×3 table on page 376 and decide at the 0.05 level of significance whether the differences in the reaction to the flavor of the new toothpaste are significant.

14.43 A sample survey, designed to show where persons living in different parts of the country buy nonprescribed medicines, yielded the following results:

	North-east	North Central	South	West
Drugstores	218	200	183	179
Grocery stores	39	52	87	62
Others	43	48	30	59

Test the null hypothesis that, so far as nonprescribed medicines are concerned, the buying habits of persons living in the given parts of the country are the same. Use the level of significance $\alpha = 0.05$.

14.44 Random samples of the records of the quality control department of a large firm show that vendor A shipped 12 parts that were rejected, 23 parts that were imperfect but acceptable, and 85 parts that were perfect; vendor B shipped 8 parts that were rejected, 14 parts that were imperfect but acceptable, and 68 parts that were perfect; and vendor C shipped 21 parts that were rejected, 30 parts that were imperfect but acceptable, and 119 parts that were perfect. Test at the level of significance $\alpha = 0.01$ whether the three vendors ship parts of equal quality.

14.45 In a study devoted to stockholder attitude toward its community relations programs, a large national manufacturer takes a random sample of 370 of its stockholders, classifies their holdings as either small, medium-sized, or large, and asks each person to rate the company programs as either good, fair, or poor. The following are the results:

	Small	Medium	Large
Good	35	50	20
Fair	60	55	35
Poor	45	55	15

At the 0.01 level of significance, is there a relationship between the size of stockholders' holdings and their attitude toward the company programs?

14.46 In a study to determine whether there is a relationship between bank employees' standard of dress and their professional advancement, a random sample of size $n = 500$ yielded the results shown in the following table:

	Speed of advancement		
	Slow	Average	Fast
Very well dressed	38	135	129
Well dressed	32	68	43
Poorly dressed	13	25	17

Use the 0.05 level of significance to test the null hypothesis that there is no real relationship between standard of dress and speed of professional advancement.

14.47 A market research firm wants to determine, on the basis of the following information, whether there exists a relationship between the size of a tube of toothpaste which a customer buys and the number of persons in the customer's household:

		Number of persons in household			
		1–2	3–4	5–6	7 or more
	Giant	22	107	76	45
Size of tube bought	Large	55	23	16	12
	Small	31	69	37	7

At the level of significance $\alpha = 0.01$, is there a relationship?

14.48 Use an argument similar to that on page 377 to show that the rule for calculating the expected cell frequencies (dividing the product of the column total and the row total by the grand total) applies also when the column totals are fixed sample sizes and do not depend on chance.

14.49 If the analysis of a contingency table shows that there is a relationship between the two variables under consideration, the strength of the relationship can be measured by the **contingency coefficient**

$$C = \sqrt{\frac{\chi^2}{\chi^2 + n}}$$

where n is the total frequency for the table. This coefficient assumes values between 0 (corresponding to independence) and a maximum value of less than 1 depending on the size of the table. For example, for a $k \times k$ table the maximum value of C is $\sqrt{(k-1)/k}$. The larger C is, the stronger the relationship is between the variables.
(a) What is the maximum value of C for a 3×3 table?
(b) Calculate C for a contingency table with $n = 300$ and $\chi^2 = 25.6$.
(c) Calculate C for the data of Exercise 14.47.

14.50 To determine whether a die is balanced, it is rolled 360 times. The following are the results: 1 showed 57 times, 2 showed 46 times, 3 showed 68 times, 4 showed 52 times, 5 showed 72 times, and 6 showed 65 times. At the 0.05 level of significance, can we reject the null hypothesis that the die is balanced?

14.51 Ten years' data show that in one Western city there were no bank robberies in 64 months, one bank robbery in 29 months, two bank robberies in 19 months, and three bank robberies in 8 months. At the level of significance $\alpha = 0.05$, does this substantiate the claim that the probabilities of 0, 1, 2, or 3 bank robberies are 0.40, 0.30, 0.20, and 0.10?

14.52 Each jar of instant coffee in a large production lot is sealed with either a "25¢-off" coupon or a "10¢-off" coupon. Subsequently, a random sample of 120 cartons, each containing 6 jars, is taken from the lot and inspected. The numbers of 25¢-off coupons sealed in the jars of the sample cartons were:

Number of 25¢-off coupons	Number of cartons
0	3
1	17
2	25
3	35
4	24
5	11
6	5

At the 0.05 level of significance, does it appear that the data may be looked upon as values of a random variable having the binomial distribution with $p = 0.50$ and $n = 6$?

14.53 The following is the distribution of the numbers of calls received at the switchboard of a government agency during 600 five-minute intervals:

Number of calls	Frequency
0	34
1	131
2	160
3	136
4	72
5	37
6	22
7	8

Test at the 0.01 level of significance the hypothesis that the underlying distribution from which this sample came is a Poisson distribution with the parameter $\lambda = 2.5$.

14.54 The following is the distribution of the times it took a random sample of 200 persons to complete a screening test required of applicants by a firm specializing in executive placement:

Time (minutes)	Number of persons
24 or less	15
25–29	50
30–34	75
35–39	40
40–44	15
45 or over	5

The mean and the standard deviation of these times, calculated before they were grouped, are $\bar{x} = 32.1$ minutes and $s = 5.6$ minutes.

(a) Find the area under a normal curve with $\mu = 32.1$ and $\sigma = 5.6$ which lies to the left of 24.5, between 24.5 and 29.5, between 29.5 and 34.5, between 34.5 and 39.5, between 39.5 and 44.5, and to the right of 44.5.

(b) Calculate the expected normal-curve frequencies for the six classes of the distribution by multiplying each of the areas found in part (a) by 200.

(c) Test, at the 0.05 level of significance, the null hypothesis that this sample might reasonably have come from a population having approximately the shape of a normal distribution. In making the χ^2 test of the goodness of fit of a normal distribution, the number of degrees of freedom is $k - 3$, where k is the number of classes, and 3 degrees of freedom are lost since the sums of the expected and observed frequencies must agree and the mean and standard deviation of the normal curve had to be estimated from the data.

14.7

Check List of Key Terms

Cell, 372
Chi-square distribution, 372
Chi-square statistic, 372
Contingency coefficient, 383
Contingency table, 376
Count data, 375
Expected cell frequencies, 377
Goodness of fit, 379

Grand total, 379
Observed cell frequencies, 371
Pooling, 369
$r \times c$ table, 376
Sample proportion, 358
Standard error of a proportion, 359
Standard error of difference between two proportions, 369

14.8

Review Exercises

14.55 Five coins are tossed 320 times, and 0, 1, 2, 3, 4, and 5 heads showed 12, 51, 88, 105, 59, and 5 times. At the 0.05 level of significance, is it reasonable to suppose that the coins are balanced?

14.56 A random sample of 200 of a professional football team's season-ticket holders shows that 176 intend to renew their tickets for the next season.
(a) Construct a 95% confidence interval for the true proportion of ticket holders who intend to renew their tickets for the next season.
(b) If we use the sample proportion, 0.88, to estimate this true proportion, what can we say with 99% confidence about the maximum size of our error?

14.57 In studying problems related to its handling of reservations, an airline takes a random sample of 80 of the complaints about reservations filed in each of four cities. If 52 of the complaints from city A, 48 of the complaints from city B, 63 of the complaints from city C, and 57 of the complaints from city D allege overbooking, test at the 0.05 level of significance whether the differences among the corresponding sample proportions are real.

14.58 A social scientist claims that at least 90 percent of all patients have to wait 25 minutes or more in their doctors' offices before being called. If, in a random sample, 10 of 14 patients had to wait this long, what can we conclude about the claim at the 0.05 level of significance?

14.59 In a random sample of 100 persons who skipped breakfast, 46 reported that they experienced midmorning fatigue, and in a random sample of 400 persons who ate breakfast, 114 reported that they experienced midmorning fatigue.

Use the z statistic to test at the 0.01 level of significance whether there is a real difference between the proportions of persons experiencing midmorning fatigue in the two populations sampled.

14.60 Use the χ^2 statistic to rework the preceding exercise, and verify that the value obtained for the χ^2 statistic equals the square of the value obtained for the z statistic.

14.61 The personnel manager of a large department store feels that at least 4 percent of the sales tickets written by part-time employees of the store are in error. What can we conclude about this claim at the level of significance $\alpha = 0.01$ if a random sample of 600 sales tickets written by part-time employees of the store includes 14 which are in error?

14.62 In a random sample of 300 television viewers in a certain area, 114 had seen a certain controversial program. Construct a 99% confidence interval for the actual percentage of television viewers in that area who saw the program.

14.63 An opinion pollster wants to determine what proportion of the voting population is opposed to operating nuclear power plants. How large a sample is required in order to be able to assert with probability 0.95 that the sample proportion will be off by less than 0.025?

14.64 With reference to the preceding exercise, how large a sample would be required if the population proportion is presumed to be at most 0.30?

14.65 The following table shows how samples of 100 residents each of three federally financed housing projects replied to a question of whether they would continue to live there if they had the choice:

	Project 1	Project 2	Project 3
Yes	83	68	65
No	17	32	35

Test at the 0.05 level of significance whether the differences among the corresponding proportions of "Yes" answers are significant.

14.66 In a random sample of 250 retired persons, 192 stated that they would prefer living in an apartment to living in a one-family home. At the 0.01 level of significance, does this refute a claim that at most 60 percent of all retired persons would prefer living in an apartment to living in a one-family home?

14.67 Based on the results of 13 random trials, we want to test the null hypothesis $p = 0.20$ against the alternative hypothesis $p > 0.20$ at the 0.05 level of significance. For what numbers of successes must the null hypothesis be rejected? What is the actual level of significance?

14.68 In an occupational survey a large university took a random sample of 385 male graduates of 10 years earlier who had been working continuously in business since their graduation. Each person was asked to state whether, in his opinion, his career progress had been better than was expected at graduation, about what was expected, or poorer than expected, and each person was ranked as having stood in the bottom third, middle third, or top third of his class. The results are shown in the following table:

	Bottom third	Middle third	Top third
Better than expected	80	45	75
About what was expected	30	20	50
Poorer than expected	30	15	40

At the 0.05 level of significance, is there a relationship between the class standing and the career progress of the men?

14.69 Tests are made on the proportion of defective castings produced by two molds. If in a random sample of 100 castings from mold A there were 13 defectives and in a random sample of 200 castings from mold B there were 35 defectives, can the null hypothesis $p_1 = p_2$ be rejected against the alternative hypothesis $p_1 < p_2$ at the 0.01 level of significance?

14.70 In a random sample of 600 voters interviewed in a large city, only 342 felt that the President was doing a good job. If we use $\frac{342}{600} = 0.57$ as an estimate of the true proportion of voters in this city who feel that the President is doing a good job, what can we say with 99% confidence about the maximum size of our error?

14.71 The following table shows how many times, Monday through Friday, a bus was late arriving at a given stop in 40 weeks:

Number of times bus was late	Number of weeks
0	4
1	11
2	15
3 or more	10

Use the 0.05 level of significance to test the null hypothesis that the bus is late 30 percent of the time, namely, the null hypothesis that the number of times the bus arrives late per week is a random variable having the binomial distribution with $n = 5$ and $p = 0.30$.

14.72 Tests of the fidelity and the selectivity in a random sample of 190 radios produced the results shown in the following table:

		Fidelity		
		Low	Average	High
	Low	7	12	31
Selectivity	Average	35	59	18
	High	15	13	0

At the 0.05 level of significance, can we conclude that there is a relationship between the fidelity and the selectivity of radios?

14.73 In a random sample of 320 car owners who reported car nonstart problems caused, they admitted, by owner neglect, 160 reported that their negligence consisted of leaving their headlights on and running their batteries down.

(a) Verify that $0.445 < p < 0.555$ is a 95% confidence interval for the actual proportion of such owners in the population sampled.

(b) Verify that

$$0.50 - 1.75(0.028) < p < 0.50 + 2.33(0.028)$$

$$0.451 < p < 0.565$$

is an alternative 95% confidence interval for the same parameter.

(c) Explain in what way the (symmetric) confidence interval of part (a) is preferable to that of part (b).

14.74 In a random sample of 100 real estate brokers drawn from among 600 licensed brokers in a certain region, 70 felt that warranty policies (providing breakdown protection on plumbing, heating, electrical, and air-conditioning systems, and on built-in appliances) offered by the sellers of used houses makes such warranted houses easier to sell. Construct a 90% confidence interval for the corresponding proportion of all the licensed brokers in that region who feel that way.

15

Decision Making: Analysis of Variance

In this chapter we shall consider the problem of deciding whether observed differences among more than two sample means can be attributed to chance, or whether there are real differences among the means of the populations sampled. For instance, we may want to decide on the basis of sample data whether there really is a difference in the effectiveness of three methods of teaching managerial accounting, we may want to compare the average monthly sales of several insurance salesmen, we may want to see whether there really is a difference in the average mileage obtained with four kinds of gasoline, we may want to judge whether there really is a difference in the durability of five kinds of carpet, and so on. The method we shall introduce for this purpose is a powerful statistical tool called analysis of variance, or ANOVA for short. A one-way analysis of variance is described in Section 15.2 and a two-way analysis of variance in Section 15.3.

Differences among k Means

Suppose that we want to compare the cholesterol contents of four competing diet foods on the basis of the following data (in milligrams per package) which were obtained for three 6-ounce packages of each of the diet foods:

Diet food A: 3.6, 4.1, 4.0
Diet food B: 3.1, 3.2, 3.9
Diet food C: 3.2, 3.5, 3.5
Diet food D: 3.5, 3.8, 3.8

The means of these four samples are 3.9, 3.4, 3.4, and 3.7, and we want to know whether the differences among them are significant or whether they can be attributed to chance.

In general, in problems like this, if $\mu_1, \mu_2, \ldots,$ and μ_k are the means of k populations from each of which a sample is drawn, we want to test the null hypothesis $\mu_1 = \mu_2 = \cdots = \mu_k$ against the alternative that these means are not all equal.[†] Evidently, this null hypothesis would be supported if the differences among the sample means were small, and the alternative hypothesis would be supported if the differences among the sample means were large. Thus, we need a precise measure of the discrepancies among the $\bar{x}$'s, and with it a rule to follow which tells us when the discrepancies are so large that the null hypothesis should be rejected. An obvious choice of such a measure is the variance of the $\bar{x}$'s, and this is the measure we shall use here. For the diet foods the mean of the four $\bar{x}$'s is

$$\frac{3.9 + 3.4 + 3.4 + 3.7}{4} = 3.6$$

and their variance is

$$s_{\bar{x}}^2 = \frac{(3.9 - 3.6)^2 + (3.4 - 3.6)^2 + (3.4 - 3.6)^2 + (3.7 - 3.6)^2}{4 - 1}$$

$$= 0.06$$

[†] In connection with work later in this chapter, it is desirable to write these means as $\mu_1 = \mu + \alpha_1, \mu_2 = \mu + \alpha_2, \ldots,$ and $\mu_k = \mu + \alpha_k$. Here

$$\mu = \frac{\mu_1 + \mu_2 + \cdots + \mu_k}{k}$$

is called the grand mean, and the α's, whose sum is zero (see Exercise 15.7 on page 399), are called the treatment effects. In this notation, the null hypothesis becomes $\alpha_1 = \alpha_2 = \cdots = \alpha_k = 0$, and the alternative hypothesis is that the α's are not all equal to zero.

where the subscript $\bar{x}$ is used to show that this is the variance of the sample means.

Let us now make two assumptions which are critical to the method of analysis we shall use: It will be assumed that (1) the populations from which we are sampling can be approximated closely with normal distributions, and (2) they all have the same variance σ^2. With these assumptions, we note that, if the null hypothesis $\mu_1 = \mu_2 = \cdots = \mu_k$ is true, we can look upon the k samples as samples from one and the same (normal) population and, hence, upon the variance of their means, $s_{\bar{x}}^2$, as an estimate of $\sigma_{\bar{x}}^2$, the square of the standard error of the mean. Now, since $\sigma_{\bar{x}} = \dfrac{\sigma}{\sqrt{n}}$ for samples from infinite populations, we can look upon $s_{\bar{x}}^2$ as an estimate of $\sigma_{\bar{x}}^2 = \left(\dfrac{\sigma}{\sqrt{n}}\right)^2 = \dfrac{\sigma^2}{n}$ and, therefore, upon $n \cdot s_{\bar{x}}^2$ as an estimate of σ^2. For instance, for our example we have $n \cdot s_{\bar{x}}^2 = 3(0.06) = 0.18$ as an estimate of σ^2, the common variance of the four populations sampled.

If σ^2 were known, we could compare $n \cdot s_{\bar{x}}^2$ with σ^2 and reject the null hypothesis that the population means are all equal if this value is much larger than σ^2. However, in most practical problems σ^2 is not known and we have no choice but to estimate it on the basis of the sample data. Having assumed that the k samples do, in fact, all come from identical populations, we could use any one of their variances, $s_1^2, s_2^2, \ldots,$ or s_k^2, as a second estimate of σ^2, and we can also use their mean. Averaging, or **pooling**, the four sample variances in our example, we get

$$
\begin{aligned}
\frac{s_1^2 + s_2^2 + s_3^2 + s_4^2}{4} = \frac{1}{4} \Bigg[& \frac{(3.6 - 3.9)^2 + (4.1 - 3.9)^2 + (4.0 - 3.9)^2}{3 - 1} \\
& + \frac{(3.1 - 3.4)^2 + (3.2 - 3.4)^2 + (3.9 - 3.4)^2}{3 - 1} \\
& + \frac{(3.2 - 3.4)^2 + (3.5 - 3.4)^2 + (3.5 - 3.4)^2}{3 - 1} \\
& + \frac{(3.5 - 3.7)^2 + (3.8 - 3.7)^2 + (3.8 - 3.7)^2}{3 - 1} \Bigg] \\
= 0.08 &
\end{aligned}
$$

and we now have two estimates of σ^2,

$$
n \cdot s_{\bar{x}}^2 = 0.18 \quad \text{and} \quad \frac{s_1^2 + s_2^2 + s_3^2 + s_4^2}{4} = 0.08
$$

If the first of two such estimates of σ^2 (which is based on the variation among the sample means) is "much" larger than the second estimate (which is based on the variation within the samples and, hence, measures variation due to chance), it stands to reason that the null hypothesis should be rejected. After all, in that case the variation among the sample means would be greater than we would expect it to be if it were due only to chance. To put the comparison of the two estimates of σ^2 on a rigorous basis, we use the statistic

Statistic for test concerning differences among means

$$F = \frac{\text{estimate of } \sigma^2 \text{ based on the variation among the } \bar{x}\text{'s}}{\text{estimate of } \sigma^2 \text{ based on the variation within the samples}}$$

which is appropriately called a **variance ratio**.

If the null hypothesis is true and if the assumptions we made are valid, the sampling distribution of this statistic is the F distribution which we introduced in Chapter 13. Since the null hypothesis will be rejected only when F is large (that is, when the variability of the $\bar{x}$'s is too great to be attributed to chance), we base our decision on the criterion of Figure 15.1. Here F_α is such that the

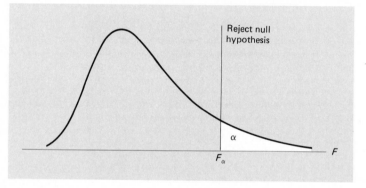

15.1

F distribution.

area under the curve to its right equals the level of significance α. For $\alpha = 0.05$ or 0.01, the critical values of F_α may be looked up in Table IV at the end of the book, and if we compare the means of k random samples of size n, we have $k - 1$ degrees of freedom for the numerator and $k(n - 1)$ degrees of freedom for the denominator.†

Returning to our example, we find that for $k = 4$ and $n = 3$ the numerator and denominator degrees of freedom are $k - 1 = 4 - 1 = 3$ and $k(n - 1) = 4(3 - 1) = 8$, and that $F_{0.05} = 4.07$. Since

$$F = \frac{0.18}{0.08} = 2.25$$

is less than this critical value, we find that the null hypothesis cannot be rejected.

---

†In connection with the numerator degrees of freedom, the numerator is $n \cdot s_{\bar{x}}^2$, and $s_{\bar{x}}^2$ is the variance of k means and, hence, has $k - 1$ degrees of freedom in accordance with the terminology introduced in the footnote to page 307. As for the denominator degrees of freedom, the denominator is the mean of k sample variances with each having $n - 1$ degrees of freedom.

Although there are what may seem to be substantial differences among the means of the samples, this, we conclude, is due to chance.

The technique we have just described is the simplest form of an analysis of variance. Although we could go ahead and perform F tests for differences among k means without further discussion, it will be instructive to look at the problem from an analysis-of-variance point of view, and we shall do so in the next section.

15.2

One-Way Analysis of Variance

The basic idea in the analysis of variance is to express a measure of the total variation in a set of data as a sum of terms, which can be attributed to specific sources, or causes, of variation. With regard to the example of the preceding section, two such sources of variation might be (1) actual differences in the average cholesterol content of the four diet foods, and (2) chance, which in problems of this kind is usually called the **experimental error**. The measure of the total variation in a set of data which we shall use is the **total sum of squares**†

$$SST = \sum_{i=1}^{k} \sum_{j=1}^{n} (x_{ij} - \bar{x}_{..})^2$$

where x_{ij} is the jth observation of the ith sample ($i = 1, 2, \ldots, k$ and $j = 1, 2, \ldots, n$), and $\bar{x}_{..}$ is the **grand mean**, the mean of all the kn measurements or observations. Note that if we divided the total sum of squares SST by $kn - 1$, we would get the variance of the data; hence, the total sum of squares of a set of data is interpreted in much the same way as its sample variance.

If we let $\bar{x}_{i.}$ denote the mean of the ith sample, we can now write the following identity, which forms the basis of a **one-way analysis of variance**:‡

Identity for one-way analysis of variance

$$SST = n \cdot \sum_{i=1}^{k} (\bar{x}_{i.} - \bar{x}_{..})^2 + \sum_{i=1}^{k} \sum_{j=1}^{n} (x_{ij} - \bar{x}_{i.})^2$$

Looking closely at the two terms into which the total sum of squares SST has been partitioned, we find that the first term is a measure of the variation among

†The use of double subscripts and double summations is explained briefly in Section 3.10.
‡This identity may be derived by writing the total sum of squares as

$$SST = \sum_{i=1}^{k} \sum_{j=1}^{n} (x_{ij} - \bar{x}_{..})^2$$

$$= \sum_{i=1}^{k} \sum_{j=1}^{n} [(\bar{x}_{i.} - \bar{x}_{..}) + (x_{ij} - \bar{x}_{i.})]^2$$

and then expanding the squares $[(\bar{x}_{i.} - \bar{x}_{..}) + (x_{ij} - \bar{x}_{i.})]^2$ by means of the binomial theorem and simplifying algebraically.

the sample means; in fact, if we divide it by $k - 1$ we get the quantity which we earlier denoted $n \cdot s_{\bar{x}}^2$. Similarly, the second term is a measure of the variation within the individual samples, and if we divided this term by $k(n - 1)$ we would get the mean of the variances of the individual samples, the quantity which we put into the denominator of F in the preceding section.

It is customary to refer to the first term, the quantity which measures the variation among the sample means, as the **treatment sum of squares** $SS(Tr)$, and to the second term, which measures the variation within the samples, as the **error sum of squares** SSE. This terminology is explained by the fact that most analysis-of-variance techniques were originally developed in connection with agricultural experiments where different fertilizers, for example, were regarded as different **treatments** applied to the soil. The word "error" in "error sum of squares" pertains to the experimental error, what we also refer to as chance. Although this may sound confusing at first, we shall refer to the four diet foods of our example as four treatments, and in other problems we may refer to three kinds of packaging as three different treatments, five kinds of advertising campaigns as five different treatments, and so on.

Before we go any further, let us verify the identity $SST = SS(Tr) + SSE$ with reference to the diet-foods example of the preceding section. Substituting into the formulas for the different sums of squares, we get

$$
\begin{aligned}
SST = \ &(3.6 - 3.6)^2 + (4.1 - 3.6)^2 + (4.0 - 3.6)^2 \\
&+ (3.1 - 3.6)^2 + (3.2 - 3.6)^2 + (3.9 - 3.6)^2 \\
&+ (3.2 - 3.6)^2 + (3.5 - 3.6)^2 + (3.5 - 3.6)^2 \\
&+ (3.5 - 3.6)^2 + (3.8 - 3.6)^2 + (3.8 - 3.6)^2 \\
= \ &1.18
\end{aligned}
$$

$$
\begin{aligned}
SS(Tr) = \ &3[(3.9 - 3.6)^2 + (3.4 - 3.6)^2 + (3.4 - 3.6)^2 \\
&+ (3.7 - 3.6)^2] \\
= \ &0.54
\end{aligned}
$$

$$
\begin{aligned}
SSE = \ &(3.6 - 3.9)^2 + (4.1 - 3.9)^2 + (4.0 - 3.9)^2 \\
&+ (3.1 - 3.4)^2 + (3.2 - 3.4)^2 + (3.9 - 3.4)^2 \\
&+ (3.2 - 3.4)^2 + (3.5 - 3.4)^2 + (3.5 - 3.4)^2 \\
&+ (3.5 - 3.7)^2 + (3.8 - 3.7)^2 + (3.8 - 3.7)^2 \\
= \ &0.64
\end{aligned}
$$

and it can be seen that

$$
SS(Tr) + SSE = 0.54 + 0.64 = 1.18 = SST
$$

To test the null hypothesis $\mu_1 = \mu_2 = \cdots = \mu_k$ (or $\alpha_1 = \alpha_2 = \cdots = \alpha_k = 0$ in the notation of the footnote to page 390) against the alternative that the treatment means are not all equal (or that the treatment effects are not all zero), we now proceed as in the preceding section and compare $SS(Tr)$ with SSE by means of an F statistic. In practice, we usually exhibit the necessary work in an **analysis-of-variance table** as follows:

Source of variation	Degrees of freedom	Sum of squares	Mean square	F
Treatments	$k-1$	$SS(Tr)$	$MS(Tr) = \dfrac{SS(Tr)}{k-1}$	$\dfrac{MS(Tr)}{MSE}$
Error	$k(n-1)$	SSE	$MSE = \dfrac{SSE}{k(n-1)}$	
Total	$kn-1$	SST		

Here the second column lists the degrees of freedom (the number of independent deviations from the mean on which the sums of squares are based), the fourth column lists the **mean squares** $MS(Tr)$ and MSE, which are obtained by dividing the corresponding sums of squares by their degrees of freedom, and the right-hand column gives the value of the F statistic as the ratio of the two mean squares. These two mean squares are, in fact, the two estimates of σ^2 referred to on page 391; the numerator and denominator degrees of freedom for the F test, $k-1$ and $k(n-1)$, are shown opposite "Treatments" and "Error" in the "Degrees of freedom" column. The significance test is the same as before; we compare F with F_α for $k-1$ and $k(n-1)$ degrees of freedom.

EXAMPLE　Construct an analysis-of-variance table for the diet-foods example.

SOLUTION　The degrees of freedom for the treatment, error, and total sums of squares are $k-1 = 4-1 = 3$, $k(n-1) = 4(3-1) = 8$, and $kn-1 = 4 \cdot 3 - 1 = 11$ (which merely provides a check as it equals the sum of the other two). The sums of squares are $SS(Tr) = 0.54$, $SSE = 0.64$, and $SST = 1.18$, so that $MS(Tr) = \dfrac{0.54}{3} = 0.18$, $MSE = \dfrac{0.64}{8} = 0.08$, and $F = \dfrac{0.18}{0.08} = 2.25$. All these results are summarized in the following table:

Source of variation	Degrees of freedom	Sum of squares	Mean square	F
Treatments	3	0.54	0.18	2.25
Error	8	0.64	0.08	
Total	11	1.18		

Since $F = 2.25$ is less than 4.07, the value of $F_{0.05}$ for 3 and 8 degrees of freedom, we find (as before) that the null hypothesis cannot be rejected at the 0.05 level of significance.

The numbers which we used in our illustration were intentionally chosen so that the calculations would be relatively simple. In actual practice, the calculation of the sums of squares can be quite tedious unless we use the following computing formulas, in which $T_{i.}$ denotes the total of the observations for the ith treatment (that is, the sum of the values in the ith sample), and $T_{..}$ denotes the grand total of all the data:†

Computing formulas for sums of squares

$$SST = \sum_{i=1}^{k} \sum_{j=1}^{n} x_{ij}^2 - \frac{1}{kn} \cdot T_{..}^2$$

$$SS(Tr) = \frac{1}{n} \cdot \sum_{i=1}^{k} T_{i.}^2 - \frac{1}{kn} \cdot T_{..}^2$$

and by subtraction

$$SSE = SST - SS(Tr)$$

EXAMPLE Use these computing formulas to calculate SST, $SS(Tr)$, and SSE for the example dealing with the four diet foods.

SOLUTION Substituting $k = 4$, $n = 3$, $T_{1.} = 11.7$, $T_{2.} = 10.2$, $T_{3.} = 10.2$, $T_{4.} = 11.1$, $T_{..} = 43.2$, and $\sum \sum x^2 = 156.70$ into the formulas, we get

$$SST = 156.70 - \frac{1}{12}(43.2)^2 = 1.18$$

$$SS(Tr) = \frac{1}{3}(11.7^2 + 10.2^2 + 10.2^2 + 11.1^2) - \frac{1}{12}(43.2)^2$$

$$= 0.54$$

and

$$SSE = 1.18 - 0.54 = 0.64$$

Of course, these results are identical with those obtained before.

The method we have discussed here applies only when each sample has the same number of observations, but minor modifications make it applicable also to situations where the sample sizes are not all equal. If there are n_i values for the ith treatment, the computing formulas for the sums of squares become

†In many instances, the calculations can also be simplified by coding, that is, by subtracting the same constant from each value and/or multiplying each value by a constant.

$$SST = \sum_{i=1}^{k} \sum_{j=1}^{n_i} x_{ij}^2 - \frac{1}{N} \cdot T_{..}^2$$

$$SS(Tr) = \sum_{i=1}^{k} \frac{T_{i.}^2}{n_i} - \frac{1}{N} \cdot T_{..}^2$$

$$SSE = SST - SS(Tr)$$

where $N = n_1 + n_2 + \cdots + n_k$. The only other change is that the total number of degrees of freedom is $N - 1$, and the degrees of freedom are $k - 1$ for treatments and $N - k$ for error.

EXAMPLE A restaurant manager wants to determine whether the sales of steak and shrimp combination dinners depend on how this entree is described on the menu. He has three kinds of menus printed, listing steak and shrimp combination dinners among the other entrees, featuring them as "Chef's Special," and as "Gourmet's Delight," and he intends to use each kind of menu on six different Saturday nights. However, only the following data, showing the numbers of steak and shrimp combination dinners sold on 15 Saturdays, are actually collected:

Listed among other entrees	221, 205, 198, 237, 206, 223
Featured as "Chef's Special"	247, 214, 222, 244, 215, 226
Featured as "Gourmet's Delight"	206, 219, 202

Use the 0.05 level of significance to test the null hypothesis that the different descriptions do not affect the sales of the dinner.

SOLUTION Substituting $n_1 = 6$, $n_2 = 6$, $n_3 = 3$, $N = 15$, $T_{1.} = 1{,}290$, $T_{2.} = 1{,}368$, $T_{3.} = 627$, $T_{..} = 3{,}285$, and $\sum\sum x^2 = 722{,}531$ into the formulas for the sums of squares, we get

$$SST = 722{,}531 - \frac{1}{15}(3{,}285)^2 = 722{,}531 - 719{,}415$$

$$= 3{,}116$$

$$SS(Tr) = \frac{1{,}290^2}{6} + \frac{1{,}368^2}{6} + \frac{627^2}{3} - 719{,}415$$

$$= 882$$

and

$$SSE = 3{,}116 - 882 = 2{,}234$$

Also, the degrees of freedom are $k - 1 = 3 - 1 = 2$, $N - k = 15 - 3 = 12$, and $N - 1 = 15 - 1 = 14$, the mean squares are $MS(Tr) = \frac{882}{2} = 441$ and $MSE = \frac{2,234}{12} = 186.2$, $F = \frac{441}{186.2} = 2.37$, and all these results are summarized in the following table:

Source of variation	Degrees of freedom	Sum of squares	Mean square	F
Treatments	2	882	441	2.37
Error	12	2,234	186.2	
Total	14	3,116		

Since $F = 2.37$ does not exceed 3.89, the value of $F_{0.05}$ for 2 and 12 degrees of freedom, we find that the null hypothesis (that the different descriptions do not affect the sales of the dinners) cannot be rejected.

EXERCISES

15.1 The following are the numbers of miles per gallon which a test driver got with four tankfuls each of five brands of gasoline:

> Brand A: 27, 21, 26, 22
> Brand B: 24, 29, 27, 28
> Brand C: 27, 27, 30, 32
> Brand D: 24, 21, 24, 23
> Brand E: 25, 20, 22, 21

(a) Use the method of Section 15.1 to test the null hypothesis that the five brands of gasoline yield the same average mileage at the 0.01 level of significance.

(b) Perform an analysis of variance, using the computing formulas for the required sums of squares, and compare the resulting value of F with that obtained in part (a).

15.2 Samples of peanut butter produced by three different manufacturers are tested for aflatoxin content (ppb), with the following results:

> Brand 1: 0.5, 6.3, 1.1, 2.7, 5.5, 4.3
> Brand 2: 2.5, 1.8, 3.6, 5.2, 1.2, 0.7
> Brand 3: 3.3., 1.5, 0.4, 4.8, 2.2, 1.0

Use the 0.05 level of significance to test whether the differences among the means of the three samples are significant.

15.3 The following are eight consecutive weeks' earnings (in dollars) of three door-to-door vacuum cleaner salespersons employed by a given firm:

$$A: 231, 209, 216, 243, 207, 200, 223, 184$$
$$B: 215, 200, 219, 230, 218, 204, 213, 207$$
$$C: 229, 213, 204, 220, 226, 208, 232, 196$$

Use the 0.05 level of significance to test the null hypothesis that on the average the three salespersons' weekly earnings are the same.

15.4 With reference to the illustration of Section 3.12, perform an analysis of variance to test, at the 0.05 level of significance, whether the differences among the mean areas covered by the three paints are significant.

15.5 To study its performance, a newly designed motorboat was timed (in minutes) over a marked course under three different wind and water conditions. Use the following data to test, at the 0.05 level of significance, the null hypothesis that the boat's performance is not affected by the differences in wind and water conditions:

Calm conditions: 20, 17, 14, 24
Moderate conditions: 21, 23, 16, 25, 18, 23
Choppy conditions: 26, 24, 23, 29, 21

15.6 The following are the numbers of words per minute which a secretary typed on several occasions on four different typewriters:

Typewriter C: 71, 75, 69, 77, 61, 72, 71, 78
Typewriter D: 68, 71, 74, 66, 69, 67, 70, 62
Typewriter E: 75, 70, 81, 73, 78, 72
Typewriter F: 62, 59, 71, 68, 63, 65, 72, 60, 64

Use the level of significance $\alpha = 0.05$ to test whether the differences among the means of the four samples can be attributed to chance.

15.7 With reference to the notation introduced in the footnote to page 390, show that the sum of the α's, the treatment effects, is equal to zero.

15.3

Two-Way Analysis of Variance

In the example we used to illustrate a one-way analysis of variance, we were unable to show that there really is a difference in the average cholesterol content of the four diet foods, even though the sample means varied from 3.4 to 3.9 milligrams per package. The results were not statistically significant because there were also considerable differences among the values within each of the samples and, hence, a large experimental error. Since this is the quantity in the denominator of the F statistic, the F ratio was not large enough to lead us to reject the null hypothesis. But, suppose now we learn something that we did not know earlier—the measurements of the cholesterol contents were performed in different laboratories. The first value of each sample, we learn, came from one laboratory, the second value came from another laboratory, and the third value came from a third laboratory. This puts the whole study in a new light and we might picture the original data as follows:

	Laboratory 1	Laboratory 2	Laboratory 3
Diet food A	3.6	4.1	4.0
Diet food B	3.1	3.2	3.9
Diet food C	3.2	3.5	3.5
Diet food D	3.5	3.8	3.8

From these data we find that the means of the cholesterol readings from the three laboratories are 3.35, 3.65, and 3.80 milligrams, and this suggests that what we called chance variation, or experimental error, in our earlier analysis may well have been caused in part by differences in laboratories.

It also suggests that we should perform a **two-way analysis of variance**, in which the total variation of the data is partitioned into one component which we ascribe to possible differences due to one variable (the different treatments), a second component which we ascribe to possible differences due to a second variable—an extraneous variable which causes variations which should not be included in the error sum of squares—and a third component which we ascribe to chance. It is customary to call the categories of such an extraneous variable—the three laboratories in our example—**blocks**, thanks again to the origin of this method in agricultural research.

To formulate the null hypotheses to be tested in a two-way analysis of variance, let us write μ_{ij} for the true mean which corresponds to the ith treatment and the jth block (in our numerical illustration, the true average cholesterol content of the ith diet food as measured by the jth laboratory) and express it as

$$\mu_{ij} = \mu + \alpha_i + \beta_j$$

As in the notation of the footnote to page 390, μ is the grand mean (the average of all the μ_{ij}) and the α_i are the treatment effects (whose sum is zero). Correspondingly, we refer to the β_j as the **block effects** (whose sum is also zero), and write the two null hypotheses we want to test as

$$\alpha_1 = \alpha_2 = \cdots = \alpha_k = 0$$

and

$$\beta_1 = \beta_2 = \cdots = \beta_n = 0$$

The alternative to the first null hypothesis (which in our illustration amounts to the hypothesis that the true average cholesterol content is the same for the four foods) is that the treatment effects α_i are not all zero. Also, the alternative to the second null hypothesis (which in our illustration amounts to the hypothesis that the use of the different laboratories in performing the measurements has no effect on the results) is that the block effects β_j are not all zero.

To test the second of the null hypotheses, we need a quantity, similar to the treatment sum of squares, which measures the variation in the different block means (3.35, 3.65, and 3.80 milligrams in our example) instead of the variation in the different treatment means. So, if we let $T_{\cdot j}$ denote the total of the values in the jth block, substitute it for $T_{i\cdot}$ in the formula for $SS(Tr)$, sum on j instead of i, and interchange n and k, we obtain, analogous to $SS(Tr)$, the **block sum of squares**

Computing formula for block sum of squares

$$SSB = \frac{1}{k} \cdot \sum_{j=1}^{n} T_{\cdot j}^2 - \frac{1}{kn} \cdot T_{\cdot\cdot}^2$$

In a two-way analysis of variance we compute SST and $SS(Tr)$ according to the formulas on page 396, SSB according to the formula immediately above, and then we get SSE by subtraction. Since

$$SST = SS(Tr) + SSB + SSE$$

we have

Error sum of squares (Two-way analysis of variance)

$$SSE = SST - SS(Tr) - SSB$$

Observe that the error sum of squares for a two-way analysis of variance does not equal the error sum of squares for a one-way analysis of variance performed on the same data, even though we denote both with the symbol SSE. In fact, we are now partitioning the error sum of squares for the one-way analysis of variance into two terms: the block sum of squares, SSB, and the remainder which is the new error sum of squares, SSE.

We can now construct the following analysis-of-variance table for a two-way analysis of variance:

Source of variation	Degrees of freedom	Sum of squares	Mean square	F
Treatments	$k - 1$	$SS(Tr)$	$MS(Tr) = \dfrac{SS(Tr)}{k - 1}$	$\dfrac{MS(Tr)}{MSE}$
Blocks	$n - 1$	SSB	$MSB = \dfrac{SSB}{n - 1}$	$\dfrac{MSB}{MSE}$
Error	$(k - 1)(n - 1)$	SSE	$MSE = \dfrac{SSE}{(k - 1)(n - 1)}$	
Total	$kn - 1$	SST		

The mean squares are again given by the sums of squares divided by their degrees of freedom, and the two F values are given by the mean squares for treatments and blocks divided by the mean square for error. Also, the degrees of freedom for blocks is $n - 1$ (like those for treatments with n substituted for k), and the degrees of freedom for error can be found by subtracting the degrees of freedom for treatments and blocks from $kn - 1$, the total number of degrees of freedom:

$$(kn - 1) - (k - 1) - (n - 1) = kn - k - n + 1$$
$$= (k - 1)(n - 1)$$

Thus, in the significance test for treatments the numerator and denominator degrees of freedom for F are $k - 1$ and $(k - 1)(n - 1)$, and in the significance test for blocks the numerator and denominator degrees of freedom for F are $n - 1$ and $(k - 1)(n - 1)$.

EXAMPLE Based on the data on page 400, test at the 0.05 level of significance whether the differences among the means obtained for the different diet foods (treatments) are significant, and also whether the differences among the means obtained for the different laboratories (blocks) are significant.

SOLUTION Substituting $k = 4$, $n = 3$, $T_{.1} = 13.4$, $T_{.2} = 14.6$, $T_{.3} = 15.2$, and $T_{..} = 43.2$ into the formula for SSB, we get

$$SSB = \frac{1}{4}(13.4^2 + 14.6^2 + 15.2^2) - \frac{1}{12}(43.2)^2$$

$$= 155.94 - 155.52$$

$$= 0.42$$

and, since $SST = 1.18$ and $SS(Tr) = 0.54$, we find that

$$SSE = 1.18 - 0.54 - 0.42$$
$$= 0.22$$

Also, the degrees of freedom are $k - 1 = 4 - 1 = 3$, $n - 1 = 3 - 1 = 2$, $(k - 1)(n - 1) = (4 - 1)(3 - 1) = 6$, and $kn - 1 = 4 \cdot 3 - 1 = 11$, the mean squares are $MS(Tr) = \frac{0.54}{3} = 0.18$, $MSB = \frac{0.42}{2} = 0.21$, and $MSE = \frac{0.22}{6} = 0.0367$, so that for treatments $F = \frac{0.18}{0.0367} = 4.90$ and for blocks $F = \frac{0.21}{0.0367} = 5.72$. All these results are summarized in the following analysis-of-variance table:

Source of variation	Degrees of freedom	Sum of squares	Mean square	F
Treatments	3	0.54	0.18	4.90
Blocks	2	0.42	0.21	5.72
Error	6	0.22	0.0367	
Total	11	1.18		

Since $F = 4.90$ exceeds 4.76, the value of $F_{0.05}$ for 3 and 6 degrees of freedom, we find that we must reject the null hypothesis that the true average cholesterol content of the four diet foods is the same. Also, since $F = 5.72$ exceeds 5.14, the value of $F_{0.05}$ for 2 and 6 degrees of freedom, we find that we must reject the null hypothesis about the laboratories; in other words, we have shown that there is a systematic difference in the results provided by the three laboratories. This may not be of any special interest, but after taking the differences among the laboratories into account, it now seems evident that there are real differences among the cholesterol contents of the four foods.

The diet foods data which we analyzed above constitute a **complete-block experiment**. It is complete in the sense that each treatment occurs in each block the same number of times—in our example, each diet food is tested once by each laboratory. In a complete-block experiment, the treatments are our primary concern while the blocks represent an extraneous variable, sometimes called a **nuisance variable**, which we introduce to eliminate the variation caused by it from the error sum of squares.

However, a two-way analysis of variance can also be used in connection with **two-factor experiments**, where both variables are of material concern. This would be the case, for example, in an analysis of the following data collected in an experiment designed to test whether or not the range of a missile flight (in miles) is affected by three different launchers and also by four different fuels (see Exercise 15.12 on page 405):

	Fuel 1	Fuel 2	Fuel 3	Fuel 4
Launcher X	45.9	57.6	52.2	41.7
Launcher Y	46.0	51.0	50.1	38.8
Launcher Z	45.7	56.9	55.3	48.1

Note that we used a different format for the table to distinguish between two-factor and complete-block experiments.

When a two-way analysis of variance is used in this way, we usually call the two variables **factors** A and B (instead of treatments and blocks) and write SSA instead of $SS(Tr)$; we still write SSB, but now B stands for factor B instead of for blocks.

EXERCISES

15.8 A laboratory technician measures the breaking strength of each of five kinds of linen threads by using four different measuring instruments, I_1, I_2, I_3, and I_4, and obtains the following results (in ounces):

	I_1	I_2	I_3	I_4
Thread 1	20.9	20.4	19.9	21.9
Thread 2	25.0	26.2	27.0	24.8
Thread 3	25.5	23.1	21.5	24.4
Thread 4	24.8	21.2	23.5	25.7
Thread 5	19.6	21.2	22.1	21.1

Perform a two-way analysis of variance, using $\alpha = 0.05$ for both tests of significance.

15.9 Four different, although supposedly equivalent, forms of a standardized achievement test are given to each of four students, and the following are their scores:

	Student C	*Student D*	*Student E*	*Student F*
Form 1	77	62	52	66
Form 2	85	63	49	65
Form 3	81	65	46	64
Form 4	88	72	55	60

Perform a two-way analysis of variance to test at the 0.01 level of significance whether it is reasonable to treat the four forms as equivalent. If the order in which each student takes the four tests is randomized by some means, we call the design of this experiment a **randomized block design**. The purpose of this randomization is to take care of such possible extraneous factors as fatigue, or perhaps the experience gained from repeatedly taking the test.

15.10 To study the effectiveness of five different kinds of packaging, a processor of a breakfast food puts each kind into five different supermarkets, J, K, L, M, and N. Perform a two-way analysis of variance on the following data representing the number of sales of the breakfast food on a given day, to test the null hypothesis that packaging has no effect on sales at the 0.05 level of significance:

	J	K	L	M	N
Packaging 1	45	32	36	32	40
Packaging 2	37	34	46	44	34
Packaging 3	35	37	48	46	35
Packaging 4	36	38	50	36	45
Packaging 5	42	39	40	45	51

15.11 To study the performance of three different detergents at three different water temperatures, the following "whiteness" readings were obtained with specially designed equipment for nine loads of washing:

	Detergent A	Detergent B	Detergent C
Cold water	45	43	55
Warm water	37	40	56
Hot water	42	44	46

Perform a two-way analysis of variance, using the level of significance $\alpha = 0.05$.

15.12 With reference to the missile-range data on page 403, perform a two-way analysis of variance to test at the 0.05 level of significance whether there are significant differences
(a) among the mean ranges for the three launchers;
(b) among the mean ranges for the four fuels.

15.4

A Word of Caution

Although the analysis of variance is a very powerful statistical tool, our example shows what can happen when we use the wrong kind of analysis, in this case, a one-way analysis when we should have used a two-way analysis. In addition to this there are many other pitfalls, for it is often difficult to determine whether the necessary assumptions are met. In the two-way analysis we had only one observation from each population, that is, one observation for each combination of diet foods and laboratories. Consequently, it is impossible in this example to determine statistically whether the populations sampled have roughly the shape of normal distributions with the same variance. To aid in such a determination, we might have taken several observations from each population, but there are many situations in which this is neither feasible nor practical.

Furthermore, it should be understood that what we have presented here is merely an introduction to some of the most basic techniques which come under the general heading of analysis of variance. An obvious generalization would be to apply the idea of analyzing the total variation of a set of data to experiments in which there are more than two variables about which we want to test hypotheses. Then there are situations in which the variables under consideration are not independent. Suppose, for instance, that a tire manufacturer is experimenting with various treads under various road conditions, and he finds that one kind is especially good for use on dirt roads while another kind is especially good for use on icy roads. If this is the case, we say that there is an **interaction** between road conditions and the designs of the treads. On the other hand, if each of the treads performed equally under all kinds of road conditions, we would say that there is no interaction and that the two variables (tread design and road conditions) are independent. Here we have studied only the case where there is assumed to be no interaction.

Finally, let us mention the problem of how to interpret the results of an analysis of variance once it has been shown that the populations sampled do not have equal means. For instance, in the example dealing with the four diet foods, A, B, C, and D, the sample means were 3.9, 3.4, 3.4, and 3.7, and the differences among them were shown to be significant, but can we conclude that in general food A contains more cholesterol on the average than the other three? Or that foods B and C are really lower in cholesterol content than food D? There exist statistical techniques which provide answers to questions of this kind; they are called **multiple-comparisons tests** and they may be found in more advanced texts.

15.5

Check List of Key Terms

15.6

Review Exercises

15.13 To compare four different golf-ball designs, *A, B, C,* and *D,* several balls of each kind were driven by a golf professional and the following are the distances (in yards) from the tee to the points where the balls came to rest:

> *Golf ball A:* 262, 245, 237, 280, 236
> *Golf ball B:* 244, 216, 251, 263, 214, 228
> *Golf ball C:* 272, 265, 244, 259
> *Golf ball D:* 250, 233, 217, 267, 258

At the 0.05 level of significance, can the differences among the means of the four samples be attributed to chance?

15.14 The following are the numbers of defective pieces produced by four workmen operating, in turn, three different machines:

	Machine 1	Machine 2	Machine 3
Workman 1	27	29	21
Workman 2	31	30	27
Workman 3	24	26	22
Workman 4	21	25	20

Perform a two-way analysis of variance and test, at the 0.05 level of significance, whether

(a) the differences among the means obtained for the four workmen can be attributed to chance;

(b) the differences among the means obtained for the three machines can be attributed to chance.

15.15 To find the best arrangement of instruments on a control panel of an airplane, three different arrangements were tested by simulating an emergency condition and observing the reaction time required to correct the condition. The reaction times (in tenths of a second) of 12 pilots (randomly assigned to the different arrangements) were as follows:

> *Arrangement 1:* 8, 15, 10, 11
> *Arrangement 2:* 16, 11, 14, 19
> *Arrangement 3:* 12, 7, 13, 8

(a) Use the method of Section 15.1 and the 0.05 level of significance to test whether the differences among the three sample means can be attributed to chance.

(b) Use the computing formulas for *SST*, *SS(Tr)*, and *SSE* to determine the values of these sums of the squares for the given data, construct an analysis-of-variance table, and compare the value of *F* with that obtained in part (a).

15.16 The sample data in the following table are the grades in a statistics test obtained by nine college students from three majors who were taught by three different instructors:

	Instructor A	Instructor B	Instructor C
Marketing	77	88	71
Finance	88	97	81
Advertising	85	95	72

Analyze this two-factor experiment using the 0.05 level of significance.

15.17 The following are the numbers of mistakes made in five successive days by four technicians working for a photographic laboratory:

Technician I: 8, 11, 7, 9, 10
Technician II: 9, 11, 6, 14, 10
Technician III: 8, 13, 11, 9, 13
Technician IV: 13, 5, 9, 10, 7

Test at the 0.01 level of significance whether the differences among the means of the four samples can be attributed to chance.

15.18 An experiment was conducted to compare three methods of teaching the programming of a certain digital computer. Random samples of size four were taken from each of three groups of students taught, respectively, by method A (straight teaching-machine instruction), method B (personal instruction and some direct experience working with the computer), and method C (personal instruction but no work with the computer itself) and the following are the grades obtained by these students on an appropriate achievement test:

Method A: 76, 80, 70, 74
Method B: 95, 85, 91, 89
Method C: 77, 82, 81, 84

Test at the 0.05 level of significance whether the differences among the three sample means are significant.

ost of the tests of hypotheses discussed in Chapters 12 through 15 require specific assumptions about the population, or populations, sampled. In many cases we must assume that the populations have roughly the shape of normal distributions, or that their variances are known or are known to be equal, or that the samples are independent. Since there are many situations where these assumptions cannot be met, statisticians have developed alternative techniques based on less stringent assumptions, which have become known as **nonparametric tests**.

Aside from the fact that nonparametric tests may be used under very general conditions, they are often easier to explain and understand than the standard tests which they replace; moreover, in many of these tests the computational burden is relatively light. For these reasons, nonparametric tests have become quite popular, and extensive literature is devoted to their theory and application.

In Sections 16.1 and 16.2 we shall present the **sign test** as a nonparametric alternative to tests concerning means and tests concerning differences between means based on paired data; in Sections 16.3 and 16.4 we shall study some methods based on **rank sums**; and in Sections 16.5 and 16.6 we shall learn how to test the **randomness** of a sample once the data have been obtained.

16

Decision Making: Nonparametric Tests

16.1

The One-Sample Sign Test

Except for the large-sample tests, all the standard tests concerning means that we have studied are based on the assumption that the populations sampled have roughly the shape of normal distributions. When in a particular case this assumption is untenable, the standard test can be replaced by one of several nonparametric alternatives, among them the sign test, which we shall study in this section and the next. The one-sample sign test applies when we sample a continuous symmetrical population, so that the probability a sample value is less than the mean and the probability a sample value is greater than the mean are both $\frac{1}{2}$. To test the null hypothesis $\mu = \mu_0$ against an appropriate alternative on the basis of a random sample of size n, we replace each sample value greater than μ_0 with a plus sign and each sample value less than μ_0 with a minus sign; then we test the null hypothesis that these plus and minus signs are values of a random variable having the binomial distribution with $p = \frac{1}{2}$. (If a sample value equals μ_0, which is possible since we usually deal with rounded data, we simply discard it.)

To perform the actual test when the sample is small, we can refer directly to tables of binomial probabilities such as Table V, and when the sample is large, we can use the normal approximation to the binomial distribution.

EXAMPLE In a given year, the average number of days that a (small) sample of 15 wholesalers of drugs and drug sundries required to convert receivables into cash were 33.9, 35.4, 37.3, 40.9, 27.8, 35.5, 34.6, 41.1, 30.0, 43.2, 33.9, 41.3, 32.0, 37.7, and 35.2 days. Use the one-sample sign test to test the null hypothesis $\mu = 32.0$ days against the alternative hypothesis $\mu > 32.0$ days at the 0.01 level of significance.

SOLUTION Replacing each value less than 32.0 with a minus sign, each value greater than 32.0 with a plus sign, and discarding the one value which actually equals 32.0, we get

$$+ + + + - + + + - + + + + +$$

and the question is whether 12 plus signs observed in 14 trials supports the null hypothesis $p = \frac{1}{2}$ or the alternative hypothesis $p > \frac{1}{2}$. From

Table V we find that for $n = 14$ and $p = \frac{1}{2}$ the probability of 12 or more successes is $0.006 + 0.001 = 0.007$, and since this is less than $\alpha = 0.01$, the null hypothesis must be rejected; it appears that for the time period and population studied, wholesalers of drugs and drug sundries took more than 32.0 days on the average to convert receivables into cash.

Since np and $n(1 - p)$ are both greater than 5 in the preceding example, we could have used the normal approximation to the binomial distribution, but we illustrate this technique next with a different example.

EXAMPLE In the beginning of Chapter 12 we gave the following data on a large industrial plant's daily emission of sulfur oxides (in tons):

17	15	20	29	19	18	22	25	27	9
24	20	17	6	24	14	15	23	24	26
19	23	28	19	16	22	24	17	20	13
19	10	23	18	31	13	20	17	24	14

Use the one-sample sign test to test the null hypothesis that the plant's true average daily emission of sulfur oxides is $\mu = 23.5$ tons against the alternative hypothesis $\mu < 23.5$ tons at the 0.05 level of significance.

SOLUTION There are 11 plus signs and 29 minus signs, and we must see whether, on the basis of this, we can reject the null hypothesis $p = \frac{1}{2}$ against the alternative hypothesis $p < \frac{1}{2}$. Substituting $x = 11$, $n = 40$, and $p_0 = \frac{1}{2}$ into the formula

$$z = \frac{x - np_0}{\sqrt{np_0(1 - p_0)}}$$

we get

$$z = \frac{11 - 40 \cdot \frac{1}{2}}{\sqrt{40 \cdot \frac{1}{2} \cdot \frac{1}{2}}} = -2.85$$

Since this is less than $-z_{0.05} = -1.645$, it follows that the null hypothesis must be rejected, and we conclude that the plant's true average daily emission of sulfur oxides is less than 23.5 tons. When n is small in problems like this, it may be desirable to use the continuity correction given in the footnote to page 366, but if we had substituted $11 + \frac{1}{2} - 20$ instead of $11 - 20$ into the formula for z, we would have obtained $z = -2.69$, and the conclusion would have been the same.

The Paired-Sample Sign Test

The sign test has important applications in problems involving paired data. In these problems, each pair of sample values can be replaced with a plus sign if the first value is greater than the second, a minus sign if the first value is smaller than the second, or be discarded if the two values are equal. Then, we proceed as in Section 16.1.

EXAMPLE To determine the effectiveness of a new traffic control system, the numbers of accidents that occurred at a random sample of eight dangerous intersections during the four weeks before and the four weeks following the installation of the new system were observed with the following results:

| 9 and 5 | 7 and 3 | 3 and 4 | 16 and 11 |
| 12 and 7 | 12 and 5 | 5 and 5 | 6 and 1 |

Use the sign test at the level of significance $\alpha = 0.10$ to test the null hypothesis that the new traffic control system is as effective as the old system against the alternative hypothesis that the new system is more effective.

SOLUTION Since there are six plus signs and one minus sign, it remains to be seen whether the probability of "6 or more successes in 7 trials" supports the alternative hypothesis $p > \frac{1}{2}$. Referring to Table V, we find that for $n = 7$ and $p = \frac{1}{2}$ the probability of "6 or more successes" is $0.055 + 0.008 = 0.063$, and it follows at the 0.10 level of significance that the null hypothesis must be rejected. Apparently, the new traffic control system is effective.

The test we have described here is only one of several nonparametric tests used in the analysis of paired data. Another popular test used for this purpose, the **Wilcoxon signed-rank test,** may be found in the books on nonparametric statistics listed in the Bibliography at the end of the book.

EXERCISES

16.1 On 12 occasions, a random sample, a woman had to wait 2, 7, 9, 3, 7, 9, 5, 7, 8, 7, 10, and 4 minutes for her bus to work. Use the sign test based on Table V and the 0.05 level of significance to test the null hypothesis $\mu = 5$ (that on the average she has to wait five minutes) against the alternative hypothesis $\mu \neq 5$.

16.2 The following random sample shows the amounts (in dollars) spent by 15 persons at a certain amusement park: 7.50, 10.00, 13.75, 9.50, 10.35, 11.45, 8.85,

9.25, 6.65, 15.60, 11.10, 8.50, 13.85, 9.85, and 10.15. Use the sign test (based on Table V) and the level of significance $\alpha = 0.05$ to test the null hypothesis that on the average a person spends $9.00 at the park against the alternative that this figure is too low.

16.3 A random sample of nine women buying new eyeglasses tried on 12, 11, 14, 15, 10, 14, 11, 8, and 12 frames before making a choice. Use the sign test at the 0.05 level of significance to test the null hypothesis $\mu = 10$ (that on the average a woman buying new eyeglasses tries on 10 frames before making a choice) against the alternative hypothesis $\mu > 10$.

16.4 Use the normal approximation to the binomial distribution to rework the example on page 410, which dealt with the number of days required by certain wholesalers to convert receivables into cash.

16.5 Twenty-four cans of floor wax, randomly selected from a large production lot, have the following net weights (in ounces): 12.0, 11.9, 12.2, 12.0, 11.9, 12.0, 12.0, 12.1, 11.8, 12.0, 12.0, 12.1, 11.9, 11.9, 12.2, 12.1, 12.0, 11.9, 11.9, 12.1, 12.0, 12.0, 11.9, and 12.0. Use the sign test and the level of significance $\alpha = 0.05$ to test the null hypothesis that the true average weight for the entire production lot of floor wax is 12.05 ounces per can against the one-sided alternative $\mu < 12.05$.

16.6 The following are the numbers of twists that were required to break the bars in a random sample of 20 forged alloy bars: 37, 29, 34, 21, 54, 38, 30, 26, 48, 37, 24, 33, 39, 51, 44, 38, 35, 29, 46, and 31. Use the sign test and the level of significance $\alpha = 0.01$ to test the null hypothesis that on the average 30 twists are required to break such a bar against the alternative hypothesis that this figure is too low.

16.7 The following are the numbers of speeding tickets issued by two policemen on a random sample of 30 days: 7 and 10, 11 and 13, 10 and 11, 14 and 14, 11 and 15, 12 and 9, 6 and 10, 9 and 13, 8 and 11, 10 and 11, 11 and 15, 13 and 11, 7 and 10, 6 and 12, 10 and 14, 8 and 8, 11 and 12, 9 and 14, 9 and 7, 10 and 12, 6 and 7, 12 and 14, 9 and 11, 12 and 10, 11 and 13, 12 and 15, 7 and 9, 10 and 9, 11 and 13, and 8 and 10. Use the sign test at the level of significance $\alpha = 0.01$ to test the null hypothesis that on the average the two policemen issue equally many speeding tickets against the alternative hypothesis that on the average the second policeman issues more speeding tickets than the first.

16.8 The following are the numbers of employees absent from two departments of a large firm in a random sample of 25 days: 2 and 4, 6 and 3, 5 and 5, 7 and 2, 3 and 1, 4 and 3, 2 and 5, 3 and 1, 4 and 3, 5 and 6, 5 and 4, 3 and 8, 6 and 4, 5 and 2, 4 and 3, 3 and 0, 2 and 5, 6 and 4, 3 and 1, 2 and 4, 5 and 2, 3 and 2, 4 and 6, 6 and 3, and 4 and 3. Use the sign test at the level of significance $\alpha = 0.05$ to test the null hypothesis that on the average there are equally many absences in the two departments against the alternative hypothesis that on the average there are more absences in the first department.

16.9 The following are the numbers of passengers carried on flight No. 136 and flight No. 137 between Chicago and Phoenix on 12 days: 232 and 189, 265 and 230, 249 and 236, 250 and 261, 255 and 249, 236 and 218, 270 and 258, 266 and 253, 249 and 251, 240 and 233, 257 and 254, and 239 and 249. Use the sign test (based on Table V) and the 0.05 level of significance to test the null hypothesis

$\mu_1 = \mu_2$ (that on the average flight No. 136 carries equally many passengers as flight No. 137) against the alternative hypothesis $\mu_1 > \mu_2$.

16.10 Use the paired-sample sign test to rework Exercise 12.55 on page 339.

16.3

Rank Sums: The U Test

In this section we shall present a nonparametric alternative to the two-sample t test for the difference between two means. It is called the U test, the **Mann–Whitney test**, or the **Wilcoxon test**, named after the statisticians who contributed to its development. With this test we will be able to test the null hypothesis $\mu_1 = \mu_2$ without having to assume that the populations sampled have roughly the shape of normal distributions; in fact, the test requires only that the populations be continuous (in order to avoid ties), and in practice it does not matter whether this assumption is satisfied or not.

To illustrate how the U test is performed, suppose that we want to compare the mean lifetimes of two kinds of 9-volt batteries on the basis of the following lifetimes (in hours):

Brand A: 6.9, 11.2, 14.0, 13.2, 9.1, 13.9, 16.1, 9.3, 2.4, 6.4, 18.0, 11.5
Brand B: 15.5, 11.1, 16.0, 15.8, 18.2, 13.7, 18.3, 9.0, 17.2, 17.8, 13.0, 15.1

The means of these two random samples are 11.0 and 15.1, and their difference seems large, but it may or may not be significant. Shortcut methods suggest that the variance in the brand A batteries is almost three times the variance in the brand B batteries, so it may be unreasonable to use the two-sample t test, which is based on the assumption that the samples come from populations with equal variability.

We begin the U test by arranging the data jointly, as if they comprise one sample, in an increasing order of magnitude. For our data we get

2.4	6.4	6.9	9.0	9.1	9.3	11.1	11.2	11.5	13.0	13.2	13.7
A	A	A	B	A	A	B	A	A	B	A	B
13.9	14.0	15.1	15.5	15.8	16.0	16.1	17.2	17.8	18.0	18.2	18.3
A	A	B	B	B	B	A	B	B	A	B	B

where we indicated for each value whether it belongs to brand A or to brand B. Assigning the data in this order the ranks 1, 2, 3, . . . , and 24, we find that the lifetimes of the brand A batteries occupy ranks 1, 2, 3, 5, 6, 8, 9, 11, 13, 14, 19, and 22, while those of brand B occupy ranks 4, 7, 10, 12, 15, 16, 17, 18, 20, 21, 23, and 24. There are no ties here between values belonging to different samples, but if there were, we would assign each of the tied observations the mean of the ranks which they jointly occupy. For instance, if the third and fourth values were the same, we would assign each the rank $\dfrac{3+4}{2} = 3.5$, and if the ninth, tenth, and eleventh values were the same, we would assign each the rank $\dfrac{9+10+11}{3} = 10$.

The null hypothesis we want to test in a problem like this is that the two samples come from identical populations. If this hypothesis is true, it seems reasonable to suppose that the means of the ranks assigned to the values of the two samples should be more or less the same. The alternative hypothesis is that the means of the populations are not equal, and if this is the case and the difference is pronounced, most of the smaller ranks will go to the values of one sample, while most of the higher ranks will go to those of the other sample.

The test of the null hypothesis that the two samples come from identical populations may either be based on W_1, the sum of the ranks of the values of the first sample, or on W_2, the sum of the ranks of the values of the second sample, and in practice it does not matter which sample we call sample 1 and which we call sample 2. (When the sample sizes are unequal, we usually let the smaller of the two be sample 1; however, this is not required for the work in this book.)

If the sample sizes are n_1 and n_2, the sum of W_1 and W_2 is simply the sum of the first $n_1 + n_2$ positive integers, which is known to be

$$\frac{(n_1 + n_2)(n_1 + n_2 + 1)}{2}$$

This formula enables us to find W_2 if we know W_1, and vice versa. For our example we get

$$W_1 = 1 + 2 + 3 + 5 + 6 + 8 + 9 + 11 + 13 + 14 + 19 + 22$$
$$= 113$$

and since the sum of the first 24 positive integers is $\dfrac{24 \cdot 25}{2} = 300$, it follows that $W_2 = 300 - 113 = 187$. (This value is the sum of the ranks 4, 7, 10, 12, 15, 16, 17, 18, 20, 21, 23, and 24.)

When the use of rank sums was first proposed as a nonparametric alternative to the two-sample t test, the decision was based on W_1 or W_2, but now the decision is usually based on either of the related statistics

U_1 and U_2 statistics

or

$$U_1 = n_1 n_2 + \frac{n_1(n_1 + 1)}{2} - W_1$$

$$U_2 = n_1 n_2 + \frac{n_2(n_2 + 1)}{2} - W_2$$

or on the statistic U, which always equals the smaller of the two. The resulting tests are equivalent to those based on W_1 or W_2, but they have the advantage that they lend themselves more readily to the construction of tables of critical values. Not only do U_1 and U_2 take on values on the interval from 0 to $n_1 n_2$—indeed, their sum is always equal to $n_1 n_2$—but their sampling distributions are symmetrical about $\frac{n_1 n_2}{2}$.

The use of U, which always equals the smaller of the values of U_1 and U_2, has the added advantage that the resulting test is one-tailed and hence easier to tabulate. Accordingly

> **We reject the null hypothesis that the two samples come from identical populations and accept the alternative hypothesis that the two populations have unequal means, if**
>
> $$U \leq U'_\alpha$$
>
> **where U'_α may be read from Table VII for values of n_1 and n_2 through 15, and $\alpha = 0.05$ and $\alpha = 0.01$.**

In the construction of Table VII, U'_α is the largest value of U for which the probability of $U \leq U'_\alpha$ is less than or equal to α, and the blank spaces indicate that the null hypothesis cannot be rejected at the given level of significance regardless of the calculated value of U. More extensive tables may be found in handbooks of statistical tables, but when n_1 and n_2 are both greater than 8, we can use instead the large-sample test described later in this section.

Returning now to our example dealing with the lifetimes of the two kinds of batteries, let us use the U test at the 0.05 level of significance to test the null hypothesis that the two samples come from identical populations against the alternative hypothesis that the two populations have unequal means. Having already shown that $W_1 = 113$ and $W_2 = 187$, we find that

$$U_1 = 12 \cdot 12 + \frac{12 \cdot 13}{2} - 113 = 109$$

and

$$U_2 = 12 \cdot 12 + \frac{12 \cdot 13}{2} - 187 = 35$$

and, hence, that $U = 35$. Note that $U_1 + U_2 = 109 + 35 = 144$, which equals $n_1 n_2 = 12 \cdot 12$.

Finally, since $U = 35$ is less than 37, the value of U'_α in Table VII for $n_1 = 12$, $n_2 = 12$, and $\alpha = 0.05$, we find that the null hypothesis must be rejected; in other words, we conclude that there is a difference in the mean lifetime of the two kinds of batteries.

Table VII, slightly modified, can also be used when the alternative hypothesis is $\mu_1 < \mu_2$ or $\mu_1 > \mu_2$. In that case the test is based on U_1 or U_2 instead of U, and this complicates matters because the same one-tailed criterion is used regardless of whether the inequality is $<$ or $>$. Thus, we shall consider one-sided alternative hypotheses here only in connection with the large-sample test, which we describe next.

The large-sample U test may be based on either U_1 or U_2 as defined on page 415, but since the resulting tests are equivalent and it does not matter how we number the samples, we shall use here the statistic U_1.

Under the null hypothesis that the two samples come from identical populations, it can be shown that the mean and the standard deviation of the sampling distribution of U_1 are†

Mean and standard deviation of U_1 statistic

$$\mu_{U_1} = \frac{n_1 n_2}{2}$$

and

$$\sigma_{U_1} = \sqrt{\frac{n_1 n_2 (n_1 + n_2 + 1)}{12}}$$

Furthermore, if n_1 and n_2 are both greater than 8, the sampling distribution of U_1 can be approximated closely with a normal curve. Thus, we base the test of the null hypothesis that the two samples come from identical populations on the statistic

Statistic for large-sample U test

$$z = \frac{U_1 - \mu_{U_1}}{\sigma_{U_1}}$$

which has approximately the standard normal distribution. If the alternative hypothesis is $\mu_1 \neq \mu_2$, we reject the null hypothesis for $z < -z_{\alpha/2}$ or $z > z_{\alpha/2}$; if the alternative hypothesis is $\mu_1 < \mu_2$, we reject the null hypothesis for $z > z_\alpha$ since large values of U_1 correspond to small values of W_1; and if the alternative hypothesis is $\mu_1 > \mu_2$, we reject the null hypothesis for $z < -z_\alpha$ since small values of U_1 correspond to large values of W_1.

EXAMPLE   The following are the weight gains (in pounds) of two random samples of young turkeys fed two different diets but otherwise kept under identical conditions:

†If there are ties in rank, these formulas provide only approximations, but if the number of ties is small, there is usually no need to make any corrections.

Diet 1: 16.3, 10.1, 10.7, 13.5, 14.9, 11.8, 14.3, 10.2
12.0, 14.7, 23.6, 15.1, 14.5, 18.4, 13.2, 14.0
Diet 2: 21.3, 23.8, 15.4, 19.6, 12.0, 13.9, 18.8, 19.2
15.3, 20.1, 14.8, 18.9, 20.7, 21.1, 15.8, 16.2

Use the large-sample U test at the 0.01 level of significance to test the null hypothesis that the two populations sampled have identical distributions against the alternative hypothesis that on the average the second diet produces a greater gain in weight.

SOLUTION Arranging the data jointly according to size, we get 10.1, 10.2, 10.7, 11.8, 12.0, 12.0, 13.2, 13.5, 13.9, 14.0, 14.3, 14.5, 14.7, 14.8, 14.9, 15.1, 15.3, 15.4, 15.8, 16.2, 16.3, 18.4, 18.8, 18.9, 19.2, 19.6, 20.1, 20.7, 21.1, 21.3, 23.6, and 23.8. Then, assigning the data in this order the ranks 1, 2, 3, ..., and 32, we find that the values of the first sample (diet 1) occupy ranks 1, 2, 3, 4, 5.5, 7, 8, 10, 11, 12, 13, 15, 16, 21, 22, and 31, while those of the second sample (diet 2) occupy ranks 5.5, 9, 14, 17, 18, 19, 20, 23, 24, 25, 26, 27, 28, 29, 30, and 32. (The 5th and 6th values are both equal to 12.0, so we assigned each the rank 5.5.) Thus,

$$W_1 = 1 + 2 + 3 + 4 + 5.5 + 7 + 8 + 10 + 11 + 12$$
$$+ 13 + 15 + 16 + 21 + 22 + 31$$
$$= 181.5$$

and

$$U_1 = 16 \cdot 16 + \frac{16 \cdot 17}{2} - 181.5$$
$$= 210.5$$

and since

$$\mu_{U_1} = \frac{16 \cdot 16}{2} = 128 \quad \text{and} \quad \sigma_{U_1} = \sqrt{\frac{16 \cdot 16 \cdot 33}{12}} = 26.53$$

it follows that

$$z = \frac{210.5 - 128}{26.53} = 3.11$$

Since $z = 3.11$ exceeds $z_{0.01} = 2.33$, the null hypothesis must be rejected; in other words, we conclude that on the average the second diet produces a greater weight gain.

16.4

Rank Sums: The *H* Test

The *H* test, or **Kruskal–Wallis test**, is a rank-sum test which serves to test the null hypothesis that k independent random samples come from identical populations against the alternative hypothesis that the means of these populations are not all equal. Unlike the standard test which it replaces, the one-way analysis of variance of Section 15.2, it does not require the assumption that the samples come from populations having roughly the shape of normal distributions.

In the *H* test the data are ranked jointly from low to high as though they constitute a single sample. Then, if R_i is the sum of the ranks assigned to the n_i values of the ith sample and $n = n_1 + n_2 + \cdots + n_k$, the *H* test is based on the statistic

Statistic for H test

$$H = \frac{12}{n(n+1)} \sum_{i=1}^{k} \frac{R_i^2}{n_i} - 3(n+1)$$

If the null hypothesis is true and each sample has at least five observations, the sampling distribution of H can be approximated closely with a chi-square distribution with $k - 1$ degrees of freedom. Consequently, we can reject the null hypothesis that $\mu_1 = \mu_2 = \cdots = \mu_k$ and accept the alternative that the μ's are not all equal at the level of significance α, if $H > \chi_\alpha^2$ for $k - 1$ degrees of freedom. If any sample has less than five items, the χ^2 approximation cannot be used, and the test must be based on special tables.

EXAMPLE A company's trainees are randomly assigned to groups which are taught a certain industrial inspection procedure by three different methods, and at the end of the instruction period they are tested for inspection performance quality. The following are their scores:

> *Method E:* 94, 87, 91, 74, 86, 97
> *Method F:* 85, 82, 79, 84, 61, 72, 80
> *Method G:* 89, 67, 72, 76, 69

Use the *H* test to determine at the 0.05 level of significance whether the three methods are equally effective.

SOLUTION After arranging the 18 scores according to size and ranking them from low to high, we find that

$$R_1 = 6 + 13 + 14 + 16 + 17 + 18 = 84$$
$$R_2 = 1 + 4.5 + 8 + 9 + 10 + 11 + 12 = 55.5$$

and

$$R_3 = 2 + 3 + 4.5 + 7 + 15 = 31.5$$

There is one tie, and the two tied observations which occupy ranks 4 and 5 are both ranked 4.5.

Then, substituting these values of R_1, R_2, and R_3 together with $n_1 = 6$, $n_2 = 7$, and $n_3 = 5$ into the formula for H, we get

$$H = \frac{12}{18 \cdot 19}\left(\frac{84^2}{6} + \frac{55.5^2}{7} + \frac{31.5^2}{5}\right) - 3 \cdot 19$$
$$= 6.67$$

Since 6.67 exceeds 5.991, the value of $\chi^2_{0.05}$ for $k - 1 = 3 - 1 = 2$ degrees of freedom, we reject the null hypothesis; that is, we conclude that the three methods are not equally effective.

EXERCISES

16.11 The following are the numbers of mistakes counted on pages randomly selected from reports typed by a company's two secretaries:

Secretary 1: 6, 14, 10, 10, 5, 11
Secretary 2: 4, 6, 7, 5, 2, 9

Use the U test at the 0.05 level of significance to test the null hypothesis that the two secretaries average equally many mistakes per page against the alternative hypothesis that they do not average equally many mistakes per page.

16.12 The following are the numbers of minutes it took random samples of 15 men and 12 women to complete a short screening test given to job applicants at a large bank:

Men: 6.5, 10.0, 7.0, 9.8, 8.5, 9.2, 9.0, 8.2, 10.8, 8.7, 6.7, 8.1, 7.9, 6.4, 8.9
Women: 8.6, 7.8, 8.3, 6.6, 10.5, 6.3, 9.3, 8.4, 9.7, 8.8, 9.9, 7.6

Use the U test (based on Table VII) at the 0.05 level of significance to test the null hypothesis that the two samples come from identical populations against the alternative hypothesis that they have unequal means.

16.13 The following are the Rockwell hardness numbers obtained for six aluminum die castings randomly selected from production lot A and for eight aluminum die castings randomly selected from production lot B:

Production lot A: 75, 56, 63, 70, 58, 74

Production lot B: 63, 85, 77, 80, 86, 76, 72, 82

Use the *U* test at the 0.05 level of significance to test the claim that the average hardness of die castings from the two production lots is the same.

16.14 An examination designed to measure basic economic knowledge was given to random samples of 25 male freshmen and 25 female freshmen entering a major university, and their grades were

Women: 75, 70, 56, 90, 85, 91, 95, 98, 89, 68, 96, 74, 95, 88, 60, 67, 88, 90, 76, 94, 99, 82, 71, 94, 78

Men: 87, 72, 43, 54, 69, 72, 98, 92, 86, 64, 60, 49, 86, 61, 61, 86, 35, 61, 73, 76, 32, 73, 61, 66, 74

Apply the *U* test at the 0.05 level of significance to test the null hypothesis that there is no difference in the average knowledge of economics between male and female freshmen entering the university.

16.15 The following are data on the breaking strength (in pounds) of random samples of two kinds of 2-inch cotton ribbons:

Type I ribbon: 143, 180, 199, 186, 168, 170, 185, 193, 198, 196, 175, 181, 132, 192, 196, 164, 179, 197, 176, 180

Type II ribbon: 176, 165, 173, 195, 177, 199, 155, 135, 172, 183, 170, 165, 186, 160, 162, 190, 171, 165, 189, 171, 180, 176, 193, 194

Use the *U* test and the level of significance $\alpha = 0.05$ to test the null hypothesis that there is no difference in the average strength of the two kinds of ribbons.

16.16 Use the large-sample *U* test to rework Exercise 16.12.

16.17 The following are the weekly food expenditures (in dollars) of families with two children chosen at random from two suburbs of a large city:

Suburb K: 88.60, 80.50, 85.38, 96.45, 77.95, 80.78, 77.89, 82.00, 74.19, 81.15

Suburb L: 72.63, 85.16, 65.35, 88.19, 81.72, 73.12, 85.91, 76.51, 73.76, 70.78

Use the large-sample *U* test at the 0.05 level of significance to test the claim that on the average such weekly food expenditures are higher in suburb *K* than in suburb *L*.

16.18 The following are the miles per gallon which a test driver got in random samples of six tankfuls each of three kinds of gasoline:

Gasoline 1: 28, 23, 26, 31, 14, 29

Gasoline 2: 21, 31, 32, 19, 27, 16

Gasoline 3: 24, 17, 21, 31, 22, 18

Use the H test at the 0.05 level of significance to test the claim that there is no difference in the true average mileage yield of the three kinds of gasoline.

16.19 Use the H test to rework Exercise 15.3 on page 398.

16.20 Use the H test to rework Exercise 15.6 on page 399.

16.21 The following are the lifetimes (in hours) of samples of four kinds of light bulbs in continuous use:

      Brand A: 603, 625, 641, 622, 585, 593, 660, 600, 633, 580, 615, 648
      Brand B: 620, 640, 646, 620, 652, 639, 590, 646, 631, 669, 610, 619
      Brand C: 587, 602, 617, 650, 588, 612, 574, 628, 617, 598, 602, 657
      Brand D: 626, 608, 596, 601, 637, 654, 601, 597, 644, 614, 582, 626

Use the H test at the 0.05 level of significance to test the null hypothesis that the true average lifetimes of the four kinds of light bulbs are all equal.

16.5

Tests of Randomness: Runs

All the methods of inference we have discussed so far are based on the assumption that we are dealing with random samples. However, there are many applications in which it is hard to decide whether this assumption is justifiable. This is true, particularly, when we have little or no control over the selection of the data. For instance, if we want to predict a department store's sales volume for a given month, we have no choice but to use sales data from previous years and, perhaps, collateral information about economic conditions in general. None of this information constitutes a random sample in the strict sense. Also, we have no choice but to rely on whatever records are available if we want to make long-range predictions of the weather, if we want to estimate the mortality rate of a disease, or if we want to study traffic accidents at a dangerous intersection.

Several methods have been developed in recent years which make it possible to judge the randomness of a sample on the basis of the order in which the observations are taken. We can thus test, after the data have been taken, whether patterns that look suspiciously nonrandom may be attributed to chance. The technique we shall describe in this and the following section is based on the **theory of runs**. A **run** is a succession of identical letters (or other kinds of symbols) which is followed and preceded by different letters or no letters at all. To illustrate, consider the following arrangement of letters typed by a secretary in the given order which do, *D*, and do not, *N*, contain erasures:

<u>*NNNN*</u><u>*DDD*</u><u>*NNNNNNN*</u><u>*DD*</u><u>*NN*</u><u>*DDDD*</u>

Using underlines to combine the letters which constitute the runs, we find that first there is a run of four N's, then a run of three D's, then a run of seven N's, then a run of two D's, then a run of two N's, and finally a run of four D's. In all, there are six runs of varying lengths.

The **total number of runs** appearing in an arrangement of this kind is often a good indication of a possible lack of randomness. If there are too few runs, we might suspect a definite grouping or clustering, or perhaps a trend; if there are too many runs, we might suspect some sort of repeated alternating pattern. In the example above there seems to be a definite clustering—the letters with erasures seem to come in groups—but it remains to be seen whether this is significant or whether it can be attributed to chance.

If there are n_1 letters of one kind, n_2 letters of another kind, and u runs, we base this kind of decision on the following criterion:

Reject the null hypothesis of randomness if

$$u \leq u'_{\alpha/2} \quad \text{or} \quad u \geq u_{\alpha/2}$$

where $u'_{\alpha/2}$ and $u_{\alpha/2}$ may be read from Table VIII for values of n_1 and n_2 through 15, and $\alpha = 0.05$ and $\alpha = 0.01$.

In the construction of Table VIII, $u'_{\alpha/2}$ is the largest value of u for which the probability of $u \leq u'_{\alpha/2}$ is less than or equal to $\alpha/2$, $u_{\alpha/2}$ is the smallest value of u for which the probability of $u \geq u_{\alpha/2}$ is less than or equal to $\alpha/2$, and the blank spaces indicate that the null hypothesis of randomness cannot be rejected for values in that tail of the sampling distribution of u regardless of the value of u. More extensive tables for the u test may be found in handbooks of statistical tables.

Returning to our illustration dealing with the letters with and without erasures, we find that $n_1 = 13$, $n_2 = 9$, and $u = 6$. For these values of n_1 and n_2 we get $u'_{0.025} = 6$ and $u_{0.025} = 17$ from Table VIII, and it follows that the null hypothesis of randomness must be rejected at the 0.05 level of significance. The letters with erasures seem to come in clusters.

Under the null hypothesis that n_1 letters of one kind and n_2 letters of another kind are arranged at random, it can be shown that the mean and the standard deviation of u are

Mean and standard deviation of u

$$\mu_u = \frac{2n_1 n_2}{n_1 + n_2} + 1$$

and

$$\sigma_u = \sqrt{\frac{2n_1 n_2 (2n_1 n_2 - n_1 - n_2)}{(n_1 + n_2)^2 (n_1 + n_2 - 1)}}$$

Furthermore, if neither n_1 nor n_2 is less than 10, the sampling distribution of u can be approximated closely with a normal curve, and we can base the test of the null hypothesis of randomness on the statistic

Statistic for large-sample u test

$$z = \frac{u - \mu_u}{\sigma_u}$$

which has approximately the standard normal distribution. If the alternative hypothesis is that the arrangement is not random, we reject the null hypothesis for $z < -z_{\alpha/2}$ or $z > z_{\alpha/2}$; if the alternative hypothesis is that there is a clustering or a trend, we reject the null hypothesis for $z < -z_\alpha$; and if the alternative hypothesis is that there is an alternating, or cyclical, pattern, we reject the null hypothesis for $z > z_\alpha$.

EXAMPLE The following is an arrangement of 30 men, *M*, and 18 women, *W*, lined up to purchase tickets for a rock concert:

$$MWMWMMMWMWMMMWWMMMMWWMWM$$
(cont.) $$MMWMMMWWWWMWMMMWMWMMMMWWM$$

Test for randomness at the 0.05 level of significance.

SOLUTION Since $n_1 = 30$, $n_2 = 18$, and $u = 27$, we get

$$\mu_u = \frac{2 \cdot 30 \cdot 18}{30 + 18} + 1 = 23.5$$

$$\sigma_u = \sqrt{\frac{2 \cdot 30 \cdot 18(2 \cdot 30 \cdot 18 - 30 - 18)}{(30 + 18)^2(30 + 18 - 1)}} = 3.21$$

and, hence,

$$z = \frac{27 - 23.5}{3.21} = 1.09$$

Since this value does not exceed $z_{0.025} = 1.96$, the null hypothesis cannot be rejected; in other words, there is no real evidence to suggest that the arrangement is not random.

16.6

Tests of Randomness:
Runs Above and Below the Median

The method of the preceding section is not limited to tests of the randomness of series of attributes (such as the *N*'s and *D*'s and the *M*'s and *W*'s of our examples). Any sample consisting of numerical measurements or observations can be treated similarly by using the letters a and b to denote values falling above, a, and below, b, the median of the sample. Numbers equal to the median are omitted. The resulting series of a's and b's (representing the data in their original order) can be tested for randomness on the basis of the total number of runs of a's and b's, the total number of runs above and below the median.

EXAMPLE On 24 successive trips between two cities, a bus carried 24, 19, 32, 28, 21, 23, 26, 17, 20, 28, 30, 24, 13, 35, 26, 21, 19, 29, 27, 18, 26, 14, 21, and 23 passengers. Use the total number of runs above and below the median and the 0.01 level of significance to test whether it is reasonable to treat these data as if they constitute a random sample.

SOLUTION The median number of passengers is 23.5, so we get the following arrangement of values above and below it

$$abaabbabbaaabaabbaababbb$$

in which there are $u = 14$ runs. For $n_1 = 12$, $n_2 = 12$, and $\alpha = 0.01$, we get $u'_{0.005} = 6$ and $u_{0.005} = 20$ from Table VIII, and since $u = 14$ falls between these two critical values, the null hypothesis of randomness cannot be rejected. In other words, there is no real evidence to suggest that the data may not be treated as if they constitute a random sample.

EXERCISES

16.22 The following sequence of C's and A's shows the order in which 25 cars with California or Arizona license plates crossed the Colorado river entering Arizona at Blythe, California:

$$CAACACCACCCAACAAAACACCACC$$

Test for randomness at the 0.05 level of significance.

16.23 The following is the order in which 22 buy gold, B, and sell gold, S, orders were received by a Swiss bank over a given period of time:

$$BBBBBBBBBSSSSSSBSSBBBBB$$

Test for randomness at the 0.01 level of significance.

16.24 A driver buys gasoline either at a Texaco station, T, or at an Arco station, A, and the following arrangement shows the order of the stations from which the driver bought over a certain period of time:

$$TTTATATAATTATATATAATAT$$

Test for randomness at the 0.05 level of significance.

16.25 The following arrangement indicates whether the price of a certain stock changed, *C*, or remained the same, *R*, in 50 trading days:

$$CCCCRCCRCCCCRRCRCCCRRRCCR$$
(cont.) $$RCCCRRCRCCCCCCRCCRRCCCRCC$$

Test for randomness at the level of significance $\alpha = 0.05$.

16.26 Representing each 0, 2, 4, 6, and 8 by the letter *E* (for even) and each 1, 3, 5, 7, and 9 by the letter *O* (for odd), test at the 0.05 level of significance whether the arrangement of the 50 digits in the first column of the random-number table on page 581 may be regarded as random.

16.27 To test whether a radio signal contains a message or constitutes random noise, an interval of time is subdivided into a number of very short intervals and for each of these it is determined whether the signal strength exceeds, *E*, or does not exceed, *N*, a certain level of background noise. Test at the level of significance $\alpha = 0.01$ whether the following arrangement, thus obtained, may be regarded as random, and hence that the signal contains no message and may be regarded as random noise:

$$NNNENENENEEENEEENEENENEE$$
(cont.) $$NEENNENEEENENNNENNENNNE$$

16.28 The following are the numbers of defective pieces turned out by a machine during 24 consecutive shifts: 15, 11, 17, 14, 16, 12, 19, 17, 21, 15, 17, 19, 21, 14, 22, 16, 19, 12, 16, 14, 18, 17, 24, and 13. Test for randomness at the level of significance $\alpha = 0.01$.

16.29 The following are the numbers of lunches that an insurance agent claimed as business deductions in 30 consecutive months: 6, 7, 5, 6, 8, 6, 8, 6, 6, 4, 3, 2, 4, 4, 3, 4, 7, 5, 6, 8, 6, 6, 3, 4, 2, 5, 4, 4, 3, and 7. Discarding the three values which equal the median, test for randomness at the 0.01 level of significance.

16.30 The total numbers of retail stores opening for business and also quitting business within the calendar years 1948–1980 in a large city were 108, 103, 109, 107, 125, 142, 147, 122, 116, 153, 144, 162, 143, 126, 145, 129, 134, 137, 143, 150, 148, 152, 125, 106, 112, 139, 132, 122, 138, 148, 155, 146, and 158; the median number is 138. Test at the 0.05 level of significance whether there is a significant upward trend in these data.

16.31 Reading successive rows across in Exercise 2.23 on page 25, test at the 0.05 level of significance whether the data on the productivity of the 100 workers may be regarded as a random sample.

16.7

A Word of Caution

Statistical methods which require no (or virtually no) assumptions about the populations from which we are sampling are usually less efficient than the corresponding standard techniques. To illustrate this point, let us refer to the calculations in Chapter 11 where we showed that, by using Chebyshev's theorem, we can assert with a probability of *at least* 0.75 that the mean of a random sample of size $n = 64$ drawn from an infinite population with $\sigma = 20$ will differ from the population mean μ by less than 5. As we also showed, however, if we can assume that the population from which we are sampling has the shape of a normal distribution, we can make the same assertion with the probability 0.9544 (instead of "at least" 0.75). To put it another way, assertions made with equal confidence require larger samples if they are made without knowledge of the form of the underlying distribution than if they are made with such knowledge. It is generally true that the more we are willing to assume, the more we can infer from a sample; however, the more we assume, the more we limit the applicability of our methods.

16.8

Check List of Key Terms

H test, 419
Kruskal–Wallis test, 419
Mann–Whitney test, 414
Nonparametric tests, 409
One-sample sign test, 410
Paired-sample sign test, 412
Rank sums, 415
Run, 422

Runs above and below the median, 425
Sign test, 410
Theory of runs, 422
Total number of runs, 423
u test, 423
U test, 414
Wilcoxon test, 412

16.9

Review Exercises

16.32 The following are the prices (in cents) charged for a certain item in a random sample of 18 discount drugstores: 57, 58, 54, 56, 57, 55, 54, 57, 57, 54, 59, 56, 60, 54, 56, 58, 58, and 58. Use the sign test at the 0.05 level of significance to test the null hypothesis that the mean item price in the population sampled is $\mu = 55$ cents against the alternative hypothesis $\mu \neq 55$ cents.

16.33 To compare four bowling balls, a professional bowler bowls five games with each ball and gets the following results:

> *Bowling ball D:* 221, 232, 207, 198, 212
> *Bowling ball E:* 202, 225, 252, 218, 226
> *Bowling ball F:* 210, 205, 189, 196, 216
> *Bowling ball G:* 229, 192, 247, 220, 208

Use the H test at the 0.05 level of significance to test the null hypothesis that on the average the bowler performs equally well with the four bowling balls.

16.34 The following are the numbers of 3-month and 6-month certificates of deposit which a major bank sold on a random sample of 32 business days: 20 and 16, 14 and 14, 12 and 9, 23 and 18, 19 and 16, 15 and 20, 11 and 11, 15 and 12, 17 and 13, 20 and 16, 10 and 14, 13 and 11, 15 and 15, 18 and 12, 22 and 20, 17 and 11, 8 and 6, 13 and 12, 15 and 8, 12 and 16, 21 and 14, 14 and 11, 9 and 10, 14 and 10, 11 and 17, 22 and 15, 17 and 10, 10 and 14, 23 and 16, 14 and 6, 7 and 7, and 13 and 8. Use the sign test at the 0.01 level of significance to test the null hypothesis that on the average the bank sells equally many 3- and 6-month certificates against the alternative hypothesis that on the average it sells more 3-month certificates.

16.35 The following are the percentages of their assets which U.S. life insurance companies invested in stocks and bonds in the years 1949–1974: 38.8, 39.6, 41.2, 42.9, 43.9, 44.2, 43.7, 43.3, 43.4, 43.8, 43.6, 43.3, 43.4, 43.1, 42.9, 42.4, 42.4, 41.6, 42.5, 43.2, 42.9, 42.7, 45.0, 47.1, 46.7, and 45.0. Use the large-sample test based on runs above and below the median and the 0.05 level of significance to test whether there is a significant trend in this series.

16.36 The following are the grades which a random sample of 17 students received on the midterm and final examinations in accounting: 65 and 78, 78 and 81, 75 and 78, 80 and 76, 62 and 68, 68 and 77, 72 and 72, 81 and 79, 63 and 66, 84 and 89, 74 and 71, 92 and 98, 70 and 75, 74 and 72, 90 and 94, 77 and 79, and 84 and 84. Use the sign test (based on Table V) and the 0.05 level of significance to test the null hypothesis $\mu_1 = \mu_2$ (that the true average midterm and final examination grades of the population sampled are equal) against the alternative hypothesis $\mu_1 < \mu_2$.

16.37 The following arrangement shows whether 50 persons interviewed consecutively in the order shown were for, F, or against, A, a proposed city rezoning ordinance:

$$AAAAAFAAFFAAAAAAAFAAAAFFF$$
$$(cont.) \quad AAAFAAFAAAAFFAAAAAAAFAAA$$

Test for randomness at the level of significance $\alpha = 0.01$.

16.38 In a random sample of six rounds of golf at the Paradise Valley Country Club, a professional golfer scored 71, 69, 72, 74, 71, and 72. Use the sign test at the 0.05 level of significance to test the null hypothesis $\mu = 70$ (that the golfer can be expected to average 70 on this course) against the alternative hypothesis $\mu > 70$.

16.39 The following is the order in which prices of cotton went up, U, or down, D, on fifteen trading days in March:

$$DDDUDUUUUDDDUDD$$

Test for randomness at the 0.05 level of significance.

16.40 In two random samples of size 14 each it was found that insecticide A was 43.0, 41.9, 46.9, 44.6, 43.9, 42.0, 44.0, 44.5, 41.0, 43.1, 39.0, 45.2, 44.6, and 42.0 percent effective in killing mosquitos, and insecticide B was 39.6, 45.7, 39.8, 42.8, 41.2, 45.0, 40.2, 40.1, 40.2, 41.7, 37.4, 38.8, 41.7, and 38.7 percent effective. Use the large-sample U test at the 0.05 level of significance to test the null hypothesis that on the average the two insecticides are equally effective against the alternative hypothesis that on the average they are not equally effective.

16.41 Rework the preceding exercise using Table VII instead of the large-sample approximation.

16.42 The following are the numbers of minutes that patients had to wait in the offices of four doctors:

> Doctor 1: 20, 25, 38, 31, 23, 22
> Doctor 2: 18, 19, 12, 21, 24, 9
> Doctor 3: 11, 8, 10, 27, 25, 14
> Doctor 4: 14, 17, 25, 28, 21, 15

Use the H test at the 0.05 level of significance to test the null hypothesis that the four samples come from identical populations against the alternative hypothesis that the means of the populations are not all equal.

16.43 Test at the 0.05 level of significance whether the following sequence of cars observing, O, and exceeding, E, the 55-mph speed limit may be regarded as random:

> OOOEEOEOOOOEEEOEOOOOEO

16.44 The following are the octane ratings measured in a random sample of 14 specimens of a certain kind of gasoline: 91.0, 89.8, 90.9, 93.6, 87.1, 90.0, 92.5, 90.5, 91.0, 88.2, 90.3, 92.6, 90.0, and 90.8. Use the sign test (based on Table V) and the 0.05 level of significance to test the null hypothesis $\mu = 90.0$ (that the true mean octane rating of the given kind of gasoline is 90.0) against the alternative hypothesis $\mu \neq 90.0$.

16.45 Rework the preceding exercise using the large-sample technique.

16.46 Before placing a large order for road flares to be carried by all company cars, a major fleet operator life tests flares of brand P and brand Q. The following data show the burning times (in minutes) observed in random samples of 12 flares of each kind:

> Brand P: 16, 18, 16, 13, 16, 15, 11, 15, 14, 16, 25, 17
> Brand Q: 17, 13, 15, 19, 18, 14, 20, 14, 19, 17, 19, 21

Use the U test (based on Table VII) and the 0.01 level of significance to test whether it is reasonable to conclude that there is no difference in the actual mean burning time of the two brands of flares.

16.47 Rework the preceding exercise using the large sample approximation to the sampling distribution of the U statistic.

16.48 The following sequence shows whether a television news program had at least 12 percent of the total viewing audience, A, or less than 12 percent, L, on 36 consecutive viewing days:

$$LLLLAALLLALLLAAAAL$$
$$(cont.) \quad ALLLAALLLLLALLLLLA$$

Test for randomness at the 0.05 level of significance.

The main objective of many statistical investigations is to establish relationships which make it possible to predict one or more variables in terms of others. For instance, studies are made to predict the future sales of a new product in terms of its price, a woman's weight loss in terms of the number of weeks she will stay on an 800-calorie-per-day diet, family expenditures for housing in terms of family income, the per capita consumption of certain food items in terms of the amount of money spent advertising them on television, and so forth.

It would be ideal, of course, if we could predict one quantity exactly in terms of another, but this is seldom possible; in most instances we must be satisfied with predicting averages or expected values. For instance, we cannot predict exactly how much money a specific college graduate will earn 10 years after graduation, but given suitable data we can predict the average earnings of all college graduates 10 years after graduation. Similarly, we can predict the average yield of a given variety of wheat in terms of the rainfall in the month of July, and we can predict the expected grade-point average of a student starting an M.B.A. program in terms of his undergraduate grade-point average. This problem of predicting the average value of one variable in terms of the known value of another variable (or the known values of other variables) is the problem of regression. This term dates back to Francis Galton, who used it first in connection with a study of the relationship between the heights of fathers and sons.

The case where predictions are based on the known value of one variable is treated in Sections 17.1 through 17.3, and the case where predictions are based on the known values of several variables is treated in Section 17.4. Related problems will be taken up in Chapter 19.

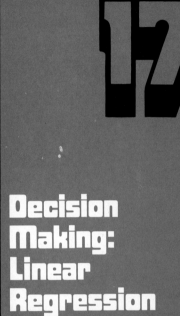

Decision Making: Linear Regression

Curve Fitting

Whenever possible, scientists strive to express, or approximate, relationships between known quantities and quantities that are to be predicted in terms of mathematical equations. In physics it is known, for example, that at a constant temperature the relationship between the volume, y, of a gas and its pressure, x, is given by the formula

$$y = \frac{k}{x}$$

where k is a numerical constant; also, in biology it has been discovered that the size of a culture of bacteria, y, can be expressed in terms of the time, x, it has been exposed to certain favorable conditions by means of the formula

$$y = a \cdot b^x$$

where a and b are numerical constants.

Business and economic statisticians have borrowed and continue to borrow liberally from the tools of the natural sciences. For instance, the first equation above, $y = k/x$, is often used to express the relationship between the demand, y, for a commodity and its price, x; and the second equation, $y = a \cdot b^x$, is often used to express, among other things, how a company's production or sales, y, grow with time, x.

In any situation where we want to use observed data to derive a mathematical equation and use it to predict the value of one variable from a given value of another—a procedure known generally as **curve fitting**—there are essentially three problems to be solved. We must decide what kind of "predicting" equation is to be used; then we must find the particular equation which is in some sense the best of its kind; and finally we must settle certain questions regarding the goodness of the particular equation, or of the predictions made from it.

With respect to the first of these problems of curve fitting, there are many different kinds of curves (and their equations) that can be used for predictive purposes. Designating the variable to be predicted (often called the "dependent" variable) by y and the predictor variable (the "independent" variable) by x, we might use a straight line having the equation $y = a + bx$, a parabola given by $y = a + bx + cx^2$, an exponential curve given by $y = a \cdot b^x$, or any one of many other mathematical equations. The choice of one of these is sometimes decided for us by theoretical considerations, but usually it is decided by direct inspection of the data. We plot the data on ordinary (arithmetic) graph paper, sometimes on special graph paper with logarithmic scales, and we decide by visual study of the plot upon the kind of curve which best describes the overall pattern of the data. There exist various methods for putting this decision on a more objective basis, but they will not be discussed in this book. So far as the second and third problems of curve fitting are concerned, we shall study the second in some detail in Section 17.2, and the third in Section 17.3.

17.2

Linear Regression

Of the various kinds of equations used to predict values of one variable, y, from associated values of another variable, x, the simplest and most widely used is the **linear equation in two unknowns**, which is of the form

$$y = a + bx$$

where a is the y intercept (the value of y at the point where $x = 0$) and b is the slope of the line (the change in y which accompanies a change of one unit in x). Ordinarily, the numerical constants a and b are estimated from sample data, and once they have been determined, we can substitute a given value of x into the equation and calculate the predicted value of y. Linear equations are useful and important not only because many relationships are actually of this form, but also because they often provide close approximations to relationships which would otherwise be difficult to describe in mathematical terms.

The term "linear equation" arises from the fact that, when plotted on ordinary graph paper (or arithmetic paper), all pairs of values of x and y which satisfy an equation of the form $y = a + bx$ fall on a straight line. Suppose, for instance, that a large mail-order firm wants to predict the number of orders it can expect to receive, y, in terms of the number of catalogs it distributes, x, and that on the basis of past experience it has derived the predicting equation

$$y = 11,400 + 1.8x$$

The graph of this linear equation is shown in Figure 17.1, and any pair of values of x and y which are such that $y = 11,400 + 1.8x$ forms a point (x, y) that falls on the line. Substituting $x = 5,000$, for example, the firm finds that when 5,000 catalogs are distributed, it can expect to receive $11,400 + (1.8)(5,000) = 20,400$ orders; and when $x = 10,000$ catalogs are distributed, it can expect to receive $11,400 + (1.8)(10,000) = 29,400$ orders.

Once we have decided to fit a straight line, we are faced with the problem of finding the equation of the particular line which in some sense provides the best possible fit to the observed data. To show how this is done, let us consider the following: An automatic vending machine company owns and controls a very large stock of vending machines of a certain kind. The machines are installed in various locations and they are of various ages. Company records show,

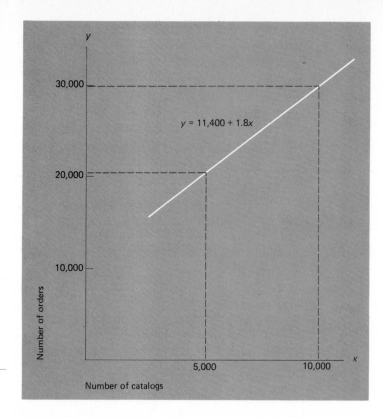

17.1

Graph of linear
equation.

among other things, for each active machine the amount of money spent on
its maintenance in its last (full) year of use. From this stock we take a random
sample of $n = 10$ machines and, recording for each its maintenance cost, y, and
its last year of use, x, we get†

Year x	Maintenance cost (dollars) y
4	148
2	128
3	133
5	154
2	118
3	145
4	143
5	159
4	142
3	127

†Here and there, for the sake of simplicity, we shall call this (somewhat loosely) the "age"
of a machine.

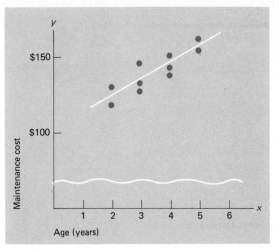

17.2

Data on age and
maintenance cost of
vending machines.

Plotting these pairs of values on arithmetic paper as in Figure 17.2, we observe that, although the points do not all fall on a straight line, the overall pattern of the relationship is reasonably well described by the white line. There is no noticeable departure from linearity in the scatter of the points, so we feel justified in deciding that a straight line is a suitable description of the underlying relationship.

We now face the problem of finding the equation of the line which in some sense provides the best fit to the data and which, it is hoped, will later yield the best possible predictions of y from x. Logically speaking, there is no limit to the number of straight lines which can be drawn on a piece of graph paper. Some of these lines would be such obviously poor fits to the data that we could not consider them seriously, but there are many lines which would seem to provide more or less "good" fits, and the problem is to find that one line which fits the data "best" in some well-defined sense. If all the points actually fell on a straight line, the criterion would be self-evident, but this is an extreme case which we rarely encounter in practice. In general, we have to be satisfied with a line having certain desirable, but not perfect, properties.

The criterion which, nowadays, is used almost exclusively for defining a "best" fit dates back to the early part of the nineteenth century and the French mathematician Adrien Legendre; it is known as the **method of least squares**. As it will be used here, this method requires that the line which we fit to our data be such that the sum of the squares of the vertical deviations of the points from the line is a minimum.

For the maintenance-cost problem, the method of least squares requires that the sum of the squares of the distances represented by the solid-line segments of Figure 17.3 be as small as possible. To explain the nature of this procedure, let us consider one of the vending machines, say, the one whose maintenance cost was $118 during its second year of use. If we mark $x = 2$ on the horizontal scale and read the corresponding value of y off the line of Figure 17.3, we find

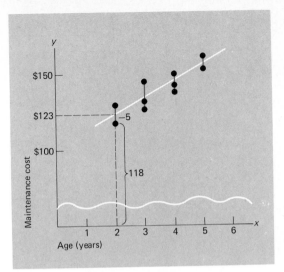

17.3

Line fitted to data on age and maintenance cost of vending machines.

that this maintenance cost is about \$123; therefore, the error in the prediction based on the line, represented by the vertical distance from the point to the line, is $118 - 123 = -5$. Altogether, there are 10 such errors in this example, and the least-squares criterion requires that we minimize the sum of their squares.

To show how a **least-squares line** is actually fitted to a set of data, let us consider n pairs of numbers $(x_1, y_1), (x_2, y_2), \ldots, (x_n, y_n)$, which might represent such things as the reading rate and reading comprehension of n financial analysts, the number of labor units per acre and crop yield per acre on n farms, the score on a paper-and-pencil personality test and vocational success of n college graduates, and product advertising expenditures and product sales of n consumer goods manufacturers.

If we write the equation of the line as $\hat{y} = a + bx$, where the symbol $\hat{y}$ ("y-hat") is used to distinguish between observed values of y and the corresponding values $\hat{y}$ on the line, the least-squares criterion requires that we minimize the sum of the squares of the differences between the y's and the $\hat{y}$'s (see Figure 17.4). This means that we must find the numerical values of the constants a and b appearing in the equation $\hat{y} = a + bx$ for which

$$\sum (y - \hat{y})^2 = \sum (y - a - bx)^2$$

is as small as possible. We shall not go through the derivation of the two equations, called the **normal equations**, which provide the solution to this problem. Instead, we merely state that minimizing $\sum (y - \hat{y})^2$ yields the following system of two linear equations in the unknowns a and b:

Normal equations

$$\sum y = na + b(\sum x)$$
$$\sum xy = a(\sum x) + b(\sum x^2)$$

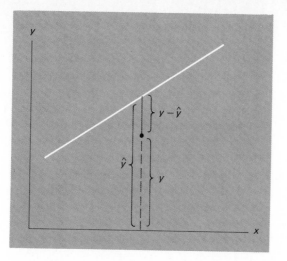

17.4

Least-squares line.

In these equations, whose solution gives the least-squares values of a and b, n is the number of pairs of observations, $\sum x$ and $\sum y$ are the sums of the observed x's and y's, $\sum x^2$ is the sum of the squares of the x's, and $\sum xy$ is the sum of the cross products of the x's and their associated observed values of y.

EXAMPLE   Fit a least-squares line to the vending machine data on page 434.

SOLUTION   We get the sums needed for substitution into the normal equations by performing the calculations shown in the following table:

Year x	Maintenance cost (dollars) y	x^2	xy
4	148	16	592
2	128	4	256
3	133	9	399
5	154	25	770
2	118	4	236
3	145	9	435
4	143	16	572
5	159	25	795
4	142	16	568
3	127	9	381
35	1,397	133	5,004

Substituting $n = 10$ and the four column totals into the two normal equations, we get

$$1{,}397 = 10a + 35b$$

$$5{,}004 = 35a + 133b$$

and we must now solve these two simultaneous linear equations for a and b. There are several ways in which this can be done; from elementary algebra we can use either the method of elimination or the method of determinants. In either case, we get $a = 101.5$ and $b = 10.9$.

As an alternative to these procedures we can use the following formulas, which result when we symbolically solve the two normal equations for a and b:

Solutions of
normal equations

$$a = \frac{(\sum y)(\sum x^2) - (\sum x)(\sum xy)}{n(\sum x^2) - (\sum x)^2}$$

$$b = \frac{n(\sum xy) - (\sum x)(\sum y)}{n(\sum x^2) - (\sum x)^2}$$

EXAMPLE   Rework the preceding example, using the formulas above for a and b.

SOLUTION   Substituting $n = 10$ and the various sums shown in the table above, we get as before

$$a = \frac{(1{,}397)(133) - (35)(5{,}004)}{10(133) - (35)^2} = \frac{10{,}661}{105} = 101.5$$

$$b = \frac{10(5{,}004) - (35)(1{,}397)}{105} = \frac{1{,}145}{105} = 10.9$$

There is still another way of finding a and b which is often used because of its convenience. We solve the first normal equation for a, getting

$$a = \frac{\sum y - b(\sum x)}{n}$$

Then we calculate b using the formula above and substitute its value into this formula for a.

Once we have determined the equation of a least-squares line, we can use it to make predictions.

EXAMPLE   Use the least-squares line $\hat{y} = 101.5 + 10.9x$ to predict the maintenance cost of one of the vending machines during its second year of use.

SOLUTION Substituting $x = 2$ into the equation, we get

$$\hat{y} = 101.5 + 10.9(2) = \$123.30$$

and this is the best estimate we can make in the least-squares sense.

In the discussion of this section we have considered only the problem of fitting a straight line to paired data. More generally, the method of least squares can also be used to fit other kinds of curves and to derive predicting equations in more than two unknowns. The problem of fitting some curves other than straight lines by the method of least squares will be discussed briefly in Chapter 19; a simple example of a predicting equation in more than two unknowns will be treated later in this chapter.

EXERCISES

17.1 The following table shows how many weeks six persons have worked at an automobile inspection station and the number of cars each one inspected between noon and 2 P.M. on a given day:

Number of weeks employed x	Number of cars checked y
5	14
1	13
7	20
2	12
12	20
9	22

(a) Solve the normal equations to find the equation of the least-squares line which will enable us to predict y in terms of x.
(b) Use the formulas on page 438 to check the values of a and b obtained in part (a).

17.2 Use the results of part (a) of the preceding exercise to estimate the number of cars we can expect someone who has been working at the inspection station for 10 weeks to inspect during this two-hour period.

17.3 The following sample data were collected in a study of the relationship between the conditioning period (in days) of a laminating glue and the storage life of the glue (in months) at 70°F:

Conditioning period x	1	2	3	4	4	5	5	6	6	7
Storage life y	7	8	9	8	9	11	10	13	14	13

Find the equation of the least-squares line from which we can predict the storage life of this glue in terms of the length of the conditioning period. Also plot the ten points representing the data, the "data points," together with the least-squares line on one diagram.

17.4 With reference to the preceding exercise, predict the storage life of the glue at 70°F when the conditioning period is $x = 5$ days.

17.5 In connection with planning servicing facilities for machines that require attention, a company wants to study the relationship between the number of machines waiting for attention at a given time and the average time required by operators to service the machines. More specifically, the company wants to know whether there is a tendency for operators to work faster (and reduce the service time) when the number of machines waiting for service is large. Accordingly, the company randomly selects eight records showing the number of machines in line at the beginning of a given time period and the number of services completed by the operator during the period. The following are the data:

Machines in line x	Number of completed services y
3	3
6	2
5	3
4	5
4	3
6	6
8	6
7	4

(a) Find the least-squares line which best fits the data and plot it together with the eight data points on one diagram.

(b) Predict the average number of services an operator will complete during a period when there are five machines in line at the beginning of the period.

17.6 The following table shows 10 weeks' sales of a downtown department store and its suburban branch (in $10,000):

Sales of downtown store	Sales of suburban store
71	49
64	31
58	24
80	68
63	30
69	40
76	62
60	22
66	35
55	16

(a) Find the equation of the least-squares line which will enable us to predict the sales of the suburban store from the sales of the downtown store.

(b) Predict the sales of the suburban store for a week in which the sales of the downtown store are $650,000.

17.7 The following data show the advertising expenses (expressed as a percentage of total expenses) and the net operating profits (expressed as a percentage of total sales) in a random sample of 12 sporting goods stores:

Advertising expenses x	Profit y
1.5	3.1
0.8	1.9
2.6	4.2
1.0	2.3
0.6	1.2
2.8	4.9
1.2	2.8
0.9	2.1
0.4	1.4
1.3	2.4
1.2	2.4
2.0	3.8

Fit a least-squares line and use it to predict the profit of such a store (expressed as a percentage of total sales) when its advertising expenses are 1.2 percent of its total expenses.

17.8 Suppose that in the preceding exercise we had wanted to predict, or estimate, how much should be spent on advertising so that there will be an expected net operating profit of 3 percent of total sales. We could substitute $\hat{y} = 3$ into the equation obtained in Exercise 17.7 and solve for x, but this would not be a prediction in the least-squares sense. To make the best possible least-squares predictions, or estimates, of advertising expenses, we denote profits x, advertising expenses y, and then fit a least-squares line to these data. Find the equation of this line and use it to estimate how much should be spent on advertising so that there will be an expected net operating profit of 3 percent of total sales.

17.9 A study is made to determine the relationship between the area burned in forest fires in U.S. counties having a formal fire-control organization (with a county ranger, smoke-chaser assistants during the fire season, fire-crew members, and so on, supported by annual fire-control appropriations) and the variable operating costs of the fire protection. (The ranger's salary is a fixed operating cost and is unrelated to the area burned. Money over and above the ranger's salary which determines how much he can travel, the equipment and facilities he can buy, and how many men he can hire—the variable operating funds—directly affect the quality of fire protection.) A random sample of the annual figures for 12 counties shows the following variable operating costs (in cents per acre) and the area burned (as a percentage of the protected area):

Variable operating costs *x*	Protected area burned *y*
2.8	0.09
1.7	0.24
1.9	0.36
2.5	0.12
2.7	0.20
0.8	0.40
1.5	0.34
2.8	0.14
2.2	0.22
1.2	0.42
2.1	0.30
0.7	0.48

(a) Fit a least-squares line to these data and plot the line together with the points representing the data on one diagram.

(b) Predict the percentage of protected area burned in a county having 2.0 cents per acre variable operating cost.

17.10 In a study devoted to determining the composition of brandy, a random sample of six brandy specimens showed the following tannin content and esters content (both in grams for 100 liters at 100 proof):

Tannin *x*	19	15	9	10	11	19
Esters *y*	31.7	32.3	8.5	14.3	14.0	17.8

(a) Find the equation of the line of best fit to these data and plot the line together with the original data on one diagram.

(b) What is the best estimate of the amount of esters in a brandy with tannin content of $x = 10$?

17.3

Regression Analysis

In the preceding section we used a least-squares line to predict the maintenance cost of a vending machine during its second year of use as $123.30, but even if we interpret the line correctly as a **regression line** (that is, treat the predictions made from it as averages, or expected values), several questions remain to be answered.

1. How good are the values we found for the constants a and b in the equation $\hat{y} = a + bx$? After all, $a = 101.5$ and $b = 10.9$ are only estimates based on a random sample, and if we based our work on another sample of ten of the vending machines, the method of least squares would surely lead to different values of a and b.

2. How good an estimate is \$123.30 of the true average maintenance cost of one of the vending machines during its second year?

Also, we might ask

3. How can we obtain limits (two numbers) and an associated degree of confidence which measure the goodness of a prediction in the same way in which a confidence interval measures the goodness of, say, an estimate of the mean of a population?

When in the first question above, we said that the numbers $a = 101.5$ and $b = 10.9$ are "only estimates based on sample data," we implied the existence of corresponding true values, usually denoted by α and β, and therefore of a true regression line $\mu_{y|x} = \alpha + \beta x$, where $\mu_{y|x}$ is read "the true mean of y for a given value of x." It is customary to refer to α and β as **regression coefficients**, and to a and b as the corresponding **estimated regression coefficients**.†

To clarify the idea of a true regression line, let us consider Figure 17.5, in which we have drawn the distributions of y for several values of x. With reference

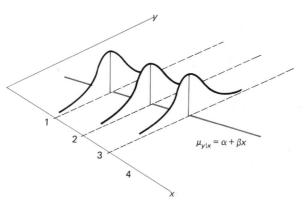

17.5

Distributions of y for given values of x.

to our example, these curves should be looked upon as the distributions of the maintenance costs of such vending machines during their first, second, and third years of use; to complete the picture, we can visualize similar distributions for all other values of x within the range of values under consideration. Note that the means of all the distributions of Figure 17.5 lie on the true regression line given by $\mu_{y|x} = \alpha + \beta x$.

In **linear regression analysis** we assume that the x's are constants, not values

†The coefficients α and β should not be confused with the probabilities α and β of committing Type I and Type II errors.

of random variables, and that for each value of x the variable to be predicted has a certain distribution (as shown in Figure 17.5) whose mean is $\alpha + \beta x$. In **normal regression analysis** we assume, furthermore, that these distributions are all normal distributions having the same standard deviation σ. In other words, the distributions pictured in Figure 17.5 as well as those we imagine, are normal distributions with the means $\alpha + \beta x$ and the standard deviation σ.

If we make all these assumptions, inferences about the regression coefficients α and β may be based on the statistics

Statistics for inferences about regression coefficients

$$t = \frac{a - \alpha}{s_e\sqrt{\dfrac{1}{n} + \dfrac{n \cdot \bar{x}^2}{n(\sum x^2) - (\sum x)^2}}}$$

$$t = \frac{b - \beta}{s_e}\sqrt{\frac{n(\sum x^2) - (\sum x)^2}{n}}$$

whose sampling distributions are t distributions with $n - 2$ degrees of freedom. Here α and β are the regression coefficients we want to estimate or test, and a and b are their estimates calculated from a given set of data by the method of least squares. Also, n, $\bar{x}$, $\sum x$, and $\sum x^2$ come from the original paired data, and s_e is an estimate of σ, the common standard deviation of the normal distributions of Figure 17.5, given by

$$s_e = \sqrt{\frac{\sum (y - \hat{y})^2}{n - 2}}$$

Here again, y is an observed value of y and $\hat{y}$ is the corresponding value on the least-squares line. We call s_e the **standard error of estimate**, and it should be observed that its square, s_e^2, is the sum of the squares of the vertical deviations from the points to the line in Figure 17.3 divided by $n - 2$. (We lose two degrees of freedom, so to speak, because α and β are replaced by estimates, a and b, in the formula for s_e.) In practice, it is easier to calculate s_e by means of the formula

Standard error of estimate

$$s_e = \sqrt{\frac{\sum y^2 - a(\sum y) - b(\sum xy)}{n - 2}}$$

The following example illustrates how the two t statistics are used to make inferences about the regression coefficients α and β.

EXAMPLE Suppose that someone claims that $\beta = 12.50$ in the vending-machine example, and that one purpose of the study was to test this claim. Now, β is the slope of the regression line; that is, it is the change in y associated with a change of one unit in x. Hence, the hypothesis asserts that, in the population of vending machines sampled, each year's use adds another \$12.50 on the average to the maintenance cost of a machine. Test this hypothesis against the alternative hypothesis $\beta \neq 12.50$ at the 0.05 level of significance.

SOLUTION Among the various quantities needed for substitution into the formula for the second of the two t statistics above, we have $n = 10$, $\sum x = 35$, $\sum x^2 = 133$, $\sum y = 1{,}397$, $\sum xy = 5{,}004$, and $b = 10.9$ from the work of the preceding section. By hypothesis $\beta = 12.5$, and to calculate s_e, the only other quantity needed, we first determine $\sum y^2 = 196{,}665$ from the original data. Now we have

$$s_e = \sqrt{\frac{196{,}665 - (101.5)(1{,}397) - (10.9)(5{,}004)}{8}}$$

$$= 6.4$$

and, hence,

$$t = \frac{10.9 - 12.5}{6.4} \sqrt{\frac{10(133) - (35)^2}{10}}$$

$$= -0.81$$

Since $t = -0.81$ falls between -2.306 and 2.306, where 2.306 is the value of $t_{0.025}$ for $n - 2 = 8$ degrees of freedom, we find that the null hypothesis cannot be rejected. In other words, the difference between $\beta = 12.50$ and $b = 10.9$ may well be due to chance.

Tests concerning the regression coefficient α are performed in the same way, except that we use the first, instead of the second, of the two t statistics above. In most practical applications, however, the regression coefficient α is not of as much interest as the regression coefficient β since α is just the y intercept. In many cases (for instance, in our example) this value has no real meaning.

To construct confidence intervals for the regression coefficients α and β, we substitute into the middle term of $-t_{\alpha/2} < t < t_{\alpha/2}$ the appropriate t statistic from page 444. Then, simple algebra leads to the confidence limit formulas

Confidence limits for regression coefficients

and

$$a \pm t_{\alpha/2} \cdot s_e \sqrt{\frac{1}{n} + \frac{n \cdot \bar{x}^2}{n(\sum x^2) - (\sum x)^2}}$$

$$b \pm \frac{t_{\alpha/2} \cdot s_e}{\sqrt{\dfrac{n(\sum x^2) - (\sum x)^2}{n}}}$$

where the degree of confidence is $1 - \alpha$ and $t_{\alpha/2}$ is the entry in Table II for $n - 2$ degrees of freedom.

Construct a 95% confidence interval for β in the vending-machine example.

Having all the quantities needed from the preceding example, we substitute them into the formula and get

$$10.9 \pm \frac{(2.306)(6.4)}{\sqrt{\dfrac{10(133) - (35)^2}{10}}}$$

This leads to the 95% confidence interval

$$6.3 < \beta < 15.5$$

which may seem rather wide. This lack of precision is due to two things—the large variation in the maintenance cost of vending machines of the same age and the small sample—but it does not mean that the estimate has no value in practice.

To answer the other two questions raised on page 443, about the "goodness" of an estimate or a prediction made from a least-squares equation, we use methods very similar to the ones just discussed. In fact, they are based on two more t statistics, which are given in Exercises 17.20 and 17.21.

EXERCISES

17.11 With reference to the vending-machines example, use the 0.05 level of significance to test the null hypothesis $\alpha = \$125.00$ against the alternative hypothesis $\alpha < \$125,000$.

17.12 With reference to the vending-machines example, construct a 95% confidence interval for the regression coefficient α.

17.13 With reference to Exercise 17.7 on page 441, test the null hypothesis $\beta = 1.6$ against the alternative hypothesis $\beta \neq 1.6$ at the 0.05 level of significance. Also, state in words the hypothesis being tested.

17.14 With reference to Exercise 17.9 on page 441, test the null hypothesis $\beta = -0.2$ against the alternative hypothesis $\beta > -0.2$ at the 0.05 level of significance.

17.15 The following table shows the assessed values and the selling prices of eight houses, constituting a random sample of all the houses sold in December 1979 in the Phoenix metropolitan area:

Assessed value (in $1,000)	Selling price (in $1,000)
40.3	63.4
72.0	118.3
32.5	55.2
44.8	74.0
27.9	48.8
51.6	81.1
80.4	123.2
58.0	92.5

Fit a least-squares line which will enable us to predict the selling price of a house in terms of its assessed value and test the null hypothesis $\beta = 1.30$ against the alternative hypothesis $\beta > 1.30$ at the 0.05 level of significance.

17.16 With reference to Exercise 17.7 on page 441, test the null hypothesis $\alpha = 0.4$ against the alternative hypothesis $\alpha \neq 0.4$ at the 0.01 level of significance. Also, state in words the hypothesis being tested.

17.17 With reference to Exercise 17.10 on page 442, construct a 95% confidence interval for the regression coefficient β.

17.18 With reference to Exercise 17.15, construct a 99% confidence interval for the regression coefficient β.

17.19 With reference to Exercise 17.9 on page 441, construct a 99% confidence interval for the regression coefficient α.

17.20 To answer the second of the three questions asked in the beginning of Section 17.3, we can use the following $(1 - \alpha)100\%$ confidence interval for $\mu_{y|x_0}$, the mean of y when $x = x_0$:

Confidence interval for mean of y when $x = x_0$

$$(a + bx_0) \pm t_{\alpha/2} \cdot s_e \sqrt{\frac{1}{n} + \frac{n(x_0 - \bar{x})^2}{n(\sum x^2) - (\sum x)^2}}$$

As before, $t_{\alpha/2}$ may be read from Table II, and the number of degrees of freedom is $n - 2$. To illustrate, let us refer again to the maintenance-cost example, and let us find a 95% confidence interval for the true average cost of maintaining such a vending machine during its second year of use. Copying the various quantities from pages 438 and 445 and substituting them, together with $x_0 = 2$, into the formula for the confidence limits, we get

$$123.3 \pm (2.306)(6.4) \sqrt{\frac{1}{10} + \frac{10(2 - 3.5)^2}{10(133) - (35)^2}}$$

$$123.3 \pm 14.76\sqrt{0.314}$$

and, hence, the 95% confidence interval

$$115.0 < \mu_{y|2} < 131.6$$

(a) With reference to Exercise 17.7 on page 441, find a 99% confidence interval for the true average net operating profits (expressed as a percentage of total sales) when advertising expenses are 1.5 percent of total expenses.

(b) With reference to Exercise 17.15, find a 95% confidence interval for the mean selling price of a house which is assessed at $60,000.

17.21 The third of the questions asked in the beginning of Section 17.3 differs from the other two in that it does not involve an estimate of a population parameter, but a prediction of a single future observation. Limits (two values) for which we can assert with a given degree of confidence that they will contain such an observation are called **limits of prediction**. Appropriate limits for predicting with $(1 - \alpha)100\%$ confidence a value of y when $x = x_0$ are given by

Limits of prediction

$$(a + bx_0) \pm t_{\alpha/2} \cdot s_e \sqrt{1 + \frac{1}{n} + \frac{n(x_0 - \bar{x})^2}{n(\sum x^2) - (\sum x)^2}}$$

Again, $t_{\alpha/2}$ may be read from Table II, and the number of degrees of freedom is $n - 2$. To illustrate, let us refer again to the maintenance-cost example, and let us find 95% limits of prediction for the maintenance cost of a particular vending machine during its second year. Noting that the only difference between the above limits and the confidence limits of Exercise 17.20 is that we add 1 to the quantity under the radical, we can immediately write the limits of prediction as

$$123.3 \pm 14.76\sqrt{1.314}$$

Thus, the 95% limits of prediction are 106.4 and 140.2.

(a) With reference to Exercise 17.7 on page 441, find 99% limits of prediction for the net operating profits (expressed as a percentage of total sales) when advertising expenses are 1.5 percent of total expenses.

(b) With reference to Exercise 17.15, find 95% limits of prediction for the selling price of a house which is assessed at $60,000.

17.4

Multiple Regression

Although there are many problems in which one variable can be predicted quite adequately in terms of another, it stands to reason that predictions should improve if one considers additional relevant information. For instance, we should be able to make better predictions of a company's newly hired salesmen's first-year sales if we consider not only their years of experience, but their sales aptitude, their education, and their personality. Also, we should be able to make better predictions of the performance of heavy equipment operators if we consider not only their years of experience, but their visual acuity, their ability to judge spatial relations, and their eye–hand coordination.

Many mathematical formulas can serve to express relationships between more than two variables, but most commonly used in statistics (partly for reasons of convenience) are linear equations of the form

$$y = b_0 + b_1x_1 + b_2x_2 + b_3x_3 + \cdots + b_kx_k$$

Here y is the variable which is to be predicted, $x_1, x_2, x_3, \ldots,$ and x_k are the k known variables on which predictions are to be based, and $b_0, b_1, b_2, b_3, \ldots,$ and b_k are numerical constants which must be determined from the observed data.

To illustrate, consider the following equation which was obtained in a study of the demand for different meats

$$\hat{y} = 3.489 - 0.090x_1 + 0.064x_2 + 0.019x_3$$

Here y denotes the total consumption of federally inspected beef and veal in millions of pounds, x_1 denotes a composite retail price of beef in cents per pound, x_2 denotes a composite retail price of pork in cents per pound, and x_3 denotes income as measured by a certain payroll index. With this equation, we can predict the total consumption of federally inspected beef and veal on the basis of specified values of x_1, x_2, and x_3.

The main problem in deriving a linear equation in more than two variables which best describes a given set of data is that of finding numerical values for b_0, b_1, b_2, b_3, ..., and b_k. This is usually done by the method of least squares; that is, we minimize the sum of squares $\sum (y - \hat{y})^2$, where, as before, the y's are the observed values and the $\hat{y}$'s are the values calculated by means of the linear equation. In principle, the problem of determining the values of b_0, b_1, b_2, b_3, ..., and b_k is the same as it is in a two-variable problem; however, manual solutions may be practically impossible because the method of least squares requires that we solve as many normal equations as there are unknown constants b_0, b_1, b_2, b_3, ..., and b_k. For instance, when there are two independent variables x_1 and x_2, and we want to fit the equation $y = b_0 + b_1 x_1 + b_2 x_2$, we must solve the three normal equations

Normal equations (two independent variables)

$$\sum y = n \cdot b_0 + b_1 (\sum x_1) + b_2 (\sum x_2)$$
$$\sum x_1 y = b_0 (\sum x_1) + b_1 (\sum x_1^2) + b_2 (\sum x_1 x_2)$$
$$\sum x_2 y = b_0 (\sum x_2) + b_1 (\sum x_1 x_2) + b_2 (\sum x_2^2)$$

Here $\sum x_1 y$ is the sum of the cross products of the given values of x_1 and their associated values of y, $\sum x_1 x_2$ is the sum of the cross products of the given values of x_1 and their associated values of x_2, and so on.

EXAMPLE   The following data show the number of bedrooms, the number of baths, and the prices at which eight one-family houses sold recently in a certain community:

Number of bedrooms x_1	Number of baths x_2	Price (dollars) y
3	2	48,800
2	1	44,300
4	3	53,800
2	1	44,200
3	2	49,700
2	2	44,900
5	3	58,400
4	2	52,900

Find a linear equation which will enable us to predict the average sale price of a one-family house in the given community in terms of the number of bedrooms and the number of baths.

SOLUTION   To get the sums needed for substitution into the three normal equations, we perform the calculations shown in the following table:

x_1	x_2	y	x_1y	x_2y	x_1^2	x_1x_2	x_2^2
3	2	48,800	146,400	97,600	9	6	4
2	1	44,300	88,600	44,300	4	2	1
4	3	53,800	215,200	161,400	16	12	9
2	1	44,200	88,400	44,200	4	2	1
3	2	49,700	149,100	99,400	9	6	4
2	2	44,900	89,800	89,800	4	4	4
5	3	58,400	292,000	175,200	25	15	9
4	2	52,900	211,600	105,800	16	8	4
25	16	397,000	1,281,100	817,700	87	55	36

Then, substituting the column totals and $n = 8$ into the normal equations, we get

$$397,000 = 8b_0 + 25b_1 + 16b_2$$
$$1,281,100 = 25b_0 + 87b_1 + 55b_2$$
$$817,700 = 16b_0 + 55b_1 + 36b_2$$

and the solution of this system of linear equations is $b_0 = 35,197$, $b_1 = 4,149$, and $b_2 = 731$ (see Exercise 17.25 on page 452). Thus, the least-squares equation is

$$\hat{y} = 35,197 + 4,149x_1 + 731x_2$$

and this tells us that, in this study, each extra bedroom adds on the average $4,149, and each bath $731, to the sale price of a house. To estimate (predict) the average sale price of three-bedroom houses with two baths, for instance, we substitute $x_1 = 3$ and $x_2 = 2$, and get

$$\hat{y} = 35,197 + (4,149)(3) + (731)(2) = \$49,106$$

or approximately $49,100.

EXERCISES   17.22 The following are data on the average weekly net profits (in $1,000) of five restaurants, their seating capacities, and the average daily traffic (in thousands of cars) which passes their locations:

Seating capacity x_1	Traffic count x_2	Weekly net profit y
120	19	23.8
200	8	24.2
150	12	22.0
180	15	26.2
240	16	33.5

(a) Fit an equation of the form $y = b_0 + b_1 x_1 + b_2 x_2$ to these data.

(b) Use the equation obtained in part (a) to predict the average weekly net profit of a restaurant with a seating capacity of 210 at a location where the daily traffic count averages 14,000 cars.

17.23 The following are data on the ages and incomes of a random sample of five executives working for a large multinational company, and the number of years each went to colllege:

Age x_1	Years college x_2	Income (dollars) y
37	4	41,200
45	0	36,800
38	5	45,000
42	2	40,300
31	4	35,400

(a) Fit an equation of the form $y = b_0 + b_1 x_1 + b_2 x_2$ to these data.

(b) Use the equation obtained in part (a) to estimate the average income of 40-year-old executives with four years of college working for this company.

17.24 Suppose that we are given the following sample data to study the relationship between the grades students get in a certain examination, their IQ's, and the number of hours they studied for the test:

Number of hours studied x_1	IQ x_2	Grade in examination y
8	98	56
5	99	44
11	118	79
13	94	72
10	109	70
5	116	54
18	97	94
15	100	85
2	99	33
8	114	65

(a) Fit an equation of the form $y = b_0 + b_1x_1 + b_2x_2$ to these data.

(b) Use the equation obtained in part (a) to predict the grade of a student with an IQ of 105 who studied 12 hours for the examination.

17.25 Use the method of elimination or determinants to verify that the solution of the three normal equations on page 449 is $b_0 = 35,197$, $b_1 = 4,149$, and $b_2 = 731$. (There will most likely be differences in the results due to rounding.)

17.5

Check List of Key Terms

Curve fitting, 432
Estimated regression coefficients, 443
Least-squares line, 436
Limits of prediction, 447
Linear equation, 433
Linear regression analysis, 443
Method of least squares, 435

Multiple regression, 448
Normal equations, 436
Normal regression analysis, 444
Regression, 431
Regression coefficients, 443
Regression line, 442
Standard error of estimate, 444

17.6

Review Exercises

17.26 The following sample data show the demand for a product (in thousands of units) and its price (in cents) charged in six different market areas:

Price x	Demand y
19	55
23	7
21	20
15	123
16	88
18	76

(a) Fit a least-squares line which will enable us to predict the demand for the product in terms of its price, and plot its graph.

(b) Use the equation obtained in part (a) to predict the demand for the product in a market area where it is priced at 15 cents.

17.27 The following are sample data provided by a moving company on the weights of six shipments, the distances they were moved, and the damage that was incurred:

Weight (1,000 pounds) x_1	Distance (1,000 miles) x_2	Damage (dollars) y
4.0	1.5	160
3.0	2.2	112
1.6	1.0	69
1.2	2.0	90
3.4	0.8	123
4.8	1.6	186

(a) Fit an equation of the form $y = b_0 + b_1x_1 + b_2x_2$ to these data.

(b) Use the equation obtained in part (a) to predict the damage when a shipment weighing 2,400 pounds is moved 1,200 miles.

17.28 The following sample data show the average annual yield of wheat (in bushels per acre) in a given county and the annual rainfall (in inches) measured from September through August:

Rainfall x	Yield of wheat y
8.8	39.6
10.3	42.5
15.9	69.3
13.1	52.4
12.9	60.5
7.2	26.7
11.3	50.2
18.6	78.6

(a) Fit a least-squares line from which we can predict the yield of wheat in this county in terms of the rainfall, and plot its graph.

(b) Use the equation obtained in part (a) and also its graph to predict the annual yield of wheat in the county when the annual rainfall is 9.0 inches.

17.29 With reference to the preceding exercise, test the null hypothesis $\beta = 5$ against the alternative hypothesis $\beta < 5$ at the 0.01 level of significance. Also state in words the hypothesis being tested.

17.30 With reference to Exercise 17.28, find a 95% confidence interval for the true average yield of wheat in the county when there are 12 inches of rain during the year.

17.31 With reference to Exercise 17.28, find 95% limits of prediction for the yield of wheat in the county when there are 12 inches of rain during the year.

17.32 With reference to Exercise 17.1 on page 439, construct
(a) a 95% confidence interval for the regression coefficient α;
(b) a 95% confidence interval for the regression coefficient β.

17.33 The following sample data were collected to determine the relationship between two processing variables and the current gain of a certain kind of transistor:

Diffusion time (hours) x_1	Sheet resistance (ohm-cm) x_2	Current gain y
1.5	66	5.3
2.5	87	7.8
0.5	69	7.4
1.2	141	9.8
2.6	93	10.8
0.3	105	9.1
2.4	111	8.1
2.0	78	7.2
0.7	66	6.5
1.6	123	12.6

Fit an equation of the form $y = b_0 + b_1 x_1 + b_2 x_2$ to these data, and use it to estimate the expected current gain corresponding to a diffusion time of 2.2 hours and a sheet resistance of 90 ohm-cm.

aving learned how to fit a least-squares line to paired data, we turn now to the problem of determining how well such a line actually fits the data. Of course, in the vending-machines example we can get some idea of this by drawing and inspecting the diagram of Figure 17.3, but in order to determine this objectively, let us refer back to the original data:

Year	Maintenance cost (dollars)
x	y
4	148
2	128
3	133
5	154
2	118
3	145
4	143
5	159
4	142
3	127

As can be seen from this table, there are substantial differences among the maintenance costs of the vending machines, the smallest being $118 and the largest being $159. However, we also see that the $118 cost is for a machine in its second year of use, while the $159 is for a machine in its fifth year, and this suggests that the differences among the maintenance costs may well be due, at least in part, to differences in the ages of the machines. These observations raise the following question, which we shall answer in this chapter: Of the total variation among the y's, how much can be attributed to the relationship between the two variables x and y (that is, to the fact that the observed y's correspond to different values of x), and how much can be attributed to chance?

In Section 18.1 we shall introduce the coefficient of correlation as a measure of the strength of the linear relationship between two variables, and in Section 18.2 we shall learn how to test its significance; the problems of rank correlation and multiple correlation will be treated in Sections 18.3 and 18.4.

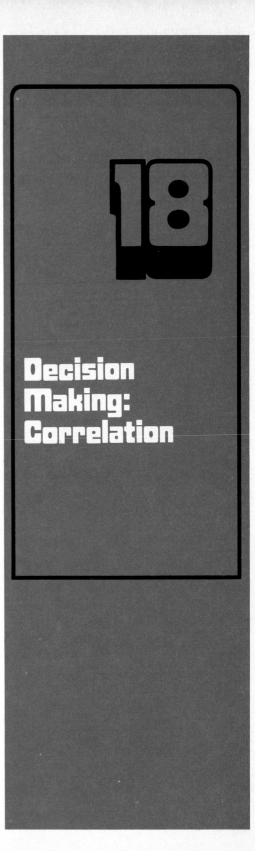

18

Decision Making: Correlation

The Coefficient of Correlation

With regard to the question raised directly above, we are faced here with an analysis-of-variance problem, and a study of Figure 18.1 will help to understand this. Referring to this figure, we see that the deviation of any observed value y from the mean of all the observed values of y, $y - \bar{y}$, can be written as the sum of two parts: $\hat{y} - \bar{y}$, the deviation of the value on the line (corresponding to an observed value of x) from the mean of the y's, and $y - \hat{y}$, the deviation of the observed value of y from the corresponding value on the line. Symbolically, we write

$$y - \bar{y} = (\hat{y} - \bar{y}) + (y - \hat{y})$$

and if we square the expressions on both sides of this identity and sum over all n values of y, we find that algebraic simplifications lead to

$$\sum (y - \bar{y})^2 = \sum (\hat{y} - \bar{y})^2 + \sum (y - \hat{y})^2$$

As our measure of the total variation of the y's we use the quantity on the left above, $\sum (y - \bar{y})^2$, called the **total sum of squares**. This measure is just $n - 1$ times the variance of the y's, and, as the equation shows, it has been partitioned into two additive components. The first of these, $\sum (\hat{y} - \bar{y})^2$, consists of the sum of the squares of the deviations of the values on the line from the mean $\bar{y}$, and it is called the **regression sum of squares**. It measures that portion of the total variation of the y's which would exist if differences in x were the only cause of differences among the y's, or in other words, if all the y's lay directly on the line.

This is hardly ever the case in practice, though, and the fact that all the points do not lie on a regression line is an indication that there are other factors than differences among the x's which affect the values of y. It is customary to lump together all these other factors—not separately considered in a study—under the general heading of "chance." Chance variation thus depends on the amounts by which the points deviate from the line, and its measure, $\sum (y - \hat{y})^2$, is the sum of the squared deviations of the observed y values from their corresponding values $\hat{y}$ on the line. This quantity, the second of the two components into which we partition the total sum of squares, is called the **residual sum of squares**.

To calculate the various sums of squares for the maintenance-cost example,

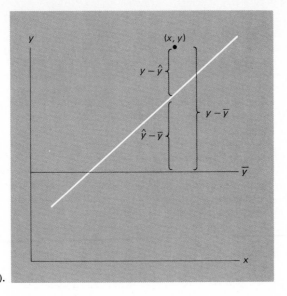

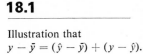

18.1

Illustration that
$y - \bar{y} = (\hat{y} - \bar{y}) + (y - \hat{y}).$

where $\bar{y} = \dfrac{1{,}397}{10} = 139.7$, we find by substitution into the formula for the total sum of squares (but see Exercise 18.15 on page 465) that

$$\sum (y - \bar{y})^2 = (148 - 139.7)^2 + (128 - 139.7)^2 + \cdots + (127 - 139.7)^2$$
$$= 1{,}504.10$$

To find the residual sum of squares by substitution into the formula (but again see Exercise 18.15 on page 465), we must first determine the value of $\hat{y} = 101.5 + 10.9x$ for each of the given values of x, and we get $101.5 + (10.9)(4) = 145.1$, $101.5 + (10.9)(2) = 123.3, \ldots$, and $101.5 + (10.9)(3) = 134.2$. Then

$$\sum (y - \hat{y})^2 = (148 - 145.1)^2 + (128 - 123.3)^2 + \cdots + (127 - 134.2)^2$$
$$= 255.53$$

Finally, the regression sum of squares may either be calculated directly from $\sum (\hat{y} - \bar{y})^2$, or by subtracting $\sum (y - \hat{y})^2$ from $\sum (y - \bar{y})^2$. Having already calculated these latter two sums of squares, we find that the regression sum of squares equals

$$\sum (y - \bar{y})^2 - \sum (y - \hat{y})^2 = 1{,}504.10 - 255.53$$
$$= 1{,}248.57$$

Comparing the regression sum of squares with the total sum of squares, we find that

$$\frac{\sum (\hat{y} - \bar{y})^2}{\sum (y - \bar{y})^2} \cdot 100 = \frac{1{,}248.57}{1{,}504.10} \cdot 100 = 83.0\%$$

of the total variation in the maintenance costs can be attributed to differences in the ages of the vending machines. It follows that the other 17.0 percent of the variation in the maintenance costs can be attributed to chance.

In general, if we divide $\sum(y-\bar{y})^2 = \sum(\hat{y}-\bar{y})^2 + \sum(y-\hat{y})^2$ through by $\sum(y-\bar{y})^2$, we get two ratios whose sum is 1, as follows:

$$1 = \frac{\sum(\hat{y}-\bar{y})^2}{\sum(y-\bar{y})^2} + \frac{\sum(y-\hat{y})^2}{\sum(y-\bar{y})^2}$$

The first of these ratios on the right is the ratio of the regression sum of squares to the total sum of squares, or the proportion of the total variation in y which is determined (accounted for) by differences in x. This ratio is called the **coefficient of determination** and designated r^2. Noting that r^2 is equal to 1 minus the second ratio on the right, we can now define the following measure r, called the **coefficient of correlation**, or more explicitly the "Pearsonian" or "product-moment" coefficient of correlation:

Coefficient of correlation

$$r = \pm\sqrt{1 - \frac{\sum(y-\hat{y})^2}{\sum(y-\bar{y})^2}}$$

where the sign given to r is the sign of b in the least-squares predicting equation. Thus, since $b = +10.9$ in the vending-machines example, we get $r = +\sqrt{0.83} = 0.91$ rounded to two decimals.

It follows from the rule for the sign of r that r is positive when the least-squares line has an upward slope; that is, when the relationship between x and y is such that small values of y tend to go with small values of x and large values of y tend to go with large values of x. Also, r is negative when the least-squares line has a downward slope; that is, when large values of y tend to go with small values of x and small values of y tend to go with large values of x. Geometrically, the ideas of a **positive correlation** and a **negative correlation** are illustrated in the first two diagrams of Figure 18.2; the third diagram illustrates the case where the least-squares line is horizontal, $r = 0$ and there is **no correlation**.

18.2

Types of correlation.

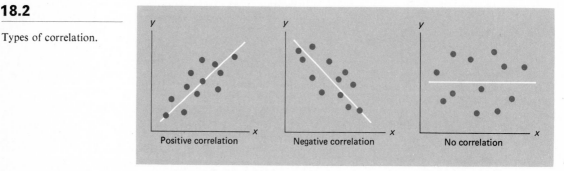

Positive correlation   Negative correlation   No correlation

Since part of the variation of the y's cannot exceed their total variation, $\sum (y - \hat{y})^2$ cannot exceed $\sum (y - \bar{y})^2$, and it follows from the formula defining r that a correlation coefficient must lie on the interval from -1 to $+1$. If all the points actually fall on a straight line, the residual sum of squares, $\sum (y - \hat{y})^2$, is zero and the resulting value of r, -1 or $+1$, indicates that the fit of the line to the paired observations is perfect. If, however, the scatter of the points is such that the least-squares line is a horizontal line coincident with $\bar{y}$ (that is, a line with slope $b = 0$ which intersects the y-axis at height $a = \bar{y}$), then $\sum (y - \hat{y})^2$ equals $\sum (y - \bar{y})^2$ and $r = 0$. In this case none of the variation of the y's can be attributed to their relationship with x, and the fit is so poor that knowledge of x is of no help in predicting y—the predicted value of y is $\bar{y}$ for any x.

The formula which defines r shows clearly the essence of the coefficient of correlation, but in actual practice it is much easier to calculate r from the following computing formula†:

Computing formula
for coefficient
of correlation

$$r = \frac{n(\sum xy) - (\sum x)(\sum y)}{\sqrt{n(\sum x^2) - (\sum x)^2}\sqrt{n(\sum y^2) - (\sum y)^2}}$$

This formula may look imposing but, with the exception of $\sum y^2$, the quantities it calls for are precisely the same ones required in calculating the coefficients a and b of a least-squares regression line. It has the added advantage that r calculated in this way always has the correct sign.

EXAMPLE   Use the computing formula for r to verify the value $r = 0.91$, which we calculated for the vending-machines data.

†This computing formula follows directly from an alternative, but equivalent, definition of the coefficient of correlation based on the sample covariance

$$s_{xy} = \frac{\sum (x - \bar{x})(y - \bar{y})}{n - 1}$$

In this formula, we add the products obtained by multiplying the deviation of each x from $\bar{x}$ by the deviation of the corresponding y from $\bar{y}$, and divide by $n - 1$. In this way we literally measure the way in which the values of x and y vary together. If the relationship between the x's and the y's is such that large values of x tend to go with large values of y, and small values of x with small values of y, the deviations $x - \bar{x}$ and $y - \bar{y}$ tend to be both positive or both negative, so that most of the products $(x - \bar{x})(y - \bar{y})$ and, hence, the covariance, are positive. On the other hand, if the relationship between the x's and the y's is such that large values of x tend to go with small values of y and vice versa, the deviations $x - \bar{x}$ and $y - \bar{y}$ tend to be of opposite sign, so most of the products and, hence, the covariance, are negative. Using the covariance and the standard deviations, s_x and s_y, of the x's and the y's, we can define the coefficient of correlation as

$$r = \frac{s_{xy}}{s_x \cdot s_y}$$

Since the sum of products $\sum (x - \bar{x})(y - \bar{y})$ in the formula for the sample covariance is called a **product moment**, this explains why r is also often called the **product-moment coefficient of correlation**.

SOLUTION Copying $n = 10$, $\sum x = 35$, $\sum x^2 = 133$, $\sum y = 1{,}397$, and $\sum xy = 5{,}004$ from page 437, and $\sum y^2 = 196{,}665$ from page 445, we find that substitution into the computing formula for r yields

$$r = \frac{10(5{,}004) - (35)(1{,}397)}{\sqrt{10(133) - (35)^2}\sqrt{10(196{,}665) - (1{,}397)^2}} = 0.91$$

This agrees, as it should, with our earlier result.

As we have defined r, the coefficient of correlation, $100r^2$ gives the percentage of the total variation of the y's which is explained by, or is due to, their relationship with x. This itself is a very important measure in the study of relationships between two variables; beyond this, $100r^2$ also permits valid comparisons to be made among several relationships of this kind. If, for instance, in one study $r = 0.80$, then 64 percent of the variation in y is accounted for by its relationship with x; if in another study $r = 0.40$, only 16 percent of the variation in y is accounted for by its relationship with x. Thus, in the sense of "percentage of variation accounted for" we can say that a correlation of 0.80 is four times as strong as a correlation of 0.40. In the same way, we say that a correlation of 0.60 is nine times as strong as a correlation of 0.20, and so on.

In interpreting a correlation coefficient it is sometimes assumed that a numerically high value of r establishes a cause-and-effect relationship running from the independent variable x to the dependent variable y. We want to make it clear that this is not so. There are cases in which changes in one variable are considered to be the "cause" of changes in another variable (called its "effect"), but only logical argument, not a high value of r, can establish such a relationship. This is discussed further in Section 18.5.

We also want to make it clear that r measures only the strength of linear relationships. The value of r calculated for the data of Figure 18.3, for example, is near zero even though the points are all very close to the white curve. There is an obvious strong curvilinear relationship between the two variables, but virtually no linear relationship.

18.3

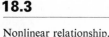

Nonlinear relationship.

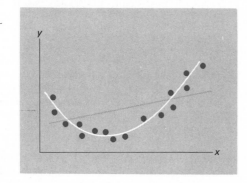

18.2

A Significance Test for r

When r is calculated on the basis of sample data, we may get a strong positive or negative correlation purely by chance, even though there is actually no linear relationship whatever between the two variables in the population from which the sample came.

Suppose, for instance, that we take two dice, one red and one green, roll them five times, and get the results shown in the following table:

Red die x	Green die y
4	5
2	2
4	6
2	1
6	4

Presumably there is no relationship between x and y, the numbers of points showing on the two dice. It is hard to see why, for example, large values of x should be associated with large values of y and small values of x with small values of y. But calculating r, we get the surprisingly high value $r = 0.66$. This raises the question of whether something is wrong with the presumption that there is no relationship between x and y, and to answer it we shall have to see whether the high value of r may be attributed to chance.

When a correlation coefficient is calculated from sample data, as in the above example, the value we obtain for r is only an estimate of a corresponding parameter, the **population correlation coefficient**, which we denote by the Greek letter ρ (rho). What r measures for a sample, ρ measures for a population.

There are several ways of testing the null hypothesis of no correlation, the null hypothesis $\rho = 0$. Here we shall make the same assumptions as in Section 17.3, except that x is also a random variable having a normal distribution. Referring to Table VI,

> We reject the null hypothesis of no correlation at the level of significance α if the value of r calculated for a set of paired data exceeds $r_{\alpha/2}$ or is less than $-r_{\alpha/2}$; otherwise, we accept the null hypothesis.

With Table VI, we can perform this test at the 0.05, 0.02, and 0.01 levels of significance.

EXAMPLE Use the 0.05 level of significance to test the null hypothesis of no correlation for the example where we rolled a pair of dice five times and got $r = 0.66$.

SOLUTION Since $r = 0.66$ does not exceed 0.878, the value of $r_{0.025}$ for $n = 5$, we find that the null hypothesis cannot be rejected; in other words, the correlation coefficient of 0.66 obtained for the pair of dice is not significant.

EXAMPLE Use the 0.01 level of significance to test the null hypothesis of no correlation for the maintenance-cost example where we had $n = 10$ and $r = 0.91$.

SOLUTION Since $r = 0.91$ exceeds 0.765, the value of $r_{0.005}$ for $n = 10$, the null hypothesis must be rejected; in fact, it appears virtually certain that the two variables—machine age and maintenance cost—are related.

EXERCISES

18.1 The following are the numbers of mistakes which 12 inspectors made during the first hour after coming to work and also during the last hour before going home:

First hour x	Last hour y
3	5
2	1
4	4
5	6
4	3
1	4
2	3
5	2
0	3
3	2
2	2
4	5

Calculate r for these data and test the null hypothesis of no correlation at the 0.05 level of significance.

18.2 The following table shows the percentages of the vote predicted by a poll for seven candidates for the U.S. Senate in different states, x, and the percentages of the vote which they actually received, y:

Poll x	Election y
42	51
34	31
59	56
41	42
53	58
40	35
55	54

Calculate r for these data.

18.3 Since r does not depend on the scales of x and y, its calculation can often be simplified by adding a suitable positive or negative number to each x, each y, or both. Rework the preceding exercise after subtracting 34 from each x and 31 from each y.

18.4 With reference to Exercise 17.3 on page 439,
  (a) calculate r for the given data;
  (b) use the 0.01 level of significance to test whether there is a real relationship between the two variables.

18.5 With reference to Exercise 17.6 on page 440,
  (a) calculate r for the given data;
  (b) use the 0.05 level of significance to test whether there is a real relationship between the two variables.

18.6 With reference to Exercise 17.7 on page 441,
  (a) calculate r for the advertising expenses and profits;
  (b) find the percentage of the total variation in the profits that is accounted for by differences in the advertising expenses;
  (c) use the 0.05 level of significance to test the null hypothesis of no correlation between the two variables.

18.7 With reference to Exercise 17.9 on page 441, what percentage of the total variation in the area burned is accounted for by differences in the variable operating costs?

18.8 If $r = 0.41$ for one set of paired data and $r = 0.29$ for another set of paired data, compare the strengths of the two relationships.

18.9 If we calculate r for each of the following sets of data, should we be surprised if we get $r = 1$ and $r = -1$? Explain your answer.

(a)

x	y
15	6
5	1

(b)

x	y
10	30
20	5

18.10 With reference to Exercise 17.23 on page 451, calculate r for each of the following pairs of variables:

(a) income and age;

(b) income and years of college;

(c) age and years of college.

18.11 Test in each case whether the value of r is significant at the 0.05 level of significance:

(a) $n = 12$ and $r = -0.53$;

(b) $n = 20$ and $r = 0.58$;

(c) $n = 15$ and $r = 0.55$.

18.12 Test in each case whether the value of r is significant at the 0.01 level of significance:

(a) $n = 14$ and $r = 0.63$;

(b) $n = 16$ and $r = -0.58$;

(c) $n = 32$ and $r = -0.47$.

18.13 Correlation methods are sometimes used to study the relationship between two (time) series of data which are recorded annually, monthly, weekly, daily, and so on. Suppose, for instance, that in the years 1963–1976 a large textile manufacturer spent 0.8, 0.5, 0.8, 1.0, 1.0, 0.9, 0.8, 1.2, 1.0, 0.9, 0.8, 1.0, 1.0, and 0.8 million dollars on research and development, and that in these years its share of the market was 20.4, 18.6, 19.1, 18.0, 18.2, 19.6, 20.0, 20.4, 19.2, 20.5, 20.8, 18.9, 19.0, and 19.8 percent. To see whether and how the company's share of the market in a given year may be related to its expenditures on research and development in prior years, let x_t denote the company's research and development expenditures and y_t its market share in the year t, and calculate

(a) the correlation coefficient for y_t and x_{t-1};

(b) the correlation coefficient for y_t and x_{t-2};

(c) the correlation coefficient for y_t and x_{t-3};

(d) the correlation coefficient for y_t and x_{t-4}.

For instance, in part (a) calculate r after pairing the 1964 percentage share of the market with the 1963 expenditures on research and development, the 1965 market share with the 1964 expenditures, and so on, . . . , and in part (d) calculate r after pairing the 1967 percentage share of the market with the 1963 expenditures on research and development, the 1968 market share with the 1964 expenditures, and so on. These time-lag correlations are called cross correlations. To continue,

(e) test, at the level of significance 0.05, whether the correlation coefficients obtained in parts (a) through (d) are significant;

(f) discuss the apparent duration of the effect of expenditures on research and development on the company's share of the market.

18.14 Correlation methods are sometimes used to study the internal structure (or systematic patterns) in series of data (time series) which are recorded on an annual, monthly, weekly, . . . , basis. Consider, for example, the diagram of Figure 18.4, which shows a line chart of a company's annual sales for the years 1958–1977. There is an obvious linear trend in the series as shown by the least-squares line, and to look for further patterns, we can study the deviations from the line, $y - \hat{y}$, which are $-2, 6, 0, 3, -2, -13, -5, -10, 1, 18, 6, 10, 1, -8,$ $-15, 0, 3, -7, 15,$ and 0 million dollars for the years 1958–1977. Letting y_t denote the deviation from the line in the year t, calculate

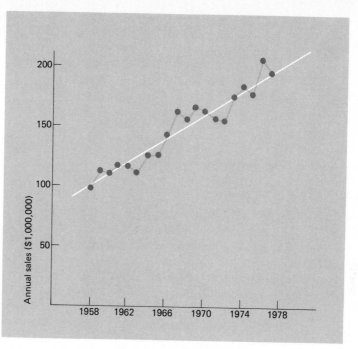

(a) the correlation coefficient for y_t and y_{t-1};
(b) the correlation coefficient for y_t and y_{t-2};
(c) the correlation coefficient for y_t and y_{t-3};
(d) the correlation coefficient for y_t and y_{t-4};
(e) the correlation coefficient for y_t and y_{t-5}.

For instance, in part (a) calculate r after pairing the 1959 deviation from the line with the 1958 deviation from the line, the 1960 deviation from the line with the 1959 deviation from the line, and so on, . . . , and in part (e) calculate r after pairing the 1963 deviation from the line with the 1958 deviation from the line, the 1964 deviation from the line with the 1959 deviation from the line, and, finally, the 1977 deviation from the line with the 1972 deviation from the line. These lag-time correlations are also called **autocorrelations**. To continue,

(f) test at the 0.05 level of significance whether the correlation coefficients obtained in parts (a) through (e) are significant;
(g) discuss the possible existence of a cyclical (or repeating) pattern in the series.

18.15 On page 457 we calculated for the maintenance-cost example the total sum of squares and the residual sum of squares by direct substitution into $\sum (y - \bar{y})^2$ and $\sum (y - \hat{y})^2$. However, it can be shown that

$$\sum (y - \bar{y})^2 = \sum y^2 - n \cdot \bar{y}^2$$

and

$$\sum (y - \hat{y})^2 = \sum y^2 - a(\sum y) - b(\sum xy)$$

where a and b are the y intercept and the slope of the least-squares line, and in most problems these two formulas will greatly simplify the calculation of these sums of squares. Use these computing formulas to recalculate the two sums of squares for the maintenance-cost example. As will be seen, substantial differences may arise due to rounding the estimated regression coefficients.

18.3

Rank Correlation

Since the significance test for r of the preceding section is based on fairly stringent assumptions, it is sometimes preferable to use a nonparametric alternative which can be applied under much more general conditions. This test of the null hypothesis of no correlation is based on the **rank-correlation coefficient** (often called **Spearman's rank-correlation coefficient**), which is essentially the coefficient of correlation for the ranks of the x's and the y's within the two samples. As we shall see, the rank-correlation coefficient has the added advantage that it is usually easier to determine than the product-moment coefficient of correlation when no calculator is available.

To calculate the rank-correlation coefficient, we first rank the x's among themselves, giving rank 1 to the largest (or smallest) value, rank 2 to the second largest (or second smallest), and so on; then we rank the y's similarly among themselves, find the sum of the squares of the differences, d, between the ranks of the x's and y's, and substitute into the formula

Rank-correlation coefficient

$$r_s = 1 - \frac{6(\sum d^2)}{n(n^2 - 1)}$$

where n is the number of pairs of observations. When there are ties in rank, we assign to each of the tied observations the mean of the ranks which they jointly occupy. For instance, if the third and fourth largest values of a variable are the same, we assign each the rank $\frac{3 + 4}{2} = 3.5$, and if the fifth, sixth, and seventh largest values of a variable are the same, we assign each the rank $\frac{5 + 6 + 7}{3} = 6$.

EXAMPLE The following are the numbers of hours which a random sample of ten students studied for an examination and the grades the students received:

Number of hours studied *x*	Grade in examination *y*
8	56
5	44
11	79
13	72
10	70
5	54
18	94
15	85
2	33
8	65

Calculate r_s.

SOLUTION Ranking the *x*'s among themselves from low to high, and also the *y*'s, we get the ranks shown in the first two columns of the following table:

Rank of x	Rank of y	d	d²
4.5	4	0.5	0.25
2.5	2	0.5	0.25
7	8	−1.0	1.00
8	7	1.0	1.00
6	6	0.0	0.00
2.5	3	−0.5	0.25
10	10	0.0	0.00
9	9	0.0	0.00
1	1	0.0	0.00
4.5	5	−0.5	0.25
			3.00

Then, determining the *d*'s and their squares, and substituting $n = 10$ and $\sum d^2 = 3.00$ into the formula for r_s, we get

$$r_s = 1 - \frac{6 \cdot 3}{10(10^2 - 1)} = 0.98$$

If we calculate *r* for these data (the original *x*'s and *y*'s), we get 0.96, so that the difference between *r* and r_s is very small in this case.

When there are no ties, r_S equals the correlation coefficient r calculated for the two sets of ranks; when ties exist, there will be a small (but usually negligible) difference. By using ranks, we naturally lose some information, but as we said, r_S is usually easier to determine than r when no calculator is available. Mainly, though, rank-correlation methods have the advantage that they can be used to measure relationships in problems where items cannot be measured but can be ranked (see, for instance, Exercises 18.23 and 18.24), and that tests of significance based on them are relatively unrestrictive.

In testing the null hypothesis of no correlation between two variables x and y, we do not have to make any assumptions about the nature of the populations from which the samples came. Under the null hypothesis of no correlation (indeed, the null hypothesis that the x's and y's are randomly matched), the sampling distribution of r_S has the mean 0 and the standard deviation†

$$\sigma_{r_S} = \frac{1}{\sqrt{n-1}}$$

Since this sampling distribution can be approximated with a normal distribution even for relatively small values of n, we base the test of the null hypothesis on the statistic

Statistic for testing significance of r_S

$$z = \frac{r_S - 0}{1/\sqrt{n-1}} = r_S\sqrt{n-1}$$

which has approximately the standard normal distribution.

EXAMPLE With reference to the preceding example where we had $n = 10$ and $r_S = 0.98$, test the significance of this value of r_S at the 0.01 level of significance.

SOLUTION Substituting $n = 10$ and $r_S = 0.98$ into the formula for z above, we get

$$z = 0.98\sqrt{10 - 1} = 2.94$$

Since this value exceeds $z_{0.005} = 2.575$, we reject the null hypothesis of no correlation and conclude at the 0.01 level of significance that there is, in fact, a real (positive) relationship between study time and grades in the population sampled.

†There exists a correction for σ_{r_S} that accounts for ties in rank, but it is seldom used unless the number of ties is large.

18.4

Multiple Correlation

In the beginning of this chapter we introduced the correlation coefficient as a measure of the goodness of the fit of a least-squares line to a set of paired data. If predictions are to be made with equations of the form

$$\hat{y} = b_0 + b_1 x_1 + b_2 x_2 + \cdots + b_k x_k$$

as in Section 17.4, we define the **multiple correlation coefficient** in the same way we originally defined r. We take the square root of the quantity

$$1 - \frac{\sum (y - \hat{y})^2}{\sum (y - \bar{y})^2}$$

which is the proportion of the total variation of the y's that can be attributed to the relationship with the x's. The only difference is that we now calculate $\hat{y}$ by means of the multiple regression equation instead of the equation $\hat{y} = a + bx$.

EXAMPLE In the example on page 450, where we derived the equation $\hat{y} = 35{,}197 + 4{,}149 x_1 + 731 x_2$, we find that $\sum (y - \hat{y})^2 = 686{,}719$ and $\sum (y - \bar{y})^2 = 185{,}995{,}000$. What is the value of the multiple correlation coefficient?

SOLUTION Since

$$1 - \frac{\sum (y - \hat{y})^2}{\sum (y - \bar{y})^2} = 1 - \frac{686{,}719}{185{,}955{,}000} = 0.9963$$

it follows that the multiple correlation coefficient is $\sqrt{0.9963} = 0.998$.

The preceding example also serves to illustrate that adding more independent variables in a correlation study is not always sufficiently productive to justify the extra work. The value of r for y and x_1 alone is actually 0.996, so very little seems to be gained by considering x_2 also. In this example, predictions based on the number of bedrooms alone are virtually as good as predictions which take into account also the number of baths. However, the situation is quite different in Exercise 18.27, where two independent variables together account for a much higher percentage of the total variation in y than does either x_1 or x_2 alone.

18.16 Calculate r_S for the data of Exercise 17.1 on page 439 and test the null hypothesis of no correlation at the 0.05 level of significance.

18.17 Calculate r_S for the data of Exercise 18.2 on page 462, and compare it with the value of r obtained in that exercise.

18.18 Calculate r_S for the data of Exercise 17.6 on page 440, and compare it with the value of r obtained in Exercise 18.5 on page 463.

18.19 Calculate r_S for the data of Exercise 17.7 on page 441, and compare it with the value of r found in Exercise 18.6 on page 463.

18.20 Use the value of r_S in the preceding exercise to test the null hypothesis of no correlation at the 0.01 level of significance.

18.21 The following table shows the 1979 final standing of the six baseball teams in the Western Division of the National League and the average salaries which they paid to their players:

	Final standing in 1979	Average salary in 1979
Los Angeles	1	$134,305
Cincinnati	2	165,144
Houston	3	73,660
San Francisco	4	120,737
San Diego	5	103,819
Atlanta	6	93,366

Calculate r_S.

18.22 Calculate r_S for the following data showing the statistics grades, x, and accounting grades, y, in a random sample of 18 students:

x	y	x	y
83	71	71	82
67	68	71	68
46	60	50	68
68	81	67	64
91	82	55	64
91	87	71	68
50	52	59	61
75	77	94	87
86	83	71	66

Also test whether the value of r_S is significant at the 0.05 level of significance.

18.23 The following table shows how a panel of nutrition experts and a panel of heads of household ranked 15 breakfast foods on their palatability:

Breakfast foods	Nutrition experts	Heads of household
A	3	5
B	7	4
C	11	8
D	9	14
E	1	2
F	4	6
G	10	12
H	8	7
I	5	1
J	13	15
K	12	9
L	2	3
M	15	10
N	6	11
O	14	13

Calculate r_S as a measure of the consistency of the two rankings.

18.24 The following are the rankings which three judges gave to the work of ten corporate accounting department trainees:

Judge A	Judge B	Judge C
6	2	7
4	5	3
2	4	1
5	8	2
9	10	10
3	1	6
1	6	4
8	9	9
10	7	8
7	3	5

Calculate r_S for each pair of rankings and determine which pair of rankings is
(a) most consistent;
(b) least consistent.

18.25 In a multiple regression problem, the residual sum of squares is $\sum (y - \hat{y})^2 = 75,240$ and the total sum of squares is $\sum (y - \bar{y})^2 = 112,550$. Find the value of the multiple correlation coefficient.

18.26 Use the least-squares equation of Exercise 17.22 on page 450 to calculate $\hat{y}$ for each of the five restaurants, determine the two sums of squares $\sum (y - \hat{y})^2$ and $\sum (y - \bar{y})^2$, and calculate the multiple correlation coefficient.

18.27 Use the least-squares equation derived in Exercise 17.23 on page 451 to calculate $\hat{y}$ for each of the five executives, determine the two sums of squares $\sum (y - \hat{y})^2$ and $\sum (y - \bar{y})^2$, and calculate the multiple correlation coefficient. Also compare the result with the values of r obtained in parts (a) and (b) of Exercise 18.10 on page 464.

18.28 Use the least-squares equation of Exercise 17.24 on page 451 to calculate $\hat{y}$ for each of the ten students, determine the total and residual sums of squares, and calculate the multiple correlation coefficient.

18.5

A Word of Caution

One must always take special care in interpreting the results of correlation studies. We call attention again to the fact that the correlation coefficient r measures only the strength of linear relationships (see Figure 18.3), and that it is possible to find a high degree of correlation in sample data when actually there is no relationship whatever in the population sampled.

We also call attention again to the danger of presuming that a high correlation coefficient implies a cause-and-effect relationship between two variables x and y. For instance, a high positive correlation has been observed in the study of the relationship between teachers' salaries and liquor consumption, and a high negative correlation has been observed in a study of the annual per capita consumption of chewing tobacco in the United States and the number of automobile thefts reported in a sample of urban areas in the same years. Moreover, in another study, a strong positive correlation was observed between the number of storks seen nesting in English villages and the number of children born in these same villages. We leave it to the reader's ingenuity to explain why there might be strong correlations observed in these instances in the absence of any cause-and-effect relationships.

Finally, we observe that there is a significantly high correlation between the advertising expenses and net operating profits of the sporting goods stores of Exercise 17.7 on page 441. But which way (if either) should we argue? That high advertising outlays lead to high profits, or that high profits lead to high advertising outlays?

18.6

Check List of Key Terms

Coefficient of correlation, 458
Coefficient of determination, 458
Multiple correlation coefficient, 469

Negative correlation, 458
Population correlation coefficient, 461
Positive correlation, 458

18.7

Review Exercises

18.29 If $r = 0.56$ for one set of paired data and $r = -0.97$ for another, compare the strengths of the two relationships.

18.30 The following are the numbers of persons who attended seminars on estate planning, x, and seminars on tax shelters, y, sponsored by an investment firm in twelve communities:

x	y
37	18
25	19
38	25
20	22
24	17
21	15
42	35
36	20
25	24
12	15
28	21
33	19

Calculate r_S.

18.31 If a set of $n = 42$ paired observations yields $r = 0.33$, test the null hypothesis of no correlation
(a) at the 0.05 level of significance;
(b) at the 0.01 level of significance.

18.32 The following are the numbers of hours which ten executives (interviewed as part of a sample survey) said they spent reading the financial pages of newspapers, x, and watching television, y, during the preceding week:

x	y
6	12
5	25
3	15
10	12
1	18
4	20
2	27
9	9
8	6
12	5

Calculate r.

18.33 For the data of the preceding exercise, test the null hypothesis of no correlation at the 0.05 level of significance.

18.34 In a multiple regression problem, the residual sum of squares is $\sum (y - \hat{y})^2 = 463$ and the total sum of squares is $\sum (y - \bar{y})^2 = 851$. Find the value of the multiple correlation coefficient.

18.35 For the following sample data

x	y
4	3
2	4
1	2
3	1

verify that $r = 0$ and that the least-squares line which fits these data best is a horizontal line coincident with $\bar{y}$.

18.36 In a random sample of ten door-to-door salespersons, the correlation between years of experience, x, and gross sales, y, is 0.72. What percentage of the differences in sales can be attributed to differences in experience?

18.37 Calculate r for the data of Exercise 17.26 on page 452, and test the null hypothesis of no correlation at the 0.05 level of significance.

18.38 Use the least-squares equation of Exercise 17.27 on page 452, to calculate $\hat{y}$ for each of the shipments, determine the two sums of squares $\sum (y - \hat{y})^2$ and $\sum (y - \bar{y})^2$, and calculate the multiple correlation coefficient.

18.39 The following are the ways in which two judges ranked ten corporate annual reports on their overall quality:

Judge A	Judge B
5	8
4	3
7	7
10	9
1	2
6	5
2	4
9	10
3	1
8	6

Calculate r_S as a measure of the consistency of the two rankings.

18.40 The following are the numbers of inquiries which a brokerage house received in eight weeks about municipal bonds, x, and money-market funds, y:

x	y
57	88
33	69
45	50
17	29
38	53
47	61
72	85
60	82

Calculate r and test the null hypothesis of no correlation at the 0.05 level of significance.

18.41 If a random sample of $n = 20$ pairs of observations yielded $r_S = 0.41$, is this rank correlation coefficient significant at the 0.05 level of significance?

Decision Making: Time Series Analysis

It has been said in many different ways that the future belongs to those who plan for it best. This is certainly true in business and economics where, in planning for an uncertain future, many important, powerful, and often necessary tools are available for help. These include, among others, the ones we have already studied in this book—index numbers, estimation, tests of hypotheses, analysis of variance, regression and correlation, and the rest. We come now to a new tool of absolutely basic importance in business planning. Called time-series analysis, it is essentially a study of the way in which such things as work force, production, sales, and profits move through time. In Section 19.1 we begin our study with some introductory remarks on looking backward and looking ahead, then in Section 19.2 we discuss in some detail the components of a time series. In Sections 19.3 through 19.7 we learn how to measure these components, that is, how to describe them statistically, and in Sections 19.8 and 19.9 we present some special applications.

Some Preliminary Remarks

Business planning is not an end in itself, but organized planning utilizing various statistical techniques (intended to assess past performance and estimate the success or failure of proposed strategies) seems to have everything in its favor. Aside from its intuitive appeal, there are the achievement records of many highly successful companies which treat planning as an organized activity and analyze exhaustively the many factors bearing on planning decisions. Marketing strategy is often planned in great detail for several years ahead, with enough flexibility built in to allow for whatever changes market conditions may require. Financial strategy is also carefully planned, so that operating plans can be carried out and a proper balance maintained between distributed earnings and retained earnings necessary for future growth. Many manufacturing companies attempt to make their long-range planning (for example, beyond a year) more effective by maintaining projections of 10 years or longer on sales, profits, cash needs, and so on, for all of their major product groups. No intelligent planning of future needs for raw materials and production facilities, for instance, can be done without predictions of such basic variables as product or service demand, production costs, and restrictions on capacity.

Predictions of the sort that involve explaining events which will occur at some future time are called **forecasts**, and the process of arriving at such explanations is called **forecasting**. There are various ways of forecasting the future values of economic variables, including the **intrinsic method**, in which the future values of variables are predicted from their past values. One important statistical technique included in the intrinsic method is **time-series analysis**. By a **time series** we mean statistical data that are collected, observed, or recorded at regular intervals of time. The term "time series," or simply "series," applies, for example, to data recorded periodically showing the total annual sales of retail stores, the total quarterly value of construction contracts awarded, the total amount of unfilled orders in durable-goods industries at the end of each month, and the daily clearings in the Chicago Clearing House. We shall restrict ourselves to the analysis of business and economic data, but neither the term "time series" nor the methods of analysis which we shall discuss are limited to these kinds of data.

Although in forecasting our concern is with the future, time-series analysis begins by looking backward. After all, it would be silly not to put relevant experience from the past to use in planning for an uncertain future. Thus, we search for observable regularities and patterns in historical series which are so persistent that they cannot be ignored. If we subsequently base our forecasts on such regularities and patterns, we are simply expressing the feeling that the future follows the past with some degree of consistency, that what has happened

in the past will, to a greater or lesser extent, continue to happen or will happen
again in the future.

19.2

The Components of a Time Series

Sometimes, when we look at the graph of a time series, we get the impression
that it has been scrawled by a small child, and it is hard to believe that any kind
of analysis could bring order into the seemingly haphazard movement of the
data through time. Nevertheless, if we make some simplifying assumptions it
becomes possible to identify, explain, and measure the fluctuations that appear
in time series. Specifically, let us assume that there are four basic types of variation in a series which, superimposed and acting in concert, account for the
observed changes over a period of time and give the series its erratic appearance.
These four **components** are

1. Secular trend.
2. Seasonal variation.
3. Cyclical variation.
4. Irregular variation.

We shall assume further that there is a multiplicative relationship between these
four components; that is, any particular value in a series is the product of
factors which can be attributed to the four components.

This is the traditional approach to time-series analysis, but it is only one of
many possible models (or schemes) which might be used in studying time series.
Although it ignores the hidden interactions and interrelationships in the data
and the entire "complex of individually small shifts and nuances," this approach
has been and continues to be widely used in practice where, in many instances,
it provides entirely satisfactory results. It is possible to construct mathematically
sophisticated forecasting models, but there is much of fundamental importance
to be learned about the movements of data through time from a study of the
traditional methods.

By the **secular trend**, or long-term trend, of a time series we mean the
smooth or regular underlying movement of a series over a fairly long period of
time. Intuitively speaking, the trend of a time series characterizes the gradual
and consistent pattern of changes in the series which are thought to result from
persistent forces affecting growth or decline (changes in population, income, and
wealth; changes in the level of education and technology; etc.) that exert their
influence more or less slowly. For example, Figure 19.1 shows the overall upward
trend in group life insurance in force in the United States, and Figure 19.2
shows the persistent downward trend in agricultural employment in the United
States.

The problem in trend analysis is to describe the underlying movement or

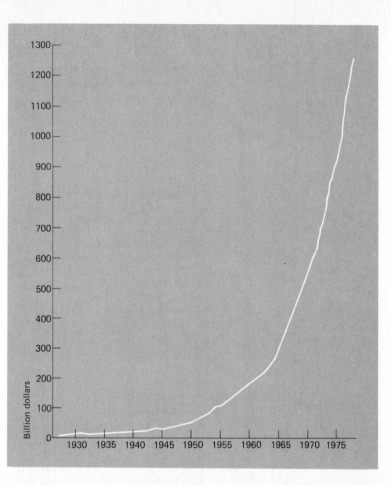

19.1

Group life insurance in force in the United States, 1927–1978.

19.2

Agricultural employment in the United States, 1939–1977.

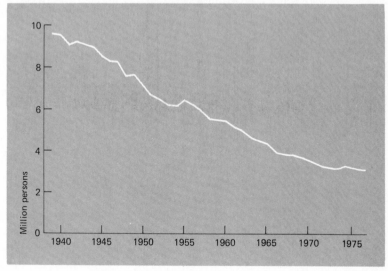

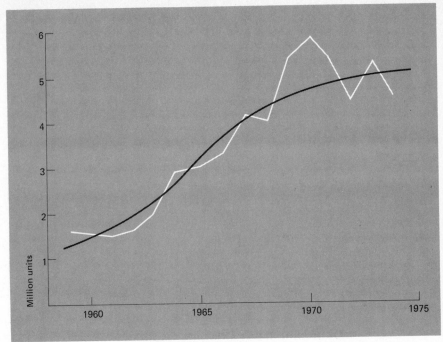

Growth curve fitted to
factory shipments of room
air conditioners.

general sweep of a time series in quantitative terms. In many series the patterns
of gradual growth or decline can be described reasonably well by means of a
straight line, but in others more complicated curves are required. For example,
the series on factory shipments of room air conditioners in the United States
shown in Figure 19.3 has the general shape of an elongated letter S, and its
trend is not well described by a straight line. The curve shown fitted to these
data is one of the so-called **growth curves**, and it reflects a type of growth that
is frequently observed in time series.

Strictly speaking, **seasonal variation** is the movements in a time series, like
those shown in Figure 19.4, which recur year after year in the same months (or
the same quarters) of the year with more or less the same intensity. Thus, the
month-to-month variation observed in retail sales and the quarter-to-quarter
variation observed in consumer installment debt are both examples of seasonal
variation in time series. Sometimes the term "seasonal variation" is also applied
to other inherently periodic movements, such as those occurring within a day
or week or month, whose period is at most one year. In any case, the movement
described is a most obvious one.

Few businesses are free of the effects of seasonal variation. The examination
of almost any series of economic data recorded on a quarterly, monthly, weekly,
daily, or hourly basis shows movements within the series which seem to occur
period after period with some definite degree of regularity. About two thirds to
three fourths of the total annual business in the jewelry trade is done in the

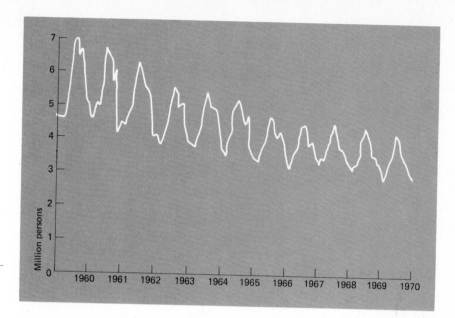

Farm employment in
the United States.

two months before Christmas; except for the holiday season, airline passenger
traffic normally drops during the winter months; electric power and natural
gas consumption is normally higher during the winter months; city traffic on
workdays is always heaviest in the early morning and late afternoon; and so on.
To businessmen responsible for realistic planning of such activities as production,
purchasing, inventory, personnel, advertising, and sales, an understanding of
seasonal patterns is of primary concern. Even in cases where the seasonal variation is not of basic concern, it must often be measured statistically in order to
facilitate the study of other types of variation.

After the trend and seasonal variation have been eliminated statistically
from a time series, in the model we are discussing here there remain the **cyclical**
and **irregular variations**. Irregular variations in time series are of two types:
(1) variations which are caused by such readily identifiable special events as
elections, wars, floods, earthquakes, strikes, and bank failures, and (2) random or chance variations whose causes cannot be definitely assigned. Most
of the time, irregular variations resulting from the occurrence of special events
can be easily recognized and identified with the phenomena which caused
them; then the data reflecting their impact can simply be eliminated before
measuring the other time-series components. As for those essentially random
kinds of fluctuations, there is little to be said except that they usually tend to
average out in the long run.

It is conceivable that in any given time series which one is studying there are
really no systematic movements of any sort, that all the observed fluctuations
in the series are in fact irregular ones. Ordinarily, before attempting to measure,
say, a trend, we like to test whether or not there actually is a significant movement of this kind in the series. There are several tests of significance which can be

applied in these cases. One of them, the test based on runs above and below the median, was described in Chapter 16.

Cyclical variation is sometimes defined as the variation which remains in a time series after the trend, seasonal, and irregular variations have been eliminated. Actually, there is much more to it than that, but in classical time-series analysis such a process of elimination is the usual way of measuring (business) cyclical variation, or **business cycles**. Generally speaking, business cycles consist of recurring up and down movements of business or economic activity which differ from seasonal variations in that they extend over longer periods of time and, supposedly, result from an entirely different set of causes. The recurring periods of prosperity, recession, depression, and recovery, which constitute the four phases of a complete business cycle, are considered to be due to factors other than the weather, social customs, and so on, which account for seasonal variations. Because of the great importance of cyclical swings of business to the economic and social life of this country, an enormous amount of effort has been spent in studying the business cycle. Many theories have been proposed to account for it, but no generally accepted explanation of this complicated phenomenon has appeared.

19.3

Linear Trends

The most widely used method of fitting trend lines to time series is the method of least squares. As we have seen, the problem of fitting a least-squares line $\hat{y} = a + bx$ is essentially that of determining values of a and b which, for a given set of data, make $\sum (y - \hat{y})^2$ as small as possible. We can find these two quantities in any problem by solving the two normal equations,

$$\sum y = na + b(\sum x)$$
$$\sum xy = a(\sum x) + b(\sum x^2)$$

or by using formulas derived from them.

In time-series analysis, however, the x's practically always refer to successive periods (usually years) and where this is the case we can simplify the work of fitting a least-squares trend line by performing a change of scale, or coding the x's, so that in the new scale the sum of the x's is zero. If the series has an odd number of years, we count from the middle of the period in units of one year, assigning $x = 0$ to the middle year and 1, 2, 3, ... to the following years and $-1, -2, -3, ...$ to the preceding years. If the series has an even number of years, though, there are two middle years, not one, and the midpoint of the series falls between them. Assigning $x = 0$ to this point in time and counting from here in units of six months (or half-years), the x's are 1, 3, 5, ... for the following years and $-1, -3, -5, ...$ for the preceding years. In either case (an odd number or an even number of years in the series) $\sum x = 0$, and substituting this into the two normal equations and solving for a and b, we get

Computing formulas for a and b (with coding)

$$a = \frac{\sum y}{n} \quad \text{and} \quad b = \frac{\sum xy}{\sum x^2}$$

The advantage of this kind of coding is evident.

EXAMPLE For the years 1971 through 1979, FMC Corporation reported annual net incomes of 38.8, 49.1, 79.2, 80.9, 108.2, 80.2, 120.6, 140.9, and 151.6 millions of dollars. Fit a least-squares trend line of the form $y = a + bx$ to this series.

SOLUTION Since we have data for nine (an odd number) of years, we set $x = 0$ opposite 1975, the middle year, and count both backward and forward from that origin in full years. We show the coded x-values in the second column below and develop the sums we need to find a and b in the third, fourth, and fifth columns.

Year	x	Net income y	xy	x^2
1971	-4	38.8	-155.2	16
1972	-3	49.1	-147.3	9
1973	-2	79.2	-158.4	4
1974	-1	80.9	-80.9	1
1975	0	108.2	0	0
1976	1	80.2	80.2	1
1977	2	120.6	241.2	4
1978	3	140.9	422.7	9
1979	4	151.6	606.4	16
	0	849.5	808.7	60

Substituting $n = 9$, $\sum y = 849.5$, $\sum xy = 808.7$, and $\sum x^2 = 60$ into the new formulas for a and b, we get

$$a = \frac{\sum y}{n} = \frac{849.5}{9} = 94.4$$

$$b = \frac{\sum xy}{\sum x^2} = \frac{808.7}{60} = 13.5$$

and hence $\hat{y} = 94.4 + 13.5x$ for the equation of the trend line.

To avoid confusion over just what a trend equation really says, it is always advisable to add a line of explanation, called a **legend**, stating precisely the origin of x and the units of both x and y. In the preceding example, the year 1975 is the origin (corresponding to zero in the x-scale), the units of x are full years, and the y's are annual net incomes in millions of dollars. So, we write the trend equation and its legend as

$$\hat{y} = 94.4 + 13.5x$$

(origin, 1975; x units, 1 year; y, annual net income in millions of dollars)

All of this makes it clear that the trend value for 1975 is \$94.4 million and that the **annual trend increment** (the year-to-year growth) in the corporation's net income is estimated to be 13.5 million for this period.

Once we have calculated a trend equation, we can use it to determine the trend value for any year simply by substituting into the equation the value of x corresponding to that year.

<hr>

EXAMPLE   Based on the results of the preceding example, find the 1971 and 1979 trend values of the corporation's net income.

SOLUTION   Substituting $x = -4$ into the trend equation we get $\hat{y} = 94.4 + 13.5(-4) = 40.4$ for 1971, and substituting $x = 4$ we get $\hat{y} = 94.4 + 13.5(4) = 148.4$ for 1979.

   If we now plot these two trend values and join them by a straight line, we get the least-squares trend line shown drawn through the original series in Figure 19.5.

It is sometimes desirable, or necessary, to modify a trend equation like the one above so that it can be used with monthly data, so that the x's refer to successive months instead of successive years, or so that the origin of x is the middle of a month instead of the middle of a year. In the three examples that follow we show how modifications such as these can be made, using for illustration the net income series of the example on page 483.

<hr>

EXAMPLE   Change the y units in the equation $\hat{y} = 94.4 + 13.5x$ from annual (net) incomes to average monthly incomes.

SOLUTION   Since the average monthly figures for the nine years are one twelfth the corresponding annual figures (that is, $\frac{y}{12}$), we must write a new equation with a and b replaced by $\frac{a}{12}$ and $\frac{b}{12}$. The modified equation and its legend are

$$\hat{y} = 7.9 + 1.1x$$

(origin, 1975; x units, 1 year; y, average monthly net income in millions of dollars)

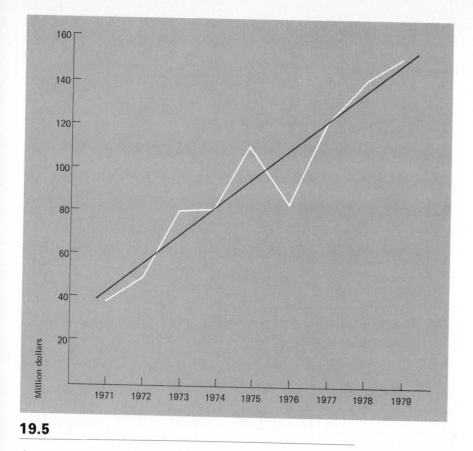

19.5

Annual net incomes of FMC Corporation, 1971–1979.

EXAMPLE Modify the trend equation obtained in the preceding example so that the x's refer to successive months instead of successive years.

SOLUTION Since b measures the trend increment, the increase or decrease of trend values corresponding to one unit of x, we must divide b by 12, changing it from an annual trend increment to a **monthly trend increment**. Leaving a unchanged, we thus get

$$\hat{y} = 7.9 + 0.1x$$

(origin, 1975; x units, 1 month; y, average monthly net income in millions of dollars)

The equation obtained in the preceding example can be used with monthly data just as it stands. The origin is at the end of June (the instant when June becomes July) 1975, but to complete the modification we often shift the origin from the middle of a year to the middle of a month.

EXAMPLE Shift the origin of the trend equation obtained in the preceding example from June–July 1975 to the middle of January 1975.

SOLUTION The middle of January 1975 is $5\frac{1}{2}$ months earlier than the middle of the year 1975. Therefore, we must subtract 5.5 monthly trend increments from the 1975 trend value of 7.9, getting $7.9 - (5.5)(0.1) = 7.4$ and, finally,

$$\hat{y} = 7.4 + 0.1x$$

(origin, January 1975; x units, 1 month; y, average monthly net income in millions of dollars)

19.4

Nonlinear Trends

When the data seem to depart more or less widely from linearity in regression or time-series analysis, we must consider fitting some curve other than a straight line. One of the most useful of these other curves is the **parabola** whose equation is

$$\hat{y} = a + bx + cx^2$$

In fitting a parabola by the method of least squares, we must determine a, b, and c so that $\sum (y - \hat{y})^2 = \sum (y - a - bx - cx^2)^2$ is a minimum. This requires that we solve the following set of three normal equations for a, b, and c:

$$\sum y = na + b(\sum x) + c(\sum x^2)$$
$$\sum xy = a(\sum x) + b(\sum x^2) + c(\sum x^3)$$
$$\sum x^2 y = a(\sum x^2) + b(\sum x^3) + c(\sum x^4)$$

When the values in a series are equally spaced, this work can be simplified appreciably by using the same coding as in the preceding section. Putting the zero of the new scale at the middle of the series and observing the conventions for coding an odd number of periods and an even number of periods will make $\sum x = 0$ and $\sum x^3 = 0$, and the normal equations reduce to

Normal equations for fitting parabola (with coding)

$$\sum y = na + c(\sum x^2)$$
$$\sum xy = b(\sum x^2)$$
$$\sum x^2 y = a(\sum x^2) + c(\sum x^4)$$

We can then find b directly from the second equation,

$$b = \frac{\sum xy}{\sum x^2}$$

and we can find a and c by solving the first and third equations simultaneously. Parabolas are also referred to as **second-degree polynomial equations**, and polynomial equations of higher degree in x than two, such as $\hat{y} = a + bx + cx^2 + dx^3$ and $\hat{y} = a + bx + cx^2 + dx^3 + ex^4$, can also be fitted by the method of least squares.

EXAMPLE

In the years 1969 through 1979 there were 318, 277, 234, 246, 254, 255, 207, 225, 224, 228, and 260 millions of wine gallons of denatured alcohol produced in the United States.
(a) Fit a parabolic trend curve of the form $\hat{y} = a + bx + cx^2$ to this series.
(b) Calculate the trend values for 1969, 1977, and 1980.
(c) Plot the graph of the parabola together with the original series of data.

SOLUTION

(a) In order to calculate a, b, and c from the reduced normal equations, we must find n, $\sum y$, $\sum xy$, $\sum x^2 y$, $\sum x^2$, and $\sum x^4$. In the second column of the table below we show the production figures (the y-values) and in the five columns to its right we show the work we do to find the required sums:

Year	Production y	x	xy	$x^2 y$	x^2	x^4
1969	318	−5	−1,590	7,950	25	625
1970	277	−4	−1,108	4,432	16	256
1971	234	−3	−702	2,106	9	81
1972	246	−2	−492	984	4	16
1973	254	−1	−254	254	1	1
1974	255	0	0	0	0	0
1975	207	1	207	207	1	1
1976	225	2	450	900	4	16
1977	224	3	672	2,016	9	81
1978	228	4	912	3,648	16	256
1979	260	5	1,300	6,500	25	625
	2,728	0	−605	28,997	110	1,958

Now, with all this done, we can find b directly by substituting into the formula, and we get

$$b = \frac{\sum xy}{\sum x^2} = \frac{-605}{110} = -5.5$$

Then, substituting $n = 11$ together with the totals of the y, x^2y, x^2 and x^4 columns into the first and third reduced normal equations, we get

$$2{,}728 = 11a + 110c$$

$$28{,}997 = 110a + 1{,}958c$$

Solving these two equations by the method of elimination or by determinants, we find that $a = 228.0$ and $c = 2.0$, both rounded to one decimal. Accordingly, we write the following parabolic trend equation and its legend:

$$\hat{y} = 228.0 - 5.5x + 2.0x^2$$

(origin, 1974; x units, 1 year; y, production of
denatured alcohol in millions of wine gallons)

In this parabolic equation, which describes the trend in the production of denatured alcohol over the 1969–1979 period, $a = 228.0$ is the trend value for 1974, $b = -5.5$ is the slope of the curve at $x = 0$ (the origin), and $2c = 4.0$ is the constant rate at which the slope changes at that point.

(b) To find the trend value on a parabolic curve for any year, we merely substitute the appropriate value of x into the trend equation. For 1969 we substitute $x = -5$ and get

$$\hat{y} = 228.0 - 5.5(-5) + 2.0(-5)^2 = 305.5$$

for 1977 we substitute $x = 3$ and get

$$\hat{y} = 228.0 - 5.5(3) + 2.0(3)^2 = 229.5$$

and for 1980 (one year beyond the end of the series) we substitute $x = 6$ and get

$$\hat{y} = 228.0 - 5.5(6) + 2.0(6)^2 = 267.0$$

(c) To plot a parabolic trend we need at least three points. Thus, using the results of part (b) together with the information that the 1974 trend value is $a = 228.0$, we obtain the curve shown in Figure 19.6.

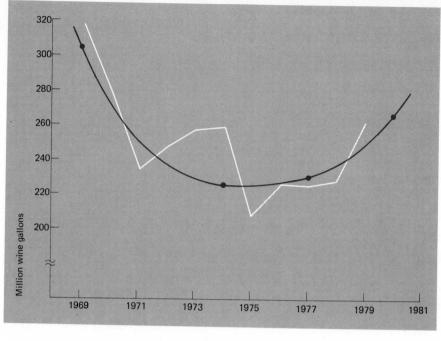

19.6

Production of denatured alcohol, 1969–1979.

Often a set of data which does not seem linear when plotted on ordinary graph paper (**arithmetic paper**) appears to "straighten out" when plotted on paper with a logarithmic vertical scale (**semilog paper** or **ratio paper**). The following series showing, for the years 1976–1981, the net profits (in thousands of dollars) of a manufacturer of components for pollution control devices is a good illustration of such data:

Year	Net profit y
1976	162
1977	201
1978	285
1979	402
1980	635
1981	920

As Figure 19.7 (the plot on arithmetic paper) shows, the path of the data is certainly not well described by a straight line. On the other hand, Figure 19.8

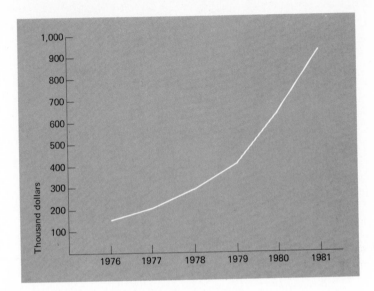

19.7

Net profits plotted on ordinary graph paper.

19.8

Net profits plotted on semilog graph paper.

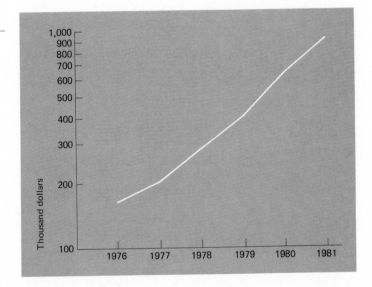

(the plot on semilog paper) shows that it is straightened out remarkably well when we use a logarithmic scale for y.

On arithmetic paper, equal intervals on the vertical scale represent equal amounts of change, and the $\hat{y}$-values calculated from the equation $\hat{y} = a + bx$ plot as a straight line on arithmetic paper. On ratio paper, however, equal intervals on the vertical scale represent equal rates of change and the $\hat{y}$-values calculated from the equation

$$\hat{y} = a \cdot b^x$$

plot as a straight line on ratio paper. This latter curve is called **exponential** since the x appears in the equation as the exponent of b, and the trends of time series which appear linear when plotted on ratio paper are called **exponential trends**.

Taking the logarithm of the expressions on both sides of the equation $\hat{y} = a \cdot b^x$, we have

$$\log \hat{y} = \log a + x \cdot \log b$$

which is a linear equation in x and $\log \hat{y}$. (Writing A, B, and Y for $\log a$, $\log b$, and $\log \hat{y}$, the equation becomes $Y = A + Bx$, which is the usual equation of a straight line.) In order to fit an exponential trend by the method of least squares (that is, to fit a straight line to the logarithms of the y values), we find numerical values for $\log a$ and $\log b$ from the formulas

Computing formulas for fitting exponential curve

$$\log a = \frac{\sum \log y}{n} \quad and \quad \log b = \frac{\sum (x \cdot \log y)}{\sum x^2}$$

provided that, by the usual change of scale, we make $\sum x = 0$; then we find a and b. The work proceeds exactly the same as in fitting a straight line to the y-values themselves, with the exception that we use $\log y$ instead of y.

EXAMPLE Fit an exponential curve to the 1976–1981 net incomes of the components manufacturer given above.

SOLUTION There is an even number of years in the period, so we code the data by setting the zero of a new x-scale at the middle of the period, where 1978 becomes 1979, and count backward and forward from here in units of a half-year. In other words, we code the years -5, -3, -1, 1, 3, and 5. Taking the logarithms needed for the work from Table XI, we develop the sums needed to find $\log a$ and $\log b$ in the following table:

x	Net profit (thousands of dollars) y	log y	$x \cdot$log y	x^2
−5	162	2.2095	−11.0475	25
−3	201	2.3032	−6.9096	9
−1	285	2.4548	−2.4548	1
1	402	2.6042	2.6042	1
3	635	2.8028	8.4084	9
5	920	2.9638	14.8190	25
		15.3383	5.4197	70

Substituting the appropriate column totals and $n = 6$ into the formulas for log a and log b, we get

$$\log a = \frac{15.3383}{6} = 2.5564$$

$$\log b = \frac{5.4197}{70} = 0.0774$$

and we write the trend equation in its logarithmic form and its legend as

$$\log \hat{y} = 2.5564 + 0.0774x$$

(origin, 1978–1979; x units, 6 months; y, annual net sales in thousands of dollars)

Table XI shows that the numbers whose logarithms are 2.5564 and 0.0774 are 360.1 and 1.195, so we write the exponential trend equation as

$$\hat{y} = 360.1(1.195)^x$$

In this form, 360.1 is the trend value for 1978–1979 and 1.195 is 1 plus the average six months' rate of growth in net income over the period. Hence, the average six months' growth rate is 0.195 or 19.5 percent.

For most practical purposes, the logarithmic form of the exponential trend equation is more convenient. For instance, to estimate the net income for 1983, we substitute $x = 9$ into the logarithmic form of the trend equation and get

$$\log \hat{y} = 2.5564 + 0.0774(9) = 3.2530$$

Turning to Table XI once again, we find that $\hat{y}$ itself is 1,790. Consequently, the company's estimated 1983 net income (based on its 1976–1981 trend) is $1,790,000.

There are various other ways trends can be described mathematically, but even though no two series may be exactly alike, most of them can be handled by the methods we have described here. In each case, our goal is to select that equation or that method of measuring trend which best describes the gradual and consistent pattern of growth.

EXERCISES

19.1 In the central area of a large city, 4, 3, 5, 7, and 7 major private construction projects were begun in the years 1977–1981.
(a) Plot the series on arithmetic paper.
(b) Fit a least-squares line to the data and plot it on the chart showing the original data.

19.2 The following are the earnings before income taxes of the IBM Corporation for the years 1976 through 1979: 4.5, 5.1, 5.3, and 5.6 billions of dollars.
(a) Plot the series on arithmetic paper.
(b) Fit a least-squares line to this series and plot it on the chart drawn in part (a).

19.3 The following are the gross incomes from sales of the IBM Corporation for the years 1975–1979: 4.5, 6.0, 7.1, 8.8, and 9.5 billions of dollars.
(a) Plot the series on arithmetic paper.
(b) Fit a least-squares line to the data and plot it on the chart drawn in part (a).
(c) Modify the trend equation by shifting the origin of x to the year 1979.

19.4 United States imports of wine from Japan for the years 1974 through 1979 were 1,173, 848, 838, 689, 682, and 579 thousands of gallons.
(a) Plot the series on arithmetic paper.
(b) Fit a least-squares trend line to these imports and plot the line on the chart drawn in part (a).
(c) Modify the equation obtained in part (b) for use with monthly data and shift the origin to January 1970.

19.5 At the ends of the years 1972–1980 a manufacturing company had the following net investments in plants and equipments: 1.5, 1.6, 1.7, 2.3, 3.1, 3.5, 3.4, 3.9, and 4.7 millions of dollars.
(a) Fit a least-squares line to the series.
(b) Modify the equation of the trend line of part (a) by shifting the origin to the middle of the year 1979.
(c) Modify the equation of the trend line of part (a) by shifting the origin to the end of the year 1974.

19.6 The equation of a least-squares trend line fit to the total amounts of U.S.-produced beverage brandy entering distribution channels in the United States for the years 1964–1974 is

$$\hat{y} = 8.35 + 0.54x$$

(origin, 1969; x units, 1 year; y, annual amounts in millions of proof gallons)

Modify this trend equation for use with monthly data and shift the origin of the equation to January 1969.

19.7 The equation of a least-squares trend line fit to the annual revenues of the Kennecott Copper Corporation for the years 1965–1974 is

$$\hat{y} = 1,039.5 + 53.6x$$

(origin, 1969–1970; x units, 6 months; y, total annual revenues in millions of dollars)

Modify this trend equation for use with monthly data and shift the origin of x to January 1971.

19.8 In order to calculate the trend values $\hat{y} = a + bx + cx^2$ of a parabolic trend, we can make up a table with columns headed *year*, x, $a + bx$, cx^2, and $\hat{y}$, then calculate $a + bx$ and cx^2 separately for each x and add them to get $\hat{y}$. Turn to the example on page 487 and calculate the trend values for the eleven years of the series showing the production of denatured alcohol in the United States.

19.9 The net sales of American Brands, Inc., for the years 1969–1979 were 2.7, 2.7, 2.8, 3.0, 3.1, 3.6, 4.1, 4.1, 4.6, 5.2, and 5.8 billions of dollars. Fit a parabolic trend to this series and plot it together with the original data on arithmetic paper.

19.10 A small technological research and development firm, founded in 1967, has received the following contract awards (in thousands of dollars) for the years 1967 through 1981: 95, 106, 93, 102, 119, 135, 210, 200, 198, 197, 245, 240, 275, 325, and 340.

(a) Plot this series on arithmetic paper.
(b) Fit a parabola to this series by the method of least squares.
(c) What is the trend value for 1974?

19.11 With reference to the preceding exercise, calculate the yearly trend values for the contract awards series. Plot them on the chart drawn in part (a) of that exercise and draw a smooth curve through them to indicate the parabolic trend.

19.12 The contract awards series of Exercise 19.10 straightens out quite well when it is plotted on ratio paper.

(a) Fit a least-squares line to the logarithms of the values of this series.
(b) From the equation of part (a) find the logarithms of the trend values for the years 1967–1981, then find the trend values themselves.
(c) Plot the trend values on the chart of Exercise 19.11, draw a smooth curve through them to indicate the exponential trend, then compare visually the parabolic and exponential trends.
(d) What was the firm's average annual growth in contracts awarded over the 15-year period under consideration?

19.13 For the years 1968–1974 Argentina's Index of Consumer Prices with 1967 = 100 stood at 116, 125, 142, 191, 303, 486, and 604. Fit an exponential curve to these data.

19.14 In both parts (a) and (b) below choose what seems to be the best mathematical description of the trend of the given short series, then fit the least-squares trend curve in the most economical way you can.

(a) In 1980 and 1981 a real estate firm sold 5 and 10 estates at prices in excess of $2 million.

(b) In 1979, 1980, and 1981 another real estate firm sold 1, 2, and 4 estates at prices in excess of $2 million.

19.15 Comment on the following statement: The use of a least-squares line eliminates all subjectivity in measuring the trend in a time series since, for a given series of data, there is only one line for which $\sum (y - \hat{y})^2$ is a minimum.

19.5

Moving Averages

A secular trend is often considered to be an indication of the "general sweep" of the development of a time series. If it is uncertain whether the trend is linear or whether it might be better described by some other kind of curve, if we are not sure whether we are actually dealing with a trend or part of a cycle, and if we are not really interested in obtaining a mathematical equation, we can describe the overall "behavior" of a time series quite well by means of an artificial series called a **moving average**. A moving average is constructed by replacing each value in a series by the mean of itself and some of the values directly preceding and directly following it. For instance, in a three-year moving average calculated for annual data, each annual figure is replaced by the mean of itself and the annual figures for the two adjacent years; in a five-year moving average each annual figure is replaced by the mean of itself, those of the two preceding years, and those of the two following years. If the averaging is done over an even number of periods, say, 4 years or 12 months, the moving average will initially fall between successive years or months. In such cases, the values are customarily brought "back in line" (or "centered") by taking a subsequent two-year (or two-month) moving average. We shall use this procedure later in measuring seasonal variation.

The basic problem in constructing a moving average is choosing an appropriate period for the average. This choice depends largely on the nature of the data and the purpose for which the average is constructed. Ordinarily, the purpose of fitting a moving average is to eliminate, insofar as possible, some sort of unwanted or distracting fluctuations in the data. In describing the trend of annual data by a moving average, for example, the main problem is to eliminate those up and down departures of the data from the basic trend which result from business cyclical influences. If all business cycles were exactly alike both in duration and amplitude, their influences could be easily removed from (averaged out of) a series because any absolutely uniform periodic movements are completely eliminated by a moving average whose period is equal to (or a multiple of) the period of the movement. This means also that, if the seasonal movements in a series of monthly data were exactly uniform, the seasonal (and also most of the irregular) variations could be removed from the series by fitting to it a 12-month moving average. However, uniformly periodic cyclical, seasonal, and irregular movements do not appear in economic time series, so the effect of fitting moving averages to series is to smooth out, but not eliminate completely, certain fluctuations in the series.

EXAMPLE Construct a five-year moving average to smooth the series consisting of the business failure rates (the number of failures per 10,000 concerns) in the United States for the years 1947 through 1978.

SOLUTION We show the original data in the second column of the table below.

Year	Business failure rate	Five-year moving totals	Five-year moving averages
1947	14.3		
1948	20.4		
1949	34.4	134.1	26.8
1950	34.3	148.5	29.7
1951	30.7	161.3	32.3
1952	28.7	168.9	33.8
1953	33.2	176.2	35.2
1954	42.0	193.5	38.7
1955	41.6	216.5	43.3
1956	48.0	239.2	47.8
1957	51.7	249.0	49.8
1958	55.9	264.4	52.9
1959	51.8	280.8	56.2
1960	57.0	289.9	58.0
1961	64.4	290.3	58.1
1962	60.8	291.7	58.3
1963	56.3	288.0	57.6
1964	53.2	275.2	55.0
1965	53.3	263.4	52.7
1966	51.6	245.7	49.1
1967	49.0	229.8	46.0
1968	38.6	220.3	44.1
1969	37.3	210.4	42.1
1970	43.8	199.7	39.9
1971	41.7	197.5	39.5
1972	38.3	198.6	39.7
1973	36.4	197.4	39.5
1974	38.4	190.5	38.1
1975	42.6	180.2	36.0
1976	34.8	167.7	33.5
1977	28.0		
1978	23.9		

In the third column we show the five-year **moving totals** which consist, for each total shown, of the sum of that year's figure and the figures for the two preceding and the two following years. We get the five-year moving averages shown in the last column by dividing each moving total by 5 (or multiplying it by $\frac{1}{5} = 0.20$).

Both the original series and the five-year moving average are shown in Figure 19.9, and it is evident from this figure that the moving average has substantially reduced the fluctuations in the series and given it a much smoother appearance. The missing values at the beginning and at the end of the artificial series are characteristic of moving averages of this sort: we lose one value at each end for a three-year moving average, two for a five-year moving average, three for a seven-year moving average, and so on. This is often of no consequence, but it may cause a problem if a series is very short or if values are needed for each year for further calculations.

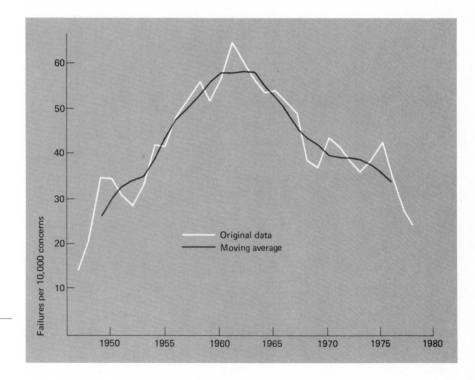

19.9

Business failure rate in the United States, 1947–1978.

19.6

Exponential Smoothing

Moving averages have been widely used for many years to smooth certain unwanted fluctuations out of time series and, hence, describe their underlying nature. A more recently developed smoothing technique—called **exponential smoothing**—is now also available for this purpose. In addition, exponential smoothing is ideally suited to short-run forecasting for inventory control; indeed, the technique was originally used in business mainly to provide an efficient and economical way to forecast item-by-item demand for items covered by automated inventory control systems. This is particularly important in large companies where, often, literally thousands and thousands of items are stocked and it is necessary to make routine forecasts of, say, the week-to-week, or month-to-month, demand for each item.

Exponential smoothing is a particular kind of moving average, but its nature is quite different from the conventional moving-average process described in the preceding section. In calculating a conventional moving average, each item in the averaging period is given a weight and all items outside the period are given no weight at all. For instance, as we proceed through the calculation of a five-year moving average using equal weights, each item in each set of five items being averaged is weighted 1 and all the other items not included in the set are weighted 0.

In exponential smoothing, the smoothed value at time t, S_t, is a weighted average of the observed value at time t, y_t, and all the other past (historical) values in the series: $y_{t-1}, y_{t-2}, \ldots,$. and y_1. This is not immediately apparent from the way in which the smoothed values are calculated. In practice, we begin by setting S_1, the first value in the series of smoothed values, equal to y_1, the first actual value in the series. Then, as we move forward, at each new time period the new smoothed value is α times the current observed value of y plus $1 - \alpha$ times the previous smoothed value of y, where α, the smoothing constant, is a fraction between 0 and 1 which we may choose at our discretion. Hence, the second smoothed value is

$$S_2 = \alpha y_2 + (1 - \alpha)S_1$$

the third smoothed value is

$$S_3 = \alpha y_3 + (1 - \alpha)S_2$$

and in general the smoothed value for time period t is

Exponential smoothing

$$S_t = \alpha y_t + (1 - \alpha)S_{t-1}$$

To show that S_t does, indeed, depend on all the past values in the series, we write $S_{t-1} = \alpha y_{t-1} + (1 - \alpha)S_{t-2}$, and, substituting into the formula for S_t above, we get

$$S_t = \alpha y_t + (1 - \alpha)[\alpha y_{t-1} + (1 - \alpha)S_{t-2}]$$
$$= \alpha y_t + \alpha(1 - \alpha)y_{t-1} + (1 - \alpha)^2 S_{t-2}$$

Then we write $S_{t-2} = \alpha y_{t-2} + (1 - \alpha)S_{t-3}$, and, substituting into the result just obtained, we get

$$S_t = \alpha y_t + \alpha(1 - \alpha)y_{t-1} + (1 - \alpha)^2[\alpha y_{t-2} + (1 - \alpha)S_{t-3}]$$
$$= \alpha y_t + \alpha(1 - \alpha)y_{t-1} + \alpha(1 - \alpha)^2 y_{t-2} + (1 - \alpha)^3 S_{t-3}$$

Repeating this process $t - 4$ more times, we finally arrive at the result that

$$S_t = \alpha y_t + \alpha(1 - \alpha)y_{t-1} + \alpha(1 - \alpha)^2 y_{t-2} + \alpha(1 - \alpha)^3 y_{t-3} + \cdots$$
$$+ (1 - \alpha)^{t-1}y_1$$

We see from this that the weight assigned to each observed value in the series decreases exponentially, and this is why we refer to the process as exponential smoothing. Depending on the value of α, the weights assigned to the earlier values in the series decrease more or less rapidly, and, if the averaging process is carried far enough forward, a time comes when the earliest values have very little effect on the current smoothed value.

At the level of this book, we cannot discuss the mathematics of choosing an optimum smoothing constant to meet a particular objective. It is clear, however, that if α is too large, we give too much weight to the current values as they occur, and we will not adequately smooth out the irregular variations. On the other hand, if α is too small, we give too little weight to the current values in the series, and the moving average is insensitive to changes that may actually be taking place.

EXAMPLE A real estate development corporation keeps voluminous statistical data on various areas and properties which seem to have above average potential for future purchase or development of one sort or another. Among these data are the annual amounts of snowfall (in feet) during the snow season at the site of a proposed winter resort. The snowfall amounts for the 26-year period 1955–1980 are shown in the second column of the table on page 500, and the graph of the series is shown in Figure 19.10. From an inspection of the data in the table and its graph no trend is apparent, but there is obviously a good deal of irregular, or random, year-to-year variation, and this is what we want to smooth (or average) out. Smooth the series using exponential smoothing, first with the smoothing constant $\alpha = 0.2$, and then with the smoothing constant $\alpha = 0.5$.

SOLUTION The third column of the table shows the exponentially smoothed values calculated with the constant $\alpha = 0.2$, and the fourth column shows the smoothed values calculated with $\alpha = 0.5$. In this example, we took the first year's snowfall, 9.9 feet, for the first value, S_1, of both smoothed

Year	Annual snow-fall, y_t	S_t ($\alpha = 0.2$)	S_t ($\alpha = 0.5$)
1955	9.9	9.9	9.9
1956	22.2	12.4	16.0
1957	11.4	12.2	13.8
1958	14.8	12.7	14.3
1959	19.7	14.1	17.0
1960	14.9	14.3	16.0
1961	15.9	14.6	16.0
1962	13.4	14.4	14.7
1963	12.0	13.9	13.4
1964	7.9	12.7	10.6
1965	12.9	12.7	11.8
1966	16.8	13.5	14.3
1967	11.6	13.1	13.0
1968	14.9	13.5	14.0
1969	13.3	13.5	13.6
1970	20.2	14.8	17.0
1971	14.3	14.7	15.6
1972	20.4	15.8	18.0
1973	13.0	15.2	15.6
1974	10.0	14.2	12.8
1975	15.9	14.5	14.4
1976	17.4	15.1	15.9
1977	18.7	15.8	17.3
1978	14.3	15.5	15.8
1979	16.2	15.6	16.0
1980	21.1	16.7	18.6

series. Then, with $\alpha = 0.2$, the second value is

$$S_2 = (0.2)(22.2) + (0.8)(9.9) = 12.4$$

the third value is

$$S_3 = (0.2)(11.4) + (0.8)(12.4) = 12.2$$

the fourth value is

$$S_4 = (0.2)(14.8) + (0.8)(12.2) = 12.7$$

and so on. The values in the series with the smoothing constant $\alpha = 0.5$ are calculated in the same way.

It can be seen from the plot of the original data and the two smoothed series in Figure 19.10 that the choice of the smoothing constant has a pronounced effect on the extent of the smoothing. The series with $\alpha = 0.2$ is a good deal

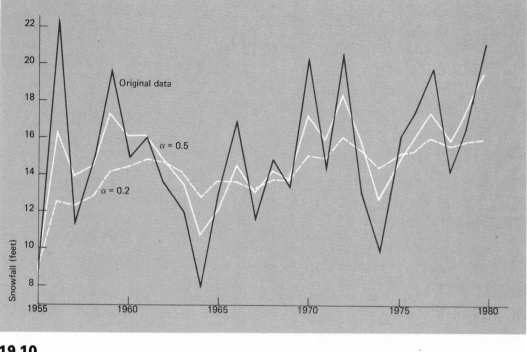

19.10

Exponentially smoothed time series.

more stable than the series with $\alpha = 0.5$, and this is what we want when our objective is to average out the year-to-year changes in the data (which we presume are merely random fluctuations) rather than to respond to them the way the series with $\alpha = 0.5$ does. Of course, when we use exponential smoothing to help predict, say, the market demand for an item at a later time, we want quick responses to whatever actual changes may occur in the level of demand (for instance, an anticipated large stepwise increase resulting from an intensive advertising campaign); this suggests that, at least for a time, we use a relatively large value of α.

EXERCISES

19.16 Construct a three-year moving average for the business failure rate series given in the example on page 496. Plot the original data, the three-year moving average, and the five-year moving average as in Figure 19.9, and compare the graphs of the two moving averages.

19.17 The following series shows (to the nearest million pounds) the annual production of acetylsalicylic acid (aspirin) in the United States for the years 1947–1976: 12, 11, 10, 11, 14, 13, 14, 14, 15, 17, 18, 21, 18, 24, 23, 27, 28, 28, 29, 34, 30, 31, 37, 35, 32, 35, 32, 33, 25, and 29.

(a) Construct a three-year moving average and draw a graph showing this moving average together with the original data.

(b) Construct a five-year moving average, add it to the graph of part (a), and compare the two moving averages.

19.18 The following series shows the annual production (to the nearest million pounds) of ethyl acetate (85%) in the United States for the years 1947–1976: 87, 62, 80, 92, 86, 72, 81, 73, 86, 91, 92, 86, 101, 107, 102, 102, 118, 118, 114, 122, 137, 179, 167, 161, 159, 222, 222, 198, 171, and 159.

(a) Calculate a three-year moving average and draw a graph showing this moving average together with the original data.

(b) Calculate a five-year moving average, add it to the graph of part (a), and compare the two moving averages.

19.19 Calculate a three-year moving average for the series of Exercise 16.35 on page 428, and draw a graph showing it together with the original data.

19.20 Calculate a five-year moving average for the series of Exercise 16.30 on page 426, and draw a graph showing it together with the original data.

19.21 Use the formula on page 499 to express S_4 in terms of α, y_4, y_3, y_2, and y_1, and use this result to verify the two smoothed 1958 values given in the table on page 500.

19.22 The following series shows the number of requests for a particular kind of service at a bank window on 12 consecutive working days: 20, 39, 18, 25, 30, 12, 21, 27, 15, 35, 30, and 17. Smooth this series using

(a) the smoothing constant $\alpha = 0.2$;

(b) the smoothing constant $\alpha = 0.5$.

19.23 Financial records of a large manufacturing company show retained net incomes of 41.6, 44.7, 37.3, 43.2, 30.5, 38.1, 40.9, 20.4, 35.8, 45.0, and 42.3 (millions of dollars) for a given 11-year period. Smooth this series using

(a) $\alpha = 0.1$;   (b) $\alpha = 0.4$.

19.24 The exponential smoothing process with smoothing constant α produces an average which is statistically similar to a $[(2/\alpha) - 1]$-term conventional moving average. For instance, if $\alpha = 0.5$, an exponentially smoothed average is statistically similar to a $(2/0.5) - 1 = 3$-term moving average, and if $\alpha = 0.01$ the exponentially smoothed average is similar to a $(2/0.01) - 1 = 199$-term moving average. This helps explain why a small value of α produces a more stable exponentially smoothed series than a larger value of α does.

(a) If $\alpha = 0.2$, what is the equivalent number of terms being averaged in a conventional moving average?

(b) For what value of α is an exponentially smoothed series equivalent to a 19-term moving average?

By statistical similarity we mean that the variabilities in the exponentially smoothed series and the moving average, not the corresponding terms in the two series, are equivalent.

19.25 When we use exponential smoothing in making short-range predictions, we take the smoothed value at period t, S_t, as the estimate of y at period $t + 1$. Designating this estimate y'_{t+1}, we have

$$y'_{t+1} = S_t = \alpha y_t + (1 - \alpha)S_{t-1}$$

and since $S_{t-1} = y'_t$,

$$y'_{t+1} = \alpha y_t + (1 - \alpha)y'_t$$

As before, we may set the smoothed value at period 1, S_1 (and consequently, the estimated value at period 2, y'_2), equal to y_1; if we prefer, we may assign the average of some recent values to S_1 and y'_2. If, for instance, we want to use an exponential smoothing formula which is statistically equivalent to a 15-term moving average, we can set S_1 and y'_2 equal to the average of the 15 most recent values of y. Estimates for succeeding periods may then be made by using either of the smoothing formulas above with the appropriate α each time the new value y_t becomes available.

(a) A large manufacturer wants to estimate at the end of each week the number of replacement parts of a particular kind which will be demanded the following week for one group of machines. No trend has been observed in the demand for the part for some time and none is anticipated in the near future, nor is there any periodic variation in the demand. The following are the numbers of parts demanded in weekly periods 1–14: 24, 20, 16, 20, 25, 24, 26, 19, 16, 22, 20, 24, 19, and 23. Choose a smoothing formula which is equivalent to a four-term moving average, set $S_1 = y'_2 = 19$ (the average demand in the four weeks preceding period 1), and estimate the demand for the remaining periods.

(b) With reference to part (a), set $S_1 = y_1 = 24$ (the actual demand in period 1) and estimate the demand for the remaining periods using $\alpha = 0.2$.

19.7

Seasonal Variation

Let us now consider the problem of measuring those movements in a time series which recur more or less regularly in the same months of successive years. We call the measures of this seasonal variation an **index of seasonal variation**, or a **seasonal index**. For monthly data a seasonal index consists of 12 numbers, one for each month, each of which expresses that particular month's activity as a percentage of the average month's activity. For instance, if the June seasonal index of sales of a wholesaler is 92, this means that June sales are typically 92 percent of sales in the average month. We use the word "typically" here because the actual percentage for a given month varies more or less widely from year to year, and 92 percent is an average of these percentages.

There are many ways in which seasonal variation can be measured or a seasonal index can be constructed. These range from rather crude measures based on very simple calculations to highly refined measures based on involved computer techniques. We shall illustrate the construction of a seasonal index by using the basic **ratio-to-moving-average**, or **percentage-of-moving-average**, method. Until certain refinements were made possible by the use of high-speed computers, this basic method was probably the most widely used and the most generally satisfactory one available. Today, it is still the best introduction to one's study of seasonal variation.

In constructing a seasonal index all our efforts are aimed at eliminating trend, cyclical, and irregular variations from the series. The way this is done in the basic ratio-to-moving-average method is relatively simple. We begin by calculating a 12-month moving average of the data in order to remove the seasonal movements from the series. Since an n-period moving average will completely eliminate any absolutely uniform n-period recurring movement, a 12-month moving average would eliminate all the seasonal movements from the series, provided these movements recurred with complete regularity year after year. Of course, in actual practice seasonal patterns vary somewhat from year to year, so the moving average cannot be expected to eliminate all of the seasonal variation. It will eliminate most of it, however, as well as most of the irregular variation, so the 12-month moving average is an estimate of the trend and cyclical components of the series. In the classical model we are discussing, each value in the original series is assumed to be the product of factors attributed to the four basic components (secular trend, seasonal variation, cyclical variation, and irregular variation). Therefore, dividing each value by the corresponding value of the 12-month moving average gives an estimate of the seasonal and irregular components in the series. In other words, dividing $T \cdot S \cdot C \cdot I$ by $T \cdot C$ leaves us with $S \cdot I$, the product of the factors attributed to seasonal and irregular variations. All that is left to do then is to eliminate, insofar as possible, the irregular fluctuations.

When one knows a good deal about the series under study, it may be possible to identify and eliminate directly the monthly $S \cdot I$ values which reflect the impact of extraordinary events (for example, a crippling nine-day tule fog, or a strike in a supplier's plant). Irregular variations of this sort, as well as those which are due to chance (nonassignable causes) can also be effectively eliminated by averaging, in some way, the $S \cdot I$ figures for the different Januaries, for the different Februaries, and so on. We can, for instance, reduce the effect of the irregular forces by using the median of the values given for each month, or perhaps by using the **modified (arithmetic) mean** which is the mean of the values remaining after the smallest and largest values have been cast out. Moving averages can also be used for smoothing out the irregular variations remaining at this stage of the calculation. In any case, by some sort of averaging process we finally arrive at an estimate of the way seasonal factors alone influence the values of a series. This estimate, consisting of the 12 monthly values, is the seasonal index.

The series we shall use to illustrate the calculation of a seasonal index is the sales by shoe stores in the United States. The original data, plotted in Figure 19.11 and given in column 1 of the table on pages 506, 507, and 508, are the monthly sales in millions of dollars for the years 1967–1971, a period which we consider to be typical of the way in which such sales vary from month to month. The pronounced seasonal pattern in the sales and an upward trend are evident from the graph.

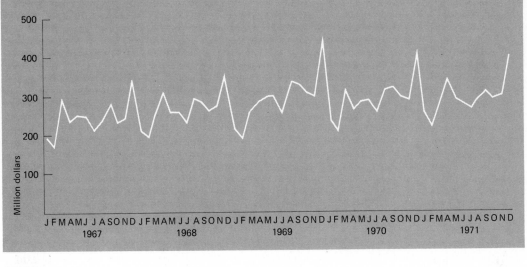

19.11

Monthly sales by shoe stores in the United States, 1967–1971.

EXAMPLE   Calculate a seasonal index for sales by shoe stores in the United States by using the basic ratio-to-moving-average method.

SOLUTION   The first step in the procedure is to calculate the 12-month moving totals shown in column 2. The first entry in this column, 2,934, is the sum of the 12 monthly sales figures for 1967, and it is recorded at the middle of the period, between June and July 1967. The second entry in this column, 2,945, is obtained by subtracting the January 1967 figure from 2,934 and adding the January 1968 figure; in other words, 2,945 is the sum of the 12 monthly sales from February 1967 through January 1968, and it is recorded at the middle of this period. The third and succeeding entries in the column are found by continuing this process of subtracting and adding monthly values.

In order to get a 12-month moving average which is centered on the original data, we next calculate two-month moving totals of the entries of column 2. These are shown in column 3, with the first entry being the sum of the first two values in column 2, the second entry being the sum of the second and third values in column 2, and so on. These entries in column 3 are recorded between those of column 2 and, hence, they are in line with (or centered on) the original data.

Since each entry of column 2 is the sum of 12 monthly figures and each entry of column 3 is the sum of two entries of column 2, or altogether the sum of 24 monthly figures, we finally get the centered 12-

SALES BY SHOE STORES
(MILLIONS OF DOLLARS)

Year	Month	Sales (1)	12-month moving total (2)	2-month moving total (3)	Centered 12-month moving average (4)	Percentages of 12-month moving average (5)
1967	January	199				
	February	168				
	March	291				
	April	236				
	May	248				
	June	247				
			2,934			
	July	209		5,879	245.0	85.3
			2,945			
	August	240		5,915	246.5	97.4
			2,970			
	September	281		5,912	246.3	114.1
			2,942			
	October	233		5,960	248.3	93.8
			3,018			
	November	242		6,047	252.0	96.0
			3,029			
	December	340		6,070	252.9	134.4
			3,041			
1968	January	210		6,109	254.5	82.5
			3,068			
	February	193		6,191	258.0	74.8
			3,123			
	March	263		6,249	260.4	101.0
			3,126			
	April	312		6,284	261.8	119.2
			3,158			
	May	259		6,351	264.6	97.9
			3,193			
	June	259		6,395	266.5	97.2
			3,202			
	July	236		6,409	267.0	88.4
			3,207			
	August	295		6,411	267.1	110.4
			3,204			

Year	Month	Sales (1)	12-month moving total (2)	2-month moving total (3)	Centered 12-month moving average (4)	Percent-ages of 12-month moving average (5)
	September	284		6,410	267.1	106.3
			3,206			
	October	265		6,382	265.9	99.7
			3,176			
	November	277		6,379	265.8	104.2
			3,203			
	December	349		6,436	268.2	130.1
			3,233			
1969	January	215		6,492	270.5	79.5
			3,259			
	February	190		6,560	273.3	69.5
			3,301			
	March	265		6,646	276.9	95.7
			3,345			
	April	282		6,732	280.5	100.5
			3,387			
	May	286		6,797	283.2	101.0
			3,410			
	June	289		6,915	288.1	100.3
			3,505			
	July	262		7,030	292.9	89.4
			3,525			
	August	337		7,074	294.6	114.4
			3,546			
	September	328		7,145	297.7	110.2
			3,599			
	October	307		7,182	299.3	102.6
			3,583			
	November	300		7,164	298.5	100.5
			3,581			
	December	444		7,163	298.5	148.7
			3,582			
1970	January	235		7,162	298.4	78.8
			3,580			
	February	211		7,142	297.6	70.9
			3,562			
	March	318		7,116	296.5	107.2
			3,554			
	April	266		7,100	295.8	89.9
			3,546			

Year	Month	Sales (1)	12-month moving total (2)	2-month moving total (3)	Centered 12-month moving average (4)	Percentages of 12-month moving average (5)
	May	284		7,083	295.1	96.2
			3,537			
	June	290		7,038	293.3	98.9
			3,501			
	July	260		7,020	292.5	88.9
			3,519			
	August	319		7,045	293.5	108.7
			3,526			
	September	320		7,009	292.0	109.6
			3,483			
	October	299		7,041	293.4	101.9
			3,558			
	November	291		7,123	296.8	98.0
			3,565			
	December	408		7,120	296.7	137.5
			3,555			
1971	January	253		7,116	296.5	85.3
			3,561			
	February	218		7,098	295.8	73.7
			3,537			
	March	275		7,069	294.5	93.4
			3,532			
	April	341		7,057	294.0	116.0
			3,525			
	May	291		7,059	294.1	98.9
			3,534			
	June	280		7,063	294.3	95.1
			3,529			
	July	266				
	August	295				
	September	315				
	October	292				
	November	300				
	December	403				

month moving average shown in column 4 by dividing each entry of column 3 by 24. These moving average values are the trend-cycle estimates, and we use them now to eliminate the $T \cdot C$ components from the original series. This is done by dividing the original $T \cdot S \cdot C \cdot I$ data month by month by the corresponding $T \cdot C$ estimates (that is, by the corresponding values of the moving average) and multiplying these ratios by 100. In this way, we arrive at the percentages of moving average shown in column 5.

All that remains to be done is to eliminate the irregular variations as best we can, and to this end we rearrange the entries of column 5 in the five dated columns of the following table:

Month	1967	1968	1969	1970	1971	Median	Seasonal index
January		82.5	79.5	78.8	85.3	81.0	81.0
February		74.8	69.5	70.9	73.7	72.3	72.3
March		101.0	95.7	107.2	93.4	98.4	98.4
April		119.2	100.5	89.9	116.0	108.2	108.2
May		97.9	101.0	96.2	98.9	98.4	98.4
June		97.2	100.3	98.9	95.1	98.0	98.0
July	85.3	88.4	89.4	88.9		88.6	88.6
August	97.4	110.4	114.4	108.7		109.6	109.6
September	114.1	106.3	110.2	109.6		109.9	109.9
October	93.8	99.7	102.6	101.9		100.8	100.8
November	96.0	104.2	100.5	98.0		99.2	99.2
December	134.4	130.1	148.7	137.5		136.0	136.0
						1,200.4	1,200

From the various ways we could use to average the figures given for each month, we choose the median here. (In this case, where we have four values for each month, the median is, in fact, equivalent to the modified mean.) The 12 medians are shown in the second column from the right. Now, since the seasonal index for each month is supposed to be a percentage of the average month, the sum of the 12 values should equal 1,200. Actually, the medians total 1,200.4, and so we adjust for this by multiplying each of the medians by 1,200/1,200.4 = 0.999667. In some cases this adjustment is substantial (see Exercise 19.26 on page 514), but here it has no effect since we are carrying the values of the index only to one decimal. Consequently, the final values of the seasonal index, shown in the last column, are the same as the 12 medians.

The interpretation of seasonal indexes is straightforward. For instance, in our example January sales by shoe stores are typically 81.0 percent of those of the average month, sales are usually low in February, and December sales are typically 36.0 percent above sales of the average month.

Observe, however, that in using a seasonal index for any purpose, we must always be mindful of its limitations. An index is based on historical (past) data, and we cannot reasonably expect seasonal patterns to remain completely constant over long periods of time. The method we have illustrated here applies to the description of constant seasonal patterns or seasonal patterns which do not change very much. If there are pronounced changes in the seasonal pattern with the passage of time, the sort of index we have discussed will not be suitable.

19.8

Seasonally Adjusted Data

Seasonal indexes are extremely important in various practical applications. We shall briefly explain two of these, the first in **deseasonalizing** data and the second in **forecasting**. Leaving the use of seasonal indexes in forecasting to the section which follows, let us now describe the process of removing seasonal influences from a given set of data, or deseasonalizing a series. Of course, when seasonal fluctuations have actually occurred, nobody knows how things would have been if a series had been uninfluenced by seasonal factors. So, when we speak of what things would have been like without seasonal fluctuations, we are speaking rather loosely. Nevertheless, the notion of a series of data free from seasonal influences is a useful way of understanding the concept.

The process of removing seasonal variation, or deseasonalizing data, consists merely of dividing each value in a series by the corresponding value of the seasonal index and multiplying the result by 100 (or by dividing by the corresponding value of the seasonal index written as a proportion). The logic of this process is quite simple: If April shoe sales are 108.2 percent of those in the average month, taking $\frac{100}{108.2} \cdot 100 = 92.4$ percent of the April sales would tell us what these sales should have been if there had been no seasonal variation.

EXAMPLE Deseasonalize the 1970 sales by shoe stores in the United States, using the seasonal index constructed above.

SOLUTION In the table which follows, the sales figures and the seasonal index are copied from pages 507, 508, and 509, with the values of the seasonal index given as proportions and rounded to two decimals. The values in the right-hand column are the deseasonalized sales, and they are computed by dividing each month's actual sales by the corresponding value of the seasonal index.

1970	Sales	Seasonal index	Deseasonalized sales
January	235	0.81	290
February	211	0.72	293
March	318	0.98	324
April	266	1.08	246
May	284	0.98	290
June	290	0.98	296
July	260	0.89	292
August	319	1.10	290
September	320	1.10	291
October	299	1.01	296
November	291	0.99	294
December	408	1.36	300

Inspecting this table we discover several interesting facts. For instance, there was an increase of $59 million in sales from July to August. This is nothing to rejoice over, however, since the deseasonalized data show that this increase is actually less than might have been expected in accordance with typical seasonal patterns. Also, there was a drop of $21 million in sales from September to October. This is no cause for alarm, though, since the deseasonalized data show that this drop is actually less than might be expected in accordance with typical seasonal patterns. The need for taking seasonal variation into account in the analysis of business and economic time series should be apparent from these observations.

The principal source of seasonally adjusted (deseasonalized) series is the federal government. In constructing these series, seasonal indexes are derived by highly sophisticated computer methods which are essentially a refinement and an extension of the basic ratio-to-moving-average method. The computational burden in this work is tremendous, and without the great arithmetic ability of digital computers, the necessary calculations could never be made for a large number of series within any reasonable limits of time and cost. The Bureau of the Census, however, has developed computer methods to the point where it can now virtually mass produce seasonally adjusted series of a high order of quality for almost all important series of data. Two variants of the Census program, X-11 (for monthly data) and X-11Q (for quarterly data), are now available to the public on a computer tape. Other computer approaches to the seasonal adjustment problem, including further refinements in the ratio-to-moving-average technique and, what is an entirely different concept, seasonal adjustments based on regression analysis, may result in still further improvements in seasonally adjusted series.

19.9

Forecasting

Now that we are familiar with time series and know how to measure some of their components, let us discuss briefly the tremendously complicated problem of business forecasting. The rationale of basing forecasts on time series is that, having observed some regularity in the movement of data through time, we are hopeful that "what has happened in the past will, to a greater or lesser extent, continue to happen or will happen again in the future." Thus, the obvious way to forecast the trend of a time series is to extrapolate from the trend equation describing the historical data. By "extrapolate" we mean extend the trend into the future so as to estimate a value that lies beyond the range of the values used to derive the trend equation.

EXAMPLE The following trend equation, calculated by the method of least squares from company sales data extending back over a long period of years, describes the long-term growth in sales of a paint manufacturer:

$$\hat{y} = 64.4 + 2.88x$$

(origin, 1980; x units, 1 year; y, total annual
sales in millions of dollars)

Estimate total sales for the year 1984, and based on a seasonal index to be given below, monthly sales for that year.

SOLUTION The 1980 trend value is $64.4 million and the growth forces are estimated to be producing a $2.88 million increase in sales per year. Substituting $x = 4$ into the trend equation, we find that, based on the long-term growth forces alone, the 1984 expected total sales are $\hat{y} = 64.4 + 2.88(4) = 75.92$, or $75,920,000.

If the manufacturer's sales were uninfluenced by seasonal fluctuations, we could estimate 1984 monthly sales in precisely the same way. Modifying the trend equation for use with monthly data by dividing a by 12 and b by 144 (by 12 and then 12 again) and then shifting the origin to January 1984, we get

$$\hat{y} = 6.22 + 0.02x$$

(origin, January 1984; x units, 1 month; y, average
monthly sales in millions of dollars)

Hence, $6.22 million is the trend value for January 1984, and substituting $x = 1, 2, 3, \ldots$, and 11 into this equation we get the trend values for the other eleven months of 1984; all these values are shown in the "Trend values" column of the following table:

Month	Trend values	Seasonal index	Predicted monthly sales for 1984
January	6.22	0.77	4.79
February	6.24	0.73	4.56
March	6.26	1.02	6.39
April	6.28	1.25	7.85
May	6.30	1.40	8.82
June	6.32	1.20	7.58
July	6.34	1.00	6.34
August	6.36	1.05	6.68
September	6.38	0.98	6.25
October	6.40	1.04	6.66
November	6.42	0.81	5.20
December	6.44	0.75	4.83

However, there is a very pronounced variation in the manufacturer's sales, which results from seasonal influences. Sales are typically quite low in November, December, January, and February and quite high in April, May, and June. In fact, a seasonal index calculated from recent historical data by the ratio-to-moving-average method shows that, for example, January sales are only 77 percent and May sales are 140 percent of what they would be if there were no seasonal variation. The complete seasonal index, written as proportions, is shown in the third column of the table.

Assuming that the seasonal pattern is not changing and is adequately described by the seasonal index, we complete the solution of this example by multiplying the trend value for each of the 12 months by the seasonal index for that month. That is, we multiply the January trend value by 0.77, the February trend value by 0.73, and so on, getting the predicted monthly sales (in millions of dollars) shown in the last column of the table. If there were no seasonal variation, we would expect May sales to be $\frac{6.30}{6.22} = 1.013$, or 1.3 percent higher than January sales because of the upward trend in sales; taking into account the seasonal influences, though, May sales are projected at $\frac{8.82}{4.79} = 1.841$, or 84.1 percent higher than January sales.

What we have done in arriving at these monthly sales predictions is precisely the opposite of deseasonalizing data. We have introduced the seasonal patterns into the data (rather than removed them from it) by multiplying the trend values by the corresponding values of the seasonal index, written as proportions. These products of a measure of the trend and of the seasonal, or $T \cdot S$, are the values we would expect if trend and seasonal forces were the only factors influencing the values of a series, and they are often called the **normal values**.

In the predicted sales values above, we have taken into account the effect of trend and seasonal patterns on the manufacturer's sales, but we have not yet considered the possible effects of cyclical and irregular influences. The latter, we have said, are essentially unpredictable; in the manufacturer's case, a large freight shipment of raw materials lost in transit for five weeks, for instance, or a fire which closes a competing manufacturer's plant for six months cannot be foreseen. The effects of such events may tend to average out in the long run, but they can substantially affect sales in particular months and cause even the most careful forecasts to go astray. In connection with the major problems of forecasting both short-run and long-run business cycles and their effect at the level of the economy as a whole and at industry and firm levels, we shall only remark that much help is available to everyone from both private and public sources. One very important aid is the Department of Commerce's monthly report *Business Conditions Digest.* Among the wealth of economic data contained in this report are various economic indicators. These are series of data which tend to turn up or down before overall economic activity does, the **leading indicators**; series which tend to move at about the same time as overall activity, the **roughly coincident indicators**; and series which tend to move somewhat behind overall activity, the **lagging indicators**, and hence to confirm or refute the earlier directional signals.

EXERCISES

19.26 In constructing a seasonal index for the sales of a branch department store by the ratio-to-moving-average method, a statistician has arrived at the following medians (in the same way we arrived at the values in the column headed "Median" on page 509): 78.4, 75.5, 93.5, 96.1, 103.9, 96.4, 83.8, 87.9, 100.6, 120.1, 111.7, and 134.3. Complete the calculation of the seasonal index.

19.27 The following data show the monthly sales (in thousands of dollars) of a manufacturer of home warm-air furnaces for the years 1971–1975, a period the manufacturer considers to be typical of these sales:

Year	Jan.	Feb.	Mar.	Apr.	May	June
1971	61.0	51.5	60.2	59.5	59.3	70.5
1972	55.1	52.2	56.5	56.1	64.0	83.6
1973	71.2	68.8	77.2	80.1	83.2	102.2
1974	59.2	61.0	63.7	67.8	69.4	88.3
1975	57.5	59.6	61.7	62.2	71.4	88.9

Year	July	Aug.	Sept.	Oct.	Nov.	Dec.
1971	71.4	102.5	130.0	114.1	77.6	43.5
1972	79.1	110.4	140.5	132.4	100.5	75.3
1973	109.3	134.7	154.1	155.6	102.5	68.8
1974	80.0	113.1	128.5	120.6	80.8	54.4
1975	85.3	110.2	129.1	133.8	80.1	66.7

Compute a seasonal index for this series by the ratio-to-moving-average method, using the median to average the percentages of moving average for the individual months.

19.28 Use the seasonal index calculated in Exercise 19.27 to deseasonalize the 1971–1975 monthly sales data shown in the exercise. Plot the deseasonalized data together with the original data on one chart.

19.29 With reference to Exercise 19.27, fit a least-squares trend line to the total annual furnace sales for the years 1971–1975. Modify this equation and use it and the seasonal index computed in Exercise 19.27 to forecast the monthly dollar sales of home furnaces in the year 1980.

19.30 The following data show the factory production of creamery butter (in millions of pounds) in the United States for the years 1969–1974:

Year	Jan.	Feb.	Mar.	Apr.	May	June
1969	106.8	96.3	103.7	109.0	116.6	110.8
1970	99.8	92.5	107.0	111.2	116.9	113.3
1971	102.6	97.7	111.2	112.3	118.2	112.6
1972	101.5	99.4	106.8	110.2	119.4	111.1
1973	96.1	84.4	90.6	93.7	100.3	87.6
1974	80.6	69.0	77.4	88.1	99.8	91.8

Year	July	Aug.	Sept.	Oct.	Nov.	Dec.
1969	92.6	76.8	67.7	76.7	71.2	90.0
1970	92.1	78.7	71.9	81.6	79.0	93.1
1971	89.4	81.1	69.5	79.9	79.3	89.9
1972	89.4	76.3	65.4	75.9	73.2	81.0
1973	69.1	58.4	51.3	62.7	60.3	67.5
1974	78.4	73.3	65.4	70.8	67.3	89.6

Compute a seasonal index for these data by the ratio-to-moving-average method; use the modified mean to average the percentages of moving average for the individual months.

19.31 Use the seasonal index of Exercise 19.30 to deseasonalize the 1974 butter production figures.

19.32 With reference to Exercise 19.30, fit a trend line by the method of least squares to the total annual butter production for the years 1969–1974. Modify this equation and use it and the seasonal index calculated in Exercise 19.30 to forecast the factory production of creamery butter for the 12 months of 1980.

19.33 Deseasonalized monthly data are often multiplied by 12 and then referred to as **annual rates**. The use of seasonally adjusted annual rates is particularly helpful in facilitating the analysis of month-to-month changes in series of data which are best understood on an annual basis (such as gross national product, new housing starts, and various other production series). Accordingly, we often see reported such statements as "Americans had more money income in June on an annual rate basis than in any other month in history." Based on the seasonally adjusted figure for March 1974 calculated in Exercise 19.31, at what annual rate was butter production then running?

19.34 A company selling swimming-pool supplies had sales of $20,000 and $24,000 in March and April of 1981. The company's seasonal index for these two months stands at 105 and 140. The president of the company expressed disappointment in the March-to-April performance and estimated that the total 1981 sales would be only $217,000, a figure substantially less than the sales manager's estimate of $264,000. Is there any real reason to be disappointed with the $4,000 sales increase from March to April? How do you suppose the sales manager and the president arrived at their estimates? Which one seems more reasonable?

19.35 A large furniture manufacturer estimates total 1982 sales to be $44,400,000. The company's seasonal index for furniture sales is, in the usual order, 78, 75, 100, 126, 138, 121, 101, 104, 99, 103, 80, and 75. Ignoring the possible existence of a trend in sales, draw up a tentative 1982 monthly sales budget for the company.

19.36 Suppose that there is, in fact, an upward trend in the sales of the manufacturer referred to in Exercise 19.35 as shown by the following trend equation: $\hat{y} = 30,000,000 + 2,880,000x$ (origin, 1977; x units, 1 year; y, total annual sales in dollars).

(a) Use this equation, suitably modified, to calculate the trend values for monthly sales for the twelve months of 1982.

(b) Draw up a revised monthly sales budget for 1982 which takes into account the trend in furniture sales.

19.37 Suppose that the equation which describes the trend in sales of a branch department store in a large metropolitan shopping center is $\hat{y} = 2,160 + 72x$ (origin, 1977; x units, 1 year; y, total annual sales in thousands of dollars). The seasonal index of sales is 80, 75, 95, 98, 107, 96, 82, 89, 104, 123, 110, and 141 for the months January–December. Draw up a monthly sales forecast for the store for 1983.

19.10

A Word of Caution

The real trouble in forecasting and planning for the future is that there are altogether too many variables which need to be taken into account. Some of these are at least quantifiable and essentially predictable, but some are not. In analyzing rates of return on various capital investments, for instance, such inputs to the problem as the net installed cost of a machine and earnings from new equipment can often be determined or estimated reasonably well. But sound capital budgeting forces companies also to take into account such things as government policies which might affect the availability and price of money, changes in leasing opportunities, and possible revisions of depreciation and depletion laws as well as future tax decisions on these rates and allowances. Waiting is not much help, either, since there is no possibility of waiting until all, or sometimes even any, of the basic questions are settled finally. The projection of past experience to the uncertain future is speculative and hazardous, but at some time decisions must be made on the basis of the available incomplete knowledge; otherwise, nothing gets done.

Except for a few irreversible decisions, however, no one is irrevocably committed to a forecast, to survive or perish with it once it has been made. For instance, a large department store in a shopping center may have to make adjustments in its sales forecasts to take account of improved area transportation facilities, the opening of a new competing store nearby, an increase in the sales tax, and other things which could not be foreseen at the time the forecasts were made. Actually, forecasts are tentative things—special kinds of hypotheses, so to speak—which can be modified or revised in response to changing conditions. When forecasts are revised in the light of new information, all those concerned must take whatever steps are necessary to translate revised production, sales, or other goals into action. Intelligent forecasting and planning demand one's continual attention to changing conditions.

Generally speaking, it seems clear that realistic forecasts, which contribute greatly both to individual success and to the stability of the economy, are the results of applying sound business experience and judgment to relevant and timely statistical analyses.

19.11

Check List of Key Terms

19.12

Review Exercises

19.38 There were 10, 9, 5, 4, 5, and 3 small manufacturing plants built in an industrial park during the years 1976–1981.
  (a) Plot the series on arithmetic paper.
  (b) Fit a least-squares trend line to the data and plot the line on the chart showing the original data.
  (c) Modify the trend equation by shifting the origin of x to the year 1980.

19.39 In calculating a seasonal index for a production series, a statistician has arrived at the following medians (in the same way we arrived at the values in the column headed "Median" on page 509): 58.1, 60.1, 73.8, 95.2, 116.8, 125.3, 120.6, 126.1, 123.1, 108.0, 103.1, and 71.8. Complete the calculation of this seasonal index.

19.40 The equation of a least-squares trend line fit to the total annual revenues of the Kennecott Copper Corporation for the years 1966–1974 is

$$\hat{y} = 1,079.9 + 117.2x$$

(origin, 1970; x units, 1 year; y, total annual revenues in millions of dollars)

Modify the trend equation for use with monthly data and shift the origin of the equation to January 1970.

19.41 The following are the numbers of gold coins sold by a bullion dealer in twelve consecutive weeks: 173, 203, 170, 233, 124, 167, 198, 179, 222, 196, 209, and 183. Smooth this series using the exponential smoothing constant $\alpha = 0.3$.

19.42 We have a series in which the raw data are the total annual sales volumes in millions of dollars of a large electronics chain, but we are interested in quantity changes alone, not price changes. Would any sort of statistical adjustment to the raw data seem to be in order before analysis, and if so, what sort of adjustment?

19.43 The equation describing the trend in the sales of a manufacturing company is $\hat{y} = 3,600 + 108x$ (origin, 1972; x units, 6 months; y, total annual unit sales). The company's seasonal index for the months of January through December is 84, 72, 101, 120, 100, 99, 88, 96, 108, 96, 96, and 139. Draw up a monthly sales forecast for the year 1983.

19.44 The following series shows the total annual revenues in thousands of dollars of an automobile supply company for the years 1965–1981: 88.1, 89.1, 88.6, 101.9, 86.7, 96.8, 112.7, 129.2, 202.0, 195.4, 192.8, 191.9, 237.4, 234.6, 270.9, 320.0, and 338.0.
  (a) Plot the series on arithmetic paper.
  (b) Fit a parabola to the data by the method of least squares and plot the curve on the graph of part (a).

(c) What is the trend value for 1973?

(d) What is the slope of the trend curve at $x = 0$?

19.45 Rounding the sales data in the preceding exercise to the nearest thousand dollars and plotting the data on semilog paper, we find that the series seems to straighten out quite well.

(a) Fit an exponential curve to these sales data rounded to the nearest thousand dollars.

(b) Plot the exponential curve on the graph of part (b) of the preceding exercise and judge by inspection whether the exponential curve or the parabola fits the data better.

(c) What is the average annual rate of growth in these sales?

19.46 Many time series of such things as income, production, and consumption are often published in per capita (for each person) form, not as totals for an entire population. Coffee and tea consumption, for instance, are often reported in pounds consumed per person in the adult population. Why do you suppose this adjustment is made to these series?

19.47 In a study of its sales of one kind of electrical motor, a manufacturer calculates the following least-squares trend equation:

$$\hat{y} = 2,800 + 200x$$

(origin, 1979; x units, 1 year; y, total number of units sold annually)

The company has physical facilities to produce only 3,600 units a year, and it believes that it is reasonable to assume that, at least for the next decade, the trend will continue as before.

(a) What is the expected annual increase in the number of units sold?

(b) By what year will the company's expected sales have equaled its present physical capacity?

(c) How much in excess of the company's present capacity is the estimated 1985 sales figure?

19.48 The following data show the monthly stocks of turkeys in cold storage (in millions of pounds) during six years:

Year	Jan.	Feb.	Mar.	Apr.	May	June	July	Aug.	Sep.	Oct.	Nov.	Dec.
First	170	150	137	109	98	92	90	102	149	241	220	177
Second	179	154	131	105	84	80	79	104	161	255	208	162
Third	160	141	112	87	68	65	67	87	134	220	183	149
Fourth	142	124	105	87	74	66	71	113	186	282	210	161
Fifth	169	152	126	108	94	106	128	189	270	382	318	263
Sixth	251	219	191	156	132	121	123	160	233	340	265	203

Compute a seasonal index for these data by the ratio-to-moving-average method, using the median to average the percentages of moving average for the individual months.

19.49 Use the seasonal index calculated in the preceding exercise to deseasonalize the data, and plot the deseasonalized data together with the original data on one chart.

19.50 The following is a trend equation fit by the method of least squares to the turkey storage data of Exercise 19.48:

$$\hat{y} = 181.48 + 1.04x$$

(origin, January of 6th year; x units, 1 month; y, average monthly stocks in millions of pounds)

Use this equation and the seasonal index computed in Exercise 19.48 to forecast the monthly stocks of turkeys in the ninth year.

19.51 The number of exploratory gas wells completed by an oil company during the years 1975–1979 were 8, 19, 19, 14, and 18. Fit a least-squares line to this series and plot the trend line together with the original series on arithmetic paper.

19.52 The following figures represent the number of employees (in thousands) in the contract construction industry in the United States for the years 1940 through 1971: 1,294, 1,790, 2,170, 1,567, 1,094, 1,132, 1,661, 1,982, 2,169, 2,165, 2,333, 2,603, 2,634, 2,623, 2,612, 2,802, 2,999, 2,923, 2,778, 2,960, 2,885, 2,816, 2,902, 2,963, 3,050, 3,186, 3,275, 3,203, 3,259, 3,411, 3,345, and 3,259. Construct a three-year moving average and draw a diagram showing the moving average together with the original data.

19.53 In addition to the moving averages we have discussed in Section 19.5 there are also **weighted moving averages**. The advantage of these is that by a suitable choice of weights we can exert a great degree of control over the extent to which a series is smoothed. The weights used in weighted moving averages are often based on binomial coefficients such as 1, 2, and 1; 1, 4, 6, 4, and 1; or 1, 6, 15, 20, 15, 6, and 1. For example, with weights of 1, 2, and 1 we get $(1 \cdot y_1 + 2 \cdot y_2 + 1 \cdot y_3)/4$ for the first value of the moving average, and so on.

(a) Use the series 130, 149, 144, 175, 175, 161, and 158 to verify that we obtain a five-year moving average with weights 1, 2, 3, 2, and 1 by taking a three-year moving average of a three-year moving average.

(b) Verify that the centered 12-month moving average we used in the ratio-to-moving-average method of calculating a seasonal index is actually a weighted 13-month moving average with weights 1, 2, 2, 2, 2, 2, 2, 2, 2, 2, 2, 2, and 1.

19.54 The following are the quarterly sales (in thousands of dollars) of a jewelry manufacturer for the years 1977–1980:

Year	QUARTER 1	2	3	4
1977	99	136	109	240
1978	110	212	136	249
1979	115	182	162	344
1980	135	190	160	308

(a) Draw a chart of these sales to verify that there is, indeed, a consistent seasonal pattern.

(b) Calculate a centered four-quarter moving average.

(c) Use the results of part (b) to compute a (quarterly) seasonal index for the sales by the ratio-to-moving-average method.

19.55 With reference to the preceding exercise, use the seasonal index obtained in part (c) to deseasonalize the sales data.

19.56 Assuming a linear trend in the sales data of Exercise 19.54, forecast the jewelry manufacturer's sales for the four quarters of 1982.

19.57 In time series analysis we sometimes deal with published series of data extending back fifty years or more. Over such long periods of time, though, changes of various kinds often occur which make the figures in a series not strictly comparable. Using chain-store sales as an example, suggest some changes in the reporting of these data down through the years which may make historical comparisons more approximate than exact.

19.58 In the classical time series model each original value in a series of monthly values is presumed to be the product of the trend, seasonal, cyclical, and irregular components, or $y = T \times S \times C \times I$. In Section 19.9 we called the products we used there, $T \times S$, the "normal values." If now we divide the original values by the normal values, we get the values $C \times I$, called the "cyclical-irregulars." Since $\dfrac{C \times I}{I} = C$, it appears that we can get a measure of the cyclical components alone for each month simply by dividing out the irregular component.

(a) Explain why it is not feasible to isolate the cyclical effect in this way.

(b) Suggest a way in which the irregular effects can be eliminated from the cyclical-irregulars leaving, as a residual, a measure of the cyclical variation alone.

Regardless of whether we merely describe things numerically or whether we generalize beyond our data, locating, assembling, or collecting data can raise many problems. This is a serious matter because access to a good supply of high-quality data is fundamental to all of statistics.

So, we shall begin this chapter with a discussion of sources of business data in Section 20.1, and in connection with this it is important to recognize that much of the published information available from government and private sources (indeed, much of the world's knowledge) is actually based on samples. Since the word "sample" is used somewhat loosely in everyday language, let us observe again that in statistics it has a very special meaning—a sample is a set of data that can reasonably serve to make generalizations, or inferences, about the population from which it came. In Sections 20.2 through 20.6 we shall continue the discussion of Section 11.1 and see how such data can be obtained, particularly in situations where simple random sampling is impractical or not feasible.

Finally, in Sections 20.7 and 20.8 we shall discuss some aspects of the design of experiments, of planning every detail, so that, with reasonable assurance, the data we get will ultimately serve their purpose.

20

Planning Business Research

20.1

Sources of Business Data

Data required to solve practical everyday business problems can come from many sources, sometimes broadly classified as **internal** and **external**. Internal data are generated from the activities within a firm; they may be taken from a firm's order book or inventory, payroll, personnel, service, inspection, or accounting records; they may be collected in experiments and tests of product quality characteristics, or they may be gathered by agents, by telephone, or by questionnaires from customers as well as suppliers. External data are obtained from sources outside the firm; they may come from records of state, local, and national governments and their agencies and regulatory bodies, from trade associations, private institutions, other firms, and so on. Some problems require the combination of internal and external data, say, when a company compares its own operating performance with that of its competitors or the industry as a whole.

External data are sometimes classified as **primary**, meaning that the organization gathering the data also publishes or releases them, or as **secondary**, meaning that the data are published by an organization other than the one which gathered them. There are many important and highly respected sources of primary as well as secondary data. Among the nongovernmental sources we find private statistical services, trade associations, trade publications, university research bureaus, commercial and financial periodicals, and specialized reporting agencies. From these sources come data on employment, farm prices and marketings, construction contracts, store sales, bank debits, and the like, often broken down on a regional, state, county, or city basis. In addition, there are reports on prices, production, sales, employment, and the like in different industries, and all this information is supplied to fill the needs of individuals and groups for reliable statistics.

By far the biggest collector and publisher of business data, and generally the most important single source of external data, is the federal government. Within the great mass of statistical material flowing from the government it is possible to find information relating to virtually every aspect of the life of the nation. For instance, the Department of Commerce publishes each year the *Statistical Abstract of the United States*, an immense storehouse of data referring to many things and gathered from many sources. Through the Bureau of the Census, the Commerce Department periodically takes and publishes the results of censuses of population, manufactures, distribution, housing, and agriculture, and also publishes monthly trade reports giving data on inventories, sales, and so on, in various wholesale and retail lines of business for the entire country and for selected cities. The Office of International Trade, the Office of Industry and Trade, and the Bureau of Foreign and Domestic Commerce, all of the Commerce Department, collect and publish data on the trade of the United States and other countries. Through the Bureau of Economic Analysis, the Department issues one of the most important of all statistical publications, the monthly *Survey of Current Business* (which is supplemented by weekly data on some of the major series of data). The S-pages of the regular monthly issues contain indicators of general business conditions and series relating to wholesale and retail commodity prices and trade, construction activity, population and employment, payrolls, wages and hours, finance, foreign trade, transportation, and so on. In addition, the Bureau publishes *Business Statistics*, a biennial supplement to the *Survey of Current Business*, which presents historical data for the series appearing in the monthly S-pages. Without much doubt, these publications constitute two of the most valuable collections of current and historical data.

Another Commerce Department publication of major importance is the monthly *Business Conditions Digest*, which contains about 500 economic indicators in a form which permits various types of analyses of current and prospective business conditions.

Other important statistics are collected and published by the Department of Labor, whose Bureau of Labor Statistics issues the *Monthly Labor Review*, the primary source for the indexes of consumer prices and wholesale prices, and data on construction contracts and costs, employment, payrolls, wages and hours, and work stoppages. In addition to the publications of many other departments, agencies, and commissions of the government, the Board of Governors of the Federal Reserve System publishes monthly the *Federal Reserve Bulletin* and the *Federal Reserve Chart Book on Financial and Business Statistics* and releases other data periodically. Besides containing a wealth of financial information, the *Bulletin* is the primary source of the Index of Industrial Production.

The problem of how to keep up with the continuing, massive flow of government data, and where to find needed information, has long bothered users of statistical data, since no single central reference source to the government's statistical output exists. In 1973, however, the Congressional Information Service began publication of a much needed master index to government statistics dealing with the American people (population and housing; demographic characteristics; consumer prices and expenditures; manpower, employment, and earnings; and so on). Called the *American Statistics Index*, it indexes the output of over 100 government agencies, congressional committees, and statistics-producing programs which produce statistics about the American people, and these data constitute half of the government's total statistical production.

Most of the data collected by the government are needed by the government itself in discharging its many responsibilities; data relative to some phenomena are not needed specifically by the government, but are collected and published in response to the needs of such large groups of individuals as to justify their collection at public expense.

EXERCISES

20.1 Determine whether the *Survey of Current Business* is a primary or a secondary source for data on:
(a) gross national product or expenditure;
(b) industrial production;
(c) industrial and commercial failures;
(d) wholesale (producer) prices;
(e) retail trade;
(f) federal government finance;
(g) petroleum, coal, and products.

20.2 Answer each of the following questions for which numerical data and other information are summarized (from various sources) in *Business Statistics*:
(a) What is the definition of "personal consumption expenditures?"
(b) What factors account for changes in the U.S. gold stock?
(c) What procedure is used in compiling export and import statistics for low-valued shipments?
(d) Does the labor force include the unemployed?
(e) How many gallons are there in a barrel of beer?

20.3 Find in the *Monthly Labor Review*
(a) the number of persons employed in mining for the past five years;
(b) the annual values of the Producer Price Index for the same period.

20.4 Find in the *Statistical Abstract of the United States* the primary source of the number of forest fires on federally protected areas of the United States.

20.5 According to the *Federal Reserve Bulletin*, total U.S. Reserve Assets (in millions of dollars) at the end of the years 1960 and 1970 were 19,359 and 14,487. What was the amount at the end of 1976?

20.6 Find in the *Handbook of Labor Statistics 1979* where the Bureau of the Census obtains its data for *Consumer Income*. Are the data completely reliable?

20.7 As of the most recent date or period for which you can locate data, determine
(a) the total amount of capital expenditures made by the paper industry for abatement of air and water pollution;
(b) member bank reserve requirements of the Federal Reserve System;
(c) mine production of recoverable zinc, by states;
(d) the Consumer Price Index for fuel oil and coal, Washington, D.C.

20.2

Samples and Sample Designs

Broadly speaking, there are two different types of samples: **probability samples** and **judgment samples**. By a probability sample from a finite population we mean a sample chosen in such a way that each element of the population has a known, though not necessarily equal, probability of being included in the sample. Simple random samples of size n drawn from populations of size N are probability samples, since each element of the population has a known probability n/N of being included in the sample. There are still other ways to take probability samples from populations and all such samples possess one great advantage: Only probability samples enable us to calculate sampling errors and, hence, judge the "goodness" of estimates or decisions to which statistical analyses lead.

In contrast to probability samples, we shall refer to samples as judgment samples if, in addition to (or instead of) chance, personal judgment plays a significant role in their selection. There are many situations where, for practical reasons, investigators use judgment samples to gain needed information. One important use of such sampling is in testing markets for new products. Because of the tremendous cost of market testing on a national scale, many products are first tested in one or a few cities. Such test cities are usually not selected at random; instead they are carefully chosen because in someone's considered judgment they are "typical" or "average" American cities. Subsequently, when generalizations are made to national markets, samples of public reaction to advertising, packaging, palatability, and the like constitute judgment samples.

Regardless of how the conclusions or actions based on judgment samples ultimately turn out, judgment samples have the undesirable feature that standard statistical theory cannot be applied to evaluate the accuracy and reliability of estimates or to calculate the probabilities of making various kinds of erroneous decisions. Whenever elements of judgment enter in the selection of a sample, the evaluation of the "goodness" of estimates or decisions based on the sample is again largely a matter of personal judgment.

There are a great many ways to select a sample from a population, and there is an extensive literature devoted to the subject of designing sampling procedures. In statistics, a **sample design** is a definite plan, completely determined before any data are collected, for obtaining a sample from a given population. Thus, a plan to take a simple random sample from among the 268 members of a trade association by using a table of random numbers in a prescribed way constitutes a sample design. Of the many ways in which a sample can be taken from a given population, some are quite simple while others are relatively involved. In what follows, we shall discuss briefly some of the most important kinds of sample designs.

20.3

Systematic Sampling

In many situations the most practical way of sampling is to select, say, every 10th voucher in a file, every 20th name on a list, every 50th piece coming off an assembly line, and so forth. Sampling of this sort is referred to as **systematic** sampling, and an element of randomness is usually introduced into this kind of sampling by using random numbers to pick the unit with which to start. Although a systematic sample may not be a random sample in accordance with our definition, it is often reasonable to treat systematic samples as if they were random samples. Whether or not this is justified depends entirely on the structure (order) of the list, or arrangement, the sample comes from. In many instances, systematic sampling actually provides an improvement over simple random sampling inasmuch as the sample is spread more evenly over the entire population.

The real danger in systematic sampling lies in the possible presence of hidden periodicities. For instance, if we inspect every 40th piece made by a particular machine, our results would be biased if, because of a regularly recurring failure, every 20th piece was blemished. Also, a systematic sample might yield biased results if we interviewed the residents of every 10th house along a certain route and it so happened that every 10th house selected was a corner house on a double lot.

20.4

Stratified Sampling

As we have seen, the goodness of a generalization or the closeness of an estimate depends largely on the standard error of the statistic being used, which, in turn, depends on both the sample size and the population variability. In theory, at least, we can increase the precision of a generalization by increasing the sample size and in practice, if the cost of sampling is largely overhead and the items to be sampled are readily at hand, it may be about as easy to take a sample of size 200 as it is to take a sample of size 50. On the other hand, if the cost of sampling is more or less proportional to the size of the sample (for instance, in destructive testing where the cost of sampling is largely the cost of the items tested), it may be very costly to increase the sample size. Moreover, a sample four times as large as another yields estimates which are only twice as reliable as those based on the other.

One relatively simple scheme for reducing the size of the standard error of a statistic is **stratification**. This is a procedure which consists of stratifying (or dividing) the population into a number of nonoverlapping subpopulations, or **strata**, then taking a sample from each stratum. If the items selected from each stratum constitute a simple random sample, the entire procedure (first stratification and then simple random sampling) is called **stratified (simple) random sampling**. Samples chosen in this way are probability samples.

Although the concept of stratifying is relatively simple, several substantial problems immediately arise: What should be the basis of stratification? How many strata should be formed? What sample sizes should be allocated to the different strata? How should the samples within the strata be taken? Stratification does not guarantee good results, but if successful and properly executed, a stratified sample will generally lead to a higher degree of precision, or reliability, than a simple random sample of the same size drawn from the whole population.

Suppose, for instance, that we want to estimate the mean weight of four persons on the basis of a sample of size 2; the weights of the four persons are 110, 130, 180, and 200 pounds, so that μ, the mean weight we want to estimate, is 155 pounds. If we take an ordinary random sample of size 2 from this population and weigh these two persons, $\bar{x}$ can vary from 120 to 190. In fact, the $\binom{4}{2} =$ 6 possible samples of size 2 that can be taken from this population have the means 120, 145, 155, 155, 165, and 190, so that $\sigma_{\bar{x}} = 21.0$ (see Exercise 20.9 on page 532).

Now suppose we know that the four persons consist of two men (who weigh 180 and 200 pounds) and two women (who weigh 110 and 130 pounds) and we stratify our sample by sex, then randomly select one of the two women and one of the two men. We now find that $\bar{x}$ varies on the much smaller interval from 145 to 165. In fact, the means of the four possible samples are 145, 155, 155, and 165, and $\sigma_{\bar{x}} = 7.1$ (see Exercise 20.9 on page 532). This illustrates how, by stratifying the sample, we are able to reduce $\sigma_{\bar{x}}$ from 21.0 to 7.1.

Essentially, the goal of stratification is to form strata in such a way that there is some relationship between being in a particular stratum and the answers sought in the statistical study, and that within the separate strata there is as much homogeneity (uniformity) as possible. In our example there is such a connection between sex and weight and there is much less variability in weight within the two groups than there is in the entire population.

In the example above we used what is called **proportional allocation** in selecting the sample, which means that the sizes of the individual samples were proportional to the sizes of the respective strata. In general, if we divide a population of size N into k strata of size $N_1, N_2, \ldots,$ and N_k, and take a sample of size n_1 from the first stratum, a sample of size n_2 from the second stratum, $\ldots,$ and a sample of size n_k from the kth stratum, we say that allocation is proportional if

$$\frac{n_1}{N_1} = \frac{n_2}{N_2} = \cdots = \frac{n_k}{N_k}$$

or if these ratios are as nearly equal as possible. In the example dealing with the weights we have $N_1 = 2$, $N_2 = 2$, $n_1 = 1$, and $n_2 = 1$, so that

$$\frac{n_1}{N_1} = \frac{n_2}{N_2} = \frac{1}{2}$$

and the allocation is, indeed, proportional.

It can easily be shown (see Exercise 20.12 on page 533) that allocation is proportional if

Sample sizes for proportional allocation

$$n_i = \frac{N_i}{N} \cdot n \qquad for \ i = 1, 2, \ldots, and \ k$$

where n is the total size of the sample, that is, $n = n_1 + n_2 + \cdots + n_k$. When necessary, we use the integers closest to the values given by this formula.

EXAMPLE A stratified sample of size $n = 60$ is to be taken from a population of size $N = 4,000$, which consists of three strata of size $N_1 = 2,000$, $N_2 = 1,200$, and $N_3 = 800$. If the allocation is to be proportional, how large a sample must we take from each stratum?

SOLUTION Substituting into the formula, we get $n_1 = \frac{2,000}{4,000} \cdot 60 = 30$, $n_2 = \frac{1,200}{4,000} \cdot 60 = 18$, and $n_3 = \frac{800}{4,000} \cdot 60 = 12$.

After we have taken a sample of size n_1 from the first stratum, a sample of size n_2 from the second stratum, . . . , we calculate the means of the k samples, $\bar{x}_1, \bar{x}_2, \ldots,$ and $\bar{x}_k$, then estimate the mean of the whole population as

$$\bar{x}_w = \frac{N_1 \bar{x}_1 + N_2 \bar{x}_2 + \cdots + N_k \bar{x}_k}{N_1 + N_2 + \cdots + N_k}$$

This is a weighted mean of the individual $\bar{x}$'s, and the weights are the sizes of the strata. Actually, this formula applies regardless of how we allocate parts of the total sample to the strata, but when allocation is proportional, we can replace the N's with the n's and $\bar{x}_w$ is simply the mean of all the data combined; for this reason proportional allocation is said to be self-weighting.

In proportional allocation the importance of different strata sizes is taken into account by the fact that the larger strata contribute relatively more items to the sample. However, strata differ not only in size but also in internal variability, and it might seem reasonable to take somewhat smaller samples from the less variable strata and somewhat larger samples from the more variable strata. We can take into account both the size and the internal variability of strata by taking from each stratum a sample whose size is proportional to the product of the stratum size and the stratum standard deviation. If we designate the standard deviations of the strata $\sigma_1, \sigma_2, \ldots,$ and σ_k, this kind of allocation requires that we make

$$\frac{n_1}{N_1 \sigma_1} = \frac{n_2}{N_2 \sigma_2} = \cdots = \frac{n_k}{N_k \sigma_k}$$

or make these ratios as nearly equal as possible. In this way, the larger and the more variable strata will contribute relatively more items to the total sample. This method of allocation is called **optimum allocation**, since for a fixed sample size, the sample chosen in this way will have the smallest possible standard error for the estimate of the population mean. To take a sample which meets the above requirement, we find the sample sizes for the different strata by using the formula

Sample sizes for optimum allocation

$$n_i = \frac{n \cdot N_i \sigma_i}{N_1 \sigma_1 + N_2 \sigma_2 + \cdots + N_k \sigma_k} \qquad for \ i = 1, 2, \ldots, and \ k$$

EXAMPLE   With reference to the preceding example, suppose that $\sigma_1 = 8$, $\sigma_2 = 15$, and $\sigma_3 = 32$. For optimum allocation, how large a sample must we take from each stratum?

SOLUTION   Substituting $n = 60$, $N_1 = 2,000$, $N_2 = 1,200$, $N_3 = 800$, $\sigma_1 = 8$, $\sigma_2 = 15$, and $\sigma_3 = 32$ into the formula for n_1, we get

$$n_1 = \frac{60 \cdot (2,000)(8)}{(2,000)(8) + (1,200)(15) + (800)(32)}$$

which gives $n_1 = 16$, and similarly we find that $n_2 = 18$ and $n_3 = 26$. Note that because of the large variability of the third stratum, n_3 is now larger than n_1 and n_2.

One problem in using optimum allocation is that we must know the standard deviations of the different strata. Another problem arises if we want to estimate several population characteristics from the same set of sample data, since what is optimum for one characteristic may not be optimum for another. In situations like these, it may be better to use proportional allocation. Indeed, the gain in reliability due to optimum allocation is often not large enough to offset the obvious practical advantages of (self-weighting) proportional allocation.

Stratification is not restricted to a single variable of classification, or characteristic, and populations are often stratified according to several characteristics. In a system-wide survey designed to determine the attitude of its students toward, say, a new tuition plan, a state college system with 17 colleges might stratify the students with respect to class standing, sex, major, and college. So, part of the sample would be allocated to sophomore women majoring in English in college A, part to senior men majoring in engineering at college N, and so on. Up to a point, stratification like this, called **cross stratification**, will often increase the precision (reliability) of estimates, and it is widely used, particularly in opinion sampling and market surveys.

20.5

Quota Sampling

In stratified sampling, the cost of taking random samples from the individual strata is often so expensive that interviewers are simply given quotas to be filled from the different strata, with few (if any) restrictions on how they are to be filled. For instance, in determining voters' attitudes toward increased tax refunds to elderly persons, an interviewer working a certain area might be told to interview 5 male self-employed homeowners under 30 years of age, 10 female wage earners in the 50–60 year bracket who live in apartments, 4 retired males over 60 who live in mobile homes, and so on, with the actual selection of the individuals being left to the interviewer's discretion. This is called **quota sampling**, and it is a convenient, relatively inexpensive, and often a necessary procedure, but as it is executed, the resulting samples are usually not probability samples.

In the absence of firm restrictions on their choice, interviewers naturally tend to select individuals who are most readily available—persons who work in the same building, shop in the same store, or perhaps reside in the same general area. Quota samples are thus essentially judgment samples, and although it may be possible to guess at sampling errors by using experience or corollary information, quota samples generally do not lend themselves to any sort of formal statistical evaluation.

20.6

Cluster sampling

To illustrate another important kind of sampling, suppose that a large foundation wants to study the changing patterns of family expenditures in the Los Angeles area. In attempting to complete schedules for 2,000 families, the foundation finds that simple random sampling is practically impossible, since suitable lists are not available and the cost of contacting families scattered over a wide area (with possibly two or three callbacks for the not-at-homes) is very high. One way in which a sample can be taken in this case is to divide the total area of interest into a number of smaller, nonoverlapping areas, say, city blocks. A number of these blocks is then randomly selected, with the ultimate sample consisting of all (or samples of) the families residing in these blocks. Generally speaking, in this kind of sampling, called **cluster sampling**, the total population is divided into a number of relatively small subdivisions, which are themselves clusters of still smaller units, and then some of these subdivisions, or clusters, are randomly selected for inclusion in the overall sample. If the clusters are geographic subdivisions, as in our example, this kind of sampling is also called **area sampling**.

To give two further illustrations of cluster sampling, suppose that the management of a large chain store organization wants to interview a sample of its employees to determine their attitudes toward a proposed pension plan. If random methods are used to select, say, five stores from the list and if some or all employees of these five stores are interviewed, the resulting sample is a cluster sample. Also, if the Dean of Students of a university wants to know how fraternity men at the school feel about a certain new regulation, he can take a cluster sample by interviewing some or all of the members of several randomly selected fraternities.

Although estimates based on cluster samples are usually not as reliable as estimates based on simple random samples of the same size (see Exercise 20.10 on page 533), they are usually more reliable per unit cost. Referring again to the survey of family expenditures in the Los Angeles area, it is easy to see that it may well be possible to take a cluster sample several times the size of a simple random sample for the same cost. It is much cheaper to visit and interview families living close together in clusters than families selected at random over a wide area.

In practice, several of the methods we have discussed may well be used in the same survey. For instance, if government statisticians wanted to study the attitude of American farmers toward marketing cooperatives, they might first stratify the country by states or some other geographic subdivision. To take a sample from each stratum, they might then use cluster sampling, subdividing each stratum into a number of smaller geographic subdivisions, and finally they might use simple random sampling or systematic sampling to select a sample of farmers within each cluster.

EXERCISES

20.8 The following are the numbers of commercial FM radio stations in operation in 1976 in the 50 states (listed in alphabetic order): 70, 7, 28, 51, 185, 44, 26, 8, 113, 83, 8, 17, 128, 98, 64, 39, 83, 55, 27, 33, 43, 107, 68, 66, 71, 19, 30, 12, 16, 34, 28, 118, 86, 13, 132, 49, 29, 122, 7, 53, 16, 69, 170, 16, 11, 72, 44, 31, 92, and 9.
   (a) List the 10 possible systematic samples of size 5 that can be taken from this list by starting with one of the first 10 numbers and then taking each tenth number on the list.
   (b) Calculate the means of the 10 samples obtained in part (a), and assuming that the starting point is randomly selected from among the first 10 numbers, show that the mean of this sampling distribution of $\bar{x}$ equals the population mean μ, the mean of the 50 numbers.

20.9 Verify that
   (a) $\sigma_{\bar{x}} = 21.0$ for the six sample means on page 528, which are assigned equal probabilities of $\frac{1}{6}$;
   (b) $\sigma_{\bar{x}} = 7.1$ for the four sample means on page 528, which are assigned equal probabilities of $\frac{1}{4}$.

20.10 Suppose that in a group of six athletes there are three tennis players whose weights are 140, 150, and 160 pounds, and three football players whose weights are 210, 220, and 230 pounds.

(a) List all possible random samples of size 2 which may be taken from this population, calculate the means of these samples, and show that $\sigma_{\bar{x}} = 22.7$.

(b) List all possible stratified random samples of size 2 which may be taken by selecting one tennis player and one football player, calculate the means of these samples, and show that $\sigma_{\bar{x}} = 5.8$.

(c) Suppose that the six athletes are divided into clusters according to their sports, each cluster is assigned a probability of $\frac{1}{2}$, and a random sample of size 2 is taken from one of the randomly chosen clusters. List all possible samples, calculate their means, and show that $\sigma_{\bar{x}} = 35.2$.

(d) Compare and discuss the results obtained for $\sigma_{\bar{x}}$ in parts (a), (b), and (c).

20.11 On the basis of the amounts of their total deposits, 60 savings banks in a state are classified into 30 small, 20 medium-sized, and 10 large banks.

(a) In how many ways can we choose a stratified 10 percent sample of the 60 banks if one third of the sample is allocated to each of the three strata?

(b) In how many ways can we choose a stratified 10 percent sample of the 60 banks if the proportion of the sample allocated to each stratum is proportional to the size of the stratum?

20.12 Verify that if the formula $n_i = \dfrac{N_i}{N} \cdot n$ is used to determine the sample sizes allocated to the strata, then

(a) the allocation is proportional (that is, the ratios n_i/N_i all equal the same constant);

(b) the sum of the n_i is equal to n.

20.13 A stratified sample of size $n = 40$ is to be taken from a population of size $N = 2,000$, which consists of four strata for which $N_1 = 500$, $N_2 = 1,200$, $N_3 = 200$, and $N_4 = 100$. If the allocation is to be proportional, how large a sample must be taken from each stratum?

20.14 A stratified sample of size $n = 200$ is to be taken from a population of size $N = 50,000$, which consists of five strata for which $N_1 = 20,000$, $N_2 = 15,000$, $N_3 = 5,000$, $N_4 = 8,000$, and $N_5 = 2,000$. If the allocation is to be proportional, how large a sample must be taken from each stratum?

20.15 A stratified sample of size $n = 100$ is to be taken from a population of size $N = 40,000$, which consists of two strata for which $N_1 = 10,000$, $N_2 = 30,000$, $\sigma_1 = 45$, and $\sigma_2 = 60$. How large a sample must be taken from each stratum if the allocation is to be

(a) proportional;   (b) optimal?

20.16 A stratified sample of size $n = 440$ is to be taken from a population of size $N = 22,000$, which consists of four strata for which $N_1 = 2,000$, $N_2 = 6,000$, $N_3 = 10,000$, $N_4 = 4,000$, $\sigma_1 = 25$, $\sigma_2 = 20$, $\sigma_3 = 15$, and $\sigma_4 = 30$. How large a sample must be taken from each stratum if the allocation is to be

(a) proportional;   (b) optimal?

20.17 Suppose that we want to estimate the mean weight of six persons on the basis of a sample of size 3. The weights of the six persons are 135, 141, 159, 165, 171, and 267 pounds, so the population mean which we want to estimate is

$\mu = 173$. If, furthermore, the first four of these weights are those of women and the other two are weights of men, and we stratify according to sex, show that

(a) proportional allocation leads to $n_1 = 2$ and $n_2 = 1$;

(b) optimum allocation leads to $n_1 = 1$ and $n_2 = 2$.

20.18 With reference to the preceding exercise show that

(a) if the allocation is proportional, there are 12 possible samples for which
$\sigma_{\bar{x}} = 16.7$ (provided that the selection within each stratum is random);

(b) if the allocation is optimal, there are only four possible samples for which
$\sigma_{\bar{x}} = 8.2$ (provided that the selection within each stratum is random).

20.19 If $\bar{x}$ is the mean of a stratified random sample of size n obtained by proportional allocation from a finite population of size N, which consists of k strata of size $N_1, N_2, \ldots,$ and N_k, then

$$\sigma_{\bar{x}}^2 = \sum_{i=1}^{k} \frac{(N - n)N_i^2}{nN^2(N_i - 1)} \cdot \sigma_i^2$$

where $\sigma_1^2, \sigma_2^2, \ldots,$ and σ_k^2 are the corresponding variances for the individual strata.

(a) Use this formula to verify the value $\sigma_{\bar{x}} = 7.1$, actually $\sqrt{50}$, given in the example on page 528.

(b) Use this formula to verify the value $\sigma_{\bar{x}} = 5.8$, actually $\sqrt{\frac{100}{3}}$, given in part (b) of Exercise 20.10.

This standard error formula, by itself, does not enable us to judge the effectiveness of stratification; this depends on whether or not the variances $\sigma_1^2, \sigma_2^2, \ldots,$ and σ_k^2 are appreciably smaller than the corresponding variance σ^2 for the entire population.

20.7

Planning Experiments

Even people who have never done any research should be able to visualize some of the problems involved in planning an experiment so that it can actually serve the purpose it is designed for. It happens all too often that an experiment intended to test one thing tests another, or that a poorly designed experiment tests nothing of any interest whatever. Suppose, for instance, that in order to compare the cleansing action of two detergents, someone has soiled 10 swatches of white cloth equally with India ink and oil and then washed 5 swatches in an agitator-type machine using a cup of detergent Q and the other 5 swatches in the same machine using a cup of detergent R. Following this, whiteness readings were made on the swatches with the following results:

Detergent Q: 76, 85, 82, 80, 77
Detergent R: 72, 58, 74, 66, 70

20.35 On each weekday, 4 of a bank's 13 employees are asked to remain in the bank during the lunch hour. Complete the following schedule, in which the employees are numbered 1–13, so that each of the 13 employees shares lunch-hour duty once with each of the other employees:

	Employees			
October 3	2	3	5	11
October 4	5	6	8	
October 5		9	11	4
October 6	11	12	1	7
October 7	1	2	4	10
October 10	4		7	13
October 11	7	8	10	
October 12	10		13	6
October 13	13	1	3	9
October 14	3	4	6	12
October 17	6		9	2
October 18	9	10	12	5
October 19	12		2	8

20.36 The following are the numbers of locksmiths in twenty cities: 18, 23, 12, 10, 20, 11, 7, 11, 116, 34, 9, 8, 19, 25, 9, 83, 28, 9, 13, and 15.
(a) List the five possible systematic samples of size $n = 4$ that can be taken from this list by starting with one of the first five numbers and then taking each fifth number on the list.
(b) Calculate the mean of each of the five samples obtained in part (a) and verify that their mean equals the average (mean) number of locksmiths in the given cities.

20.37 A food processor considers three spices for use individually or together in a new vegetable soup. If A, B, and C denote that the three spices are used, and a, b, and c denote that they are not used, then AbC, for example, denotes that the first and third spices are used and the second spice is not used. List the eight possible ways in which the food processor may or may not use the three spices.

20.38 A stratified sample of size $n = 90$ is to be taken from among 2,000 real estate listings, of which 600 are priced under $75,000, 800 are priced between $75,000 and $150,000, 400 are priced between $150,000 and $200,000, and 200 are priced over $200,000. If the allocation is to be proportional, what part of the sample should be allocated to each of the four strata?

20.39 Use the formula of Exercise 20.19 on page 534 to verify the value $\sigma_{\bar{x}} = 16.7$, actually $\sqrt{\frac{836}{3}}$, given in part (a) of Exercise 20.18 on page 534.

20.40 Verify symbolically that for stratified sampling with proportional allocation the weighted mean given by the formula on page 529 equals the mean of the combined data obtained for all the strata.

20.41 The records of a casualty insurance company show that among 3,800 claims filed against the company over a period of time, 2,600 were minor claims (under $200), while the other 1,200 were major claims ($200 or more). To estimate the average size of these claims, the company takes a 1 percent sample, proportionally allocated to the two strata, with the following results (rounded to the nearest dollar):

Minor claims: 42, 115, 63, 78, 45, 148, 195, 66, 18, 73, 55, 89, 170, 41, 92, 103, 22, 138, 49, 62, 88, 113, 29, 71, 58, 83

Major claims: 246, 355, 872, 649, 253, 338, 491, 860, 755, 502, 488, 311

(a) Find the means of these two samples and then determine their weighted mean, using as weights the respective sizes of the two strata.

(b) Verify that the result of part (a) equals the ordinary mean of the 38 claims; that is, verify for this example that proportional allocation is self-weighting.

Bibliography

A

Statistics for the Layman

CAMPBELL, S.K., *Flaws and Fallacies in Statistical Thinking*. Englewood Cliffs, N.J.: Prentice-Hall, Inc., 1974.

FEDERER, W. T., *Statistics and Society*. New York: Marcel Dekker, Inc., 1973.

HUFF, D., *How to Lie with Statistics*. New York: W. W. Norton & Company, Inc., 1954.

HUFF, D., AND GEIS, I., *How to Take a Chance*. New York: W. W. Norton & Company, Inc., 1959.

KIMBLE, G. A., *How to Use (and Misuse) Statistics*. Englewood Cliffs, N.J.: Prentice-Hall, Inc., 1978.

LARSEN, R. J., AND STROUP, D. F., *Statistics in the Real World*. New York: Macmillan Publishing Co., Inc., 1976.

LEVINSON, H. C., *Chance, Luck, and Statistics*. New York: Dover Publications, Inc., 1963.

MORONEY, M. J., *Facts from Figures*. New York: Penguin Books, 1956.

MOSTELLER, F., PIETERS, R. S., KRUSKAL, W. H., RISING, G. R., LINK, R. F., CARLSON, R., AND ZELINKA, M., *Statistics by Example*. Reading, Mass.: Addison-Wesley Publishing Company, Inc., 1973.

RUNYON, R. P., *Winning with Statistics*. Reading, Mass.: Addison-Wesley Publishing Company, Inc., 1977.

TANUR, J. M. (ed.), *Statistics: A Guide to Business and Economics*. San Francisco: Holden-Day, Inc., 1976.

B

Some Books on the Theory
of Probability and Statistics

FREUND, J. E., *Introduction to Probability*. Encino, Calif.: Dickenson Publishing Co., Inc., 1973.

FREUND, J. E., AND WALPOLE, R. E., *Mathematical Statistics*, 3rd ed. Englewood Cliffs, N.J.: Prentice-Hall, Inc., 1980.

GOLDBERG, S., *Probability—An Introduction*. Englewood Cliffs, N.J.: Prentice-Hall, Inc., 1960.

HODGES, J. L., AND LEHMANN, E. L., *Elements of Finite Probability*. San Francisco: Holden-Day, Inc., 1965.

HOEL, P., *Introduction to Mathematical Statistics*, 4th ed. New York: John Wiley & Sons, Inc., 1971.

MENDENHALL, W., AND SCHAEFFER, R. L., *Mathematical Statistics with Applications*. North Scituate, Mass.: Duxbury Press, 1973.

MOSTELLER, F., ROURKE, R. E. K., AND THOMAS, G. B., *Probability with Statistical Applications*, 2nd ed. Reading, Mass.: Addison-Wesley Publishing Company, Inc., 1970.

SCHAEFFER, R. L., AND MENDENHALL, W., *Introduction to Probability: Theory and Applications*. North Scituate, Mass.: Duxbury Press, 1975.

C

Some Books Dealing
with Special Topics

BOX, G. E. P., HUNTER, W. G., AND HUNTER, J. S., *Statistics for Experimenters*. New York: John Wiley & Sons, Inc., 1978.

CHATTERJEE, S., AND PRICE, B., *Regression Analysis by Example*. New York: John Wiley & Sons, Inc., 1977.

CHILDRESS, R. L., *Mathematics for Managerial Decisions*. Englewood Cliffs, N.J.: Prentice-Hall, Inc., 1974.

COCHRAN, W. G., *Sampling Techniques*, 2nd ed. New York: John Wiley & Sons, Inc., 1963.

DEMING, W. E., *Sample Design in Business Research*. New York: John Wiley & Sons, Inc., 1960.

DRESHER, M., *Games of Strategy: Theory and Applications*. Englewood Cliffs, N.J.: Prentice-Hall, Inc., 1961.

DUNCAN, A. J., *Quality Control and Industrial Statistics*, 3rd ed. Homewood, Ill.: Richard D. Irwin, Inc., 1965.

FERBER, R., AND VERDOORN, P. J., *Research Methods in Economics and Business*. New York: Macmillan Publishing Co., Inc., 1962.

GIBBONS, J. D., *Nonparametric Statistical Inference*. New York: McGraw-Hill Book Company, 1971.

GRANT, E. L., AND LEAVENWORTH, R., *Statistical Quality Control*, 4th ed. New York: McGraw-Hill Book Company, 1972.

HICKS, C. R., *Fundamental Concepts in the Design of Experiments*. New York: Holt, Rinehart and Winston, Inc., 1964.

JEFFREY, R. C., *The Logic of Decision*. New York: McGraw-Hill Book Company, 1965.

LEHMANN, E. L., *Nonparametrics: Statistical Methods Based on Ranks*. San Francisco: Holden-Day, Inc., 1975.

McRAE, T. W., *Statistical Sampling for Audit and Control*. New York: John Wiley & Sons, Inc., 1974.

MOORE, P. G., *Risk in Business Decision*. New York: John Wiley & Sons, Inc., 1972.

MORGAN, B. W., *An Introduction to Bayesian Statistical Decision Processes*. Englewood Cliffs, N.J.: Prentice-Hall, Inc., 1968.

MOSTELLER, F., AND ROURKE, R. E. K., *Sturdy Statistics, Nonparametrics and Order Statistics*. Reading, Mass.: Addison-Wesley Publishing Company, Inc., 1973.

MUDGETT, B. D., *Index Numbers*. New York: John Wiley & Sons, Inc., 1951.

NELSON, C. R., *Applied Time Series Analysis for Managerial Forecasting*. San Francisco: Holden-Day, Inc., 1973.

NOETHER, G. E., *Introduction to Statistics: A Nonparametric Approach*. Boston: Houghton Mifflin Company, 1976.

SCHLAIFER, R., *Probability and Statistics for Business Decisions*. New York: McGraw-Hill Book Company, 1961.

SCHMIDT, S. A., *Measuring Uncertainty, An Elementary Introduction to Bayesian Statistics*. Reading, Mass.: Addison-Wesley Publishing Company, Inc., 1969.

SIEGEL, S., *Nonparametric Statistics for the Behavioral Sciences*. New York: McGraw-Hill Book Company, 1956.

STUART, A., *Basic Ideas of Scientific Sampling*. New York: Hafner Press, 1962.

WILLIAMS, W. H., *A Sampler on Sampling*. New York: John Wiley & Sons, Inc., 1978.

D

Some General Reference Works and Tables

COMAN, E. T., JR., *Sources of Business Information*, 2nd ed. Berkeley, Calif.: University of California Press, 1964.

FREUND, J. E., AND WILLIAMS, F. J., *Dictionary/Outline of Basic Statistics*. New York: McGraw-Hill Book Company, 1966.

HAUSER, P. M., AND LEONARD, W. R., *Government Statistics for Business Use*, 2nd ed. New York: John Wiley & Sons, Inc., 1956.

KENDALL, M. G., AND BUCKLAND, W. R., *A Dictionary of Statistical Terms*, 3rd ed. New York: Hafner Press, 1971.

NATIONAL BUREAU OF STANDARDS, *Tables of the Binomial Probability Distribution*. Washington, D.C.: Government Printing Office, 1950.

OWEN, D. B., *Handbook of Statistical Tables*. Reading, Mass.: Addison-Wesley Publishing Company, Inc., 1962.

PEARSON, E. S., AND HARTLEY, H. O., *Biometrika Tables for Statisticians*, 3rd ed. Cambridge: Cambridge University Press, 1966.

RAND CORPORATION, *A Million Random Digits with* 100,000 *Normal Deviates*. New York: The Free Press, 1955.

ROMIG, H. G., 50–100 *Binomial Tables*. New York: John Wiley & Sons, Inc., 1953.

WASSERMAN, P., et al. (eds.), *Statistical Sources*, 3rd ed. Detroit, Mich.: Gale Research Co., 1965.

E

Some Sources of Statistical Data

Agricultural Statistics. Washington, D.C.: United States Department of Agriculture.

Commodity Yearbook. New York: Commodity Research Bureau, Inc.

Demographic Yearbook. New York: Statistical Office of the United Nations.

The Economic Almanac. New York: National Industrial Conference Board.

Industrial Marketing (Annual Market Data and Directory Number). Chicago: Advertising Publications, Inc.

The Market Guide. New York: Editor and Publishing Co., Inc.

Statistical Abstract of the United States. Washington, D.C.: United States Bureau of the Census.

Statistical Yearbook. New York: Statistical Office of the United Nations.

Survey of Buying Power. New York: Sales Management.

The World Almanac. New York: Newspaper Enterprise Association, Inc.

Statistical Tables

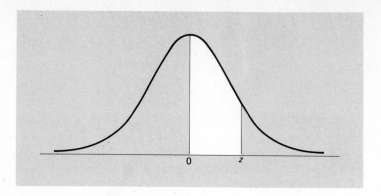

The entries in Table I are the probabilities that a random variable having the standard normal distribution will take on a value between 0 and z; they are given by the area of the white region under the curve in the figure shown above.

TABLE I Normal-curve Areas

z	.00	.01	.02	.03	.04	.05	.06	.07	.08	.09
0.0	.0000	.0040	.0080	.0120	.0160	.0199	.0239	.0279	.0319	.0359
0.1	.0398	.0438	.0478	.0517	.0557	.0596	.0636	.0675	.0714	.0753
0.2	.0793	.0832	.0871	.0910	.0948	.0987	.1026	.1064	.1103	.1141
0.3	.1179	.1217	.1255	.1293	.1331	.1368	.1406	.1443	.1480	.1517
0.4	.1554	.1591	.1628	.1664	.1700	.1736	.1772	.1808	.1844	.1879
0.5	.1915	.1950	.1985	.2019	.2054	.2088	.2123	.2157	.2190	.2224
0.6	.2257	.2291	.2324	.2357	.2389	.2422	.2454	.2486	.2517	.2549
0.7	.2580	.2611	.2642	.2673	.2704	.2734	.2764	.2794	.2823	.2852
0.8	.2881	.2910	.2939	.2967	.2995	.3023	.3051	.3078	.3106	.3133
0.9	.3159	.3186	.3212	.3238	.3264	.3289	.3315	.3340	.3365	.3389
1.0	.3413	.3438	.3461	.3485	.3508	.3531	.3554	.3577	.3599	.3621
1.1	.3643	.3665	.3686	.3708	.3729	.3749	.3770	.3790	.3810	.3830
1.2	.3849	.3869	.3888	.3907	.3925	.3944	.3962	.3980	.3997	.4015
1.3	.4032	.4049	.4066	.4082	.4099	.4115	.4131	.4147	.4162	.4177
1.4	.4192	.4207	.4222	.4236	.4251	.4265	.4279	.4292	.4306	.4319
1.5	.4332	.4345	.4357	.4370	.4382	.4394	.4406	.4418	.4429	.4441
1.6	.4452	.4463	.4474	.4484	.4495	.4505	.4515	.4525	.4535	.4545
1.7	.4554	.4564	.4573	.4582	.4591	.4599	.4608	.4616	.4625	.4633
1.8	.4641	.4649	.4656	.4664	.4671	.4678	.4686	.4693	.4699	.4706
1.9	.4713	.4719	.4726	.4732	.4738	.4744	.4750	.4756	.4761	.4767
2.0	.4772	.4778	.4783	.4788	.4793	.4798	.4803	.4808	.4812	.4817
2.1	.4821	.4826	.4830	.4834	.4838	.4842	.4846	.4850	.4854	.4857
2.2	.4861	.4864	.4868	.4871	.4875	.4878	.4881	.4884	.4887	.4890
2.3	.4893	.4896	.4898	.4901	.4904	.4906	.4909	.4911	.4913	.4916
2.4	.4918	.4920	.4922	.4925	.4927	.4929	.4931	.4932	.4934	.4936
2.5	.4938	.4940	.4941	.4943	.4945	.4946	.4948	.4949	.4951	.4952
2.6	.4953	.4955	.4956	.4957	.4959	.4960	.4961	.4962	.4963	.4964
2.7	.4965	.4966	.4967	.4968	.4969	.4970	.4971	.4972	.4973	.4974
2.8	.4974	.4975	.4976	.4977	.4977	.4978	.4979	.4979	.4980	.4981
2.9	.4981	.4982	.4982	.4983	.4984	.4984	.4985	.4985	.4986	.4986
3.0	.4987	.4987	.4987	.4988	.4988	.4989	.4989	.4989	.4990	.4990

Also, for $z = 4.0$, 5.0, and 6.0, the areas are 0.49997, 0.4999997, and 0.499999999.

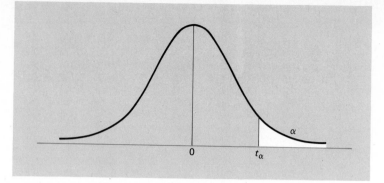

The entries in Table II are values for which the area to their right under the t distribution with given degrees of freedom (the white area in the figure shown above) is equal to α.

TABLE II   Values of t†

d.f.	$t_{.100}$	$t_{.050}$	$t_{.025}$	$t_{.010}$	$t_{.005}$	d.f.
1	3.078	6.314	12.706	31.821	63.657	1
2	1.886	2.920	4.303	6.965	9.925	2
3	1.638	2.353	3.182	4.541	5.841	3
4	1.533	2.132	2.776	3.747	4.604	4
5	1.476	2.015	2.571	3.365	4.032	5
6	1.440	1.943	2.447	3.143	3.707	6
7	1.415	1.895	2.365	2.998	3.499	7
8	1.397	1.860	2.306	2.896	3.355	8
9	1.383	1.833	2.262	2.821	3.250	9
10	1.372	1.812	2.228	2.764	3.169	10
11	1.363	1.796	2.201	2.718	3.106	11
12	1.356	1.782	2.179	2.681	3.055	12
13	1.350	1.771	2.160	2.650	3.012	13
14	1.345	1.761	2.145	2.624	2.977	14
15	1.341	1.753	2.131	2.602	2.947	15
16	1.337	1.746	2.120	2.583	2.921	16
17	1.333	1.740	2.110	2.567	2.898	17
18	1.330	1.734	2.101	2.552	2.878	18
19	1.328	1.729	2.093	2.539	2.861	19
20	1.325	1.725	2.086	2.528	2.845	20
21	1.323	1.721	2.080	2.518	2.831	21
22	1.321	1.717	2.074	2.508	2.819	22
23	1.319	1.714	2.069	2.500	2.807	23
24	1.318	1.711	2.064	2.492	2.797	24
25	1.316	1.708	2.060	2.485	2.787	25
26	1.315	1.706	2.056	2.479	2.779	26
27	1.314	1.703	2.052	2.473	2.771	27
28	1.313	1.701	2.048	2.467	2.763	28
29	1.311	1.699	2.045	2.462	2.756	29
inf.	1.282	1.645	1.960	2.326	2.576	inf.

†Abridged by permission of Macmillan Publishing Co., Inc. from STATISTICAL METHODS FOR RESEARCH WORKERS, 14th Edition, by R. A. Fisher. Copyright © 1970 University of Adelaide.

d.f
n-l   30 and below

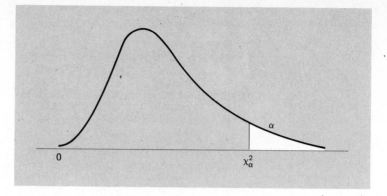

The entries in Table III are values for which the area to their right under the chi-square distribution with given degrees of freedom (the white area in the figure shown above) is equal to α.

TABLE III Values of χ^2†

d.f.	$\chi^2_{.995}$	$\chi^2_{.99}$	$\chi^2_{.975}$	$\chi^2_{.95}$	$\chi^2_{.05}$	$\chi^2_{.025}$	$\chi^2_{.01}$	$\chi^2_{.005}$	d.f.
1	.0000393	.000157	.000982	.00393	3.841	5.024	6.635	7.879	1
2	.0100	.0201	.0506	.103	5.991	7.378	9.210	10.597	2
3	.0717	.115	.216	.352	7.815	9.348	11.345	12.838	3
4	.207	.297	.484	.711	9.488	11.143	13.277	14.860	4
5	.412	.554	.831	1.145	11.070	12.832	15.086	16.750	5
6	.676	.872	1.237	1.635	12.592	14.449	16.812	18.548	6
7	.989	1.239	1.690	2.167	14.067	16.013	18.475	20.278	7
8	1.344	1.646	2.180	2.733	15.507	17.535	20.090	21.955	8
9	1.735	2.088	2.700	3.325	13.919	19.023	21.666	23.589	9
10	2.156	2.558	3.247	3.940	18.307	20.483	23.209	25.188	10
11	2.603	3.053	3.816	4.575	19.675	21.920	24.725	26.757	11
12	3.074	3.571	4.404	5.226	21.026	23.337	26.217	28.300	12
13	3.565	4.107	5.009	5.892	22.362	24.736	27.688	29.819	13
14	4.075	4.660	5.629	6.571	23.685	26.119	29.141	31.319	14
15	4.601	5.229	6.262	7.261	24.996	27.488	30.578	32.801	15
16	5.142	5.812	6.908	7.962	26.296	28.845	32.000	34.267	16
17	5.697	6.408	7.564	8.672	27.587	30.191	33.409	35.718	17
18	6.265	7.015	8.231	9.390	28.869	31.526	34.805	37.156	18
19	6.844	7.633	8.907	10.117	30.144	32.852	36.191	38.582	19
20	7.434	8.260	9.591	10.851	31.410	34.170	37.566	39.997	20
21	8.034	8.897	10.283	11.591	32.671	35.479	38.932	41.401	21
22	8.643	9.542	10.982	12.338	33.924	36.781	40.289	42.796	22
23	9.260	10.196	11.689	13.091	35.172	38.076	41.638	44.181	23
24	9.886	10.856	12.401	13.848	36.415	39.364	42.980	45.558	24
25	10.520	11.524	13.120	14.611	37.652	40.646	44.314	46.928	25
26	11.160	12.198	13.844	15.379	38.885	41.923	45.642	48.290	26
27	11.808	12.879	14.573	16.151	40.113	43.194	46.963	49.645	27
28	12.461	13.565	15.308	16.928	41.337	44.461	48.278	50.993	28
29	13.121	14.256	16.047	17.708	42.557	45.722	49.588	52.336	29
30	13.787	14.953	16.791	18.493	43.773	46.979	50.892	53.672	30

†Based on Table 8 of *Biometrika Tables for Statisticians*, Vol. I (Cambridge: Cambridge University Press, 1954) by permission of the *Biometrika* trustees.

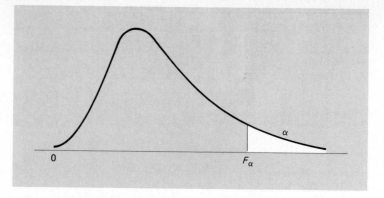

The entries in Table IV are values for which the area to their right under the F distribution with given degrees of freedom (the white area in the figure shown above) is equal to α.

TABLE IV Values of $F_{0.05}$†

Degrees of freedom for numerator

	1	2	3	4	5	6	7	8	9	10	12	15	20	24	30	40	60	120	∞
1	161	200	216	225	230	234	237	239	241	242	244	246	248	249	250	251	252	253	254
2	18.5	19.0	19.2	19.2	19.3	19.3	19.4	19.4	19.4	19.4	19.4	19.4	19.4	19.5	19.5	19.5	19.5	19.5	19.5
3	10.1	9.55	9.28	9.12	9.01	8.94	8.89	8.85	8.81	8.79	8.74	8.70	8.66	8.64	8.62	8.59	8.57	8.55	8.53
4	7.71	6.94	6.59	6.39	6.26	6.16	6.09	6.04	6.00	5.96	5.91	5.86	5.80	5.77	5.75	5.72	5.69	5.66	5.63
5	6.61	5.79	5.41	5.19	5.05	4.95	4.88	4.82	4.77	4.74	4.68	4.62	4.56	4.53	4.50	4.46	4.43	4.40	4.37
6	5.99	5.14	4.76	4.53	4.39	4.28	4.21	4.15	4.10	4.06	4.00	3.94	3.87	3.84	3.81	3.77	3.74	3.70	3.67
7	5.59	4.74	4.35	4.12	3.97	3.87	3.79	3.73	3.68	3.64	3.57	3.51	3.44	3.41	3.38	3.34	3.30	3.27	3.23
8	5.32	4.46	4.07	3.84	3.69	3.58	3.50	3.44	3.39	3.35	3.28	3.22	3.15	3.12	3.08	3.04	3.01	2.97	2.93
9	5.12	4.26	3.86	3.63	3.48	3.37	3.29	3.23	3.18	3.14	3.07	3.01	2.94	2.90	2.86	2.83	2.79	2.75	2.71
10	4.96	4.10	3.71	3.48	3.33	3.22	3.14	3.07	3.02	2.98	2.91	2.85	2.77	2.74	2.70	2.66	2.62	2.58	2.54
11	4.84	3.98	3.59	3.36	3.20	3.09	3.01	2.95	2.90	2.85	2.79	2.72	2.65	2.61	2.57	2.53	2.49	2.45	2.40
12	4.75	3.89	3.49	3.26	3.11	3.00	2.91	2.85	2.80	2.75	2.69	2.62	2.54	2.51	2.47	2.43	2.38	2.34	2.30
13	4.67	3.81	3.41	3.18	3.03	2.92	2.83	2.77	2.71	2.67	2.60	2.53	2.46	2.42	2.38	2.34	2.30	2.25	2.21
14	4.60	3.74	3.34	3.11	2.96	2.85	2.76	2.70	2.65	2.60	2.53	2.46	2.39	2.35	2.31	2.27	2.22	2.18	2.13
15	4.54	3.68	3.29	3.06	2.90	2.79	2.71	2.64	2.59	2.54	2.48	2.40	2.33	2.29	2.25	2.20	2.16	2.11	2.07
16	4.49	3.63	3.24	3.01	2.85	2.74	2.66	2.59	2.54	2.49	2.42	2.35	2.28	2.24	2.19	2.15	2.11	2.06	2.01
17	4.45	3.59	3.20	2.96	2.81	2.70	2.61	2.55	2.49	2.45	2.38	2.31	2.23	2.19	2.15	2.10	2.06	2.01	1.96
18	4.41	3.55	3.16	2.93	2.77	2.66	2.58	2.51	2.46	2.41	2.34	2.27	2.19	2.15	2.11	2.06	2.02	1.97	1.92
19	4.38	3.52	3.13	2.90	2.74	2.63	2.54	2.48	2.42	2.38	2.31	2.23	2.16	2.11	2.07	2.03	1.98	1.93	1.88
20	4.35	3.49	3.10	2.87	2.71	2.60	2.51	2.45	2.39	2.35	2.28	2.20	2.12	2.08	2.04	1.99	1.95	1.90	1.84
21	4.32	3.47	3.07	2.84	2.68	2.57	2.49	2.42	2.37	2.32	2.25	2.18	2.10	2.05	2.01	1.96	1.92	1.87	1.81
22	4.30	3.44	3.05	2.82	2.66	2.55	2.46	2.40	2.34	2.30	2.23	2.15	2.07	2.03	1.98	1.94	1.89	1.84	1.78
23	4.28	3.42	3.03	2.80	2.64	2.53	2.44	2.37	2.32	2.27	2.20	2.13	2.05	2.01	1.96	1.91	1.86	1.81	1.76
24	4.26	3.40	3.01	2.78	2.62	2.51	2.42	2.36	2.30	2.25	2.18	2.11	2.03	1.98	1.94	1.89	1.84	1.79	1.73
25	4.24	3.39	2.99	2.76	2.60	2.49	2.40	2.34	2.28	2.24	2.16	2.09	2.01	1.96	1.92	1.87	1.82	1.77	1.71
30	4.17	3.32	2.92	2.69	2.53	2.42	2.33	2.27	2.21	2.16	2.09	2.01	1.93	1.89	1.84	1.79	1.74	1.68	1.62
40	4.08	3.23	2.84	2.61	2.45	2.34	2.25	2.18	2.12	2.08	2.00	1.92	1.84	1.79	1.74	1.69	1.64	1.58	1.51
60	4.00	3.15	2.76	2.53	2.37	2.25	2.17	2.10	2.04	1.99	1.92	1.84	1.75	1.70	1.65	1.59	1.53	1.47	1.39
120	3.92	3.07	2.68	2.45	2.29	2.18	2.09	2.02	1.96	1.91	1.83	1.75	1.66	1.61	1.55	1.50	1.43	1.35	1.25
∞	3.84	3.00	2.60	2.37	2.21	2.10	2.01	1.94	1.88	1.83	1.75	1.67	1.57	1.52	1.46	1.39	1.32	1.22	1.00

Degrees of freedom for denominator

†This table is reproduced from M. Merrington and C. M. Thompson, "Tables of percentage points of the inverted beta (F) distribution," *Biometrika*, Vol. 33 (1943), by permission of the *Biometrika* trustees.

TABLE IV Values of $F_{0.01}$†

Degrees of freedom for numerator

Denom.	1	2	3	4	5	6	7	8	9	10	12	15	20	24	30	40	60	120	∞
1	4,052	5,000	5,403	5,625	5,764	5,859	5,928	5,982	6,023	6,056	6,106	6,157	6,209	6,235	6,261	6,287	6,313	6,339	6,366
2	98.5	99.0	99.2	99.2	99.3	99.3	99.4	99.4	99.4	99.4	99.4	99.4	99.4	99.5	99.5	99.5	99.5	99.5	99.5
3	34.1	30.8	29.5	28.7	28.2	27.9	27.7	27.5	27.3	27.2	27.1	26.9	26.7	26.6	26.5	26.4	26.3	26.2	26.1
4	21.2	18.0	16.7	16.0	15.5	15.2	15.0	14.8	14.7	14.5	14.4	14.2	14.0	13.9	13.8	13.7	13.7	13.6	13.5
5	16.3	13.3	12.1	11.4	11.0	10.7	10.5	10.3	10.2	10.1	9.89	9.72	9.55	9.47	9.38	9.29	9.20	9.11	9.02
6	13.7	10.9	9.78	9.15	8.75	8.47	8.26	8.10	7.98	7.87	7.72	7.56	7.40	7.31	7.23	7.14	7.06	6.97	6.88
7	12.2	9.55	8.45	7.85	7.46	7.19	6.99	6.84	6.72	6.62	6.47	6.31	6.16	6.07	5.99	5.91	5.82	5.74	5.65
8	11.3	8.65	7.59	7.01	6.63	6.37	6.18	6.03	5.91	5.81	5.67	5.52	5.36	5.28	5.20	5.12	5.03	4.95	4.86
9	10.6	8.02	6.99	6.42	6.06	5.80	5.61	5.47	5.35	5.26	5.11	4.96	4.81	4.73	4.65	4.57	4.48	4.40	4.31
10	10.0	7.56	6.55	5.99	5.64	5.39	5.20	5.06	4.94	4.85	4.71	4.56	4.41	4.33	4.25	4.17	4.08	4.00	3.91
11	9.65	7.21	6.22	5.67	5.32	5.07	4.89	4.74	4.63	4.54	4.40	4.25	4.10	4.02	3.94	3.86	3.78	3.69	3.60
12	9.33	6.93	5.95	5.41	5.06	4.82	4.64	4.50	4.39	4.30	4.16	4.01	3.86	3.78	3.70	3.62	3.54	3.45	3.36
13	9.07	6.70	5.74	5.21	4.86	4.62	4.44	4.30	4.19	4.10	3.96	3.82	3.66	3.59	3.51	3.43	3.34	3.25	3.17
14	8.86	6.51	5.56	5.04	4.70	4.46	4.28	4.14	4.03	3.94	3.80	3.66	3.51	3.43	3.35	3.27	3.18	3.09	3.00
15	8.68	6.36	5.42	4.89	4.56	4.32	4.14	4.00	3.89	3.80	3.67	3.52	3.37	3.29	3.21	3.13	3.05	2.96	2.87
16	8.53	6.23	5.29	4.77	4.44	4.20	4.03	3.89	3.78	3.69	3.55	3.41	3.26	3.18	3.10	3.02	2.93	2.84	2.75
17	8.40	6.11	5.19	4.67	4.34	4.10	3.93	3.79	3.68	3.59	3.46	3.31	3.16	3.08	3.00	2.92	2.83	2.75	2.65
18	8.29	6.01	5.09	4.58	4.25	4.01	3.84	3.71	3.60	3.51	3.37	3.23	3.08	3.00	2.92	2.84	2.75	2.66	2.57
19	8.19	5.93	5.01	4.50	4.17	3.94	3.77	3.63	3.52	3.43	3.30	3.15	3.00	2.92	2.84	2.76	2.67	2.58	2.49
20	8.10	5.85	4.94	4.43	4.10	3.87	3.70	3.56	3.46	3.37	3.23	3.09	2.94	2.86	2.78	2.69	2.61	2.52	2.42
21	8.02	5.78	4.87	4.37	4.04	3.81	3.64	3.51	3.40	3.31	3.17	3.03	2.88	2.80	2.72	2.64	2.55	2.46	2.36
22	7.95	5.72	4.82	4.31	3.99	3.76	3.59	3.45	3.35	3.26	3.12	2.98	2.83	2.75	2.67	2.58	2.50	2.40	2.31
23	7.88	5.66	4.76	4.26	3.94	3.71	3.54	3.41	3.30	3.21	3.07	2.93	2.78	2.70	2.62	2.54	2.45	2.35	2.26
24	7.82	5.61	4.72	4.22	3.90	3.67	3.50	3.36	3.26	3.17	3.03	2.89	2.74	2.66	2.58	2.49	2.40	2.31	2.21
25	7.77	5.57	4.68	4.18	3.86	3.63	3.46	3.32	3.22	3.13	2.99	2.85	2.70	2.62	2.53	2.45	2.36	2.27	2.17
30	7.56	5.39	4.51	4.02	3.70	3.47	3.30	3.17	3.07	2.98	2.84	2.70	2.55	2.47	2.39	2.30	2.21	2.11	2.01
40	7.31	5.18	4.31	3.83	3.51	3.29	3.12	2.99	2.89	2.80	2.66	2.52	2.37	2.29	2.20	2.11	2.02	1.92	1.80
60	7.08	4.98	4.13	3.65	3.34	3.12	2.95	2.82	2.72	2.63	2.50	2.35	2.20	2.12	2.03	1.94	1.84	1.73	1.60
120	6.85	4.79	3.95	3.48	3.17	2.96	2.79	2.66	2.56	2.47	2.34	2.19	2.03	1.95	1.86	1.76	1.66	1.53	1.38
∞	6.63	4.61	3.78	3.32	3.02	2.80	2.64	2.51	2.41	2.32	2.18	2.04	1.88	1.79	1.70	1.59	1.47	1.32	1.00

Degrees of freedom for denominator

†This table is reproduced from M. Merrington and C. M. Thompson, "Tables of percentage points of the inverted beta (F) distribution," *Biometrika*, Vol. 33 (1943), by permission of the *Biometrika* trustees.

TABLE V Binomial Probabilities†

n	x	0.05	0.1	0.2	0.3	0.4	0.5	0.6	0.7	0.8	0.9	0.95
2	0	0.902	0.810	0.640	0.490	0.360	0.250	0.160	0.090	0.040	0.010	0.002
	1	0.095	0.180	0.320	0.420	0.480	0.500	0.480	0.420	0.320	0.180	0.095
	2	0.002	0.010	0.040	0.090	0.160	0.250	0.360	0.490	0.640	0.810	0.902
3	0	0.857	0.729	0.512	0.343	0.216	0.125	0.064	0.027	0.008	0.001	
	1	0.135	0.243	0.384	0.441	0.432	0.375	0.288	0.189	0.096	0.027	0.007
	2	0.007	0.027	0.096	0.189	0.288	0.375	0.432	0.441	0.384	0.243	0.135
	3		0.001	0.008	0.027	0.064	0.125	0.216	0.343	0.512	0.729	0.857
4	0	0.815	0.656	0.410	0.240	0.130	0.062	0.026	0.008	0.002		
	1	0.171	0.292	0.410	0.412	0.346	0.250	0.154	0.076	0.026	0.004	
	2	0.014	0.049	0.154	0.265	0.346	0.375	0.346	0.265	0.154	0.049	0.014
	3		0.004	0.026	0.076	0.154	0.250	0.346	0.412	0.410	0.292	0.171
	4			0.002	0.008	0.026	0.062	0.130	0.240	0.410	0.656	0.815
5	0	0.774	0.590	0.328	0.168	0.078	0.031	0.010	0.002			
	1	0.204	0.328	0.410	0.360	0.259	0.156	0.077	0.028	0.006		
	2	0.021	0.073	0.205	0.309	0.346	0.312	0.230	0.132	0.051	0.008	0.001
	3	0.001	0.008	0.051	0.132	0.230	0.312	0.346	0.309	0.205	0.073	0.021
	4			0.006	0.028	0.077	0.156	0.259	0.360	0.410	0.328	0.204
	5				0.002	0.010	0.031	0.078	0.168	0.328	0.590	0.774
6	0	0.735	0.531	0.262	0.118	0.047	0.016	0.004	0.001			
	1	0.232	0.354	0.393	0.303	0.187	0.094	0.037	0.010	0.002		
	2	0.031	0.098	0.246	0.324	0.311	0.234	0.138	0.060	0.015	0.001	
	3	0.002	0.015	0.082	0.185	0.276	0.312	0.276	0.185	0.082	0.015	0.002
	4		0.001	0.015	0.060	0.138	0.234	0.311	0.324	0.246	0.098	0.031
	5			0.002	0.010	0.037	0.094	0.187	0.303	0.393	0.354	0.232
	6				0.001	0.004	0.016	0.047	0.118	0.262	0.531	0.735
7	0	0.698	0.478	0.210	0.082	0.028	0.008	0.002				
	1	0.257	0.372	0.367	0.247	0.131	0.055	0.017	0.004			
	2	0.041	0.124	0.275	0.318	0.261	0.164	0.077	0.025	0.004		
	3	0.004	0.023	0.115	0.227	0.290	0.273	0.194	0.097	0.029	0.003	
	4		0.003	0.029	0.097	0.194	0.273	0.290	0.227	0.115	0.023	0.004
	5			0.004	0.025	0.077	0.164	0.261	0.318	0.275	0.124	0.041
	6				0.004	0.017	0.055	0.131	0.247	0.367	0.372	0.257
	7					0.002	0.008	0.028	0.082	0.210	0.478	0.698
8	0	0.663	0.430	0.168	0.058	0.017	0.004	0.001				
	1	0.279	0.383	0.336	0.198	0.090	0.031	0.008	0.001			
	2	0.051	0.149	0.294	0.296	0.209	0.109	0.041	0.010	0.001		
	3	0.005	0.033	0.147	0.254	0.279	0.219	0.124	0.047	0.009		
	4		0.005	0.046	0.136	0.232	0.273	0.232	0.136	0.046	0.005	
	5			0.009	0.047	0.124	0.219	0.279	0.254	0.147	0.033	0.005
	6			0.001	0.010	0.041	0.109	0.209	0.296	0.294	0.149	0.051
	7				0.001	0.008	0.031	0.090	0.198	0.336	0.383	0.279
	8					0.001	0.004	0.017	0.058	0.168	0.430	0.663

†All values omitted in this table are 0.0005 or less.

TABLE V   Binomial Probabilities (*continued*)

n	x	0.05	0.1	0.2	0.3	0.4	0.5	0.6	0.7	0.8	0.9	0.95
9	0	0.630	0.387	0.134	0.040	0.010	0.002					
	1	0.299	0.387	0.302	0.156	0.060	0.018	0.004				
	2	0.063	0.172	0.302	0.267	0.161	0.070	0.021	0.004			
	3	0.008	0.045	0.176	0.267	0.251	0.164	0.074	0.021	0.003		
	4	0.001	0.007	0.066	0.172	0.251	0.246	0.167	0.074	0.017	0.001	
	5		0.001	0.017	0.074	0.167	0.246	0.251	0.172	0.066	0.007	0.001
	6			0.003	0.021	0.074	0.164	0.251	0.267	0.176	0.045	0.008
	7				0.004	0.021	0.070	0.161	0.267	0.302	0.172	0.063
	8					0.004	0.018	0.060	0.156	0.302	0.387	0.299
	9						0.002	0.010	0.040	0.134	0.387	0.630
10	0	0.599	0.349	0.107	0.028	0.006	0.001					
	1	0.315	0.387	0.268	0.121	0.040	0.010	0.002				
	2	0.075	0.194	0.302	0.233	0.121	0.044	0.011	0.001			
	3	0.010	0.057	0.201	0.267	0.215	0.117	0.042	0.009	0.001		
	4	0.001	0.011	0.088	0.200	0.251	0.205	0.111	0.037	0.006		
	5		0.001	0.026	0.103	0.201	0.246	0.201	0.103	0.026	0.001	
	6			0.006	0.037	0.111	0.205	0.251	0.200	0.088	0.011	0.001
	7			0.001	0.009	0.042	0.117	0.215	0.267	0.201	0.057	0.010
	8				0.001	0.011	0.044	0.121	0.233	0.302	0.194	0.075
	9					0.002	0.010	0.040	0.121	0.268	0.387	0.315
	10						0.001	0.006	0.028	0.107	0.349	0.599
11	0	0.569	0.314	0.086	0.020	0.004						
	1	0.329	0.384	0.236	0.093	0.027	0.005	0.001				
	2	0.087	0.213	0.295	0.200	0.089	0.027	0.005	0.001			
	3	0.014	0.071	0.221	0.257	0.177	0.081	0.023	0.004			
	4	0.001	0.016	0.111	0.220	0.236	0.161	0.070	0.017	0.002		
	5		0.002	0.039	0.132	0.221	0.226	0.147	0.057	0.010		
	6			0.010	0.057	0.147	0.226	0.221	0.132	0.039	0.002	
	7			0.002	0.017	0.070	0.161	0.236	0.220	0.111	0.016	0.001
	8				0.004	0.023	0.081	0.177	0.257	0.221	0.071	0.014
	9				0.001	0.005	0.027	0.089	0.200	0.295	0.213	0.087
	10					0.001	0.005	0.027	0.093	0.236	0.384	0.329
	11							0.004	0.020	0.086	0.314	0.569
12	0	0.540	0.282	0.069	0.014	0.002						
	1	0.341	0.377	0.206	0.071	0.017	0.003					
	2	0.099	0.230	0.283	0.168	0.064	0.016	0.002				
	3	0.017	0.085	0.236	0.240	0.142	0.054	0.012	0.001			
	4	0.002	0.021	0.133	0.231	0.213	0.121	0.042	0.008	0.001		
	5		0.004	0.053	0.158	0.227	0.193	0.101	0.029	0.003		
	6			0.016	0.079	0.177	0.226	0.177	0.079	0.016		
	7			0.003	0.029	0.101	0.193	0.227	0.158	0.053	0.004	
	8			0.001	0.008	0.042	0.121	0.213	0.231	0.133	0.021	0.002
	9				0.001	0.012	0.054	0.142	0.240	0.236	0.085	0.017
	10					0.002	0.016	0.064	0.168	0.283	0.230	0.099
	11						0.003	0.017	0.071	0.206	0.377	0.341
	12							0.002	0.014	0.069	0.282	0.540

TABLE V Binomial Probabilities (*continued*)

| | | | | | | p | | | | | | | |
n	x	0.05	0.1	0.2	0.3	0.4	0.5	0.6	0.7	0.8	0.9	0.95
13	0	0.513	0.254	0.055	0.010	0.001						
	1	0.351	0.367	0.179	0.054	0.011	0.002					
	2	0.111	0.245	0.268	0.139	0.045	0.010	0.001				
	3	0.021	0.100	0.246	0.218	0.111	0.035	0.006	0.001			
	4	0.003	0.028	0.154	0.234	0.184	0.087	0.024	0.003			
	5		0.006	0.069	0.180	0.221	0.157	0.066	0.014	0.001		
	6		0.001	0.023	0.103	0.197	0.209	0.131	0.044	0.006		
	7			0.006	0.044	0.131	0.209	0.197	0.103	0.023	0.001	
	8			0.001	0.014	0.066	0.157	0.221	0.180	0.069	0.006	
	9				0.003	0.024	0.087	0.184	0.234	0.154	0.028	0.003
	10				0.001	0.006	0.035	0.111	0.218	0.246	0.100	0.021
	11					0.001	0.010	0.045	0.139	0.268	0.245	0.111
	12						0.002	0.011	0.054	0.179	0.367	0.351
	13							0.001	0.010	0.055	0.254	0.513
14	0	0.488	0.229	0.044	0.007	0.001						
	1	0.359	0.356	0.154	0.041	0.007	0.001					
	2	0.123	0.257	0.250	0.113	0.032	0.006	0.001				
	3	0.026	0.114	0.250	0.194	0.085	0.022	0.003				
	4	0.004	0.035	0.172	0.229	0.155	0.061	0.014	0.001			
	5		0.008	0.086	0.196	0.207	0.122	0.041	0.007			
	6		0.001	0.032	0.126	0.207	0.183	0.092	0.023	0.002		
	7			0.009	0.062	0.157	0.209	0.157	0.062	0.009		
	8			0.002	0.023	0.092	0.183	0.207	0.126	0.032	0.001	
	9				0.007	0.041	0.122	0.207	0.196	0.086	0.008	
	10				0.001	0.014	0.061	0.155	0.229	0.172	0.035	0.004
	11					0.003	0.022	0.085	0.194	0.250	0.114	0.026
	12					0.001	0.006	0.032	0.113	0.250	0.257	0.123
	13						0.001	0.007	0.041	0.154	0.356	0.359
	14							0.001	0.007	0.044	0.229	0.488
15	0	0.463	0.206	0.035	0.005							
	1	0.366	0.343	0.132	0.031	0.005						
	2	0.135	0.267	0.231	0.092	0.022	0.003					
	3	0.031	0.129	0.250	0.170	0.063	0.014	0.002				
	4	0.005	0.043	0.188	0.219	0.127	0.042	0.007	0.001			
	5	0.001	0.010	0.103	0.206	0.186	0.092	0.024	0.003			
	6		0.002	0.043	0.147	0.207	0.153	0.061	0.012	0.001		
	7			0.014	0.081	0.177	0.196	0.118	0.035	0.003		
	8			0.003	0.035	0.118	0.196	0.177	0.081	0.014		
	9			0.001	0.012	0.061	0.153	0.207	0.147	0.043	0.002	
	10				0.003	0.024	0.092	0.186	0.206	0.103	0.010	0.001
	11				0.001	0.007	0.042	0.127	0.219	0.188	0.043	0.005
	12					0.002	0.014	0.063	0.170	0.250	0.129	0.031
	13						0.003	0.022	0.092	0.231	0.267	0.135
	14							0.005	0.031	0.132	0.343	0.366
	15								0.005	0.035	0.206	0.463

TABLE VI   Values of r

n	$r_{.025}$	$r_{.010}$	$r_{.005}$	n	$r_{.025}$	$r_{.010}$	$r_{.005}$
3	0.997			18	0.468	0.543	0.590
4	0.950	0.980	0.990	19	0.456	0.529	0.575
5	0.878	0.934	0.959	20	0.444	0.516	0.561
6	0.811	0.882	0.917	21	0.433	0.503	0.549
7	0.754	0.833	0.875	22	0.423	0.492	0.537
8	0.707	0.789	0.834	27	0.381	0.445	0.487
9	0.666	0.750	0.798	32	0.349	0.409	0.449
10	0.632	0.715	0.765	37	0.325	0.381	0.418
11	0.602	0.685	0.735	42	0.304	0.358	0.393
12	0.576	0.658	0.708	47	0.288	0.338	0.372
13	0.553	0.634	0.684	52	0.273	0.322	0.354
14	0.532	0.612	0.661	62	0.250	0.295	0.325
15	0.514	0.592	0.641	72	0.232	0.274	0.302
16	0.497	0.574	0.623	82	0.217	0.256	0.283
17	0.482	0.558	0.606	92	0.205	0.242	0.267

TABLE VII Values of U†

Values of $U'_{0.05}$

n_1 \ n_2	2	3	4	5	6	7	8	9	10	11	12	13	14	15
2							0	0	0	0	1	1	1	1
3				0	1	1	2	2	3	3	4	4	5	5
4			0	1	2	3	4	4	5	6	7	8	9	10
5		0	1	2	3	5	6	7	8	9	11	12	13	14
6		1	2	3	5	6	8	10	11	13	14	16	17	19
7		1	3	5	6	8	10	12	14	16	18	20	22	24
8	0	2	4	6	8	10	13	15	17	19	22	24	26	29
9	0	2	4	7	10	12	15	17	20	23	26	28	31	34
10	0	3	5	8	11	14	17	20	23	26	29	33	36	39
11	0	3	6	9	13	16	19	23	26	30	33	37	40	44
12	1	4	7	11	14	18	22	26	29	33	37	41	45	49
13	1	4	8	12	16	20	24	28	33	37	41	45	50	54
14	1	5	9	13	17	22	26	31	36	40	45	50	55	59
15	1	5	10	14	19	24	29	34	39	44	49	54	59	64

Values of $U'_{0.01}$

n_1 \ n_2	3	4	5	6	7	8	9	10	11	12	13	14	15
3							0	0	0	1	1	1	2
4				0	0	1	1	2	2	3	3	4	5
5			0	1	1	2	3	4	5	6	7	7	8
6		0	1	2	3	4	5	6	7	9	10	11	12
7		0	1	3	4	6	7	9	10	12	13	15	16
8		1	2	4	6	7	9	11	13	15	17	18	20
9	0	1	3	5	7	9	11	13	16	18	20	22	24
10	0	2	4	6	9	11	13	16	18	21	24	26	29
11	0	2	5	7	10	13	16	18	21	24	27	30	33
12	1	3	6	9	12	15	18	21	24	27	31	34	37
13	1	3	7	10	13	17	20	24	27	31	34	38	42
14	1	4	7	11	15	18	22	26	30	34	38	42	46
15	2	5	8	12	16	20	24	29	33	37	42	46	51

†This table is based on Table 11.4 of D. B. Owen, *Handbook of Statistical Tables*, © 1962, U. S. Department of Energy. Published by Addison-Wesley Publishing Company, Inc., Reading, Mass. Reprinted with permission of the publishers.

TABLE VIII   Values of u†

n_1 \ n_2	\multicolumn{12}{c}{Values of $u_{0.025}$}

n_1 \ n_2	4	5	6	7	8	9	10	11	12	13	14	15
4		9	9									
5	9	10	10	11	11							
6	9	10	11	12	12	13	13	13	13			
7		11	12	13	13	14	14	14	14	15	15	15
8		11	12	13	14	14	15	15	16	16	16	16
9			13	14	14	15	16	16	16	17	17	18
10			13	14	15	16	16	17	17	18	18	18
11			13	14	15	16	17	17	18	19	19	19
12			13	14	16	16	17	18	19	19	20	20
13				15	16	17	18	19	19	20	20	21
14				15	16	17	18	19	20	20	21	22
15				15	16	18	18	19	20	21	22	22

n_1 \ n_2	\multicolumn{14}{c}{Values of $u'_{0.025}$}

n_1 \ n_2	2	3	4	5	6	7	8	9	10	11	12	13	14	15
2											2	2	2	2
3					2	2	2	2	2	2	2	2	2	3
4			2	2	2	3	3	3	3	3	3	3	3	3
5		2	2	3	3	3	3	3	4	4	4	4	4	4
6		2	2	3	3	3	3	4	4	4	4	5	5	5
7		2	2	3	3	3	4	4	5	5	5	5	5	6
8		2	3	3	3	4	4	5	5	5	6	6	6	6
9		2	3	3	4	4	5	5	5	6	6	6	7	7
10		2	3	3	4	5	5	5	6	6	7	7	7	7
11		2	3	4	4	5	5	6	6	7	7	7	8	8
12	2	2	3	4	4	5	6	6	7	7	7	8	8	8
13	2	2	3	4	5	5	6	6	7	7	8	8	9	9
14	2	2	3	4	5	5	6	7	7	8	8	9	9	9
15	2	3	3	4	5	6	6	7	7	8	8	9	9	10

†This table is adapted, by permission, from F. S. Swed and C. Eisenhart, "Tables for testing randomness of grouping in a sequence of alternatives," *Annals of Mathematical Statistics*, Vol. 14.

TABLE VIII   Values of u (*continued*)

n_1 \ n_2	5	6	7	8	9	10	11	12	13	14	15
					Values of $u_{0.005}$						
5		11									
6	11	12	13	13							
7		13	13	14	15	15	15				
8		13	14	15	15	16	16	17	17	17	
9			15	15	16	17	17	18	18	18	19
10			15	16	17	17	18	19	19	19	20
11			15	16	17	18	19	19	20	20	21
12				17	18	19	19	20	21	21	22
13				17	18	19	20	21	21	22	22
14				17	18	19	20	21	22	23	23
15					19	20	21	22	22	23	24

n_1 \ n_2	3	4	5	6	7	8	9	10	11	12	13	14	15
						Values of $u'_{0.005}$							
3										2	2	2	2
4						2	2	2	2	2	2	2	3
5				2	2	2	2	3	3	3	3	3	3
6			2	2	2	3	3	3	3	3	3	4	4
7			2	2	3	3	3	3	4	4	4	4	4
8		2	2	3	3	3	3	4	4	4	5	5	5
9		2	2	3	3	3	4	4	5	5	5	5	6
10		2	3	3	3	4	4	5	5	5	5	6	6
11		2	3	3	4	4	5	5	5	6	6	6	7
12	2	2	3	3	4	4	5	5	6	6	6	7	7
13	2	2	3	3	4	5	5	5	6	6	7	7	7
14	2	2	3	4	4	5	5	6	6	7	7	7	8
14	2	3	3	4	4	5	6	6	7	7	7	8	8

TABLE IX   Binomial Coefficients

n	$\binom{n}{0}$	$\binom{n}{1}$	$\binom{n}{2}$	$\binom{n}{3}$	$\binom{n}{4}$	$\binom{n}{5}$	$\binom{n}{6}$	$\binom{n}{7}$	$\binom{n}{8}$	$\binom{n}{9}$	$\binom{n}{10}$
0	1										
1	1	1									
2	1	2	1								
3	1	3	3	1							
4	1	4	6	4	1						
5	1	5	10	10	5	1					
6	1	6	15	20	15	6	1				
7	1	7	21	35	35	21	7	1			
8	1	8	28	56	70	56	28	8	1		
9	1	9	36	84	126	126	84	36	9	1	
10	1	10	45	120	210	252	210	120	45	10	1
11	1	11	55	165	330	462	462	330	165	55	11
12	1	12	66	220	495	792	924	792	495	220	66
13	1	13	78	286	715	1287	1716	1716	1287	715	286
14	1	14	91	364	1001	2002	3003	3432	3003	2002	1001
15	1	15	105	455	1365	3003	5005	6435	6435	5005	3003
16	1	16	120	560	1820	4368	8008	11440	12870	11440	8008
17	1	17	136	680	2380	6188	12376	19448	24310	24310	19448
18	1	18	153	816	3060	8568	18564	31824	43758	48620	43758
19	1	19	171	969	3876	11628	27132	50388	75582	92378	92378
20	1	20	190	1140	4845	15504	38760	77520	125970	167960	184756

If necessary, use the identity $\binom{n}{k} = \binom{n}{n-k}$.

TABLE X   Values of e^{-x}

x	e^{-x}	x	e^{-x}	x	e^{-x}	x	e^{-x}
0.0	1.000	2.5	0.082	5.0	0.0067	7.5	0.00055
0.1	0.905	2.6	0.074	5.1	0.0061	7.6	0.00050
0.2	0.819	2.7	0.067	5.2	0.0055	7.7	0.00045
0.3	0.741	2.8	0.061	5.3	0.0050	7.8	0.00041
0.4	0.670	2.9	0.055	5.4	0.0045	7.9	0.00037
0.5	0.607	3.0	0.050	5.5	0.0041	8.0	0.00034
0.6	0.549	3.1	0.045	5.6	0.0037	8.1	0.00030
0.7	0.497	3.2	0.041	5.7	0.0033	8.2	0.00028
0.8	0.449	3.3	0.037	5.8	0.0030	8.3	0.00025
0.9	0.407	3.4	0.033	5.9	0.0027	8.4	0.00023
1.0	0.368	3.5	0.030	6.0	0.0025	8.5	0.00020
1.1	0.333	3.6	0.027	6.1	0.0022	8.6	0.00018
1.2	0.301	3.7	0.025	6.2	0.0020	8.7	0.00017
1.3	0.273	3.8	0.022	6.3	0.0018	8.8	0.00015
1.4	0.247	3.9	0.020	6.4	0.0017	8.9	0.00014
1.5	0.223	4.0	0.018	6.5	0.0015	9.0	0.00012
1.6	0.202	4.1	0.017	6.6	0.0014	9.1	0.00011
1.7	0.183	4.2	0.015	6.7	0.0012	9.2	0.00010
1.8	0.165	4.3	0.014	6.8	0.0011	9.3	0.00009
1.9	0.150	4.4	0.012	6.9	0.0010	9.4	0.00008
2.0	0.135	4.5	0.011	7.0	0.0009	9.5	0.00008
2.1	0.122	4.6	0.010	7.1	0.0008	9.6	0.00007
2.2	0.111	4.7	0.009	7.2	0.0007	9.7	0.00006
2.3	0.100	4.8	0.008	7.3	0.0007	9.8	0.00006
2.4	0.091	4.9	0.007	7.4	0.0006	9.9	0.00005

TABLE XI Logarithms

N	0	1	2	3	4	5	6	7	8	9
10	0000	0043	0086	0128	0170	0212	0253	0294	0334	0374
11	0414	0453	0492	0531	0569	0607	0645	0682	0719	0755
12	0792	0828	0864	0899	0934	0969	1004	1038	1072	1106
13	1139	1173	1206	1239	1271	1303	1335	1367	1399	1430
14	1461	1492	1523	1553	1584	1614	1644	1673	1703	1732
15	1761	1790	1818	1847	1875	1903	1931	1959	1987	2014
16	2041	2068	2095	2122	2148	2175	2201	2227	2253	2279
17	2304	2330	2355	2380	2405	2430	2455	2480	2504	2529
18	2553	2577	2601	2625	2648	2672	2695	2718	2742	2765
19	2788	2810	2833	2856	2878	2900	2923	2945	2967	2989
20	3010	3032	3054	3075	3096	3118	3139	3160	3181	3201
21	3222	3243	3263	3284	3304	3324	3345	3365	3385	3404
22	3424	3444	3464	3483	3502	3522	3541	3560	3579	3598
23	3617	3636	3655	3674	3692	3711	3729	3747	3766	3784
24	3802	3820	3838	3856	3874	3892	3909	3927	3945	3962
25	3979	3997	4014	4031	4048	4065	4082	4099	4116	4133
26	4150	4166	4183	4200	4216	4232	4249	4265	4281	4298
27	4314	4330	4346	4362	4378	4393	4409	4425	4440	4456
28	4472	4487	4502	4518	4533	4548	4564	4579	4594	4609
29	4624	4639	4654	4669	4683	4698	4713	4728	4742	4757
30	4771	4786	4800	4814	4829	4843	4857	4871	4886	4900
31	4914	4928	4942	4955	4969	4983	4997	5011	5024	5038
32	5051	5065	5079	5092	5105	5119	5132	5145	5159	5172
33	5185	5198	5211	5224	5237	5250	5263	5276	5289	5302
34	5315	5328	5340	5353	5366	5378	5391	5403	5416	5428
35	5441	5453	5465	5478	5490	5502	5514	5527	5539	5551
36	5563	5575	5587	5599	5611	5623	5635	5647	5658	5670
37	5682	5694	5705	5717	5729	5740	5752	5763	5775	5786
38	5798	5809	5821	5832	5843	5855	5866	5877	5888	5899
39	5911	5922	5933	5944	5955	5966	5977	5988	5999	6010
40	6021	6031	6042	6053	6064	6075	6085	6096	6107	6117
41	6128	6138	6149	6160	6170	6180	6191	6201	6212	6222
42	6232	6243	6253	6263	6274	6284	6294	6304	6314	6325
43	6335	6345	6355	6365	6375	6385	6395	6405	6415	6425
44	6435	6444	6454	6464	6474	6484	6493	6503	6513	6522
45	6532	6542	6551	6561	6571	6580	6590	6599	6609	6618
46	6628	6637	6646	6656	6665	6675	6684	6693	6702	6712
47	6721	6730	6739	6749	6758	6767	6776	6785	6794	6803
48	6812	6821	6830	6839	6848	6857	6866	6875	6884	6893
49	6902	6911	6920	6928	6937	6946	6955	6964	6972	6981
50	6990	6998	7007	7016	7024	7033	7042	7050	7059	7067
51	7076	7084	7093	7101	7110	7118	7126	7135	7143	7152
52	7160	7168	7177	7185	7193	7202	7210	7218	7226	7235
53	7243	7251	7259	7267	7275	7284	7292	7300	7308	7316
54	7324	7332	7340	7348	7356	7364	7372	7380	7388	7396

TABLE XI   Logarithms (*continued*)

N	0	1	2	3	4	5	6	7	8	9
55	7404	7412	7419	7427	7435	7443	7451	7459	7466	7474
56	7482	7490	7497	7505	7513	7520	7528	7536	7543	7551
57	7559	7566	7574	7582	7589	7597	7604	7612	7619	7627
58	7634	7642	7649	7657	7664	7672	7679	7686	7694	7701
59	7709	7716	7723	7731	7738	7745	7752	7760	7767	7774
60	7782	7789	7796	7803	7810	7818	7825	7832	7839	7846
61	7853	7860	7868	7875	7882	7889	7896	7903	7910	7917
62	7924	7931	7938	7945	7952	7959	7966	7973	7980	7987
63	7993	8000	8007	8014	8021	8028	8035	8041	8048	8055
64	8062	8069	8075	8082	8089	8096	8102	8109	8116	8122
65	8129	8136	8142	8149	8156	8162	8169	8176	8182	8189
66	8195	8202	8209	8215	8222	8228	8235	8241	8248	8254
67	8261	8267	8274	8280	8287	8293	8299	8306	8312	8319
68	8325	8331	8338	8344	8351	8357	8363	8370	8376	8382
69	8388	8395	8401	8407	8414	8420	8426	8432	8439	8445
70	8451	8457	8463	8470	8476	8482	8488	8494	8500	8506
71	8513	8519	8525	8531	8537	8543	8549	8555	8561	8567
72	8573	8579	8585	8591	8597	8603	8609	8615	8621	8627
73	8633	8639	8645	8651	8657	8663	8669	8675	8681	8686
74	8692	8698	8704	8710	8716	8722	8727	8733	8739	8745
75	8751	8756	8762	8768	8774	8779	8785	8791	8797	8802
76	8808	8814	8820	8825	8831	8837	8842	8848	8854	8859
77	8865	8871	8876	8882	8887	8893	8899	8904	8910	8915
78	8921	8927	8932	8938	8943	8949	8954	8960	8965	8971
79	8976	8982	8987	8993	8998	9004	9009	9015	9020	9025
80	9031	9036	9042	9047	9053	9058	9063	9069	9074	9079
81	9085	9090	9096	9101	9106	9112	9117	9122	9128	9133
82	9138	9143	9149	9154	9159	9165	9170	9175	9180	9186
83	9191	9196	9201	9206	9212	9217	9222	9227	9232	9238
84	9243	9248	9253	9258	9263	9269	9274	9279	9284	9289
85	9294	9299	9304	9309	9315	9320	9325	9330	9335	9340
86	9345	9350	9355	9360	9365	9370	9375	9380	9385	9390
87	9395	9400	9405	9410	9415	9420	9425	9430	9435	9440
88	9445	9450	9455	9460	9465	9469	9474	9479	9484	9489
89	9494	9499	9504	9509	9513	9518	9523	9528	9533	9538
90	9542	9547	9552	9557	9562	9566	9571	9576	9581	9586
91	9590	9595	9600	9605	9609	9614	9619	9624	9628	9633
92	9638	9643	9647	9652	9657	9661	9666	9671	9675	9680
93	9685	9689	9694	9699	9703	9708	9713	9717	9722	9727
94	9731	9736	9741	9745	9750	9754	9759	9763	9768	9773
95	9777	9782	9786	9791	9795	9800	9805	9809	9814	9818
96	9823	9827	9832	9836	9341	9845	9850	9854	9859	9863
97	9868	9872	9877	9881	9886	9890	9894	9899	9903	9908
98	9912	9917	9921	9926	9930	9934	9939	9943	9948	9952
99	9956	9961	9965	9969	9974	9978	9983	9987	9991	9996

Table XII contains the square roots of the numbers from 1.00 to 9.99, and also the square roots of these numbers multiplied by 10, spaced at intervals of 0.01. The square roots are all rounded to four decimals. To find the square root of any positive number rounded to three significant digits, we use the following rule in deciding whether to take the entry of the $\sqrt{n}$ or $\sqrt{10n}$ column:

> **Move the decimal point an even number of places to the right or to the left until a number greater than or equal to 1 but less than 100 is reached. If the resulting number is less than 10 go to the $\sqrt{n}$ column; if it is 10 or more go to the $\sqrt{10n}$ column.**

Thus, to find the square roots of 12,800, 379, and 0.0812, we go to the $\sqrt{n}$ column since the decimal point has to be moved, respectively, four places to the left, two places to the left, and two places to the right, to give 1.28, 3.79, and 8.12. Similarly, to find the square roots of 5,240, 0.281, and 0.0000259 we go to the $\sqrt{10n}$ column since the decimal point has to be moved, respectively, two places to the left, two places to the right, and six places to the right, to give 52.4, 28.1, and 25.9.

After we locate a square root in the appropriate column of Table XII, we must be sure to get the decimal point in the right place. Here it will help to use the following rule:

> **Having previously moved the decimal point an even number of places to the left or right to get a number greater than or equal to 1 but less than 100, move the decimal point of the entry of the appropriate column in Table XII half as many places in the opposite direction.**

For example, to determine the square root of 12,800, we first note that the decimal point has to be moved *four places to the left* to give 1.28. We then take the entry of the $\sqrt{n}$ column corresponding to 1.28, move its decimal point *two places to the right*, and get $\sqrt{12,800} = 113.14$. Similarly, to find the square root of 0.0000259, we note that the decimal point has to be moved *six places to the right* to give 25.9. We thus take the entry of the $\sqrt{10n}$ column corresponding to 2.59, move the decimal point *three places to the left*, and get $\sqrt{0.0000259} = 0.0050892$. In actual practice, if a number whose square root we want to find is rounded, the square root will have to be rounded to as many significant digits as the original number.

TABLE XII Square Roots

n	$\sqrt{n}$	$\sqrt{10n}$	n	$\sqrt{n}$	$\sqrt{10n}$	n	$\sqrt{n}$	$\sqrt{10n}$
1.00	1.0000	3.1623	1.50	1.2247	3.8730	2.00	1.4142	4.4721
1.01	1.0050	3.1780	1.51	1.2288	3.8859	2.01	1.4177	4.4833
1.02	1.0100	3.1937	1.52	1.2329	3.8987	2.02	1.4213	4.4944
1.03	1.0149	3.2094	1.53	1.2369	3.9115	2.03	1.4248	4.5056
1.04	1.0198	3.2249	1.54	1.2410	3.9243	2.04	1.4283	4.5166
1.05	1.0247	3.2404	1.55	1.2450	3.9370	2.05	1.4318	4.5277
1.06	1.0296	3.2558	1.56	1.2490	3.9497	2.06	1.4353	4.5387
1.07	1.0344	3.2711	1.57	1.2530	3.9623	2.07	1.4387	4.5497
1.08	1.0392	3.2863	1.58	1.2570	3.9749	2.08	1.4422	4.5607
1.09	1.0440	3.3015	1.59	1.2610	3.9875	2.09	1.4457	4.5717
1.10	1.0488	3.3166	1.60	1.2649	4.0000	2.10	1.4491	4.5826
1.11	1.0536	3.3317	1.61	1.2689	4.0125	2.11	1.4526	4.5935
1.12	1.0583	3.3466	1.62	1.2728	4.0249	2.12	1.4560	4.6043
1.13	1.0630	3.3615	1.63	1.2767	4.0373	2.13	1.4595	4.6152
1.14	1.0677	3.3764	1.64	1.2806	4.0497	2.14	1.4629	4.6260
1.15	1.0724	3.3912	1.65	1.2845	4.0620	2.15	1.4663	4.6368
1.16	1.0770	3.4059	1.66	1.2884	4.0743	2.16	1.4697	4.6476
1.17	1.0817	3.4205	1.67	1.2923	4.0866	2.17	1.4731	4.6583
1.18	1.0863	3.4351	1.68	1.2961	4.0988	2.18	1.4765	4.6690
1.19	1.0909	3.4496	1.69	1.3000	4.1110	2.19	1.4799	4.6797
1.20	1.0954	3.4641	1.70	1.3038	4.1231	2.20	1.4832	4.6904
1.21	1.1000	3.4785	1.71	1.3077	4.1352	2.21	1.4866	4.7011
1.22	1.1045	3.4928	1.72	1.3115	4.1473	2.22	1.4900	4.7117
1.23	1.1091	3.5071	1.73	1.3153	4.1593	2.23	1.4933	4.7223
1.24	1.1136	3.5214	1.74	1.3191	4.1713	2.24	1.4967	4.7329
1.25	1.1180	3.5355	1.75	1.3229	4.1833	2.25	1.5000	4.7434
1.26	1.1225	3.5496	1.76	1.3266	4.1952	2.26	1.5033	4.7539
1.27	1.1269	3.5637	1.77	1.3304	4.2071	2.27	1.5067	4.7645
1.28	1.1314	3.5777	1.78	1.3342	4.2190	2.28	1.5100	4.7749
1.29	1.1358	3.5917	1.79	1.3379	4.2308	2.29	1.5133	4.7854
1.30	1.1402	3.6056	1.80	1.3416	4.2426	2.30	1.5166	4.7958
1.31	1.1446	3.6194	1.81	1.3454	4.2544	2.31	1.5199	4.8062
1.32	1.1489	3.6332	1.82	1.3491	4.2661	2.32	1.5232	4.8166
1.33	1.1533	3.6469	1.83	1.3528	4.2778	2.33	1.5264	4.8270
1.34	1.1576	3.6606	1.84	1.3565	4.2895	2.34	1.5297	4.8374
1.35	1.1619	3.6742	1.85	1.3601	4.3012	2.35	1.5330	4.8477
1.36	1.1662	3.6878	1.86	1.3638	4.3128	2.36	1.5362	4.8580
1.37	1.1705	3.7014	1.87	1.3675	4.3243	2.37	1.5395	4.8683
1.38	1.1747	3.7148	1.88	1.3711	4.3359	2.38	1.5427	4.8785
1.39	1.1790	3.7283	1.89	1.3748	4.3474	2.39	1.5460	4.8888
1.40	1.1832	3.7417	1.90	1.3784	4.3589	2.40	1.5492	4.8990
1.41	1.1874	3.7550	1.91	1.3820	4.3704	2.41	1.5524	4.9092
1.42	1.1916	3.7683	1.92	1.3856	4.3818	2.42	1.5556	4.9193
1.43	1.1958	3.7815	1.93	1.3892	4.3932	2.43	1.5588	4.9295
1.44	1.2000	3.7947	1.94	1.3928	4.4045	2.44	1.5620	4.9396
1.45	1.2042	3.8079	1.95	1.3964	4.4159	2.45	1.5652	4.9497
1.46	1.2083	3.8210	1.96	1.4000	4.4272	2.46	1.5684	4.9598
1.47	1.2124	3.8341	1.97	1.4036	4.4385	2.47	1.5716	4.9699
1.48	1.2166	3.8471	1.98	1.4071	4.4497	2.48	1.5748	4.9800
1.49	1.2207	3.8601	1.99	1.4107	4.4609	2.49	1.5780	4.9900

TABLE XII Square Roots (*continued*)

n	$\sqrt{n}$	$\sqrt{10n}$	n	$\sqrt{n}$	$\sqrt{10n}$	n	$\sqrt{n}$	$\sqrt{10n}$
2.50	1.5811	5.0000	3.00	1.7321	5.4772	3.50	1.8708	5.9161
2.51	1.5843	5.0100	3.01	1.7349	5.4863	3.51	1.8735	5.9245
2.52	1.5875	5.0200	3.02	1.7378	5.4955	3.52	1.8762	5.9330
2.53	1.5906	5.0299	3.03	1.7407	5.5045	3.53	1.8788	5.9414
2.54	1.5937	5.0398	3.04	1.7436	5.5136	3.54	1.8815	5.9498
2.55	1.5969	5.0498	3.05	1.7464	5.5227	3.55	1.8841	5.9582
2.56	1.6000	5.0596	3.06	1.7493	5.5317	3.56	1.8868	5.9666
2.57	1.6031	5.0695	3.07	1.7521	5.5408	3.57	1.8894	5.9749
2.58	1.6062	5.0794	3.08	1.7550	5.5498	3.58	1.8921	5.9833
2.59	1.6093	5.0892	3.09	1.7578	5.5588	3.59	1.8947	5.9917
2.60	1.6125	5.0990	3.10	1.7607	5.5678	3.60	1.8974	6.0000
2.61	1.6155	5.1088	3.11	1.7635	5.5767	3.61	1.9000	6.0083
2.62	1.6186	5.1186	3.12	1.7664	5.5857	3.62	1.9026	6.0166
2.63	1.6217	5.1284	3.13	1.7692	5.5946	3.63	1.9053	6.0249
2.64	1.6248	5.1381	3.14	1.7720	5.6036	3.64	1.9079	6.0332
2.65	1.6279	5.1478	3.15	1.7748	5.6125	3.65	1.9105	6.0415
2.66	1.6310	5.1575	3.16	1.7776	5.6214	3.66	1.9131	6.0498
2.67	1.6340	5.1672	3.17	1.7804	5.6303	3.67	1.9157	6.0581
2.68	1.6371	5.1769	3.18	1.7833	5.6391	3.68	1.9183	6.0663
2.69	1.6401	5.1865	3.19	1.7861	5.6480	3.69	1.9209	6.0745
2.70	1.6432	5.1962	3.20	1.7889	5.6569	3.70	1.9235	6.0828
2.71	1.6462	5.2058	3.21	1.7916	5.6657	3.71	1.9261	6.0910
2.72	1.6492	5.2154	3.22	1.7944	5.6745	3.72	1.9287	6.0992
2.73	1.6523	5.2249	3.23	1.7972	5.6833	3.73	1.9313	6.1074
2.74	1.6553	5.2345	3.24	1.8000	5.6921	3.74	1.9339	6.1156
2.75	1.6583	5.2440	3.25	1.8028	5.7009	3.75	1.9365	6.1237
2.76	1.6613	5.2536	3.26	1.8055	5.7096	3.76	1.9391	6.1319
2.77	1.6643	5.2631	3.27	1.8083	5.7184	3.77	1.9416	6.1400
2.78	1.6673	5.2726	3.28	1.8111	5.7271	3.78	1.9442	6.1482
2.79	1.6703	5.2820	3.29	1.8138	5.7359	3.79	1.9468	6.1563
2.80	1.6733	5.2915	3.30	1.8166	5.7446	3.80	1.9494	6.1644
2.81	1.6763	5.3009	3.31	1.8193	5.7533	3.81	1.9519	6.1725
2.82	1.6793	5.3104	3.32	1.8221	5.7619	3.82	1.9545	6.1806
2.83	1.6823	5.3198	3.33	1.8248	5.7706	3.83	1.9570	6.1887
2.84	1.6852	5.3292	3.34	1.8276	5.7793	3.84	1.9596	6.1968
2.85	1.6882	5.3385	3.35	1.8303	5.7879	3.85	1.9621	6.2048
2.86	1.6912	5.3479	3.36	1.8330	5.7966	3.86	1.9647	6.2129
2.87	1.6941	5.3572	3.37	1.8358	5.8052	3.87	1.9672	6.2209
2.88	1.6971	5.3666	3.38	1.8385	5.8138	3.88	1.9698	6.2290
2.89	1.7000	5.3759	3.39	1.8412	5.8224	3.89	1.9723	6.2370
2.90	1.7029	5.3852	3.40	1.8439	5.8310	3.90	1.9748	6.2450
2.91	1.7059	5.3944	3.41	1.8466	5.8395	3.91	1.9774	6.2530
2.92	1.7088	5.4037	3.42	1.8493	5.8481	3.92	1.9799	6.2610
2.93	1.7117	5.4129	3.43	1.8520	5.8566	3.93	1.9824	6.2690
2.94	1.7146	5.4222	3.44	1.8547	5.8652	3.94	1.9849	6.2769
2.95	1.7176	5.4314	3.45	1.8574	5.8737	3.95	1.9875	6.2849
2.96	1.7205	5.4406	3.46	1.8601	5.8822	3.96	1.9900	6.2929
2.97	1.7234	5.4498	3.47	1.8628	5.8907	3.97	1.9925	6.3008
2.98	1.7263	5.4589	3.48	1.8655	5.8992	3.98	1.9950	6.3087
2.99	1.7292	5.4681	3.49	1.8682	5.9076	3.99	1.9975	6.3166

TABLE XII   Square Roots (*continued*)

n	$\sqrt{n}$	$\sqrt{10n}$	n	$\sqrt{n}$	$\sqrt{10n}$	n	$\sqrt{n}$	$\sqrt{10n}$
4.00	2.0000	6.3246	4.50	2.1213	6.7082	5.00	2.2361	7.0711
4.01	2.0025	6.3325	4.51	2.1237	6.7157	5.01	2.2383	7.0781
4.02	2.0050	6.3403	4.52	2.1260	6.7231	5.02	2.2405	7.0852
4.03	2.0075	6.3482	4.53	2.1284	6.7305	5.03	2.2428	7.0922
4.04	2.0100	6.3561	4.54	2.1307	6.7380	5.04	2.2450	7.0993
4.05	2.0125	6.3640	4.55	2.1331	6.7454	5.05	2.2472	7.1063
4.06	2.0149	6.3718	4.56	2.1354	6.7528	5.06	2.2494	7.1134
4.07	2.0174	6.3797	4.57	2.1378	6.7602	5.07	2.2517	7.1204
4.08	2.0199	6.3875	4.58	2.1401	6.7676	5.08	2.2539	7.1274
4.09	2.0224	6.3953	4.59	2.1424	6.7750	5.09	2.2561	7.1344
4.10	2.0248	6.4031	4.60	2.1448	6.7823	5.10	2.2583	7.1414
4.11	2.0273	6.4109	4.61	2.1471	6.7897	5.11	2.2605	7.1484
4.12	2.0298	6.4187	4.62	2.1494	6.7971	5.12	2.2627	7.1554
4.13	2.0322	6.4265	4.63	2.1517	6.8044	5.13	2.2650	7.1624
4.14	2.0347	6.4343	4.64	2.1541	6.8118	5.14	2.2672	7.1694
4.15	2.0372	6.4420	4.65	2.1564	6.8191	5.15	2.2694	7.1764
4.16	2.0396	6.4498	4.66	2.1587	6.8264	5.16	2.2716	7.1833
4.17	2.0421	6.4576	4.67	2.1610	6.8337	5.17	2.2738	7.1903
4.18	2.0445	6.4653	4.68	2.1633	6.8411	5.18	2.2760	7.1972
4.19	2.0469	6.4730	4.69	2.1656	6.8484	5.19	2.2782	7.2042
4.20	2.0494	6.4807	4.70	2.1679	6.8557	5.20	2.2804	7.2111
4.21	2.0518	6.4885	4.71	2.1703	6.8629	5.21	2.2825	7.2180
4.22	2.0543	6.4962	4.72	2.1726	6.8702	5.22	2.2847	7.2250
4.23	2.0567	6.5038	4.73	2.1749	6.8775	5.23	2.2869	7.2319
4.24	2.0591	6.5115	4.74	2.1772	6.8848	5.24	2.2891	7.2388
4.25	2.0616	6.5192	4.75	2.1794	6.8920	5.25	2.2913	7.2457
4.26	2.0640	6.5269	4.76	2.1817	6.8993	5.26	2.2935	7.2526
4.27	2.0664	6.5345	4.77	2.1840	6.9065	5.27	2.2956	7.2595
4.28	2.0688	6.5422	4.78	2.1863	6.9138	5.28	2.2978	7.2664
4.29	2.0712	6.5498	4.79	2.1886	6.9210	5.29	2.3000	7.2732
4.30	2.0736	6.5574	4.80	2.1909	6.9282	5.30	2.3022	7.2801
4.31	2.0761	6.5651	4.81	2.1932	6.9354	5.31	2.3043	7.2870
4.32	2.0785	6.5727	4.82	2.1954	6.9426	5.32	2.3065	7.2938
4.33	2.0809	6.5803	4.83	2.1977	6.9498	5.33	2.3087	7.3007
4.34	2.0833	6.5879	4.84	2.2000	6.9570	5.34	2.3108	7.3075
4.35	2.0857	6.5955	4.85	2.2023	6.9642	5.35	2.3130	7.3144
4.36	2.0881	6.6030	4.86	2.2045	6.9714	5.36	2.3152	7.3212
4.37	2.0905	6.6106	4.87	2.2068	6.9785	5.37	2.3173	7.3280
4.38	2.0928	6.6182	4.88	2.2091	6.9857	5.38	2.3195	7.3348
4.39	2.0952	6.6257	4.89	2.2113	6.9929	5.39	2.3216	7.3417
4.40	2.0976	6.6332	4.90	2.2136	7.0000	5.40	2.3238	7.3485
4.41	2.1000	6.6408	4.91	2.2159	7.0071	5.41	2.3259	7.3553
4.42	2.1024	6.6483	4.92	2.2181	7.0143	5.42	2.3281	7.3621
4.43	2.1048	6.6558	4.93	2.2204	7.0214	5.43	2.3302	7.3689
4.44	2.1071	6.6633	4.94	2.2226	7.0285	5.44	2.3324	7.3756
4.45	2.1095	6.6708	4.95	2.2249	7.0356	5.45	2.3345	7.3824
4.46	2.1119	6.6783	4.96	2.2271	7.0427	5.46	2.3367	7.3892
4.47	2.1142	6.6858	4.97	2.2293	7.0498	5.47	2.3388	7.3959
4.48	2.1166	6.6933	4.98	2.2316	7.0569	5.48	2.3409	7.4027
4.49	2.1190	6.7007	4.99	2.2338	7.0640	5.49	2.3431	7.4095

TABLE XII Square Roots (*continued*)

n	$\sqrt{n}$	$\sqrt{10n}$	n	$\sqrt{n}$	$\sqrt{10n}$	n	$\sqrt{n}$	$\sqrt{10n}$
5.50	2.3452	7.4162	6.00	2.4495	7.7460	6.50	2.5495	8.0623
5.51	2.3473	7.4229	6.01	2.4515	7.7524	6.51	2.5515	8.0685
5.52	2.3495	7.4297	6.02	2.4536	7.7589	6.52	2.5534	8.0747
5.53	2.3516	7.4364	6.03	2.4556	7.7653	6.53	2.5554	8.0808
5.54	2.3537	7.4431	6.04	2.4576	7.7717	6.54	2.5573	8.0870
5.55	2.3558	7.4498	6.05	2.4597	7.7782	6.55	2.5593	8.0932
5.56	2.3580	7.4565	6.06	2.4617	7.7846	6.56	2.5612	8.0994
5.57	2.3601	7.4632	6.07	2.4637	7.7910	6.57	2.5632	8.1056
5.58	2.3622	7.4699	6.08	2.4658	7.7974	6.58	2.5652	8.1117
5.59	2.3643	7.4766	6.09	2.4678	7.8038	6.59	2.5671	8.1179
5.60	2.3664	7.4833	6.10	2.4698	7.8102	6.60	2.5690	8.1240
5.61	2.3685	7.4900	6.11	2.4718	7.8166	6.61	2.5710	8.1302
5.62	2.3707	7.4967	6.12	2.4739	7.8230	6.62	2.5729	8.1363
5.63	2.3728	7.5033	6.13	2.4759	7.8294	6.63	2.5749	8.1425
5.64	2.3749	7.5100	6.14	2.4779	7.8358	6.64	2.5768	8.1486
5.65	2.3770	7.5166	6.15	2.4799	7.8422	6.65	2.5788	8.1548
5.66	2.3791	7.5233	6.16	2.4819	7.8486	6.66	2.5807	8.1609
5.67	2.3812	7.5299	6.17	2.4839	7.8549	6.67	2.5826	8.1670
5.68	2.3833	7.5366	6.18	2.4860	7.8613	6.68	2.5846	8.1731
5.69	2.3854	7.5432	6.19	2.4880	7.8677	6.69	2.5865	8.1792
5.70	2.3875	7.5498	6.20	2.4900	7.8740	6.70	2.5884	8.1854
5.71	2.3896	7.5565	6.21	2.4920	7.8804	6.71	2.5904	8.1915
5.72	2.3917	7.5631	6.22	2.4940	7.8867	6.72	2.5923	8.1976
5.73	2.3937	7.5697	6.23	2.4960	7.8930	6.73	2.5942	8.2037
5.74	2.3958	7.5763	6.24	2.4980	7.8994	6.74	2.5962	8.2098
5.75	2.3979	7.5829	6.25	2.5000	7.9057	6.75	2.5981	8.2158
5.76	2.4000	7.5895	6.26	2.5020	7.9120	6.76	2.6000	8.2219
5.77	2.4021	7.5961	6.27	2.5040	7.9183	6.77	2.6019	8.2280
5.78	2.4042	7.6026	6.28	2.5060	7.9246	6.78	2.6038	8.2341
5.79	2.4062	7.6092	6.29	2.5080	7.9310	6.79	2.6058	8.2401
5.80	2.4083	7.6158	6.30	2.5100	7.9373	6.80	2.6077	8.2462
5.81	2.4104	7.6223	6.31	2.5120	7.9436	6.81	2.6096	8.2523
5.82	2.4125	7.6289	6.32	2.5140	7.9498	6.82	2.6115	8.2583
5.83	2.4145	7.6354	6.33	2.5159	7.9561	6.83	2.6134	8.2644
5.84	2.4166	7.6420	6.34	2.5179	7.9624	6.84	2.6153	8.2704
5.85	2.4187	7.6485	6.35	2.5199	7.9687	6.85	2.6173	8.2765
5.86	2.4207	7.6551	6.36	2.5219	7.9750	6.86	2.6192	8.2825
5.87	2.4228	7.6616	6.37	2.5239	7.9812	6.87	2.6211	8.2885
5.88	2.4249	7.6681	6.38	2.5259	7.9875	6.88	2.6230	8.2946
5.89	2.4269	7.6746	6.39	2.5278	7.9937	6.89	2.6249	8.3006
5.90	2.4290	7.6811	6.40	2.5298	8.0000	6.90	2.6268	8.3066
5.91	2.4310	7.6877	6.41	2.5318	8.0062	6.91	2.6287	8.3126
5.92	2.4331	7.6942	6.42	2.5338	8.0125	6.92	2.6306	8.3187
5.93	2.4352	7.7006	6.43	2.5357	8.0187	6.93	2.6325	8.3247
5.94	2.4372	7.7071	6.44	2.5377	8.0250	6.94	2.6344	8.3307
5.95	2.4393	7.7136	6.45	2.5397	8.0312	6.95	2.6363	8.3367
5.96	2.4413	7.7201	6.46	2.5417	8.0374	6.96	2.6382	8.3427
5.97	2.4434	7.7266	6.47	2.5436	8.0436	6.97	2.6401	8.3487
5.98	2.4454	7.7330	6.48	2.5456	8.0498	6.98	2.6420	8.3546
5.99	2.4474	7.7395	6.49	2.5475	8.0561	6.99	2.6439	8.3606

TABLE XII Square Roots (*continued*)

n	$\sqrt{n}$	$\sqrt{10n}$	n	$\sqrt{n}$	$\sqrt{10n}$	n	$\sqrt{n}$	$\sqrt{10n}$
7.00	2.6458	8.3666	7.50	2.7386	8.6603	8.00	2.8284	8.9443
7.01	2.6476	8.3726	7.51	2.7404	8.6660	8.01	2.8302	8.9499
7.02	2.6495	8.3785	7.52	2.7423	8.6718	8.02	2.8320	8.9554
7.03	2.6514	8.3845	7.53	2.7441	8.6776	8.03	2.8337	8.9610
7.04	2.6533	8.3905	7.54	2.7459	8.6833	8.04	2.8355	8.9666
7.05	2.6552	8.3964	7.55	2.7477	8.6891	8.05	2.8373	8.9722
7.06	2.6571	8.4024	7.56	2.7495	8.6948	8.06	2.8390	8.9778
7.07	2.6589	8.4083	7.57	2.7514	8.7006	8.07	2.8408	8.9833
7.08	2.6608	8.4143	7.58	2.7532	8.7063	8.08	2.8425	8.9889
7.09	2.6627	8.4202	7.59	2.7550	8.7121	8.09	2.8443	8.9944
7.10	2.6646	8.4261	7.60	2.7568	8.7178	8.10	2.8460	9.0000
7.11	2.6665	8.4321	7.61	2.7586	8.7235	8.11	2.8478	9.0056
7.12	2.6683	8.4380	7.62	2.7604	8.7293	8.12	2.8496	9.0111
7.13	2.6702	8.4439	7.63	2.7622	8.7350	8.13	2.8513	9.0167
7.14	2.6721	8.4499	7.64	2.7641	8.7407	8.14	2.8531	9.0222
7.15	2.6739	8.4558	7.65	2.7659	8.7464	8.15	2.8548	9.0277
7.16	2.6758	8.4617	7.66	2.7677	8.7521	8.16	2.8566	9.0333
7.17	2.6777	8.4676	7.67	2.7695	8.7579	8.17	2.8583	9.0388
7.18	2.6796	8.4735	7.68	2.7713	8.7636	8.18	2.8601	9.0443
7.19	2.6814	8.4794	7.69	2.7731	8.7693	8.19	2.8618	9.0499
7.20	2.6833	8.4853	7.70	2.7749	8.7750	8.20	2.8636	9.0554
7.21	2.6851	8.4912	7.71	2.7767	8.7807	8.21	2.8653	9.0609
7.22	2.6870	8.4971	7.72	2.7785	8.7864	8.22	2.8671	9.0664
7.23	2.6889	8.5029	7.73	2.7803	8.7920	8.23	2.8688	9.0719
7.24	2.6907	8.5088	7.74	2.7821	8.7977	8.24	2.8705	9.0774
7.25	2.6926	8.5147	7.75	2.7839	8.8034	8.25	2.8723	9.0830
7.26	2.6944	8.5206	7.76	2.7857	8.8091	8.26	2.8740	9.0885
7.27	2.6963	8.5264	7.77	2.7875	8.8148	8.27	2.8758	9.0940
7.28	2.6981	8.5323	7.78	2.7893	8.8204	8.28	2.8775	9.0995
7.29	2.7000	8.5381	7.79	2.7911	8.8261	8.29	2.8792	9.1049
7.30	2.7019	8.5440	7.80	2.7928	8.8318	8.30	2.8810	9.1104
7.31	2.7037	8.5499	7.81	2.7946	8.8374	8.31	2.8827	9.1159
7.32	2.7055	8.5557	7.82	2.7964	8.8431	8.32	2.8844	9.1214
7.33	2.7074	8.5615	7.83	2.7982	8.8487	8.33	2.8862	9.1269
7.34	2.7092	8.5674	7.84	2.8000	8.8544	8.34	2.8879	9.1324
7.35	2.7111	8.5732	7.85	2.8018	8.8600	8.35	2.8896	9.1378
7.36	2.7129	8.5790	7.86	2.8036	8.8657	8.36	2.8914	9.1433
7.37	2.7148	8.5849	7.87	2.8054	8.8713	8.37	2.8931	9.1488
7.38	2.7166	8.5907	7.88	2.8071	8.8769	8.38	2.8948	9.1542
7.39	2.7185	8.5965	7.89	2.8089	8.8826	8.39	2.8965	9.1597
7.40	2.7203	8.6023	7.90	2.8107	8.8882	8.40	2.8983	9.1652
7.41	2.7221	8.6081	7.91	2.8125	8.8938	8.41	2.9000	9.1706
7.42	2.7240	8.6139	7.92	2.8142	8.8994	8.42	2.9017	9.1761
7.43	2.7258	8.6197	7.93	2.8160	8.9051	8.43	2.9034	9.1815
7.44	2.7276	8.6255	7.94	2.8178	8.9107	8.44	2.9052	9.1869
7.45	2.7295	8.6313	7.95	2.8196	8.9163	8.45	2.9069	9.1924
7.46	2.7313	8.6371	7.96	2.8213	8.9219	8.46	2.9086	9.1978
7.47	2.7331	8.6429	7.97	2.8231	8.9275	8.47	2.9103	9.2033
7.48	2.7350	8.6487	7.98	2.8249	8.9331	8.48	2.9120	9.2087
7.49	2.7368	8.6545	7.99	2.8267	8.9387	8.49	2.9138	9.2141

TABLE XII Square Roots (*continued*)

n	$\sqrt{n}$	$\sqrt{10n}$	n	$\sqrt{n}$	$\sqrt{10n}$	n	$\sqrt{n}$	$\sqrt{10n}$
8.50	2.9155	9.2195	9.00	3.0000	9.4868	9.50	3.0822	9.7468
8.51	2.9172	9.2250	9.01	3.0017	9.4921	9.51	3.0838	9.7519
8.52	2.9189	9.2304	9.02	3.0033	9.4974	9.52	3.0854	9.7570
8.53	2.9206	9.2358	9.03	3.0050	9.5026	9.53	3.0871	9.7622
8.54	2.9223	9.2412	9.04	3.0067	9.5079	9.54	3.0887	9.7673
8.55	2.9240	9.2466	9.05	3.0083	9.5131	9.55	3.0903	9.7724
8.56	2.9257	9.2520	9.06	3.0100	9.5184	9.56	3.0919	9.7775
8.57	2.9275	9.2574	9.07	3.0116	9.5237	9.57	3.0935	9.7826
8.58	2.9292	9.2628	9.08	3.0133	9.5289	9.58	3.0952	9.7877
8.59	2.9309	9.2682	9.09	3.0150	9.5341	9.59	3.0968	9.7929
8.60	2.9326	9.2736	9.10	3.0166	9.5394	9.60	3.0984	9.7980
8.61	2.9343	9.2790	9.11	3.0183	9.5446	9.61	3.1000	9.8031
8.62	2.9360	9.2844	9.12	3.0199	9.5499	9.62	3.1016	9.8082
8.63	2.9377	9.2898	9.13	3.0216	9.5551	9.63	3.1032	9.8133
8.64	2.9394	9.2952	9.14	3.0232	9.5603	9.64	3.1048	9.8184
8.65	2.9411	9.3005	9.15	3.0249	9.5656	9.65	3.1064	9.8234
8.66	2.9428	9.3059	9.16	3.0265	9.5708	9.66	3.1081	9.8285
8.67	2.9445	9.3113	9.17	3.0282	9.5760	9.67	3.1097	9.8336
8.68	2.9462	9.3167	9.18	3.0299	9.5812	9.68	3.1113	9.8387
8.69	2.9479	9.3220	9.19	3.0315	9.5864	9.69	3.1129	9.8438
8.70	2.9496	9.3274	9.20	3.0332	9.5917	9.70	3.1145	9.8489
8.71	2.9513	9.3327	9.21	3.0348	9.5969	9.71	3.1161	9.8539
8.72	2.9530	9.3381	9.22	3.0364	9.6021	9.72	3.1177	9.8590
8.73	2.9547	9.3434	9.23	3.0381	9.6073	9.73	3.1193	9.8641
8.74	2.9563	9.3488	9.24	3.0397	9.6125	9.74	3.1209	9.8691
8.75	2.9580	9.3541	9.25	3.0414	9.6177	9.75	3.1225	9.8742
8.76	2.9597	9.3595	9.26	3.0430	9.6229	9.76	3.1241	9.8793
8.77	2.9614	9.3648	9.27	3.0447	9.6281	9.77	3.1257	9.8843
8.78	2.9631	9.3702	9.28	3.0463	9.6333	9.78	3.1273	9.8894
8.79	2.9648	9.3755	9.29	3.0480	9.6385	9.79	3.1289	9.8944
8.80	2.9665	9.3808	9.30	3.0496	9.6437	9.80	3.1305	9.8995
8.81	2.9682	9.3862	9.31	3.0512	9.6488	9.81	3.1321	9.9045
8.82	2.9698	9.3915	9.32	3.0529	9.6540	9.82	3.1337	9.9096
8.83	2.9715	9.3968	9.33	3.0545	9.6592	9.83	3.1353	9.9146
8.84	2.9732	9.4021	9.34	3.0561	9.6644	9.84	3.1369	9.9197
8.85	2.9749	9.4074	9.35	3.0578	9.6695	9.85	3.1385	9.9247
8.86	2.9766	9.4128	9.36	3.0594	9.6747	9.86	3.1401	9.9298
8.87	2.9783	9.4181	9.37	3.0610	9.6799	9.87	3.1417	9.9348
8.88	2.9799	9.4234	9.38	3.0627	9.6850	9.88	3.1432	9.9398
8.89	2.9816	9.4287	9.39	3.0643	9.6902	9.89	3.1448	9.9448
8.90	2.9833	9.4340	9.40	3.0659	9.6954	9.90	3.1464	9.9499
8.91	2.9850	9.4393	9.41	3.0676	9.7005	9.91	3.1480	9.9549
8.92	2.9866	9.4446	9.42	3.0692	9.7057	9.92	3.1496	9.9599
8.93	2.9883	9.4499	9.43	3.0708	9.7108	9.93	3.1512	9.9649
8.94	2.9900	9.4552	9.44	3.0725	9.7160	9.94	3.1528	9.9700
8.95	2.9917	9.4604	9.45	3.0741	9.7211	9.95	3.1544	9.9750
8.96	2.9933	9.4657	9.46	3.0757	9.7263	9.96	3.1559	9.9800
8.97	2.9950	9.4710	9.47	3.0773	9.7314	9.97	3.1575	9.9850
8.98	2.9967	9.4763	9.45	3.0790	9.7365	9.98	3.1591	9.9900
8.99	2.9983	9.4816	9.49	3.0806	9.7417	9.99	3.1607	9.9950

TABLE XIII   Random Numbers†

04433	80674	24520	18222	10610	05794	37515
60298	47829	72648	37414	75755	04717	29899
67884	59651	67533	68123	17730	95862	08034
89512	32155	51906	61662	64130	16688	37275
32653	01895	12506	88535	36553	23757	34209
95913	15405	13772	76638	48423	25018	99041
55864	21694	13122	44115	01601	50541	00147
35334	49810	91601	40617	72876	33967	73830
57729	32196	76487	11622	96297	24160	09903
86648	13697	63677	70119	94739	25875	38829
30574	47609	07967	32422	76791	39725	53711
81307	43694	83580	79974	45929	85113	72268
02410	54905	79007	54939	21410	86980	91772
18969	75274	52233	62319	08598	09066	95288
87863	82384	66860	62297	80198	19347	73234
68397	71708	15438	62311	72844	60203	46412
28529	54447	58729	10854	99058	18260	38765
44285	06372	15867	70418	57012	72122	36634
86299	83430	33571	23309	57040	29285	67870
84842	68668	90894	61658	15001	94055	36308
56970	83609	52098	04184	54967	72938	56834
83125	71257	60490	44369	66130	72936	69848
55503	52423	02464	26141	68779	66388	75242
47019	76273	33203	29608	54553	25971	69573
84828	32592	79526	29554	84580	37859	28504
68921	08141	79227	05748	51276	57143	31926
36458	96045	30424	98420	72925	40729	22337
95752	59445	36847	87729	81679	59126	59437
26768	47323	58454	56958	20575	76746	49878
42613	37056	43636	58085	06766	60227	96414
95457	30566	65482	25596	02678	54592	63607
95276	17894	63564	95958	39750	64379	46059
66954	52324	64776	92345	95110	59448	77249
17457	18481	14113	62462	02798	54977	48349
03704	36872	83214	59337	01695	60666	97410
21538	86497	33210	60337	27976	70661	08250
57178	67619	98310	70348	11317	71623	55510
31048	97558	94953	55866	96283	46620	52087
69799	55380	16498	80733	96422	58078	99643
90595	61867	59231	17772	67831	33317	00520
33570	04981	98939	78784	09977	29398	93896
15340	93460	57477	13898	48431	72936	78160
64079	42483	36512	56186	99098	48850	72527
63491	05546	67118	62063	74958	20946	28147
92003	63868	41034	28260	79708	00770	88643
52360	46658	66511	04172	73085	11795	52594
74622	12142	68355	65635	21828	39539	18988
04157	50079	61343	64315	70836	82857	35335
86003	60070	66241	32836	27573	11479	94114
41268	80187	20351	09636	84668	42486	71303

†Based on parts of *Table of 105,000 Random Decimal Digits*, (Washington, D. C.: Interstate Commerce Commission, Bureau of Transport Economics and Statistics).

TABLE XIII   Random Numbers (*continued*)

48611	62866	33963	14045	79451	04934	45576
78812	03509	78673	73181	29973	18664	04555
19472	63971	37271	31445	49019	49405	46925
51266	11569	08697	91120	64156	40365	74297
55806	96275	26130	47949	14877	69594	83041
77527	81360	18180	97421	55541	90275	18213
77680	58788	33016	61173	93049	04694	43534
15404	96554	88265	34537	38526	67924	40474
14045	22917	60718	66487	46346	30949	03173
68376	43918	77653	04127	69930	43283	35766
93385	13421	67957	20384	58731	53396	59723
09858	52104	32014	53115	03727	98624	84616
93307	34116	49516	42148	57740	31198	70336
04794	01534	92058	03157	91758	80611	45357
86265	49096	97021	92582	61422	75890	86442
65943	79232	45702	67055	39024	57383	44424
90038	94209	04055	27393	61517	23002	96560
97283	95943	78363	36498	40662	94188	18202
21913	72958	75637	99936	58715	07943	23748
41161	37341	81838	19389	80336	46346	91895
23777	98392	31417	98547	92058	02277	50315
59973	08144	61070	73094	27059	69181	55623
82690	74099	77885	23813	10054	11900	44653
83854	24715	48866	65745	31131	47636	45137
61980	34997	41825	11623	07320	15003	56774
99915	45821	97702	87125	44488	77613	56823
48293	86847	43186	42951	37804	85129	28993
33225	31280	41232	34750	91097	60752	69783
06846	32828	24425	30249	78801	26977	92074
32671	45587	79620	84831	38156	74211	82752
82096	21913	75544	55228	89796	05694	91552
51666	10433	10945	55306	78562	89630	41230
54044	67942	24145	42294	27427	84875	37022
66738	60184	75679	38120	17640	36242	99357
55064	17427	89180	74018	44865	53197	74810
69599	60264	84549	78007	88450	06488	72274
64756	87759	92354	78694	63638	80939	98644
80817	74533	68407	55862	32476	19326	95558
39847	96884	84657	33697	39578	90197	80532
90401	41700	95510	61166	33757	23279	85523
78227	90110	81378	96659	37008	04050	04228
87240	52716	87697	79433	16336	52862	69149
08486	10951	26832	39763	02485	71688	90936
39338	32169	03713	93510	61244	73774	01245
21188	01850	69689	49426	49128	14660	14143
13287	82531	04388	64693	11934	35051	68576
53609	04001	19648	14053	49623	10840	31915
87900	36194	31567	53506	34304	39910	79630
81641	00496	36058	75899	46620	70024	88753
19512	50277	71508	20116	79520	06269	74173

TABLE XIII   Random Numbers (*continued*)

24418	23508	91507	76455	54941	72711	39406
57404	73678	08272	62941	02349	71389	45605
77644	98489	86268	73652	98210	44546	27174
68366	65614	01443	07607	11826	91326	29664
64472	72294	95432	53555	96810	17100	35066
88205	37913	98633	81009	81060	33449	68055
98455	78685	71250	10329	56135	80647	51404
48977	36794	56054	59243	57361	65304	93258
93077	72941	92779	23581	24548	56415	61927
84533	26564	91583	83411	66504	02036	02922
11338	12903	14514	27585	45068	05520	56321
23853	68500	92274	87026	99717	01542	72990
94096	74920	25822	98026	05394	61840	83089
83160	82362	09350	98536	38155	42661	02363
97425	47335	69709	01386	74319	04318	99387
83951	11954	24317	20345	18134	90062	10761
93085	35203	05740	03206	92012	42710	34650
33762	83193	58045	89880	78101	44392	53767
49665	85397	85137	30496	23469	42846	94810
37541	82627	80051	72521	35342	56119	97190
22145	85304	35348	82854	55846	18076	12415
27153	08662	61078	52433	22184	33998	87436
00301	49425	66682	25442	83668	66236	79655
43815	43272	73778	63469	50083	70696	13558
14689	86482	74157	46012	97765	27552	49617
16680	55936	82453	19532	49988	13176	94219
86938	60429	01137	86168	78257	86249	46134
33944	29219	73161	46061	30946	22210	79302
16045	67736	18608	18198	19468	76358	69203
37044	52523	25627	63107	30806	80857	84383
61471	45322	35340	35132	42163	69332	98851
47422	21296	16785	66393	39249	51463	95963
24133	39719	14484	58613	88717	29289	77360
67253	67064	10748	16006	16767	57345	42285
62382	76941	01635	35829	77516	98468	51686
98011	16503	09201	03523	87192	66483	55649
37366	24386	20654	85117	74078	64120	04643
73587	83993	54176	05221	94119	20108	78101
33583	68291	50547	96085	62180	27453	18567
02878	33223	39199	49536	56199	05993	71201
91498	41673	17195	33175	04994	09879	70337
91127	19815	30219	55591	21725	43827	78862
12997	55013	18662	81724	24305	37661	18956
96098	13651	15393	69995	14762	69734	89150
97627	17837	10472	18983	28387	99781	52977
40064	47981	31484	76603	54088	91095	00010
16239	68743	71374	55863	22672	91609	51514
58354	24913	20435	30965	17453	65623	93058
52567	65085	60220	84641	18273	49604	47418
06236	29052	91392	07551	83532	68130	56970

TABLE XIII   Random Numbers (*continued*)

94620	27963	96478	21559	19246	88097	44926
60947	60775	73181	43264	56895	04232	59604
27499	53523	63110	57106	20865	91683	80688
01603	23156	89223	43429	95353	44662	59433
00815	01552	06392	31437	70385	45863	75971
83844	90942	74857	52419	68723	47830	63010
06626	10042	93629	37609	57215	08409	81906
56760	63348	24949	11859	29793	37457	59377
64416	29934	00755	09418	14230	62887	92683
63569	17906	38076	32135	19096	96970	75917
22693	35089	72994	04252	23791	60249	83010
43413	59744	01275	71326	91382	45114	20245
09224	78530	50566	49965	04851	18280	14039
67625	34683	03142	74733	63558	09665	22610
86874	12549	98699	54952	91579	26023	81076
54548	49505	62515	63903	13193	33905	66936
73236	66167	49728	03581	40699	10396	81827
15220	66319	13543	14071	59148	95154	72852
16151	08029	36954	03891	38313	34016	18671
43635	84249	88984	80993	55431	90793	62603
30193	42776	85611	57635	51362	79907	77364
37430	45246	11400	20986	43996	73122	88474
88312	93047	12088	86937	70794	01041	74867
98995	58159	04700	90443	13168	31553	67891
51734	20849	70198	67906	00880	82899	66065
88698	41755	56216	66852	17748	04963	54859
51865	09836	73966	65711	41699	11732	17173
40300	08852	27528	84648	79589	95295	72895
02760	28625	70476	76410	32988	10194	94917
78450	26245	91763	73117	33047	03577	62599
50252	56911	62693	73817	98693	18728	94741
07929	66728	47761	81472	44806	15592	71357
09030	39605	87507	85446	51257	89555	75520
56670	88445	85799	76200	21795	38894	58070
48140	13583	94911	13318	64741	64336	95103
36764	86132	12463	28385	94242	32063	45233
14351	71381	28133	68269	65145	28152	39087
81276	00835	63835	87174	42446	08882	27067
55524	86088	00069	59254	24654	77371	26409
78852	65889	32719	13758	23937	90740	16866
11861	69032	51915	23510	32050	52052	24004
67699	01009	07050	73324	06732	27510	33761
50064	39500	17450	18030	63124	48061	59412
93126	17700	94400	76075	08317	27324	72723
01657	92602	41043	05686	15650	29970	95877
13800	76690	75133	60456	28491	03845	11507
98135	42870	48578	29036	69876	86563	61729
08313	99293	00990	13595	77457	79969	11339
90974	83965	62732	85161	54330	22406	86253
33273	61993	88407	69399	17301	70975	99129

Answers To Odd-Numbered Exercises

In exercises involving extensive calculations, the reader may well get answers differing somewhat from those given here due to rounding at various intermediate stages.

Chapter 2

2.1 There are various possibilities; one is $160.00–$179.99, $180.00–$199.99, $200.00–$219.99, $220.00–$239.99, $240.00–$259.99, and $260.00–$279.99.

2.3 (a) The class boundaries are -0.005, 24.995, 49.995, 99.995, and 199.995; (b) the class marks are 12.495, 37.495, 74.995, and 149.995; (c) the class intervals are 25, 25, 50, and 100.

2.5 The class limits are 11.0–12.9, 13.0–14.9, 15.0–16.9, 17.0–18.9, 19.0–20.9, and 21.0–22.9; the class marks are 11.95, 13.95, 15.95, 17.95, 19.95, and 21.95.

2.7 (a) The class frequencies are 4, 9, 18, 33, 38, 14, 3, and 1; (b) the cumulative "less than" frequencies are 0, 4, 13, 31, 64, 102, 116, 119, and 120.

2.9 (a) The class frequencies are 3, 5, 7, 11, 15, 7, and 2; (b) the percentages are 6, 10, 14, 22, 30, 14, and 4.

2.11

Maneuverability of the car	Number of drivers
Excellent	4
Very good	10
Good	22
Fair	7
Poor	1
Very poor	1

2.13 The cumulative "less than" frequencies are 0, 13, 77, 142, 186, and 200.

2.15 The cumulative "or more" frequencies are 80, 77, 72, 61, 46, 26, 18, 11, 6, 2, and 0.

2.17 The central angles are 262.8°, 46.8°, 28.8°, and 21.6°.

2.19 (b) The central angles are 32°, 80°, 176°, 56°, 8°, and 8°.

2.21 (a) Yes; (b) no; (c) no; (d) yes.

2.23 (a) The class frequencies are 4, 12, 18, 26, 20, 14, and 6.

2.25 (a) The class boundaries are 99.5, 139.5, 179.5, 219.5, 259.5, and 299.5; (b) the class marks are 119.5, 159.5, 199.5, 239.5, and 279.5; (c) the class interval is 40.

2.29 A year with 11 rainy days in June is not accommodated, and a year with 23 rainy days in June falls into both of the overlapping classes, 18–23 and 23–30.

2.31

Means of transportation	Frequency
Car	20
Train	9
Plane	15
Bus	6

Chapter 3

3.3 46 years.

3.5 (a) Yes; (b) no; (c) yes.

3.7 The means are 33, 27.75, and 20.625; the medians are 33, 28, and 20; and the modes are 32, 33, and 35; 30; and 20.

3.9 The mean is 115.2%, the median is 98%, and the modes are 80%, 98%, and 100%; the median is most suitable, since the mean is unduly affected by the one very large value and the mode is not unique.

3.11 (a) 140 minutes; (b) 142 minutes.

3.15 8.7%.

3.17 58 points.

3.19 63.7.

3.21 $70.04.

3.23 (a) 14; (b) 5; (c) 4; (d) 131.2%.

3.25 $\sigma = 1.71$.

3.27 (a) 0.4 ounce; (b) 0.18 ounce.

3.29 (a) 325 and $s = 113.35$ (or $\sigma = 107.54$);
(b) 265 and $s = 106.32$ (or $\sigma = 95.10$).

3.31 For protein, $\bar{x} = 72.28$, range $= 3.5$, and $s = 1.38$; for moisture, $\bar{x} = 4.56$, range $= 3.0$, and $s = 1.11$; for ash, $\bar{x} = 22.46$, range $= 3.1$, and $s = 1.17$; and for fat, $\bar{x} = 0.70$, range $= 0.5$, and $s = 0.20$.

3.33 Range $= 51$ and $s = 13.64$.

3.35 (b) $s = 0.678$.

3.37 $z = 1.68$ and $z = 3.00$; stock B is relatively more variable than stock A.

3.39 (a) $\frac{8}{9}$; (b) between 44 and 84.

3.41 (a) 5.85%; (b) 2.82%; (c) the first set of measurements is relatively more variable.

3.43 (a) $\tilde{x} = 6.24$; (b) mode $= 5.95$ (midpoint of the modal class).

3.45 $\bar{x} = 32.65$ and $s = 6.91$.

3.47 0.282.

3.49 (a) $\bar{x} = 66.94$; (b) $\tilde{x} = 66.0$; (c) $s = 10.48$; (d) $SK = 0.269$.

3.51 (a) $\bar{x} = 68.395$ and $\tilde{x} = 68.230$; (b) $s = 19.73$; (c) 52.66 and 83.84; (d) 42.66 and 95.38; (e) 59.33 and 77.05.

3.53 (a) 32.35, 4.93, and 15.24%; (b) 68.25, 15.59, and 22.84%; (c) 56.6, 10.2, and 18.0%.

3.57 (a) $x_1 + x_2 + \cdots + x_9$; (b) $(x_1 - k) + (x_2 - k) + \cdots + (x_6 - k)$;
(c) $x_3 y_3 + x_4 y_4 + x_5 y_5 + x_6 y_6 + x_7 y_7$; (d) $y_4^2 + y_5^2 + y_6^2 + y_7^2 + y_8^2$;
(e) $3x_2^2 + 3x_3^2 + 3x_4^2 + 3x_5^2$; (f) $(x_1 - z_1) + (x_2 - z_2) + \cdots + (x_n - z_n)$.

3.59 (a) -7; (b) 31; (c) -15; (d) -10; (e) 86; (f) -33.

3.63 (a) $\bar{x} = \dfrac{\sum\limits_{i=1}^{n} x_i}{n}$; (b) $\bar{x}_w = \dfrac{\sum\limits_{i=1}^{n} w_i x_i}{\sum\limits_{i=1}^{n} w_i}$; (c) $\bar{\bar{x}} = \dfrac{\sum\limits_{i=1}^{k} n_i \bar{x}_i}{\sum\limits_{i=1}^{k} n_i}$.

3.65 (a) 75%; (b) 93.75%; (c) 99%.

3.67 $2.95.

3.69 (a) The mean and the median can both be found; (b) the median can be found, but not the mean; (c) neither the mean nor the median can be found.

3.71 15.8 minutes.

3.73 0.456.

3.77 (a) $4.41; (b) $2.68.

3.79 The consensus opinion, or mode, is 3.

3.81 $0.80 per tape.

3.83 $\bar{x} = 39.6$, $\tilde{x} = 36.7$, mode $= 37$, and $s = 20.1$.

3.85 (a) 280 gallons; (b) 1.79 mpg.

3.87 $\bar{x} = 12.46\%$, $\tilde{x} = 12.0\%$, and the range is 7.2%.

3.89 (a) 15.6; (b) 14.0; (c) 38.64.

3.91 (a) 2/3; (b) $45,000.

Chapter 4

4.1 100.0, 107.3, 123.3, and 226.1.

4.3 (a) 98.1; (b) 100.3.

4.5 (a) 96.6; (b) 96.6.

4.7 (a) 63.8, 74.4, 131.2, 45.8, and 77.7; (b) 57.6.

4.9 (b) 88.7; (c) 103.6.

4.13 133.2.

4.15 (a) 93.0 and 148.2; (b) 146.1; (c) 146.9; (d) 148.3; (e) 147.8; (f) 122.0.

4.17 104.3.

4.25 (a) 100.0, 96.1, 92.7, 97.8, and 94.9; (b) 105.4, 101.3, 97.7, 103.1, and 100.0.

4.27 (a) 310.25, 319.60, 318.77, 323.10, and 324.90; (b) 100.0, 94.6, 87.7, 82.0, and 73.7; (c) 42.0%, 4.7%, 35.6%, and 26.3%.

4.29 $162.71.

4.33 100.0, 99.5, 99.0, and 98.3; 101.7, 101.2, 100.7, and 100.0.

4.35 No, wages increased by 18% while consumer prices increased by 25%.

4.37 (a) 72.4; (b) 75.3.

Chapter 5

5.1 (a) 3; (b) 2.

5.7 6.

5.9 (a) 343; (b) 9; (c) 20.

5.11 120.

5.13 32,768.

5.15 120.

5.17 43,680.

5.19 24.

5.21 (b) 120; (c) 6.

5.23 (a) 1,260; (b) 20; (c) $\dfrac{n!}{r_1!r_2!r_3!}$, 90; (d) 83,160.

5.25 1,001.

5.27 7,920.

5.29 (a) 286; (b) 156; (c) 13.

5.31 1, 6, 15, 20, 15, 6, and 1; 1, 7, 21, 35, 35, 21, 7, and 1.

5.33 (a) 2/52; (b) 12/52; (c) 26/52; (d) 16/52.

5.35 1/8, 3/8, 3/8, and 1/8.

5.37 (a) 1/3; (b) 1/2.

5.39 3/95.

5.41 (a) 1/4; (b) 1/22.

5.43 0.85.

5.45 0.25.

5.47 20 cents.

5.49 $2.50.

5.51 (a) $16,000 and $16,000; (b) $18,000 and $14,000.

5.53 $0.82.

5.55 1.43.

5.57 No, the mathematical expectation is only $3.50.

5.59 $p > 0.50$.

5.61 (a) The first job; (b) the second job.

5.63 504.

5.65 720.

5.67 In 5 ways.

5.69 The probability is 0.16.

5.71 11,880.

5.73 $126,400.

5.75 1/12.

5.77 15.

5.79 40,320.

5.81 840.

Chapter 6

6.1 (a) $B' = \{5, 6, 7, 8, 9, 10\}$ is the event that a person will pick a number greater than 4; (b) $A \cup C = \{3, 4, 5, 6, 7, 8, 9\}$ is the event that a person will pick a number from 3 to 9; (c) $A \cap B = \{3, 4\}$ is the event that a person will pick 3 or 4; (d) $B' \cap C' = \{5, 10\}$ is the event that a person will pick 5 or 10.

6.3 (a) The complaint is about baggage handling or carry-on facilities; (b) the complaint is about seats or leg room, or carry-on facilities; (c) the complaint is about baggage handling, ticketing and boarding, or seats or leg room.

6.5 (a) The two contractors get equally many jobs; (b) between them, the two contractors get both jobs; (c) the first contractor does not get either job.

6.7 (b) Q is the event that, between them, the two helicopters carry at least five passengers; R is the event that there are equally many passengers in the two helicopters; T is the event that the smaller helicopter carries more passengers than the larger helicopter; U is the event that, between them, the two helicopters carry three passengers.

6.9 (b) $K = \{(2, 2), (3, 2), (3, 3)\}$, $L = \{(1, 0), (1, 1)\}$, and $M = \{(0, 0), (1, 1), (2, 2), (3, 3)\}$; (c) $M' = \{(1, 0), (2, 0), (2, 1), (3, 0), (3, 1), (3, 2)\}$ is the event that not all of the sales persons at work are busy with customers; $M \cap L = \{(1, 1)\}$ is the event that only one sales person is at work and busy with a customer; (d) only K and L are mutually exclusive.

6.11 (a) Not mutually exclusive; (b) mutually exclusive; (c) not mutually exclusive; (d) mutually exclusive.

6.13 Region 1 represents the event that the unemployment rate will go down and the inflation rate will go up; region 2 represents the event that the unemployment rate will go down and the inflation rate will not go up; region 3 represents the event that the unemployment rate will not go down and the inflation rate will go up; region 4 represents the event that the unemployment rate will not go down and the inflation rate will not go up.

6.15 Region 1 represents the event that a person arrested for car theft can afford to pay a lawyer and is found guilty of the crime; region 2 represents the event that a person arrested for car theft can afford to pay a lawyer and is not found guilty of the crime; region 3 represents the event that a person arrested for car theft cannot afford to pay a lawyer and is found guilty of the crime; region 4 represents the event that a person arrested for car theft cannot afford to pay a lawyer and is not found guilty of the crime.

6.17 (a) The engine's gasoline consumption will be low, its maintenance cost will be low, but it cannot be sold at a profit; (b) the engine's gasoline consumption will be low and it can be sold at a profit; (c) the engine's maintenance cost will be low, but it cannot be sold at a profit; (d) the engine's gasoline consumption will be low or it can be sold at a profit, but its maintenance cost will not be low; (e) the engine cannot be sold at a profit.

6.19 (a) The students will see the PR director and the R&D department and get very tired; (b) the students will see the R&D department but not the PR director and they will not get very tired; (c) the students will see neither the PR director nor the R&D department, nor will they get very tired; (d) the students will see the R&D department but not the PR director; (e) the students will see the PR director and get very tired; (f) the students will see neither the PR director nor the R&D department.

6.21 (a) The probability that a power plant does not use coal; (b) the probability that a power plant is not able to provide enough electricity; (c) the probability that a power plant uses coal, is able to provide enough electricity, or both; (d) the probability that a power plant uses coal and is able to provide enough electricity; (e) the probability that a power plant does not use coal, is able to provide enough electricity, or both; (f) the probability that a power plant does not use coal and is not able to provide enough electricity.

6.23 (a) The sum of the probabilities exceeds 1; (b) the sum of the probabilities is less than 1; (c) the second probability cannot be less than the first; (d) the second probability cannot be greater than the first; (e) the sum of the probabilities is not equal to 1.

6.25 (a) 0.66; (b) 0.42; (c) 0.58.

6.27 0.83.

6.29 (a) 0.35; (b) 0.40; (c) 0.80.

6.31 (a) 0.260; (b) 0.985; (c) 0.361.

6.33 0.95.

6.35 (a) 0.13; (b) 0.98; (c) 0.11.

6.37 (a) 0.80; (b) 0.98.

6.39 (a) Odds are 3 to 2; (b) odds are 3 to 13; (c) odds are 21 to 11.

6.41 Odds are 11 to 5.

6.43 1/6.

6.45 Less than 1/4.

6.47 Greater than 0.88 but at most 0.90.

6.49 Odds are 11 to 9.

6.51 (a) The probability that an applicant who is employed will have his application approved; (b) the probability that an applicant who has his application approved is married; (c) the probability that an applicant who is not married will have his application approved; (d) the probability that an applicant who is not employed will not have his application approved; (e) the probability that an applicant who is employed and married will have his application approved;

(f) the probability that an applicant who is not married will be employed and will have his application approved.

6.53 (a) 0.25; (b) 0.10; (c) 0.55; (d) 0.45.

6.57 (a) 188/250; (b) 50/250; (c) 22/250; (d) 28/250; (e) 228/250; (f) 22/50; (g) 160/200; (h) 40/62; (i) 28/188.

6.59 0.915.

6.63 3/4.

6.65 (a) 297/632; (b) 60/632; (c) 275/632.

6.67 1/64.

6.69 (a) 2/81; (b) 80/81.

6.71 (a) Odds are 11 to 2; (b) odds are 1 to 3.

6.73 D and F are not mutually exclusive; E and G are mutually exclusive.

6.75 0.00105.

6.77 Independent.

6.79 (a) $K' = \{5, 6\}$; (b) $K \cup L = \{1, 2, 3, 4, 5\}$; (c) $K \cap L = \{2, 3, 4\}$; (d) $K \cap L' = \{1\}$.

6.81 0.43.

6.83 3/8.

6.85 (a) 3/5; (b) 7/10; (c) 1/5; (d) 3/10; (e) 1/3; (f) 3/7.

6.87 (a) 0.65; (b) 0.39; (c) 0.96; (d) 0.04.

6.89 (a) The team is rated among top twenty by AP and by UPI; (b) the team is rated among top twenty by AP but not by UPI; (c) the team is rated among top twenty by UPI but not by AP; (d) the team is not rated among top twenty by either AP or UPI.

6.91 0.34.

6.93 1/4,096.

6.95 (a) Visits customers in Blythe, but not in Palm Springs or El Centro; (b) visits customers in Blythe or Palm Springs; (c) visits customers in Blythe but not in El Centro; (d) visits customers in Blythe or El Centro (or both), but not in Palm Springs; (e) visits customers in neither Palm Springs nor El Centro.

Chapter 7

7.1 (a)

	Straight fee	Contingent fee
Wins	4,000	20,000
Loses	4,000	0

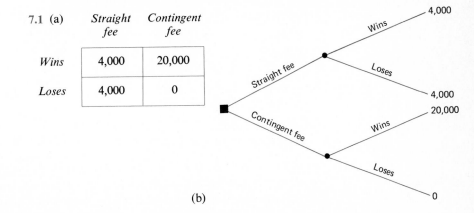

(b)

7.3 (a)

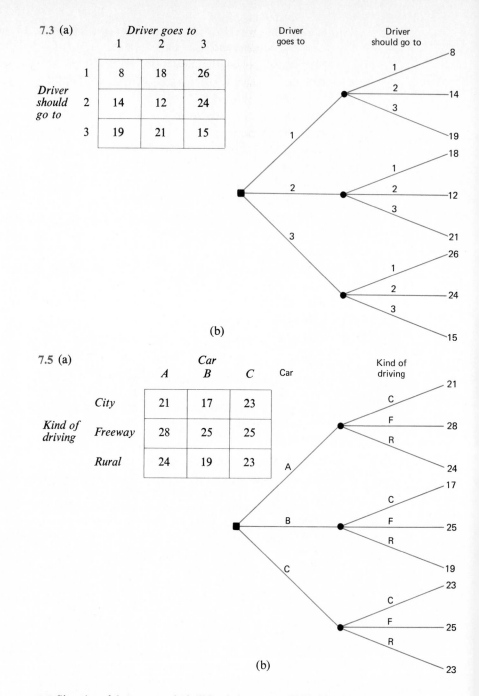

	Driver goes to		
	1	2	3
Driver should go to — 1	8	18	26
2	14	12	24
3	19	21	15

(b)

7.5 (a)

	Car		
	A	B	C
Kind of driving — City	21	17	23
Freeway	28	25	25
Rural	24	19	23

(b)

7.7 Sites 1 and 3 are not admissible; it is not possible to make a second choice.

7.9 (a) Supermarket **B** is not admissible; (b) supermarket **C**.

7.11 (a) Straight fee; (b) contingent fee.

7.13 (a) 1; (b) 3; (c) 2.

7.15 (a) Inspect both; (b) inspect neither.

7.17 (a) A_1 and E_2; (b) A_1 and E_1; (c) A_2 and E_1; (d) A_3 and E_2.

7.19 (a) A_1 and E_2; (b) A_1 and E_3, A_1 and E_4, A_3 and E_3, and A_3 and E_4; in each case the value is 5.

7.23 (a) 5/11 and 6/11; (b) 4/11 and 7/11.

7.25 1/7, 0, and 6/7.

7.27 (a) First station owner should lower his price; (b) they might take turns lowering their prices on alternate days.

7.29 0.05.

7.31 1/12 or 0.083.

7.33 (a) 2; (b) 3.

7.35 (a) $3; (b) $3.

7.37 (a) Yes; (b) yes.

7.39 (a) $pb + (1 - p)c$; (c) $p(b - a)$ and $(1 - p)(c - d)$.

7.41 $16.50.

7.43 $U = 4$.

7.45 Approximately 1.5.

7.49 Delay expansion.

7.51

	Goes to	
	LJ	*MB*
LJ	11	13
MB	15	9

Should go to (row labels for LJ and MB)

7.53 (a) La Jolla; (b) Mission Beach; (c) does not matter.

7.55 (a)

	Build	*Do not build*
Good team	240,000	120,000
Poor team	−40,000	20,000

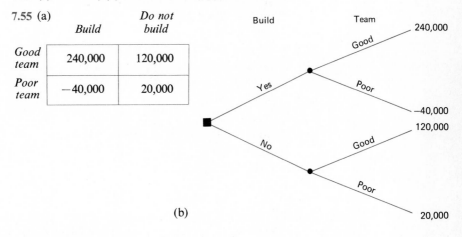

(b)

7.57 Use old stadium.

7.59 $U > 6$.

7.61 (a)

	Accept	*Reject*
Good risk	4,112	0
Bad risk	−655	0

(b) reject the applicant; (c) \$411.20; (d) yes.

7.63 (a) 6/11 and 5/11; (b) 7/11 and 4/11.

7.65 \$34.

Chapter 8

8.1 (a) No, sum of the probabilities is less than 1; (b) no, sum of the probabilities is greater than 1; (c) yes; (d) no, f(3) is negative; (e) no, sum of the probabilities is less than 1.

8.3 (a) 0.2916; (b) 0.292.

8.5 (a) 0.2835; (b) 0.283.

8.7 75/216, 15/216, 1/216, and 125/216; −7.9 cents.

8.9 1/3.

8.11 (a) 0.196; (b) 0.017; (c) 0.151; (d) 0.301.

8.13 0, 0, 0, 0, 0, 0, 0, 0, 0, 0, 0.004, 0.026, 0.123, 0.359, and 0.488.

8.15 (a) 0.0864; (b) 0.0791.

8.17 0.379.

8.19 0.758.

8.21 0.363, 0.484, 0.145, and 0.009.

8.23 (a) 0.1167; (b) 0.1148; error is 0.0019.

8.25 0.225.

8.27 0.125.

8.29 0.135, 0.270, and 0.270.

8.31 0.049.

8.33 $\mu = 1.0$ and $\sigma^2 = 1.0$.

8.35 (a) $\sigma^2 = 1.0$; (b) $\sigma = 1.37$.

8.37 (a) 1; (b) 1.

8.39 (a) $\mu = 0.498$ and $\sigma = 0.666$; (b) $\mu = 0.5$ and $\sigma = 0.67$.

8.41 (a) $\mu = 288$ and $\sigma = 12$; (b) $\mu = 67.5$ and $\sigma = 7.5$; (c) $\mu = 340$ and $\sigma = 7.14$; (d) $\mu = 96$ and $\sigma = 9.6$; (e) $\mu = 240$ and $\sigma = 12.96$.

8.43 0.918.

8.45 $\mu = 0.799$.

8.47 (a) $\mu = 3.997$; (b) $\sigma = 2.0007$.

8.49 At least 0.96.

8.53 (a) 0.238; (b) 0.262.

8.55 $\mu = 1.25$.

8.57 0.063.

8.59 (a) $\frac{1,001}{4,845}, \frac{2,184}{4,845}, \frac{1,365}{4,845}, \frac{280}{4,845}$, and $\frac{15}{4,845}$; (b) 1.2.

8.61 (a) $\mu = 3$ and $\sigma^2 = 1.5$; (b) $\mu = 3$ and $\sigma^2 = 1.5$.

8.63 (a) The probability is at least 80/81 that between 66 and 202 marriage licenses will be issued; (b) at least 63/64.

8.65 (a) 0.007; (b) 0.668; (c) 0.748.

8.67 At least 15/16.

8.69 (a) 0.107; (b) 0.008; (c) 0.0001; (d) 0.115.

8.71 (a) 10/21, 10/21, and 1/21; (b) 4/7; (c) 20 to 1.

Chapter 9

9.1 0.532.

9.3 0.758.

9.5 (a) 19/60; (b) 18/19; (c) 14/41.

9.7 88/179.

9.9 Purposeful action is the most likely cause.

9.11 (a) 0.50, 0.45, and 0.05; (b) 0.72, 0.18, and 0.10.

9.13 0.12, 0.05, 0.04, 0.11, and 0.68.

9.15 (a) 0.760, 0.173, and 0.967.

9.17 (a) Buy; (b) buy; (c) $30.96.

9.19 (a) Reject; (b) accept.

9.21 0.391.

9.23 0.72.

9.25 (a) Probability that economic conditions will remain good increases and the probability that there will be a recession decreases; (b) expand now.

9.27 Second assessment becomes nine times as likely as first assessment.

9.29 0.75, 0.18, and 0.07.

9.31 (a) Buy; $2.92; (b) buy.

Chapter 10

10.1 (a) 1/4; (b) 5/8; (c) 0.7625.

10.3 (a) 0.4176; (b) 0.8102; (c) 0.1192; (d) 0.0835.

10.5 (a) 1.63; (b) 2.22; (c) -1.00; (d) 0.44; (e) -0.50; (f) 2.17.

10.9 (a) 0.5987; (b) 0.1498.

10.11 (a) 0.330, 0.549, and 0.148; (b) 0.393 and 0.085; (c) 0.369.

10.13 (a) 27.76%; (b) 37.45%; (c) 34.94%; (d) 17.01 ounces; (e) 20.77 ounces.

10.15 Supplier A.

10.17 (a) 0.0037; (b) 0.0078.

10.19 (a) 0.1210; (b) 0.0735.

10.21 0.2266.

10.23 0.0427.

10.25 0.0465.   .8664

10.27 (a) 0.0030; (b) 0.1587; (c) 0.9644.

10.29 (a) 0.3289; (b) 0.9938; (c) 0.7734; (d) 0.0749; (e) 0.2266.

10.31 (a) 2.05; (b) 1.28.

10.33 (a) 1/16; (b) 3/4; (c) 1/4.

10.35 0.2029.

10.37 (a) 0.3944; (b) 0.5820; (c) 0.1298; (d) 0.8002; (e) 0.0212.

10.39 (a) 0.53 or -0.53; (b) -1.18; (c) 1.83; (d) 0.34 or -0.34.

11.1 (a) 10; (b) 66; (c) 300.

11.3 *pq, pr, ps, pt, qr, qs, qt, rs, rt,* and *st*.

11.5 44, 326, 558, 353, 577, 305, 24, 189, 285, 442, 569, and 555.

11.7 5605, 1451, 2582, 0935, 2431, 0574, 5804, 3534, 6107, 0113, 1860, 2562, 3534, 1678, 1448, 1074, 0163, 0920, 2056, and 5417.

11.11 (b) 6 and 7, 6 and 8, 6 and 9, 6 and 10, 6 and 11, 7 and 8, 7 and 9, 7 and 10, 7 and 11, 8 and 9, 8 and 10, 8 and 11, 9 and 10, 9 and 11, and 10 and 11; means are 6.5, 7, 7.5, 8, 8.5, 7.5, 8, 8.5, 9, 8.5, 9, 9.5, 9.5, 10, and 10.5;

(c)

Mean	Probability
6.5	1/15
7.0	1/15
7.5	2/15
8.0	2/15
8.5	3/15
9.0	2/15
9.5	2/15
10.0	1/15
10.5	1/15

(d) 8.5 and 7/6.

11.13 (a) 6 and 6, 6 and 7, 6 and 8, 6 and 9, 6 and 10, 6 and 11, 7 and 6, 7 and 7, 7 and 8, 7 and 9, 7 and 10, 7 and 11, 8 and 6, 8 and 7, 8 and 8, 8 and 9, 8 and 10, 8 and 11, 9 and 6, 9 and 7, 9 and 8, 9 and 9, 9 and 10, 9 and 11, 10 and 6, 10 and 7, 10 and 8, 10 and 9, 10 and 10, 10 and 11, 11 and 6, 11 and 7, 11 and 8, 11 and 9, 11 and 10, and 11 and 11; (b) the probabilities of 6.0, 6.5, 7.0, 7.5, 8.0, 8.5, 9.0, 9.5, 10.0, 10.5, and 11.0 are 1/36, 2/36, 3/36, 4/36, 5/36, 6/36, 5/36, 4/36, 3/36, 2/36, and 1/36; (c) 8.5 and $\sqrt{35/24}$.

11.15 11, 12, 13, 14, 15, 16, 17, 18, 19, 20, and 21 occur 1, 1, 4, 4, 12, 12, 5, 8, 1, 1, and 1 times; their standard deviation is 1.97.

11.17 $n = 100$.

11.19 (a) Divided by 3; (b) divided by 1.5; (c) multiplied by 5.

11.23 0.95.

11.25 (a) 0.5762; (b) 0.9722.

11.35 (a) 2 and 2, 2 and 4, 2 and 6, 2 and 8, 4 and 2, 4 and 4, 4 and 6, 4 and 8, 6 and 2, 6 and 4, 6 and 6, 6 and 8, 8 and 2, 8 and 4, 8 and 6, and 8 and 8; means are 2, 3, 4, 5, 3, 4, 5, 6, 4, 5, 6, 7, 5, 6, 7, and 8;

Mean	Probability
2	1/16
3	2/16
4	3/16
5	4/16
6	3/16
7	2/16
8	1/16

(b) 5 and $\sqrt{5/2}$.

11.37 (a) 0.870; (b) 0.981.

11.39 (a) Divided by 5; (b) multiplied by 3.5.

11.41 (a) 3,060; (b) 27,405; (c) 3,921,225.

11.43 A+ and A, A+ and A−, A+ and B+, A+ and B, A+ and B−, A and A−, A and B+, A and B, A and B−, A− and B+, A− and B, A− and B−, B+ and B, B+ and B−, and B and B−.

11.45 (a) Divided by 1.5; (b) multiplied by 2.5.

11.49 010, 245, 039, 076, 065, 173, 070, 103, 233, and 087.

Chapter 12

12.1 0.76 minute.

12.3 (a) $417.81 < \mu <$ $447.31; (b) $19.38.

12.5 $13.58 < \mu < 15.82$.

12.7 (a) 25.33; (b) $236.85 < \mu < 243.15$.

12.9 711.

12.11 41.

12.13 $62.94 < \mu < 66.66$.

12.15 (a) 0.47 pound; (b) $23.72 < \mu < 25.28$.

12.17 0.9273.

12.19 0.9608.

12.21 We would commit a Type I error if we erroneously reject the hypothesis that the device is effective; we would commit a Type II error if we erroneously accept the hypothesis that the device is effective.

12.25 (a) 0.0456; (b) 0.0228.

12.29 (a) 0.0164; (b) 0.0179, 0.2743, and 0.8159.

12.31 (a) The hypothesis that the clerk averages as many mistakes as all the clerks and the alternative that the clerk averages more; (b) the hypothesis that the clerk averages as many mistakes as all the clerks and the alternative that the clerk averages fewer.

12.35 $z = -1.94$; reject the rental agent's claim.

12.37 (a) $z = -3.50$; reject the null hypothesis against the alternative $\mu < 9$.

12.39 $z = 2.70$; reject the null hypothesis against the alternative $\mu > 9.5$.

12.41 (a) $\bar{x} < 14.23$; (b) 0.3974.

12.43 $t = -3.04$; null hypothesis cannot be rejected.

12.45 $z = 4.00$; the lot does not meet the manufacturer's guarantee.

12.47 $t = -3.54$; reject the null hypothesis against the alternative that the machine is underfilling cups.

12.49 $t = -2.17$; the difference is significant.

12.51 $t = 1.11$; difference is not significant.

12.53 $t = 1.29$; difference is not significant.

12.55 $t = 2.44$; the safety program is effective.

12.57 (a) Central line is at 3.0, the LCL is at 2.973, and the UCL is at 3.027; (b) the process was out of control at the times of samples 2, 8, 9, 10, 11, and 14.

12.59 (a) $\mu_1 = 70.2$ and $\sigma_1 = 0.85$; (b) 0.5778.

12.61 $z = -2.42$; null hypothesis cannot be rejected.

12.63 $\bar{x} < 3.975$.

12.65 $t = 2.11$; null hypothesis cannot be rejected; even though the difference between $\bar{x}$ and μ has increased, the estimate of the standard deviation in the denominator has been "inflated" by the error.

12.67 $t = 2.23$; difference is not significant.

12.69 0.67 microgram.

12.71 (a) $\mu < 20$ and make the modification only if the null hypothesis can be rejected; (b) $\mu > 20$ and make the modification unless the null hypothesis can be rejected.

12.73 $n = 250$.

12.75 (a) $\bar{x} > 102.47$; (b) 0.6217, 0.1539, and 0.0094.

12.77 $t = -2.68$; claim must be rejected.

Chapter 13

13.1 $1.10 < \sigma < 2.92$.

13.3 $1.25 < \sigma^2 < 19.28$.

13.5 $55.93 < \sigma < 84.51$.

13.7 (a) 1.98 ($s = 1.79$); (b) 0.43 ($s = 0.38$).

13.9 $\chi^2 = 43.74$; reject the null hypothesis.

13.11 $\chi^2 = 45.36$; reject the null hypothesis.

13.13 $z = 4.53$; reject the null hypothesis.

13.15 $F = 2.86$; reject the null hypothesis.

13.17 $F = 1.40$; cannot reject the null hypothesis.

13.19 $13.1 < \sigma < 28.2$.

13.21 (a) 613.5; (b) 656.8.

13.23 $F = 1.33$; cannot reject the null hypothesis.

Chapter 14

14.1 $0.41 < p < 0.55$.

14.3 $0.33 < p < 0.53$.

14.5 $0.67 < p < 0.75$.

14.7 $0.14 < p < 0.26$.

14.9 Can assert with 80% confidence that error is less than 0.039.

14.11 $n = 1,037$.

14.13 $n = 1,509$.

14.15 (a) $0.29 < p < 0.39$; (b) $0.18 < p < 0.42$.

14.17 At least 10.

14.19 At least 3.

14.21 At least 6.

14.23 $z = -1.45$; cannot reject the claim.

14.25 $z = 3.20$; reject the null hypothesis.

14.27 $z = -0.11$; there is no evidence of a lack of randomness.

14.29 (a) Central line is at 0.10, LCL is at 0.027, and UCL is at 0.173; (b) process was out of control at the times of the 11th and 12th samples.

14.31 $z = -3.22$; the more expensive one is more effective.

14.33 $z = 2.32$; the two machines are not equally good.

14.35 $z = -6.27$; reject the null hypothesis.

14.37 $\chi^2 = 7.49$; cannot reject the null hypothesis.

14.39 $\chi^2 = 9.32$; differences among sample proportions are significant.

14.43 $\chi^2 = 35.17$; reject the null hypothesis.

14.45 $\chi^2 = 7.037$; there is no relationship.

14.47 $\chi^2 = 94.92$; there is a relationship.

14.49 (a) $\sqrt{2/3} = 0.82$; (b) 0.28; (c) 0.40.

14.51 $\chi^2 = 9.06$; reject the claim.

14.53 $\chi^2 = 9.23$; there is a good fit.

14.55 $\chi^2 = 6.23$; yes.

14.57 $\chi^2 = 7.33$; differences may well be due to chance.

14.59 $z = 3.36$; there is a real difference.

14.61 $z = -2.08$; cannot reject the claim.

14.63 $n = 1,537$.

14.65 $\chi^2 = 9.23$; differences among sample proportions are significant.

14.67 At least 6; actual level of significance is 0.03.

14.69 $z = -1.00$; cannot reject the null hypothesis.

14.71 $\chi^2 = 4.32$; null hypothesis cannot be rejected.

14.73 (c) The confidence interval of part (a) is narrower than that of part (b).

Chapter 15

15.1 (a) $F = 6.54$; reject the null hypothesis; (b) same results.

15.3 $F = 0.08$; cannot reject the null hypothesis.

15.5 $F = 3.19$; cannot reject the null hypothesis.

15.9 For forms, $F = 1.24$, which is not significant; for students, $F = 45.62$, which is significant.

15.11 For detergents, $F = 5.74$, which is not significant; for temperatures, $F = 0.64$, which is not significant.

15.13 $F = 1.52$; differences may well be due to chance.

15.15 (a) $F = 2.93$; differences may well be due to chance; (b) same result.

15.17 $F = 0.68$; differences may well be due to chance.

Chapter 16

16.1 $x = 8$; cannot reject the null hypothesis.

16.3 $x = 7$; null hypothesis must be rejected.

16.5 $z = -2.45$; null hypothesis must be rejected.

16.7 $z = -3.40$; reject the null hypothesis.

16.9 $x = 9$; null hypothesis cannot be rejected.

16.11 $U = 6$; null hypothesis cannot be rejected.

16.13 $U = 5.5$; null hypothesis must be rejected.

16.15 $z = -1.34$; null hypothesis cannot be rejected.

16.17 $z = -1.51$; null hypothesis cannot be rejected.

16.19 $H = 0.245$; null hypothesis cannot be rejected.

16.21 $H = 4.86$; null hypothesis cannot be rejected.

16.23 $u = 5$; arrangement is not random.

16.25 $z = -0.14$; null hypothesis cannot be rejected.

16.27 $z = 1.46$; arrangement appears to be random.

16.29 $u = 5$; reject the null hypothesis of randomness.

16.31 $z = -0.51$; null hypothesis of randomness cannot be rejected.

16.33 $H = 4.51$; null hypothesis cannot be rejected.

16.35 $z = -4$; there is a significant trend.

16.37 $z = -0.89$; hypothesis of randomness cannot be rejected.

16.39 $u = 7$; arrangement appears to be random.

16.41 $U = 41$; null hypothesis must be rejected.

16.43 $u = 11$; hypothesis of randomness cannot be rejected.

16.45 $z = 1.73$; null hypothesis cannot be rejected.

16.47 $z = 1.33$; conclusion is reasonable.

Chapter 17

17.1 (a) $\hat{y} = 11.445 + 0.898x$.

17.3 $\hat{y} = 5.190 + 1.165x$.

17.5 (a) $\hat{y} = 1.836 + 0.403x$; (b) 3.851.

17.7 $\hat{y} = 0.73 + 1.46x$; 2.48.

17.9 (a) $\hat{y} = 0.576 - 0.157x$; (b) 0.26.

17.11 $t = -3.26$; reject the null hypothesis.

17.13 $t = -1.98$; null hypothesis cannot be rejected.

17.15 $\hat{y} = 6.026 + 1.493x$; $t = 3.8$; reject the null hypothesis.

17.17 $-0.60 < \beta < 3.73$.

17.19 $0.44 < \alpha < 0.71$.

17.21 (a) 2.34 and 3.51; (b) \$89,151 and \$102,061.

17.23 $\hat{y} = -6,279 + 961x_1 + 2,976x_2$; \$44,065.

17.27 (a) $\hat{y} = 14.29 + 30.15x_1 + 12.26x_2$; \$101.36. (Answer may be strongly affected by rounding.)

17.29 $t = -1.48$; null hypothesis cannot be rejected.

17.31 41.27 and 61.27.

17.33 $\hat{y} = 2.27 + 0.22x_1 + 0.062x_2$; 8.3.

Chapter 18

18.1 $r = 0.32$; null hypothesis cannot be rejected.

18.3 $r = 0.885$.

18.5 (a) $r = 0.988$; (b) there is a relationship.

18.7 84.64%.

18.11 (a) Not significant; (b) significant; (c) significant.

18.13 (a) 0.13; (b) 0.81; (c) 0.61; (d) -0.18; (e) only those of parts (b) and (c) are significant; (f) the impact seems to appear after two years and then taper off after the third year.

18.17 $r_S = 0.89$.

18.19 $r_S = 0.98$ $(r = 0.98)$.

18.21 $r_S = -0.60$.

18.23 $r_S = 0.75$.

18.25 0.58.

18.27 0.999.

18.29 Second is three times as strong as first.

18.31 (a) Significant; (b) not significant.

18.33 Correlation is significant.

18.37 $r = -0.98$; significant.

18.39 $r_S = 0.84$.

18.41 Not significant.

Chapter 19

19.1 (b) $\hat{y} = 5.2 + 1.0x$ (origin, 1979; x units, 1 year; y, annual number of projects begun).

19.3 (b) $\hat{y} = 7.18 + 1.28x$ (origin, 1977; x units, 1 year; y, annual gross incomes in billions of dollars); (c) $\hat{y} = 9.74 + 1.28x$ (origin, 1979; x units, 1 year; y, annual gross incomes in billions of dollars).

19.5 (a) $\hat{y} = 2.856 + 0.405x$ (origin, 1976–77; x units, 1 year; y, net investments in millions of dollars); (b) $\hat{y} = 3.868 + 0.405x$; (c) $\hat{y} = 2.046 + 0.405x$.

19.7 $\hat{y} = 95.925 + 0.744x$ (origin, January 1971; x units, 1 month; y, average monthly revenues in millions of dollars).

19.9 $\hat{y} = 3.51 + 0.31x + 0.028x^2$ (origin, 1974; x units, 1 year; y, annual net sales in billions of dollars).

19.11 87.7, 96.8, 107.3, 119.1, 132.2, 146.6, 162.4, 179.5, 198.0, 217.8, 238.9, 261.4, 285.2, 310.3, and 336.8.

19.13 $\log \hat{y} = 2.3634 + 0.1306x$ or $\hat{y} = 231(1.35)^x$ (origin, 1971; x units, 1 year; y, values of index).

19.17 (a) 11.0, 10.7, 11.7, 12.7, 13.7, 13.7, 14.3, 15.3, 16.7, 18.7, 19.0, 21.0, 21.7, 24.7, 26.0, 27.7, 28.3, 30.3, 31.0, 31.7, 32.7, 34.3, 34.7, 34.0, 33.0, 33.3, 30.0, and 29.0; (b) 11.6, 11.8, 12.4, 13.2, 14.0, 14.6, 15.6, 17.0, 17.8, 19.6, 20.8, 22.6, 24.0, 26.0, 27.0, 29.2, 29.8, 30.4, 32.2, 33.4, 33.0, 34.0, 34.2, 33.4, 31.4, and 30.8.

19.19 39.9, 41.2, 42.7, 43.7, 43.9, 43.7, 43.5, 43.5, 43.6, 43.6, 43.4, 43.3, 43.1, 42.8, 42.6, 42.1, 42.2, 42.4, 42.9, 42.9, 43.5, 44.9, 46.3, and 46.3.

19.21 $S_4 = \alpha y_4 + \alpha(1 - \alpha)y_3 + \alpha(1 - \alpha)^2 y_2 + (1 - \alpha)^3 y_1$.

19.23 (a) 41.6, 41.9, 41.4, 41.6, 40.5, 40.3, 40.4, 38.4, 38.1, 38.8, and 39.2; (b) 41.6, 42.8, 40.6, 41.6, 37.2, 37.6, 38.9, 31.5, 33.2, 37.9, and 39.7.

19.25 (a) 19, 19.4, 18.0, 18.8, 21.3, 22.4, 23.8, 21.9, 19.5, 20.5, 20.3, 21.8, 20.7, and 21.6; (b) 24, 23.2, 21.8, 21.4, 22.1, 22.5, 23.2, 22.4, 21.1, 21.3, 21.0, 21.6, 21.1, and 21.5.

19.27 71.6, 69.8, 74.5, 77.1, 82.6, 104.4, 96.7, 137.0, 159.3, 151.5, 105.8, and 69.7.

19.29 68.4, 66.8, 71.4, 74.0, 79.4, 100.5, 93.2, 132.2, 153.9, 146.6, 102.5, and 67.6.

19.31 74.6, 68.1, 68.3, 75.0, 79.0, 77.7, 81.8, 88.2, 89.6, 83.3, 81.1, and 94.0.

19.33 819.6 million pounds.

19.35 2,886,000, 2,775,000, 3,700,000, 4,662,000, 5,106,000, 4,447,000, 3,737,000, 3,848,000, 3,663,000, 3,811,000, 2,960,000, and 2,775,000.

19.37 170.60, 160.31, 203.54, 210.46, 230.32, 207.12, 177.32, 192.91, 225.94, 267.83, 240.08, and 308.44.

19.39 59.0, 61.0, 74.9, 96.7, 118.6, 127.2, 122.4, 128.0, 125.0, 109.6, 104.7, and 72.9.

19.41 173, 182, 178.4, 194.8, 173.6, 171.6, 179.5, 179.4, 192.2, 193.3, 198.0, and 193.5.

19.43 411.4, 353.7, 497.7, 593.1, 495.8, 492.3, 438.9, 480.2, 541.9, 483.1, 484.6, and 703.7.

19.45 (a) $\log \hat{y} = 2.1952 + 0.0406x$ or $\hat{y} = 157(1.10)^x$; (c) 10%.

19.47 (a) 200; (b) 1983; (c) 400.

19.49

J	F	M	A	M	J	J	A	S	O	N	D
143	146	158	158	175	163	160	136	127	130	143	145
150	150	151	153	150	142	140	139	137	137	135	133
134	137	129	126	121	115	119	116	114	118	119	122
119	120	121	126	132	117	126	151	159	152	137	132
142	148	145	157	168	188	227	252	231	205	207	216
211	213	220	227	236	214	218	214	199	183	172	167

19.51 $\hat{y} = 15.6 + 1.5x$ (origin, 1977; x units, 1 year; y, number of wells completed annually).

19.55

1st quarter	2nd quarter	3rd quarter	4th quarter
148	138	139	154
165	215	173	160
172	184	206	221
202	192	203	198

Chapter 20

20.1 (a) Primary; (b) secondary; (c) secondary; (d) secondary; (e) primary; (f) primary; (g) secondary.

20.11 (a) 3,719,250; (b) 7,714,000.

20.13 10, 24, 4, and 2.

20.15 (a) 25 and 75; (b) 20 and 80.   **20.21** 180.

20.23 (One possibility) (a)

	a	b	c
I	A	B	C
II	C	A	B
III	B	C	A

(b) IaA, IbB, IcC, IIaC, IIbA, IIcB, IIIaB, IIIbC, and IIIcA; (c) assign IaA, IIcB, and IIIbC to P; assign IbB, IIaC, and IIIcA to Q; assign IcC, IIbA, and IIIaB to R.

20.25 Republican.

20.27 (b) Green—Personnel
Allen—Research
Eaton—Insurance or Purchasing
Frost—Purchasing or Insurance.

20.29 (a) 855,000; (b) 5,450,625.

20.31 (a) Secondary; (b) primary; (c) secondary; (d) primary; (e) secondary.

20.33 (a) Unemployment for 15 consecutive weeks or longer; (b) no; (c) yes.

20.35 2, 3, 5, and 11 on October 3; 5, 6, 8, and 1 on October 4; 8, 9, 11, and 4 on October 5; 11, 12, 1, and 7 on October 6; 1, 2, 4, and 10 on October 7; 4, 5, 7, and 13 on October 10; 7, 8, 10, and 3 on October 11; 10, 11, 13, and 6 on October 12; 13, 1, 3, and 9 on October 13; 3, 4, 6, and 12 on October 14; 6, 7, 9, and 2 on October 17; 9, 10, 12, and 5 on October 18; 12, 13, 2, and 8 on October 19.

20.37 ABC, ABc, AbC, Abc, aBC, aBc, abC, and abc.

20.41 (a) 216.47; (b) 216.47.

Index

Multiplication rules of probability,
156, 157
Mutually exclusive events, 132

Negative correlation, **458**
Nonparametric tests, 409
Normal distribution, 258
  and binomial distribution, 270
  standard, 260
  table, 553
Normal equations, 436, 438, 449
Normal regression analysis, 444
Normal values in time series analysis,
513
Null hypothesis, 320
Numerical distribution, 9

Observed frequencies, **371**
Odds, 149
  betting, 149
  and probabilities, 150, 151
Ogive, 17
One-sample sign test, 410
One-sided alternative, 322
One-sided test, 324
One-tailed test, 324
One-way analysis of variance, 393
Open class, 10
Operating characteristic curve
    (OC-curve), 318, 319
Opportunity loss, 176
  expected (EOL), 193
Opportunity-loss table, 176
Optimum allocation, 530
Optimum choices, 181
Outcome of experiment, 130
Outliers, 33

Paasche's index, **80**
Paired-sample sign test, 412
Parabola, 486
Parameter, 32
Pascal's triangle, 114
Payoff, 168
Payoff matrix, 168
Payoff table, 168
Pearsonian coefficient of skewness,
58
Percentage distribution, 14
Percentage-of-moving-average
    method, 503
Percentile, 56
Permutations, 106
Personal probability, 119
Pictogram, 17
Pie chart, 17
Point estimate, 302

Poisson distribution, 221, 222
  mean, 227
  variance, 233
Pooling, 369, 391
Population, 30
  correlation coefficient, 461
  finite, 278
  infinite, 278
  mean, 32
  size, 30
  standard deviation, 44
  variance, 44
Positive correlation, 458
Posterior analysis, 246
Posterior distribution, 309
Posterior mean, 309
Posterior standard deviation,
    310
Postulates of probability, 139,
    140
Power function, 327
Prediction, limits of, 447
Price relatives, 78
Prior analysis, 190
Prior distribution, 309
Prior mean, 309
Prior probabilities, 190, 243
Prior standard deviation, 309
Primary data, 523
Probability:
  addition rules, 146
  classical concept, 115
  conditional, 153, 155
  consistency criterion, 152
  equiprobable events, 145
  frequency interpretation, 117
  independent events, 156
  multiplication rules, 156, 157
  and odds, 150, 151
  personal, 119
  posterior, 244
  postulates, 139, 140
  prior, 190, 243
  subjective, 119
Probability density, 256
Probability distribution, 209
  binomial, 213, 214
  geometric, 218
  hypergeometric, 220
  mean, 226
  multinomial, 223
  Poisson, 221, 222
  standard deviation, 228
  variance, 228
Probability sample, 525
Probable error, 293
Producer Price Index, 89
Product moment, 459
Product-moment coefficient of
    correlation, 459

Proportion:
  confidence interval, 359
    finite population, 363
  sample, 358
  standard error, 359
  test, 363, 365
Proportional allocation, 528
Proportions, differences among, 372
Proportions, differences between, 370
Purchasing power of the dollar, 92
Pure decision procedure, 186
p-value, 329

Qualitative distribution, **9**
Quality control, 340, 368
Quantiles, 56
Quantitative distribution, 9
Quantity index, 79
Quartiles, 56
Quartile variation, coefficient of, 61
Quota sampling, 531

r (*see* Coefficient of correlation)
Random digits, 279
Random numbers, 279
  table, 579
Random sample:
  finite population, 279
  infinite population, 280
  simple, 278
Random selection, 116
Random variable, 209
  discrete, 211, 269
Randomized block design, 404
Randomized decision procedure, 186
Randomizing, 537
Randomness, tests of, 422
Range, 43
Range, semi-interquartile, 61
Rank-correlation coefficient, 466
  significance test, 468
  Spearman's, 466
Ratio paper, 489
Ratio-to-moving-average method, 503
Raw data, 9
Real class limits, 12
Regression, 431
  linear, 443
  multiple, 448
Regression analysis, 442
  linear, 443
  normal, 444
Regression coefficients, 443
  confidence intervals, 445
  tests, 443
Regression line, 442
Regression sum of squares, 456
Regrets, 176

PROBLEMS OF ESTIMATION

Mean (large sample, σ known or estimated by s)

$$\bar{x} - z_{\alpha/2} \cdot \frac{\sigma}{\sqrt{n}} < \mu < \bar{x} + z_{\alpha/2} \cdot \frac{\sigma}{\sqrt{n}}$$

Mean (small sample)

$$\bar{x} - t_{\alpha/2} \cdot \frac{s}{\sqrt{n}} < \mu < \bar{x} + t_{\alpha/2} \cdot \frac{s}{\sqrt{n}}$$

Proportion (large sample)

$$\frac{x}{n} - z_{\alpha/2} \sqrt{\frac{\frac{x}{n}\left(1 - \frac{x}{n}\right)}{n}} < p < \frac{x}{n} + z_{\alpha/2} \sqrt{\frac{\frac{x}{n}\left(1 - \frac{x}{n}\right)}{n}}$$

Standard deviation (large sample)

$$\frac{s}{1 + \frac{z_{\alpha/2}}{\sqrt{2n}}} < \sigma < \frac{s}{1 - \frac{z_{\alpha/2}}{\sqrt{2n}}}$$

Estimation of mean

$$E = z_{\alpha/2} \cdot \frac{\sigma}{\sqrt{n}}$$

Estimation of proportion

$$E = z_{\alpha/2} \sqrt{\frac{\frac{x}{n}\left(1 - \frac{x}{n}\right)}{n}}$$

Estimation of mean

$$n = \left[\frac{z_{\alpha/2} \cdot \sigma}{E}\right]^2$$

Estimation of proportion

$$n = p(1 - p)\left[\frac{z_{\alpha/2}}{E}\right]^2 \quad \text{or} \quad n = \frac{1}{4}\left[\frac{z_{\alpha/2}}{E}\right]^2$$